IUTAM SYMPOSIUM ON DISCRETIZATION METHODS IN STRUCTURAL MECHANICS

SOLID MECHANICS AND ITS APPLICATIONS
Volume 68

Series Editor: G.M.L. GLADWELL
Solid Mechanics Division, Faculty of Engineering
University of Waterloo
Waterloo, Ontario, Canada N2L 3G1

Aims and Scope of the Series

The fundamental questions arising in mechanics are: *Why?*, *How?*, and *How much?* The aim of this series is to provide lucid accounts written by authoritative researchers giving vision and insight in answering these questions on the subject of mechanics as it relates to solids.

The scope of the series covers the entire spectrum of solid mechanics. Thus it includes the foundation of mechanics; variational formulations; computational mechanics; statics, kinematics and dynamics of rigid and elastic bodies; vibrations of solids and structures; dynamical systems and chaos; the theories of elasticity, plasticity and viscoelasticity; composite materials; rods, beams, shells and membranes; structural control and stability; soils, rocks and geomechanics; fracture; tribology; experimental mechanics; biomechanics and machine design.

The median level of presentation is the first year graduate student. Some texts are monographs defining the current state of the field; others are accessible to final year undergraduates; but essentially the emphasis is on readability and clarity.

For a list of related mechanics titles, see final pages.

IUTAM Symposium on

Discretization Methods in Structural Mechanics

Proceedings of the IUTAM Symposium held in Vienna, Austria, 2–6 June 1997

Edited by

H. A. MANG

Institute for Strength of Materials,
Vienna University of Technology,
Austria

and

F. G. RAMMERSTORFER

Institute of Lightweight Structures and Aerospace Engineering,
Vienna University of Technology,
Austria

KLUWER ACADEMIC PUBLISHERS
DORDRECHT / BOSTON / LONDON

A C.I.P. Catalogue record for this book is available from the Library of Congress.

ISBN 0-7923-5591-1

Published by Kluwer Academic Publishers,
P.O. Box 17, 3300 AA Dordrecht, The Netherlands

Sold and distributed in North, Central and South America
by Kluwer Academic Publishers,
101 Philip Drive, Norwell, MA 02061, U.S.A.

In all other countries, sold and distributed
by Kluwer Academic Publishers
P.O. Box 322, 3300 AH Dordrecht, The Netherlands

Printed on acid-free paper

Printed in Belgium

CONTENTS

Dynamics

Novel Alternative Numerical Methods

PREFACE

The IUTAM/IACM Symposium on Discretization Methods in Structural Mechanics was held in Vienna, Austria, from 2nd to 6th June 1997. The site of the Symposium was the "Theatersaal" of the Austrian Academy of Sciences. The Symposium was attended by 71 persons from 23 countries. In addition, several Austrian graduate students and research associates participated in the meeting.

In the 5-day Symposium a total of 48 papers were presented. All of them were invited and accorded equal weight in the programme. The following topics were covered:

- Error-controlled adaptivity of finite element methods
- Large deformations and buckling, including inelastic deformations
- Inelastic brittle or ductile localization, phase transition and system failure, resulting from monotonic, cyclic or impact loading
- Sensitivity analysis and inverse problems with special emphasis on identification of material parameters
- Development of linear and nonlinear finite element methods for thin-walled structures and composites
- Implicit integration schemes for nonlinear dynamics
- Coupling of rigid and deformable structures; fluid-structures and acoustic-structure interaction
- Competitive numerical methods (finite element methods, boundary element methods, coupling of these two methods)
- Identification of material and structural data.

Comments on details of the treatment of these topics are contained in the Concluding Remarks. The Editors would like to express their appreciation to E. Stein who has prepared these Concluding Remarks.

The lectures and their discussions clearly showed the remarkable evolution of "Discrete Methods in Structural Mechanics" since the first IUTAM/IACM Symposium on this topic in 1989. Several new aspects have come into focus since that time. Many of them were stimulated by challenging requirements resulting from high-tech applications. It was particularly rewarding to find that researchers both from the area of constitutive modeling and from the field of computational structural mechanics came to a consensus concerning the need of a better synthesis of efforts in both areas for the ultimate purpose of improving the quality of structures. It was no less rewarding to observe the rapidly growing acceptance of quality control of numerical analyses by means of adaptive finite element methods.

This Volume contains 41 papers, reflecting only the content of the respective lectures. Much to the regret of the Scientific Committee some manuscripts were not submitted. All the papers contained herein have been reviewed to the standard of leading scientific journals. The Editors would like to acknowledge the great efforts on behalf of both the authors and the reviewers.

The Editors particularly wish to thank the Bureau of IUTAM, the Executive Council of IACM, and the International Scientific Committee. Part of the success of the Symposium is a consequence of the excellent facilities provided by the Austrian Academy of Sciences. The organizers would like to express their appreciation to M. Milletich from the Academy and to her staff for the great efforts related to the conference facilities.

The smooth running of the Symposium owes much to the initiative and the organizational skills of Th. Huemer, R. Lackner, and G. Meschke from the Institute for Strength of Materials (TU Vienna), and G. Liebenauer, R. Frohner, and H.J. Böhm from the Institute of Lightweight Structures and Aerospace Engineering (TU Vienna).

Finally, the Editors would like to express their gratitude to the sponsoring organizations who have supported the Symposium financially, namely, the International Union of Theoretical and Applied Mechanics (IUTAM), the International Association for Computational Mechanics (IACM), the Austrian Federal Ministry of Science and Transport, the Austrian Academy of Sciences, Creditanstalt Wien Rilkeplatz, the Institute for Strength of Materials, and the Institute of Lightweight Structures and Aerospace Engineering, both of the Vienna University of Technology.

Vienna, August 1998 H.A. Mang and F.G. Rammerstorfer

International Scientific Committee

T. Kawai (Japan)

G. Kuhn (Germany)

H.A. Mang (Austria)

J.T. Oden (USA)

E. Oñate (Spain)

E. Stein (Germany)

F.G. Rammerstorfer (Austria)

F. Ziegler (Austria)

Sponsors

International Union of Theoretical and Applied Mechanics (IUTAM)

International Association for Computational Mechanics (IACM)

IUTAM/IACM Working Party on Computational Mechanics

Austrian Academy of Sciences

Austrian Federal Ministry of Science and Transport

Creditanstalt Wien Rilkeplatz, Austria

Institute for Strength of Materials
Vienna University of Technology, Austria

Institute of Lightweight Structures and Aerospace Engineering
Vienna University of Technology, Austria

OPENING ADDRESS
by the former Secretary General of IUTAM

Mr. Chairman, Mr. President of IACM,
distinguished Colleagues, Ladies and Gentlemen,

On behalf of the International Union of Theoretical and Applied Mechanics it is my duty as the *ex officio* member of the Scientific Committee, and my pleasure to welcome you at the IUTAM/IACM joint Symposium on Discretization Methods in Structural Mechanics II.

The proposal to arrange this follow-up joint Symposium was accepted in August 1994 by the General Assembly of IUTAM in Amsterdam upon recommendation by the Symposia Panel on Solid Mechanics. Since that time both chairmen, Professor Herbert A. Mang and Professor Franz G. Rammerstorfer, have done an excellent job in preparing the scientific program. The chairmen have been advised by the Scientific Committee in selecting and inviting participants from all over the world. Sometimes there are some misunderstandings concerning the appointment of the Scientific Committee by the IUTAM Bureau: If you look back to the first IUTAM/IACM Symposium on Discretization Methods in Structural Mechanics held in Vienna in 1989, to mention just the Symposium on exactly the same subject, you see different members of the Scientific Committees with relatively little overlap with the current one: Prof. T. Kawai, Prof. G. Kuhn, Prof. H.A. Mang, Prof. J.T. Oden, Prof. E. Oñate, Prof. F.G. Rammerstorfer, Prof. E. Stein, Prof. F. Ziegler (*ex officio*), and Prof. O.C. Zienkiewicz.

During the 19th ICTAM in Kyoto in 1996 a booklet "Fifty Years of Impulse to Mechanics", published by Kluwer, was made available, shedding new light on the 50 years of existence of IUTAM. However, the quadrennial Congresses span already over three quarters of this century. It was the International Council of Scientific Unions (ICSU) that suggested to transform the Congress of Applied Mechanics into a Union. It was Professor Jan Burger who sent out this invitation. In its meeting during the VIth Congress in Paris in September 1946 the president, Professor H. Villat, and the French delegates (including the famous Professor Maurice Roy) developed much sympathy with this plan, so that a decision was reached. Since that time mechanics plays an "independent" role in the concert of sciences. Presently, we are eager to find distinguished colleagues who, being convinced of the importance of mechanics, are willing to promote the Union even more rigorously.

The value and quality of IUTAM is best reflected by eight to ten annual Symposia. IUTAM has recently established a Working Party on „Computational Mechanics“ in co-operation with its largest and most important affiliated organization, namely IACM, affiliated since 1984. We are very happy to have this series of meetings and proceedings in the field of computational structural mechanics. There are many more meetings referring to this important subject. Some people even speak of an inflation. However, the quality of IUTAM-sponsored Symposia is hardly reached elsewhere. I refer in this aspect also to the forthcoming IUTAM-Symposium on Computational Method in Unbounded Domains, to be held in Boulder in early August 1997. The IUTAM Newsletter lists all Symposia. It is available free of charge from the Secretary General by e-mail: iutam@ucd.ie. Information on the published Proceedings of IUTAM-Symposia is contained in the Annual Report of IUTAM, a bound volume, which is also available free of charge. Despite there is a strict rule that only papers presented orally are eligible for publication in the Symposium Proceedings, a review process of the final papers has become mandatory.

We look forward to a well-balanced Scientific Program consisting of 16 Sessions. I am especially happy to see sessions devoted to evolving fields and those gaining importance in practical applications such as FE/BE Coupling, Sensitivity Analysis, and Optimization. The Session Chairmen are invited to stimulate the discussions. There are 30 minutes allocated to each presentation, and at least 5 minutes must be reserved for a discussion. The success of the Symposium depends on the presence of the invited participants during the whole tenure of the Symposium. In the past it was clearly understood that this is implied by the formulation "Participation on invitation only".

Computational Mechanics will play an adequate role at the 20th ICTAM, Chicago, 27 August - 2 September 2000, where not only the turn of the century will be celebrated but also joint efforts of IUTAM and IACM will be promoted. I take this opportunity to call upon you personally to submit suggestions for Minisymposia and for Prenominated Sessions to one of the members of the Working Party WP6, Professors E.R. de Arantes e Oliveira (IUTAM), J.T. Oden (IACM), H.A. Mang, T. Belytschko and P. Ladevéze.

The next General Assembly of IUTAM will be held on 29/30 August 1998 in Stuttgart. Decisions concerning future IUTAM-Symposia for the years 2000/2001 will be made depending on the quality of the proposals received and the recommendations of the two IUTAM-Symposia Panels. It would be a good idea to consider the future of joint Symposia in the field of Computational Mechanics in the tenure of our Symposium.

We shall start with an exciting session with the invited lecture given by one of the prominent founders of Computational Mechanics, Professor Zienkiewicz, after concluding the Opening Session with the address by Professor Samuelsson, President of IACM.

Thank you for your kind attention.

Franz Ziegler

RECOVERY PROCEDURES IN ERROR ESTIMATION AND ADAPTIVITY

O.C. Zienkiewicz,
B. Boroomand
University of Wales, Swansea, U.K.
J.Z. Zhu
U.E.S. Inc. Annapolis, USA

1. Introduction

Two types of procedures are currently available for deriving error estimators. They are either *Residual based* or *Recovery based.*

The residual based error estimators were first introduced by Babuška and Rheinboldt in 1978 [1] and have been since used very effectively and further developed by many others. Here substantial progress was made as recently as 1993 with the introduction of so called residual equilibration by Ainsworth and Oden [2].

The recovery based error estimators are, on the other hand, more recent having been first introduced by Zienkiewicz and Zhu in 1987 [3]. Again these were extensively improved by the introduction of new recovery processes. Here in particular the , so called, SPR (or Superconvergent Patch Recovery) method introduced in 1992 by the same authors [4] has produced a very significant improvement of performance of the Recovery based methods. In this paper we shall concentrate entirely on the *Recovery based method* of error estimation. The reasons of this are straightforward:

(i) the concept is simple to grasp as the approximation of the error is identified as the difference between the recovered solution $\mathbf{u}^*$ and the numerical solution $\mathbf{u}_h$; thus the estimate

$$||e^*|| = ||\mathbf{u}^* - \mathbf{u}_h|| \tag{1}$$

in any norm is achieved simply by assuming that the exact solution $\mathbf{u}$ can be replaced by the recovered one.

(ii) as some recovery process is invariably attached to numerical codes to present more accurate and plausible solutions, little additional computation is involved;

H. A. Mang and F. G. Rammerstorfer (eds.),
IUTAM Symposium on Discretization Methods in Structural Mechanics, 1–12.

(iii) if the recovery process itself is superconvergent, it can be shown [4] that the estimator will always be asymptotically exact;

(iv) numerical comparisons on bench mark problems and more recently by a 'patch test' procedure introduced by Babuška *et al* [6, 7] have shown that the recovery procedures are extremely accurate and robust. In all cases they appear to give a superior accuracy of estimation than that achievable by *Residual based methods.*

It is of interest to remark that in many cases it is possible to devise a Residual method which has an identical performance to a particular recovery process. In a recent paper Zhu [8] shows that:

(v) for every *Residual based* estimator there exists a corresponding *Recovery based* process. However the reverse in not true. Indeed the Recovery based methods with optimal performance appear not to have an equivalent Residual process and hence, of course, the possibilities offered by Recovery methods are greater.

In this paper we shall describe in detail the SPR based recovery as well as a new alternative REP process which appears to be comparable in performance.

With error estimation achieved the question of adaptive refinement needs to be addressed. Here we discuss some procedures of arriving at optimal mesh size distribution necessary to achieve prescribed error.

2. Solution Recovery and Error Estimation

In what follows we shall be in general concerned with the numerical solution of problems in which a differential equation of the form given as:

$$\mathbf{S}^T\mathbf{D}\mathbf{S}\mathbf{u} + \mathbf{b} = \mathbf{0} \tag{2}$$

has to be solved in a domain Ω with suitable boundary condition on $\partial\Omega = \Gamma$. In above $\mathbf{S}$ is a differential operator usually defining stresses or fluxes as

$$\boldsymbol{\sigma} = \mathbf{D}\mathbf{S}\mathbf{u} \tag{3}$$

where $\mathbf{D}$ is a matrix of physical parameters.

The solution error is defined as the difference between the exact solution and the numerical one. Thus for instance the displacement and stress error are:

$$\mathbf{e}_u = \mathbf{u} - \mathbf{u}_h \qquad \mathbf{e}_\sigma = \boldsymbol{\sigma} - \boldsymbol{\sigma}_h \tag{4}$$

at all points of the domain. It is, however, usual to define the error in terms of a suitable norm which can be written as a scalar value

$$e = ||\mathbf{e}_u|| = ||\mathbf{u} - \mathbf{u}_h|| \tag{5}$$

for any specific domain Ω. With the Recovery process we devise a procedure which gives , by suitable post-processing of $\mathbf{u}_h$ and $\boldsymbol{\sigma}_h$, the values of $\mathbf{u}^*$ and / or $\boldsymbol{\sigma}^*$ which are (hopefully) more accurate. Thus in energy norm we define the error:

$$||e^*||^E = \left[\int_\Omega (\boldsymbol{\sigma}^* - \boldsymbol{\sigma}_h)^T \mathbf{D}^{-1} (\boldsymbol{\sigma}^* - \boldsymbol{\sigma}_h) d\Omega\right]^{\frac{1}{2}} \tag{6}$$

The effectivity index of any error estimator is defined as:

$$\theta = \frac{||e||_{est}}{||e||} \quad \text{generally} \qquad \text{or} \qquad \theta^* = \frac{||e^*||}{||e||} \quad \text{for recovery based ones} \tag{7}$$

A theorem proposed by Zienkiewicz and Zhu [4] shows that for all estimators based on recovery we can establish the following bound for the effectivity:

$$1 - \frac{||\tilde{e}||}{||e||} < \theta^* < 1 + \frac{||\tilde{e}||}{||e||} \tag{8}$$

In above e is the actual error and $\tilde{e}$ is the error of the recovered solution i.e.

$$||\tilde{e}|| = ||\mathbf{u} - \mathbf{u}^*|| \tag{9}$$

The proof of the above theorem is given in reference [4]. Two important conclusions follow:

(1) that *any recovery process* which result in reduced error will give a reasonable error estimator and, more importantly,

(2) if the recovered solution converges at a higher rate than the finite element solution we shall always have asymptotically exact estimation.

3. The SPR Recovery Procedure

The SPR recovery is based on the assumption that in all element points fixed in space exist for which superconvergence occurs.

For instance it can be shown that for many elements the unknown u is superconvergent at nodal points and that the derivatives $\boldsymbol{\sigma}$ are superconvergent on

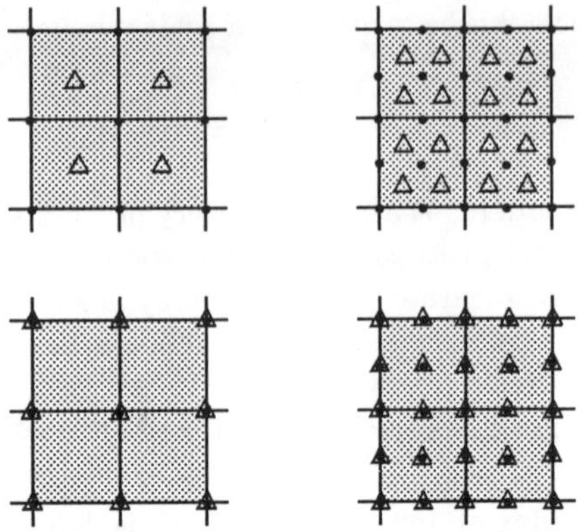

Figure 1: Superconvergent recovery of (a) $\boldsymbol{\sigma}$ and (b) $\mathbf{u}$ for bilinear and biquadratic elements in regular patches

Gauss integration points [9, 10, 11]. These statements are not in general true for distorted, isoparametric, elements but hold well for simple, regular, one or two dimensional elements.

The superconvergent recovery first establishes finite element patches at fully connected, corner nodes and specifies in each element the superconvergent points.

Figure 1 shows an example of two dimensional bi-linear and bi-quadratic elements on which superconvergent points for gradients $\boldsymbol{\sigma}$ (stresses), and displacement $\mathbf{u}$ are shown.

In the SPR procedure we approximate within each patch, each component of the gradient, by a polynomial of order $p+1$ for $\mathbf{u}$ and p for σ and make this approximation fit the superconvergent values in a least square sense.

For the bi-linear elements we have *four* superconvergent points in two dimensions and a least square fit has to be used to find *three* constants of the linear expansion for $\boldsymbol{\sigma}$.

For the bi-quadratic elements we have similarly *sixteen* superconvergent points in two dimensions and for $\boldsymbol{\sigma}$ fit, in a least square sense, *six* constants of a quadratic expansion for each component are necessary.

However we can proceed differently assuming superconvergence of the function $\mathbf{u}$ itself at nodes. Now we have to pass the $p+1$ polynomial which again is available for both bi-linear and bi-quadratic elements. Differentiation of the superconvergent $\mathbf{u}$ expansion will, presumably, yields again superconvergent expansion of σ. However, if the energy norm itself includes $\mathbf{u}$ values as well as derivatives, it is necessary to use the $\mathbf{u}$ expansion directly. This type of problems is typical of dynamics where acceleration has to be modified.

It is important to note that if the points which are being matched are exactly superconvergent then the approximation will again be superconvergent and hence solution giving gradients one order higher be automatically satisfied. This, as we shall see later, gives the fully optimal effectivity index as $\theta^* = 1$ asymptotically.

4. The REP Recovery

Although the SPR recovery has proved to work well generally, its logic for some application where superconvergent points do not in fact exist is doubtful. We have however sought to determine viable recovery alternatives. One of these, known by the acronym REP (Recovery by Equilibrium of Patches) will be here described. This procedure is presented in references [13, 14] with the second one describing a better variant.

To some extent the motivation follows that of Ladevèze [15] who sought to establish (for somewhat different reasons) a fully equilibrating stress field which can replace that of the finite element approximation. However we believe that the process derived in [14] is simpler though equilibration is only approximate.

The starting point is the governing equation, Equation (2), written with the substitution of Equation (3) and in the finite element, Galerkin, approximation sense as:

$$\int_{\Omega} \mathbf{B}^T \boldsymbol{\sigma}_h d\Omega - \int_{\Omega} \mathbf{N}^T \mathbf{b} d\Omega - \int_{\Gamma} \mathbf{N}^T \mathbf{t} d\Omega = \mathbf{0} \tag{10}$$

where the last term comes from the tractions on the boundary of the domain Ω which can, of course, represent the whole of the problem, an element patch or only a single element.

As is well known the stresses $\boldsymbol{\sigma}_h$ which result from the finite element analysis will in general be discontinuous and we shall seek to replace them in *every element patch* by a recovered system which is smooth and continuous.

To achieve the recovery we proceed in exactly an analogous way to that used in the SPR procedure, *first* approximating the stress in each patch by a polynomial of appropriate order $\boldsymbol{\sigma}^*$, *second* using these to obtain nodal values of $\bar{\boldsymbol{\sigma}}^*$ and finally interpolating these by standard shape-functions. The stress or gradient is always a vector of appropriate components $\boldsymbol{\sigma} = [\sigma_1, \sigma_2, \sigma_3]^T$ with $\sigma_1 = \sigma_x$, $\sigma_2 = \sigma_y$ and $\sigma_3 = \tau_{xy}$ in elastic two dimensional analysis for instance.

For equivalence we shall attempt always to ensure that the total smoothed stress $\boldsymbol{\sigma}^*$ satisfies in the least square sense the same patch equilibrium conditions as the finite element solution; i.e that

$$\int_{\Omega_p} \mathbf{B}^T \boldsymbol{\sigma}_h d\Omega \approx \int_{\Omega_p} \mathbf{B}^T \boldsymbol{\sigma}^* d\Omega \tag{11}$$

where for convenient we write

$$\boldsymbol{\sigma}^* = \boldsymbol{\sigma}_1^* + \boldsymbol{\sigma}_2^* + \boldsymbol{\sigma}_3^* \qquad \boldsymbol{\sigma}_i^* = (\mathbf{1}^i)^T \sigma_i^* \qquad \mathbf{1}^2 = [0,1,0] \quad \text{etc} \tag{12}$$

and

$$\sigma_i^* = \mathbf{P}\mathbf{a}^i \tag{13}$$

where $\mathbf{P}$ is polynomial terms and $\mathbf{a}^i$ is the set of unknowns coefficients for describing the ith stress component. A similar decomposition is used for $\boldsymbol{\sigma}^h$.

It has been found in practice [13, 14] that the constraint condition provided by Equation (11) are often not sufficient to ensure a non singular minimization to obtain the full set of coefficients $\mathbf{a}^i$ we therefore proposed in [14] that the Equation (11) be replaced by

$$\int_{\Omega_p} \mathbf{B}^T \boldsymbol{\sigma}_i^h d\Omega \approx \int_{\Omega_p} \mathbf{B}^T \boldsymbol{\sigma}_i^* d\Omega = \left(\int_{\Omega_p} \mathbf{B}^T (\mathbf{1}^i)^T \mathbf{P} d\Omega \right) \mathbf{a}^i \tag{14}$$

As above ensures also satisfaction of Equation (11) and merely enlarges the constraints set ensuring that each component of $\boldsymbol{\sigma}_h$ is equilibrated by the corresponding component of $\boldsymbol{\sigma}^*$. Further the imposition of the approximate equation of (14) allows each set of coefficients $\mathbf{a}^i$ to be solved independently reducing considerably the algorithm cost (and here repeating the procedure used with success in SPR.)

We minimize thus:

$$\Pi = \left(\mathbf{H}^i \mathbf{a}^i - \mathbf{F}_p^i\right)^T \left(\mathbf{H}^i \mathbf{a}^i - \mathbf{F}_p^i\right) \qquad \mathbf{H}^i = \int_{\Omega_p} \mathbf{B}^T (\mathbf{1}^i)^T \mathbf{P} d\Omega \tag{15}$$

to determine $\mathbf{a}^i$ (for details viz ref. [14]).

The REP procedure follows precisely the detail of SPR near boundaries and gives overall an approximation which does not require the knowledge of any superconvergent points. The accuracy of both processes is comparable but we quote the new method here as it is our opinion that many other alternative recovery procedures are still possible.

5. Asymptotic Behavior and Robustness of Error Estimators

It is well known that elements in which polynomials of order p are used to represent the unknown will reproduce exactly any problem for which the exact solution is also defined by such a polynomial. Indeed the verification of this behavior is the essential part of the 'patch test' which has to be satisfied by all elements to ensure convergence [9].

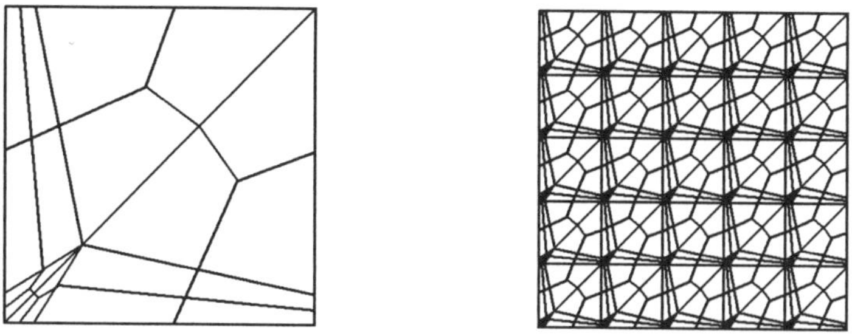

Figure 2: A repeatable patch of arbitrary quadrilateral elements

Thus if we are attempting to determine error in a general smooth solution we will find that this is dominated by terms of the order $(p+1)$. The response of the patch to exact solution of this order will therefore determine the asymptotic behaviour when the size of the patch and at all the elements tends to zero. If that patch is assumed to be one of repeatable kind, the behaviour of the patch when subjected to exact solution of that order will show the exact finite element solution and the corresponding numerical solution and establish the asymptotic error exactly. Thus any estimator can be compared with this exact value. and the asymptotic effectivity index can be established. Figure 2 show such repeatable patch of quadrilateral elements which evaluates the performance for quite irregular meshes.

In the papers presented recently by Babuška *et al* [6, 7] the procedure of dealing with such repeatable patches for various patterns of elements is developed. We shall not describe the details of the procedure here in which certain simplification are introduced to avoid some computational labour. However the basic process consists simply of

(1) subjecting the patches to individual terms of the exact solution of $(p+1)$ order

(2) computing the corresponding approximate, finite element, solutions and evaluating the *exact* error of these solution (by comparing it with the exact polynomial)

(3) applying any error estimator we wish to test to the finite element solution and evaluating its effectivity index θ

(4) determining the lowest (θ_L) and highest (θ_U) values of the effectivity *for all possible combinations of the expansion terms used in the exact solution.* This is simply accomplished by specifying a suitable eigenvalue problem.

These bounds of the effectivity index are very useful for comparing various error estimators and their behaviour for different mesh and element patterns. However, a single parameter called the *robustness index* has also been devised [6]

$$R = max\left(|1+\theta_L| + |1-\theta_U|, |1-\frac{1}{\theta_L}| + |1+\frac{1}{\theta_U}| \right) \tag{16}$$

A large value of this index obviously indicates a poor performance. Conversely the best behavior is that in which

$$\theta_L = \theta_U = 1 \tag{17}$$

and this gives $R = 0$.

In the series of tests reported in reference [6, 7] various estimators have been compared. Table 1 below shows the performances of an Equilibrium residual based error estimator with the SPR recovery error estimator for a set of particular patches of triangular elements [7]

Table 1. Robustness index for the Equilibrated Residuals (ERpB) and SPR (ZZ-discrete) estimators for a verity of anisotropic situations and element patterns $p = 2$.

Estimator	Robustness Index
ERpB	10.21
SPR (ZZ-discrete)	0.02

Table 2. Effectivity bounds and robustness of SPR and REP recovery estimator for regular patterns of triangles or rectangles with linear and quadratic shape function (applied to heat conduction and elasticity problems).

LINEAR TRIANGLES AND RECTANGLES (BOTH HEAT CONDUCTION / ELASTICITY)						
Aspect Ratio	SPR			REP		
	θ_L	θ_U	R	θ_L	θ_U	R
1/1	1.0000	1.0000	0.0000	1.0000	1.0000	0.0000
1/64	1.0000	1.0000	0.0000	1.0000	1.0000	0.0000
QUADRATIC RECTANGLES (HEAT CONDUCTION)						
1/1	1.0000	1.0000	0.0000	1.0000	1.0000	0.0000
1/64	1.0000	1.0000	0.0000	1.0000	1.0000	0.0000
QUADRATIC RECTANGLES (ELASTICITY)						
1/1	1.0000	1.0000	0.0000	0.9991	1.0102	0.0111
1/64	1.0000	1.0000	0.0000	0.9945	1.0000	0.0055
QUADRATIC TRIANGLES (ELASTICITY)						
1/1	0.9966	1.0929	0.0963	0.9562	1.0503	0.0940
1/64	0.9965	1.0947	0.0982	0.9516	1.0688	0.1172

This performance comparison is quite remarkable and it seems that in all the tests quoted by Babuška *at al* [6, 7] the SPR recovery estimator performs best. Indeed we observe that in many cases of regular subdivision when full super-convergence occur the ideal, asymptotically exact solution characterized by the parameter of Equation (17) is available.

In Table 2 we show some results obtained for regular meshes of triangles and rectangles. In the rectangular elements used for problems of heat conduction type superconvergent points are exact and the ideal result is expected for both linear and quadratic elements. It is surprising that this also occur in elasticity where the proof of superconvergent points is lacking. Further, the REP procedure also seem to yield superconvergence (except for elasticity with quadratic elements which we believe is due to round-off errors encountered.)

For regular meshes of quadratic triangles generally superconvergence is not expected and it does not occur for neither heat conduction nor elasticity. However the robustness index has very small values ($R < 0.09$ for SPR $R < 0.1$ for REP) and the estimators are therefore very good.

Similar results are also obtained for irregular meshes [14].

6. Adaptive Refinement

Most error estimations can yield reasonable evaluation of the global error and of the error contribution of individual elements. However, once again, the local estimations are substantially improved by recovery procedures and the knowledge of such local error is particularly important during adaptive refinement (and derefinement) if the desired result is to be reached in a small number of solutions.

The degree and indeed the nature of the refinement will depend much on the criteria we wish to satisfy. The most commonly used procedure aims at achieving a specific percentage error η in a global norm such as the well known energy norm and here we shall seek to achieve this in a manner we outlined originally in 1987 [3]. Many of the examples studied by the authors and others aim at satisfying;

$$\eta = \frac{||\mathbf{e}||}{||\mathbf{u}||} \leq \bar{\eta} \tag{18}$$

where $\bar{\eta}$ is the specified permissible error.

As in general we shall *aim* at equal distribution of error between all elements and if n is the number of elements in a current analysis the permissible error in each element is:

$$\bar{\eta}||\mathbf{u}_h||/n^{\frac{1}{2}} \tag{19}$$

Now we can find the approximate size of the new element vis a vis the old element by assuming $O(h^p)$ convergence and thus:

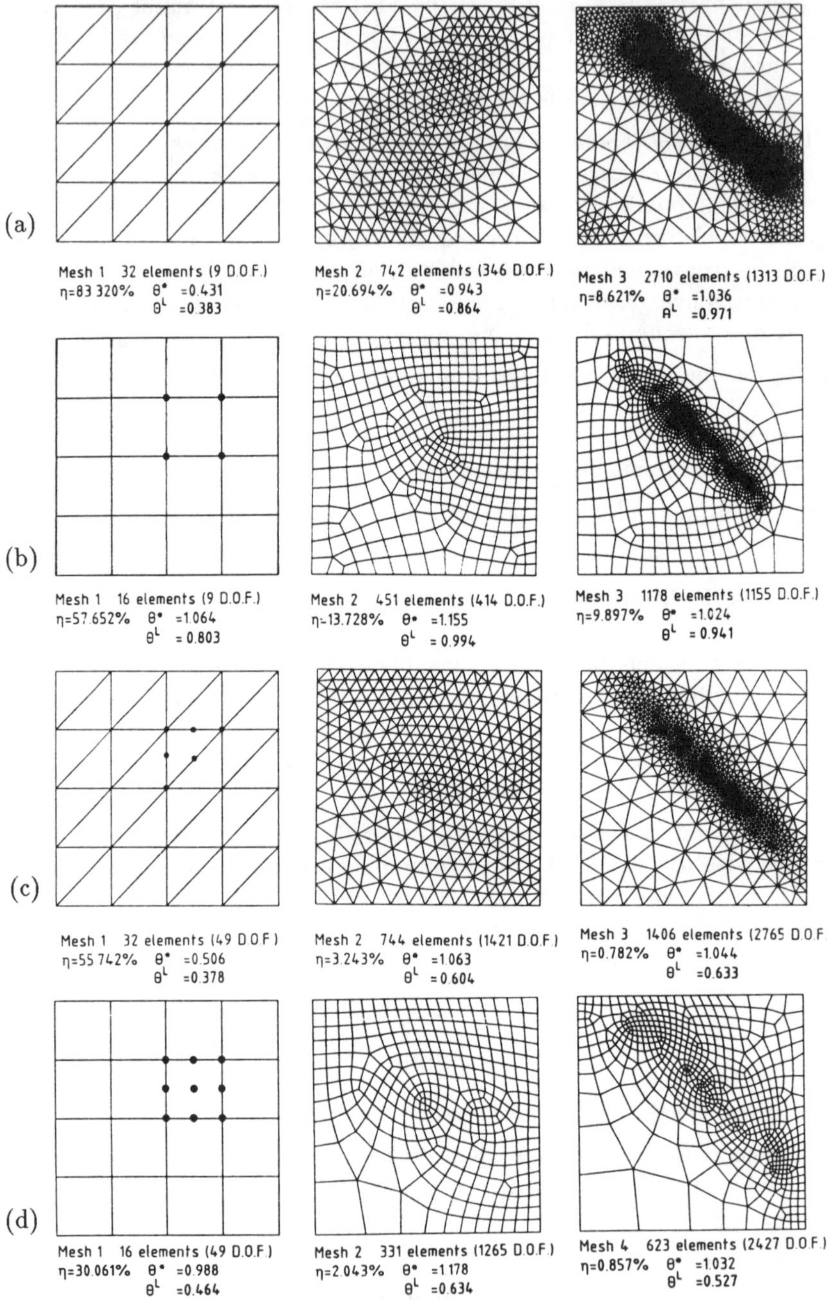

Figure 3: Adaptive mesh refinement for a Poisson type problem, triangular and quadrilateral elements (a)(b) linear (c)(d) quadratic. Aim 10% error and 1% error respectively.

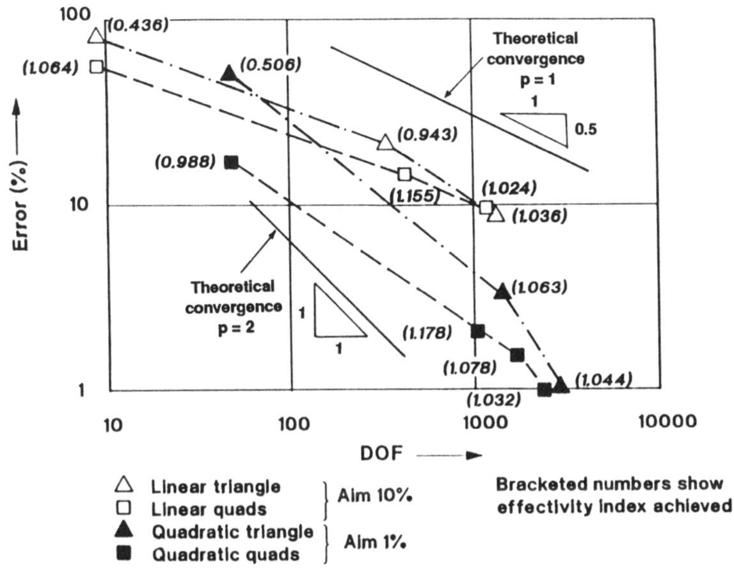

Figure 4: Convergence of adaptive refinement in example of Figure 3

$$\frac{h_{new}}{h_{old}} = \left(\frac{\bar{\eta}||\mathbf{u}_h||/n^{\frac{1}{2}}}{||e||_{\Omega_e}} \right)^{\frac{1}{p}} \tag{20}$$

Although the procedure is very crude (as the subdivisions of course changes the element number) in convergence, it is generally rapid and only one or two resolutions are generally needed to achieve final results.

Two main procedures exist to obtain the refined mesh; these are either complete remeshing or subdivision and the former is more useful generally. Figure 3 show the convergence plot for various types of elements used in solving a Poisson type problem (from reference [4]) in a square domain. The details of the problem and of the refinement are given in reference [4] and we only show the progressive steps of adaptive refinements. In Figure 4 the convergence of the error is shown versus the degree of freedoms. The reader will observe the better performance of quadrilaterals in both linear and quadratic versions.

References

[1] Babuška, I., Rheinboldt, W.C.: A posteriori error estimates for the finite element method, *Int. J. Num. Meth. Eng.*, **12** (1978), 1597-1615.

[2] Ainsworth, M., Oden, J.T.: A unified approach to a-posteriori error estimation using element residual methods, *Numer. math.*, **65** (1993), 23-50.

[3] Zienkiewicz, O.C., Zhu, J.Z.: A simple error estimator and adaptive procedure for practical engineering analysis, *Int. J. Num. Meth. Eng.*, **24** (1987), 337-357.

[4] Zienkiewicz, O.C., Zhu, J.Z.: The superconvergent patch recovery and *a posteriori* error estimates. Part 1: The recovery technique, Part 2: Error estimates and adaptivity *Int. J. Num. Meth. Eng.*, **33** (1992), 1331-1364, 1365-1382.

[5] Zienkiewicz, O.C., Zhu, J.Z.: Superconvergent patch recovery (SPR) and adaptive finite element refinement, *Com. Meth. Appl. Mech. Eng.*, **101** (1992), 207-224.

[6] Babuška, I., Strouboulis, T., Upadhyay, C.S., Gangaraj, S.K. and Copps, K.: Validation of a posteriori error estimators by numerical approach, *Int. J. Num. Meth. Eng.*, **37** (1994), 1073-1123.

[7] Babuška, I., Strouboulis, T., Upadhyay, C.S., Gangaraj, S.K., Copps, K.: An Objective criterion for assessing the reliability of A-Posteriori error estimators in finite element computations, *U.S.A.C.M Bulleten*, No. 7 (1994), 4-16.

[8] Zhu, J.Z.: A posteriori error estimation- The relationship between different procedures, *Com. Meth. Appl. Mech. Eng.* To be published

[9] Zienkiewicz, O.C., Taylor, R.L.: *The Finite Element Method* , IV Edition, **I** (1989), **II** (1991)

[10] Zlamal, M.: Superconvergence and reduced integration in the finite element method, *Math. Comput.* , **32** (1980), 663-685.

[11] Mackinnon, R.J., Carey, G.F.: Superconvergent derivatives: A Taylor series anlysis, *Int. J. Num. Meth. Eng.*, **28** (1989), 489-509.

[12] Oden, J.T., Brauchli, H.J.: On the calculation of consistant stress distribution in finite element applications, *Int. J. Num. Meth. Eng.*, **3** (1971), 317-325.

[13] Boroomand, B., Zienkiewicz, O.C.: Recovery by equilibrium in patches (REP), *Int. J. Num. Meth. Eng.*, **40** (1997), 137-154.

[14] Boroomand, B., Zienkiewicz, O.C.: An improved REP recovery and the effectivity robustness test, *Int. J. Num. Meth. Eng.* To be published.

[15] Ladevèze, P., Leguillon, D.: Error estimate procedure in the finite element method and applications, *SIAM J. Num. Anal.*, **20**, No. 3 (1983), 485-509.

Integrated Solution- and Model-Adaptivity in Finite-Element-Analysis of Plates and Shells

E. Stein & S. Ohnimus
Institute for Structural and Computational Mechanics
University of Hannover
Appelstraße 9A, D–30167 Hannover, Germany
e-mail: stein@leibniz.ibnm.uni-hannover.de

Dedicated to Professor Dr. techn. Dr. h.c. Franz Ziegler
on the occasion of his 65. birthday

ABSTRACT: Modeling of elastic thin–walled beams, plates and shells as 1D–and 2D– boundary value problems is valid in undisturbed subdomains. Disturbances near supports and free edges, in the vicinity of concentrated loads and at thickness jumps cannot be described in a sufficient way by 1D– and 2D–BVPs. In these disturbed subdomains dimensional (d)–adaptivity and model (m–)adaptivity have to be performed coupled with h– and/or p–adaptivity using hierarchically expanded test spaces in order to guarantee reliable and efficient overall results. The expansion strategy is applied for enhancing the spatial dimension and the mathematical model which is more efficient and evident for engineers than the reduction method. We present an error estimator of the dual problem, using a posterior equilibrium method (PEM) for calculation of the interface tractions on local patches with Neumann boundary conditions, using orthogonality conditions. These tractions are equilibrated with respect to the global equilibrium conditions of the stress resultants. An upper bound error estimator is presented, based on differences between the new tractions and the discontinuous tractions calculated from the stresses of the current finite element solution. The introduction of new element boundary tractions yields a method which can be regarded as a stepwise hybrid displacement method or as a Trefftz method for local Neumann problems on element patches.
An important advantage of PEM is the coupled adaptive computation of local anisotropic discretization, as well as dimension– and model errors in an additive split.

1 Introduction

The deformations and stresses of elastic beams, plates and shells in bending, modelled approximately by 1D– and 2D– theories via kinematic hypotheses, show considerable deviations from the solutions of 3D–theory of elasticity in disturbed subdomains, like boundary layers, haunched thicknesses or concentrated loads. It is possible to establish a coupled adaptive process by hierarchical projections

H. A. Mang and F. G. Rammerstorfer (eds.),
IUTAM Symposium on Discretization Methods in Structural Mechanics, 13–26.

from a more general theory, e.g. of 3D–elasticity theory, see [1, 2], and thus to improve not only the numerical solution of an approximated theory by hp–adaptivity and d–adaptivity (d=dimension), e.g. by higher shear–elastic deformation modes instead of the Kirchhoff–Love or Reissner–Mindlin hypotheses, but also the mathematical model in disturbed subdomains by m–adaptivity (m=model), e.g. extension to complete 3D–elasticity or –elastoplasticity. In this way one can start with simple 2D–elastic theories for plates and shells, e.g. with Kirchhoff–Love hypothesis, and proceed to the complete Lamé equations (expansion method), while the reverse procedure (reduction method) starts with the 3D–elasticity and ends up with reduced formulation of elasticity, e.g. 2D–elastic theory, see also [3, 4, 5, 6, 7, 8, 9, 10, 11]. These new strategies yield a quality jump of FEM regarding real engineering problems.
A rather new idea for error estimation is to formulate a local boundary value problem based on an improved stress calculation. It is possible to compute C^0–continuous stresses in the whole system with a so–called posterior equilibrium method, see [12, 13, 14]. It is also possible to compute improved stress fields by solving subproblems on patches, see [15, 16, 17, 12, 10]. With these improved local element boundary tractions it is possible to formulate a local variational problem for estimating the model–error with respect to a hierarchically expanded model, from 3D–elasticity to elastoplasticity with hardening, which is a main topic of our current research.

2 Classification of error estimators for the discretization (solution) and the model within elliptic boundary value problems

In this section a classification of different error estimators is given.

(A) Algebraic **Defect Correction Method** including error estimation for solution-, dimension- and model- adaptivity, (e.g.: Hackbusch 1989, [18]). Problems are: (1) Global solution and model error estimation are necessary; (2) Special solution techniques are necessary for defect correction. This method of interactive error controlled computation of the system matrix and the of iterative solution of the system of equations is gaining growing importance due to higher effectivity, especially by admitting controlled truncation errors in the iterative process.

(B) **Residual error estimation on patches** for solution– and dimension–adaptivity (e.g.: Babuška, Rheinboldt 1978, [19, 20]; Johnson, Hansbo 1992, [3]; Stein, Ohnimus 1992, e.g. [4, 7]). Discretization– and model– error estimation, including robust and effective calculation of model residuals is known for hierarchical reduction techniques applied to plates in bending without haunches, [1, 21]. In case of expansion method, residual errors are not useful for model adaptivity because of strong thickness locking effects for plates and shells as a class of parameter depending or non–robust problems. This residual method of the primal problem – using locally higher p– and h–testspaces in the three coordinate directions for Dirichlet–problems of element patches – controls the error of the discrete approximation space including dimensional adaptivity. It is not possible to get a workable additive split of the discretization error and the model error within the expansion method. Tests with computed model errors of enhanced plate models showed strong thickness locking phenomena. This is caused by introducing expanded constitutive equations into the approximated solution of the current model.

(C) Error estimation by **Posterior Equilibrium Method (PEM) on patches** for solution-, dimension- and model- adaptivity (e.g.: Bufler, Stein 1970 [12]; Stein, Ahmad 1973,1977, [13]; Stein [14]; Bank, Weiser 1985 [16]; Ladevèze, Leguillon 1983 [15]; Ladevèze, Maunder 1996 [22]; Ainsworth, Oden 1992 [17]; Oden, Wu, Ainsworth 1994 [23], Stein, Ohnimus 1996 [10]). Moving from the current displacement approach with stress jumps at element interfaces to the dual problem with new equilibrated continuous tractions, it is possible to get an error estimator with an upper bound for the discretization and an additive split of the solution– and the model error. The analysis occurs on patches with Neumann boundary conditions, using locally higher anisotropic testspaces. The mathematical analysis is still in progress, and the choice of adequate norms for local error estimation needs further research.

(D) **L_2–Projection** of $\boldsymbol{\sigma}_h$ (smoothing and error in the dual energy norm), Zienkiewicz, Zhu 1987, [24]. The coupled hp–adaptivity is hard to realize, the regularity of the solution has to be known and it seems not to be applicable for model adaptivity.

Superconvergent patch recovery (SPR)–techniques Zienkiewicz, Zhu 1992 [25, 26]. This method has certain relations to PEM being under research.

In the next sections we explain the basic idea of the new error estimation by posterior equilibrium method (PEM) on patches for solution–, dimension- and model– adaptivity (C).

3 (C) Posterior Equilibrium Method (PEM) for error estimation on patches

The posterior equilibrium method is based on the local calculation of improved stress tractions $\overset{\star}{\mathbf{t}}$ along the internal boundaries with C^0–continuity condition in normal directions. This introduction of new tractions is a method which can be explained as a stepwise hybrid displacement method or as a Trefftz method for Neumann problems of element patches, see also [12, 13, 15, 22, 10]. We formulate regularized variational problems on patches which equilibrate these new tractions with the known nodal forces from the previous finite element solution, using additional regularizations. With these new equilibrated tractions it is possible to introduce a new anisotropic error estimator for the discretization error and especially for the model error. The unknown locally equilibrated boundary tractions $\overset{\star}{\mathbf{t}}_h$ on patch k, figure 1, have to fulfill the weak equilibrium conditions (a hat denotes nodal quantities)

$$\int_{\partial\Omega_k} \overset{\star T}{\mathbf{t}_h} \; \mathbf{v}_h \, dO \overset{!}{=} (\, \widehat{\mathbf{p}}_{he}^T(\mathbf{u}_h)\, \widehat{\mathbf{v}}_h)_k \qquad \forall\, \mathbf{v}_h \in \mathbf{V}_h \in \mathbf{V}(\Omega_e)\,;\; \overset{\star}{\mathbf{t}}_h \in \mathbf{L}_2(\Gamma_e) \tag{1}$$

related to the previous algebraic equation system for $\boldsymbol{u}_h$. The resulting element nodal forces $\widehat{\mathbf{p}}_{he}$, the test functions (virtual displacements) $\mathbf{v}_h$ and the unknown locally equilibrated boundary tractions $\overset{\star}{\mathbf{t}}_h$ are given as

$$\widehat{\mathbf{p}}_{he}(\mathbf{u}_h) = \mathbf{K}_e \;\; \widehat{\mathbf{u}}_h \quad ; \quad \mathbf{v}_h = \mathbf{N}_v \; \widehat{\mathbf{v}}_h \;. \quad ; \quad \overset{\star}{\mathbf{t}}_h = \overline{\mathbf{N}}_t \; \widehat{\mathbf{t}}_h \;. \tag{2}$$

From equ. 1 – 2 we get

$$\widehat{\mathbf{t}}_h^T \underbrace{\int_{\partial\Omega_e} \overline{\mathbf{N}}_t^T \; \mathbf{N}_v \, dO}_{\mathbf{T}_e} \;\; \widehat{\mathbf{v}}_h \; = \; \widehat{\mathbf{p}}_{he}^T(\mathbf{u}_h)\, \widehat{\mathbf{v}}_h \qquad \forall\, \widehat{\mathbf{v}}_h \in \widehat{\mathbf{V}}_h \tag{3}$$

which results in $\mathbf{T}_e^T \, \widehat{\mathbf{t}}_h = \widehat{\mathbf{p}}_{he}(\mathbf{u}_h)$.Different from Ladevèze and Leguillon (1983) [15], Brink (1997) [27] and Brink & Stein [28], eqn. 1 and 2 fulfill the element equilibrium exactly, such that anisotropic error estimation is possible.

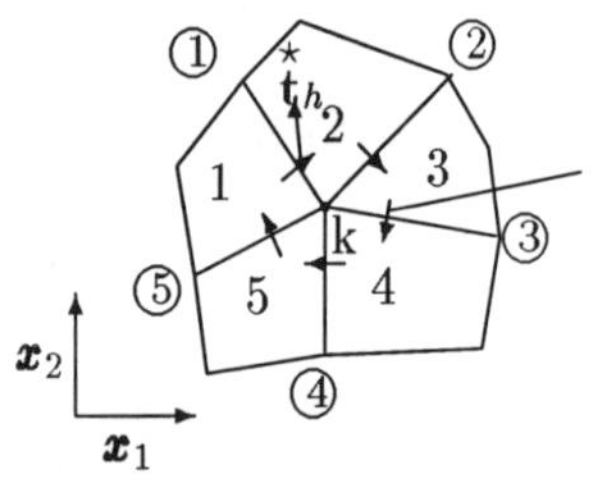

Figure 1: Patch k for node k; element edges are described by edge numbers ⓘ =① –⑤ . A priori defined normal vectors at element edge ⓘ are plotted

Only C^0–continuity conditions for $\overset{\star}{\mathbf{t}}_h$ in the normal direction of element surfaces are necessary, and due to $\overset{\star}{\mathbf{t}}_h \in \mathbf{L}_2(\Gamma_k)$ it is sufficient to describe the equilibrated tractions without C^0–continuity conditions in tangential direction of element surfaces. Consequently, a patch–wise calculation is possible, see Ainsworth (1992) [17].

To avoid global coupling effects of $\overset{\star}{\mathbf{t}}$ it is necessary to describe the new tractions with orthogonal shape functions $\overline{\mathbf{N}}_t$ with respect to the shape functions $\mathbf{N}_v$ of the test space, see equ. 3. $\mathbf{N}_v$ and $\overline{\mathbf{N}}_t$ are different bases of the same approximation space. The orthogonality condition reads

$$\int_{\partial\Omega_e} \overline{N}_t^i \, N_{v\,j} \, dO \; = \; \delta_j^i, \tag{4}$$

thus $\mathbf{N}_v$ forms a covariant and $\overline{\mathbf{N}}_t$ the corresponding contravariant basis.
The weak equilibrium conditions, equ. 3, for a patch k in figure 1 are given in each of the direction $\boldsymbol{x}_1$ and $\boldsymbol{x}_2$ by

$$\text{Element}\quad \left.\begin{array}{rlcl} 1: & \widehat{t}_{h①} - \widehat{t}_{h⑤} & = & \widehat{p}_{h1}(\mathbf{u}_h) \\ 2: & \widehat{t}_{h②} - \widehat{t}_{h①} & = & \widehat{p}_{h2}(\mathbf{u}_h) \\ 3: & \widehat{t}_{h③} - \widehat{t}_{h②} & = & \widehat{p}_{h3}(\mathbf{u}_h) \\ 4: & \widehat{t}_{h④} - \widehat{t}_{h③} & = & \widehat{p}_{h4}(\mathbf{u}_h) \\ 5: & \widehat{t}_{h⑤} - \widehat{t}_{h④} & = & \widehat{p}_{h5}(\mathbf{u}_h) \end{array}\right\} \text{for direction } \boldsymbol{x}_1 \text{ and } \boldsymbol{x}_2\,. \tag{5}$$

In eq. 5 $\widehat{p}_{h1}(\mathbf{u}_h)$ is the contribution of element 1 of the energy equivalent finite element nodal force in node k for each of the direction $\boldsymbol{x}_1$ and $\boldsymbol{x}_2$. Due to the orthogonality in eq. 4 the C^0–continuity of $\overset{\star}{\mathbf{t}}_h$ in normal direction only, the integration of $\overset{\star}{\mathbf{t}}_h$ along element edges results directly in the difference of nodal forces of adjacent element, i.e. the patchwise decoupled, e.g. for element 1:
$\int_{\Gamma_{①}} \overset{\star}{\mathbf{t}}_h \cdot \mathbf{v}_h\, dx - \int_{\Gamma_{⑤}} \overset{\star}{\mathbf{t}}_h \cdot \mathbf{v}_h\, dx = \widehat{p}_{h1} \cdot \widehat{v}_h \quad \forall \widehat{v}_h \rightarrow \widehat{t}_{h①} \cdot \widehat{v}_h - \widehat{t}_{h⑤} \cdot \widehat{v}_h =$
$\widehat{p}_{h1} \cdot \widehat{v}_h \quad \forall \widehat{v}_h \rightarrow \widehat{t}_{h①} - \widehat{t}_{h⑤} = \widehat{p}_{h1}.$
Eq. 5 result in

$$\overbrace{\begin{bmatrix} +1 & & & & -1 \\ -1 & +1 & & & \\ & -1 & +1 & & \\ & & -1 & +1 & \\ & & & -1 & +1 \end{bmatrix}}^{\boldsymbol{T}} \overbrace{\begin{bmatrix} \widehat{t}_{h①} \\ \widehat{t}_{h②} \\ \widehat{t}_{h③} \\ \widehat{t}_{h④} \\ \widehat{t}_{h⑤} \end{bmatrix}}^{\widehat{\mathbf{t}}_h} = \overbrace{\begin{bmatrix} \widehat{p}_{h1}(\mathbf{u}_h) \\ \widehat{p}_{h2}(\mathbf{u}_h) \\ \widehat{p}_{h3}(\mathbf{u}_h) \\ \widehat{p}_{h4}(\mathbf{u}_h) \\ \widehat{p}_{h5}(\mathbf{u}_h) \end{bmatrix}}^{\widehat{\mathbf{p}}_h(\mathbf{u}_h)} \tag{6}$$

where $\mathbf{T}$ can be interpreted as a topology matrix. Equ. 6 can be written as

$$\boldsymbol{T}\,\widehat{\boldsymbol{t}}_h = \bigcup_{e\in k} \boldsymbol{T}_e\,\widehat{\boldsymbol{t}}_h = \bigcup_{e\in k} \widehat{\mathbf{p}}_{he}(\boldsymbol{u}_h) = \widehat{\mathbf{p}}_h(\boldsymbol{u}_h) \tag{7}$$

We introduce covariant Lagrangean base functions and their contravariant base functions

$$\mathbf{N}_v(r,s) = \mathbf{N}_{\mathcal{L}}(r,s) = \mathbf{N}_{\mathcal{L}}(r) \otimes \mathbf{N}_{\mathcal{L}}(s)\,, \tag{8}$$

$$\overline{\mathbf{N}}_t(r,s) = \frac{\overline{\mathbf{N}}_{\mathcal{L}}(r,s)}{det\left(J(r,s)\right)_{\partial\Omega_e}} = \frac{\overline{\mathbf{N}}_{\mathcal{L}}(r) \otimes \overline{\mathbf{N}}_{\mathcal{L}}(s)}{det\left(J(r,s)\right)_{\partial\Omega_e}}, \tag{9}$$

or the hierarchical Legendre base functions, respectively. The co– and contravariant Lagrange base functions for p=1,2 are

$$p=1 \qquad N^1_{\mathcal{L}0} = \tfrac{1}{2}(1-r) \qquad \overline{N}^{10}_{\mathcal{L}} = \tfrac{1}{2}(1-3r)$$
$$N^1_{\mathcal{L}1} = \tfrac{1}{2}(1+r) \qquad \overline{N}^{11}_{\mathcal{L}} = \tfrac{1}{2}(1+3r)\,, \tag{10}$$

$$p=2 \qquad N^2_{\mathcal{L}0} = -\tfrac{1}{2}r(1-r) \qquad \overline{N}^{20}_{\mathcal{L}} = \tfrac{3}{8}(-2-4r+10r^2)$$
$$N^2_{\mathcal{L}1} = 1-r^2 \qquad \overline{N}^{21}_{\mathcal{L}} = \tfrac{3}{8}(3-5r^2) \tag{11}$$
$$N^2_{\mathcal{L}2} = \tfrac{1}{2}r(1+r) \qquad \overline{N}^{22}_{\mathcal{L}} = \tfrac{3}{8}(-2+4r+10r^2)\,,$$

with

$$\delta^j_i = \int_{-1}^{+1} N^p_{\mathcal{L}_i}(r)\; \overline{N}^{pj}_{\mathcal{L}}(r)\, dr\,. \tag{12}$$

The topology matrix $\mathbf{T}$, see equ. 2,7, is not regular and repeated as equ. (i)

$$(i) \qquad \mathbf{T}\,\widehat{\mathbf{t}}_h = \widehat{\mathbf{p}}_h(\mathbf{u}_h) \quad ; \quad \det\,(\boldsymbol{T}^T\,\boldsymbol{T}) \;= 0. \tag{13}$$

In 2D–problems there are one or more zero eigenvalues of $\mathbf{T}^T\mathbf{T}$, and in 3D–problems there are five or more zero eigenvalues.
The regularization of the local equation system for patch k is given by two additional conditions.
The **first additional condition** is posed on Neumann boundaries Γ_{et} with given traction $\bar{\mathbf{t}}$ where equilibrium with the FE–nodal forces is satisfied explicitly as

$$(ii) \qquad \widehat{\mathbf{p}}_h(\bar{\mathbf{t}}) = \widehat{\mathbf{t}}_h \qquad \text{on} \quad \boldsymbol{\Gamma}_t\,. \tag{14}$$

The **second additional condition** is gained with FE– stresses $\boldsymbol{\sigma}_h(\mathbf{u}_h)$ in Ω_e by postulating that the boundary tractions from the previous finite element solution $\mathbf{u}_h$ are approximately equal to the new tractions in the weak sense.

$$(iii) \qquad \int_{\partial\Omega_e} (\boldsymbol{\sigma}_h(\mathbf{u}_h)\;\cdot\;\mathbf{n})^T\;\mathbf{v}_h dO \simeq \int_{\partial\Omega_e} \overset{\star}{\mathbf{t}}{}_h^{\;T}\;\;\mathbf{v}_h dO \quad \forall\,\mathbf{v}_h \in \mathbf{V}_h\,, \tag{15}$$

$$\underbrace{\int_{\partial\Omega_e} (\boldsymbol{\sigma}_h(\mathbf{u}_h)\;\cdot\;\mathbf{n})^T\;\mathbf{N}_v dO\;\widehat{\mathbf{v}}_h}_{\widehat{\mathbf{p}}_h(\boldsymbol{\sigma}_h)} \simeq \widehat{\mathbf{t}}_h^T \int_{\partial\Omega_e} \underbrace{\overline{\mathbf{N}}_t^T}_{\frac{\overline{\mathbf{N}}_{\mathcal{L}}^T}{\det(J(r,s))_{\partial\Omega_e}}} \;\underbrace{\mathbf{N}_v}_{\mathbf{N}_{\mathcal{L}}} dO\;\widehat{\mathbf{v}}_h \tag{16}$$

$$\widehat{\mathbf{p}}_h(\boldsymbol{\sigma}_h)^T\;\widehat{\mathbf{v}}_h \simeq \widehat{\mathbf{t}}_h^T \underbrace{\int_{\partial I} \frac{1}{\det(J(r,s))_{\partial\Omega_e}} \overline{\mathbf{N}}_{\mathcal{L}}^T\,\mathbf{N}_{\mathcal{L}}\;\det(J(r,s))_{\partial\Omega_e} dO}_{\mathbf{I}}\;\widehat{\mathbf{v}}_h \tag{17}$$

$$\Rightarrow \quad \widehat{\mathbf{p}}_h(\boldsymbol{\sigma}_h) = \widehat{\mathbf{t}}_h\,. \tag{18}$$

This results in a least–squares–approximation

$$(iii) \qquad {}^1/_2(\widehat{\mathbf{p}}_h(\boldsymbol{\sigma}_h)-\widehat{\mathbf{t}}_h)^2 \rightarrow Min \quad . \tag{19}$$

Summarizing the calculation of tractions $\overset{\star}{\mathbf{t}}_h$:

$$\begin{array}{lll} \text{Nodal forces} & \mathbf{T}\,\widehat{\mathbf{t}}_h - \widehat{\mathbf{p}}_h(\mathbf{u}_h) = \mathbf{0} & \text{exactly fulfilled} \\ \text{Boundary tractions} & \widehat{\mathbf{t}}_h - \widehat{\mathbf{p}}_h(\mathbf{t}) = \mathbf{0} & \text{conditions (i), (ii)} \\ \text{Regularization} & {}^1/_2(\widehat{\mathbf{p}}_h(\boldsymbol{\sigma}_h)-\widehat{\mathbf{t}}_h)^2 \rightarrow Min & \text{weakly fulfilled conditions (iii)} \end{array} \tag{20}$$

Remark: Further developments are in progress to extend condition (iii) to the weak form of inelastic field equations. To avoid locking effects the local approximation space for stress calculation has to be expanded with further internal deformation modes, e.g. bubble modes.

3.1 SOLUTION– AND MODEL–ERROR ESTIMATION

After calculation of the equilibrated tractions the discretization error can be calculated by the difference of the equilibrated tractions and the previous FE–tractions in the L_2–norm or better by the energy norm of the difference of the new solution $\boldsymbol{u}_{h+}$ of the Neumann problem (for given tractions $\boldsymbol{t}_h$) minus the previous solution $\boldsymbol{u}_h$, see also Ladevèze, Leguillon 1983 [15] and Bank, Weiser 1985 [16].
For this, the local data of the current model with the approximation space $\mathbf{V}_h$ are:

$\widehat{\mathbf{u}}_h$	:	local nodal displacement for each element in $\Omega_e; \mathbf{u}_h = \mathbf{N}_v \widehat{\mathbf{u}}_h$
$\mathbf{K}_e$	:	local stiffness matrix for each element e
$\widehat{\mathbf{p}}_h = \mathbf{K}_e\,\widehat{\mathbf{u}}_h$	:	local nodal forces for each element e
$\widehat{\mathbf{p}}_h(\boldsymbol{\sigma}_h\mathbf{n})$	:	local nodal boundary tractions, calculated with $\boldsymbol{\sigma}_h\mathbf{n}; \boldsymbol{\sigma}_h = \boldsymbol{\sigma}_h(\mathbf{u}_h)$
$\overset{\star}{\mathbf{t}}_h = \overline{\mathbf{N}}_t \widehat{\mathbf{t}}_h$	:	local boundary tractions, calculated with PEM
$\widehat{\mathbf{t}}_h$	:	local nodal boundary tractions, calculated with PEM

Introducing the locally expanded test space $\mathbf{v}_{h+} \in \mathbf{V}_{h+} \subset \mathbf{V}$ and the locally expanded solution space $\mathbf{u}_{h+} \in \mathbf{V}_{h+} \subset \mathbf{V}$, the local variational problem is formulated as

$$a(\mathbf{u}_{h+}, \mathbf{v}_{h+})_{\Omega_e} = L(\mathbf{v}_{h+})_{\partial\Omega_e} \quad \forall\, \mathbf{v}_{h+} \in \mathbf{V}_{h+}\ ;\ \mathbf{u}_{h+} \in \mathbf{V}_{h+} \tag{21}$$

with the locally enhanced test space

$$\mathbf{V}_{h+} := \{\mathbf{v}_{h+} \in H^1(\Omega_e)\ ;\ \mathbf{v}_{h+} = \mathbf{0} \text{ on } \Gamma_{eu}\ ;\ \mathbf{v}_{h+} \text{ without rigid body modes}\} \tag{22}$$

and

$$L(\mathbf{v}_{h+})_{\partial\Omega_e} = \int_{\partial\Omega_e} \mathbf{v}_{h+}^T \overset{\star}{\mathbf{t}}_h dO\ . \tag{23}$$

The local tractions $\overset{\star}{\mathbf{t}}_h$ are gained from PEM. Mechanically, equ. 21 describes the weak form of equilibrium of the tractions $\overset{\star}{\mathbf{t}}_h$ and yields the displacements $\mathbf{u}_{h+}$ without rigid body modes. The **"equilibrium" discretization error estimator** η_{De} in Ω_e is given by the difference of the current FE–approximation $\mathbf{u}_h$ and the locally improved approximation $\mathbf{u}_{h+}$ in the energy norm by

$$\eta_{De}^2 = a((\mathbf{u}_{h+} - \mathbf{u}_h), (\mathbf{u}_{h+} - \mathbf{u}_h))_{\Omega_e} . \tag{24}$$

For model error estimation we introduce the local variational form a_M for an hierarchically expanded model, connected with the reduced model by a monomorphic transformation, and the corresponding discretized solution $\tilde{\mathbf{u}}_{h+}$ using again enhanced local test spaces

$$a_M(\tilde{\mathbf{u}}_{h+}, \mathbf{v}_{h+})_{\Omega_e} = L(\mathbf{v}_{h+})_{\partial\Omega_e} \quad , \quad \forall\, \mathbf{v}_{h+} \in \mathbf{V}_{h+} \; ; \; \tilde{\mathbf{u}}_{h+} \in \mathbf{V}_{h+} \tag{25}$$

and yielding $\tilde{\mathbf{u}}_{h+}$ with the right–hand side

$$L(\mathbf{v}_{h+})_{\partial\Omega_e} = \int\limits_{\partial\Omega_e} \mathbf{v}_{h+}^T \overset{\star}{\mathbf{t}}_h dO . \tag{26}$$

Equ. 25 describes the weak form of equilibrium of the new boundary tractions $\overset{\star}{\mathbf{t}}_h$ (of an element patch k) for the expanded model within element Ω_e. Note that $\mathbf{u}_{h+}$ and $\mathbf{u}_h$ have the same dimension.
Then, the **"equilibrium" model error estimator** η_{Me} in the energy norm within Ω_e is given by

$$\eta_{Me}^2 = a_M(\tilde{\mathbf{u}}_{h+} - \mathbf{u}_{h+}, \tilde{\mathbf{u}}_{h+} - \mathbf{u}_{h+})_{\Omega_e} \; ; \; \tilde{\mathbf{u}}_{h+} \in \mathbf{V}_{h+} \; ; \; \mathbf{u}_{h+} \in \mathbf{V}_{h+} . \tag{27}$$

This model error is defined by the difference between the locally expanded approximation $\mathbf{u}_{h+}$ (using the reduced $2\frac{1}{2}$D–elastic model) and the locally expanded approximation $\tilde{\mathbf{u}}_{h+}$ (using the full 3D– elastic model) in the energy norm. This model error estimator, equ. 27, is subject to locking phenomena for thin walled plates. In the next section several other possibilities of error norms are given.
To avoid substantial influences of the descretization error on the model error, the global discretization error has to be smaller than the global model error, $\eta_M^2 \geq \eta_D^2$.

3.2 MODEL–ERROR–ESTIMATORS

The model error estimator can be gained in different ways; in section 3.1 it was calculated in the energy norm by using the error of the displacements, equ. 27. The estimator of the model displacement error, derived from equ. 27, is

$$\eta_{Mu_e}^2 = ||\tilde{\mathbf{u}}_{h+} - \mathbf{u}_{h+}||_{E_M(\Omega_e)}^2 . \tag{28}$$

This error has strong locking effects due to neglecting of deformation modes in the reduced model. Next we introduce the model error estimator in the dual energy norm using the stresses of the current and the expanded model. Herein, the locking phenomena do not appear because the reduced model (e.g. for a thin plate in bending) is introduced to get fairly accurate stresses without regarding deformation modes in thickness direction. This estimator reads

$$\eta^2_{M\sigma_e} = ||\tilde{\boldsymbol{\sigma}}_{h+} - \boldsymbol{\sigma}_{h+}||^2_{E_M(\Omega_e)} \ . \tag{29}$$

A relative model error in the L_2– norm, taking into account the neglected deformation modes of the reduced model is given by

$$\zeta^2_{Mu_e} = ||\tilde{\mathbf{u}}_{h+} - \mathbf{u}_{h+}||^2_{L_2(\Omega_e)} \tag{30}$$

which is useful for the expansion of 2D–plate and –shell models. For a pure displacement method an energy oriented estimator

$$\eta^2_{MEe} = ||\tilde{\mathbf{u}}_{h+}||^2_{E_M(\Omega_e)} - ||\mathbf{u}_{h+}||^2_{E(\Omega_e)} \tag{31}$$

can be used, too. But in most investigated cases, this indicator tends to zero which is known from the good–minded convergence of the energy.
The examples of a long and a short bar with homogeneous tension, treated first as 1D–tension problems and then as 3D–problems show that the model error given in the dual energynorm, eq. 29, is adequate w.r.t. the model reduction and hence numerically effective, see [29].

3.3 LOCAL ERROR ESTIMATOR USING HIGHER ANISOTROPIC TEST FUNCTIONS

A suitable anisotropic error estimator starts with a local problem (Neumann– or Dirichlet– boundary value problem)

$$\eta_e^2 = a(\boldsymbol{e}, \boldsymbol{e}) = a(\boldsymbol{e}, \boldsymbol{e}_+) = \int_{\Omega_e} \boldsymbol{R} \cdot \boldsymbol{e}_+ d\boldsymbol{x} + \frac{1}{2} \int_{\partial\Omega_e} \boldsymbol{J} \cdot \boldsymbol{e}_+ ds \tag{32}$$

with the unknown updated error $\boldsymbol{e}_+ \in \boldsymbol{V}_+ := \boldsymbol{V} \ominus \boldsymbol{V}_h$ of the displacements. The strategy is now to choose the next higher possible degree of polynomials in the approximation space or a suitable refinement for every direction $z \in r, s, t$ of the parameter coordinates of the unit cube, in detail see [7, 29]. The approximation space is adjusted to the anisotropic refinement (directions of the anisotropic coordinates r, s, t) with $\boldsymbol{N}^z_+(r, s, t)$ denoting the refined shape functions

$$\boldsymbol{e}^z_+ = \boldsymbol{N}^z_+(r, s, t) \cdot \hat{\boldsymbol{e}}^z_+ \in \boldsymbol{V}_+ \subset \boldsymbol{V} \ ; \ z = r, s, t \quad \text{and} \tag{33}$$

$$\boldsymbol{e}^z = \boldsymbol{N}^z(r, s, t) \cdot \hat{\boldsymbol{e}}^z \in \boldsymbol{V} \quad . \tag{34}$$

This yields an approximated error, which is the sum of the error components in the directions r, s, t

$$\eta_z^2 = \sum_e \left\{ \int_{\Omega_e} \boldsymbol{R} \cdot \boldsymbol{e}^z_+ d\boldsymbol{x} + \frac{1}{2} \int_{\partial\Omega_e} \boldsymbol{J} \cdot \boldsymbol{e}^z_+ ds \right\} \quad \text{and} \quad \eta_e^2 = \sum_{z=r,s,t} \eta_z^2. \tag{35}$$

In total, local estimators are available on all unit patch which test the three directions with, e.g., $p_+ = 1; 2$ or $h_+ = \frac{h}{2}; \frac{h}{4}$. These general results are mapped in an energy–invariant way to the real patches in order to control h– and p– adaptivity locally, [7, 29]. Additionally, the pollution error has to be considered containing the global influences onto the local error. For this purpose Greens's functions of the given operator is used.

4 Example of a clamped plate in bending

A clamped plate, see Fig. 2 – 3, is adaptively calculated by the anisotropic PEM. The dimensional and model adaptivity from 2 1/2 D–Reissner–Mindlin theorie to 3D–elasticity theory starts at the boundaries and grows into the domain. This effect is correct due to the singularities at the top and bottom edge of the plate. As 2D–3D transition elements, we use special Lagrange–multiplier techniques for avoiding inconsistencies and locking effects.

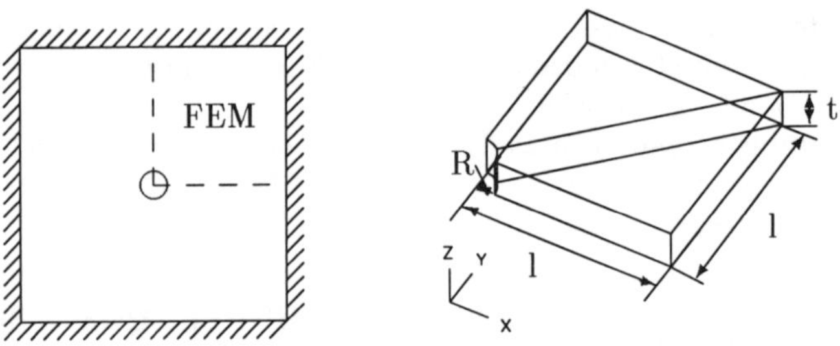

Figure 2: Example of a plate in bending with a hole. Calculation of a quarter of the system due to symmetry conditions. The plate is fully clamped at all sides and has uniform loading. Geometry: $\frac{l}{t} = 10$; $\frac{l}{R} = 10$ E= 30 000 MPa, $\nu = 0.3$

5 Conclusions

A new generation of adaptive numerical methods in structural mechanics was outlined, namely the full integration of solution–, dimensional– and model– errors into well–known hp-adaptive concepts for boundary value problems of stiffened plates and shells.
The resulting anisotropic error indicators admit optional h- and p-adaptive processes but moreover d– and m– adaptivity for plates and shells, i.e. dimensional expansions of the approximation space and even the update of the mathematical model, i.e. of constitutive and geometrical equations.
It is evident that the expansion method – developing from lower to higher approximation spaces and also from reasonably simple to more complicated mathematical models in disturbed subdomains – is numerically more efficient than the reduction method, which starts with the highest model and searchs for simpler models in subdomains, e.g. by means of kinematic hypotheses. This holds especially for complicated stiffened thin–walled structures. And last but not least this expansion method corresponds to the typical engineering way of thinking.

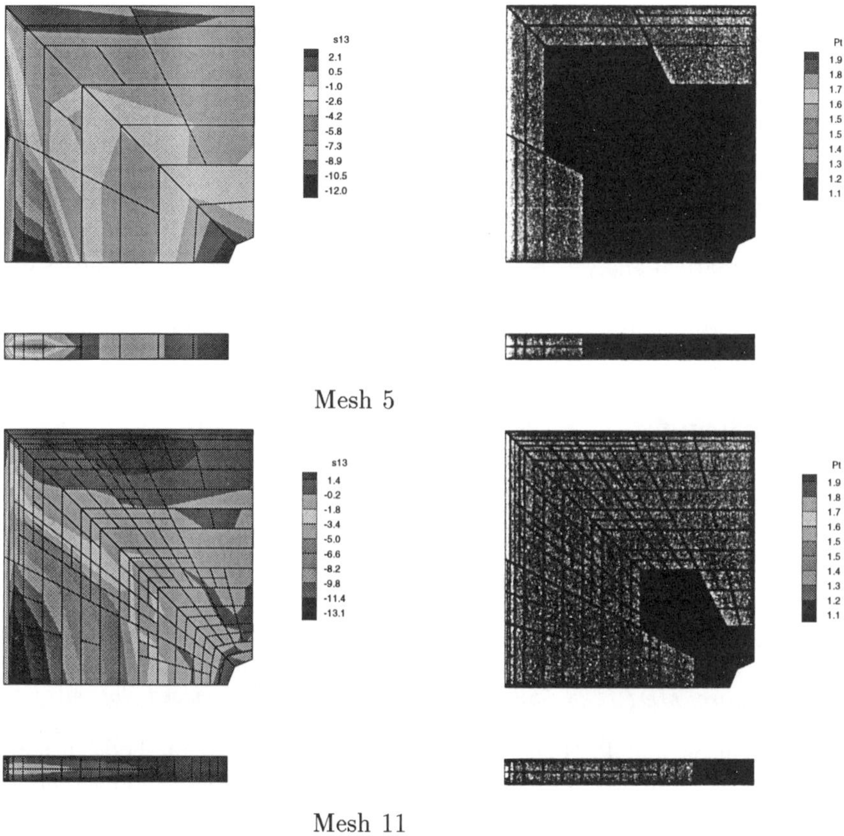

Figure 3: Example of anisotropic solution– and model adaptivity for a plate in bending with a hole; at the right side shear–stresses are given in the middle plane for mesh 5 and 11 , and at the right side the model distribution is shown for the same meshes;
2D–domain (dark region): biquadratic inplane shape functions, linear shape function in thickness direction, 2–order higher integration for equilibration, plane stress material equation;
3D–domain (light region): triquadratic shape functions, 2–order higher integration for equilibration, 3D–elastic material equation;

Residual error estimators on patches with enhanced anisotropic test spaces are beneficial for solution and dimension adaptivity.
Residual error estimators are insufficient for model adaptivity (especially the constitutive equations) because the residual model error becomes very large (due to too large stresses (locking) in enhanced models).
The posterior equilibrium method (PEM) yields physically consistent orthogonalized boundary tractions $\mathbf{t}_h$ for the expanded model, and it provides solution– and model–error estimators from local variational problems on patches which can be interpreted as a regularized Trefftz method on element patch boundaries.
In order to get optimal adaptivity for local quantities, the pollution error from outside has to be considered which needs additional global solutions, [30]. In this paper, adaptivity is scheduled in a global sense such that the pollution error is contained in the error–density function.
The mathematical analysis is still in progess, and further developments of PEM and related error estimators, including more complex material behaviour like plasticity or deformations of reinforced concrete in state 2, are in progress.

REFERENCES

[1] I. Babuška and C. Schwab. A posteriori error estimation for hierarchic models of elliptic boundary value problems on thin domains. Technical report, Technical Note BN 1148, May 1993, Institute for Physical Science and Technology, University of Maryland College Park, MD20740, USA, 1993.

[2] S. Jensen. Adaptive dimensional reduction for scalar boundary value problems. *Department of Mathematics, University of Maryland Baltimore County, Baltimore, MD 21228-5398 U.S.A., Januar 7, 1991, revised November 8 1991.*

[3] C. Johnson and P. Hansbo. Adaptive finite element methods in computational mechanics. *Computer Methods in Applied Mechanics and Engineering*, 101:143–181, 1992.

[4] E. Stein, W. Rust, and S. Ohnimus. h– and d– adaptive FE element methods for two–dimensional structural problems including postbuckling of shells, paper from the "2nd Reliability Colloquium", Crakov, 1991. *Computer Methods in Applied Mechanics and Engineering*, 101:315–354, 1992.

[5] E. Stein and S. Ohnimus. Concept and realisation of integrated adaptive finite element methods in solid– and structural–mechanics. *Numerical Methods in Engineering '92, Proceedings of the First European Conference on Numerical Methods in Engineering of ECCOMAS September 1992, Brussels, Belgium, Elsevier Science Publishers B.V.*, pages 163–170, 1992.

[6] E. Stein, B. Seifert, S. Ohnimus, and C. Carstensen. Adaptive finite element analysis of geometrically non–linear plates and shells, especially Buckling . *International Journal for Numerical Methods in Engineering*, 37:2631–2655, 1994.

[7] E. Stein and S. Ohnimus. Dimensional adaptivity in linear elasticity with hierarchical test–spaces for h– and p– refinement processes. *Engineering with Computers*, 12:107–119, 1996.

[8] E. Stein and S. Ohnimus. Integrated solution– and dimensional adaptivity of fem by hierarchical projections. *Proceedings of the Conference OFEA-1995 in St. Petersburg, Journal of mathematical physics and numerical mathematics*, 9:95–117, 1995.

[9] E. Stein and S. Ohnimus. Expansion method for the integrated solution and model adaptivity within the fe analysis of plates and shells. *Advances in Finite Element Technology, Ed. by N.-E. Wiberg, CIMNE, Barcelona 1995*, pages 150–173, 1995.

[10] E. Stein and S. Ohnimus. Equilibrium method for postprocessing and error estimation in the finite elemente method. *in First international workshop on Trefftz Method, 30. May - 1. June, 1996, Cracov, Poland, Computer Assisted Mechanics and Engineering Sciences*, 4:645–666, 1997.

[11] S. Ohnimus. *Theorie und Numerik dimensions– und modelladaptiver Finite-Elemente-Methoden von Flächentragwerken.* Berichte des Instituts für Baumechanik und Numerische Mechanik der Universität Hannover, F96/6,1996.

[12] H. Bufler and E. Stein. Zur Plattenberechnung mittels finiter Elemente. *Ing. Archiv*, 39:248–260, 1970.

[13] E. Stein and R. Ahmad. An equilibrium method for stress calculation using finite element displacements models. *Computer Methods in Applied Mechanics and Engineering*, 10:175–198, 1977.

[14] E. Stein. The practical treatment of stress concentration and singularities within finite element displacement algorithms. *Lecture Notes in Mathematics, Singularities and Constructive Methods for Their Treatment, Proceedings, Oberwolfach 1983, Edited by P. Grisvard, W. Wendland and J.R. Whiteman, Springer–Verlag Berlin Heidelberg New York Tokyo*, pages 278–299, 1983.

[15] P. Ladevèze and D. Leguillon. Error estimate procedure in the finite element method and applications. *SIAM J. Num. Anal.*, 20:485–509, 1983.

[16] R. E. Bank and A. Weiser. Some a posteriori error estimators for elliptic partial differential equations. *Mathematics of Computation*, 44:283–301, 1985.

[17] M. Ainsworth and J.T. Oden. A procedure for a posteriori error estimation for h–p finite element methods. *Computer Methods in Applied Mechanics and Engineering*, 101:73–96, 1992.

[18] W. Hackbusch. Multi–grid method and applications. *Springer Series in Computational Mathematics, Springer–Verlag, Berlin Heidelberg New York Tokyo*, 4, 1985.

[19] I. Babuška and W.C. Rheinboldt. A–posteriori error estimates for the finite element method. *International Journal for Numerical Methods in Engineering*, 12:1597–1615, 1978.

[20] I. Babuška and W.C. Rheinboldt. Error estimates for adaptive finite element computations. *SIAM Journal on Numerical Analysis*, 15:736–754, 1978.

[21] C. Schwab. A posteriori error estimation for hierarchic plate models. Technical report, Institute for Supercomputing and Applied Mathematics, University of Maryland, IBM Scientific Centre, Heidelberg.

[22] P. Ladevèze and E.A.W. Maunder. A general methodology for recovering equilibration finite element tractions and stress fields for plate and solid elements. *in First international workshop on Trefftz Method, 30. May - 1. June, 1996, Cracov, Poland, Archives of Mechanics*, pages 34–35, 1995.

[23] J.T. Oden, W. Wu, and M. Ainsworth. An a posteriori error estimator for finite element approximation of the Navier–Stokes equation. *Computer Methods in Applied Mechanics and Engineering*, 111:185–202, 1994.

[24] O.C. Zienkiewicz and J.Z. Zhu. A simble error estimator and adaptive procedure for practical engineering analysis. *International Journal for Numerical Methods in Engineering*, 24:337–357, 1987.

[25] O.C. Zienkiewicz and J.Z. Zhu. The superconvergent patch recovery and a posteriori error estimators, part 1: The recovery technique. *International Journal for Numerical Methods in Engineering*, 33:1333–1364, 1992.

[26] O.C. Zienkiewicz and J.Z. Zhu. The superconvergent patch recovery and a posteriori error estimators, part 2: Error estimators and adaptivity. *International Journal for Numerical Methods in Engineering*, 33:1365–1382, 1992.

[27] U. Brink. Adaptive gemischte finite Elemente in der linearen finiten Elastostatik und deren Kopplung mit Randelementen. *Institut für Baumechanik und Numerische Mechanik, Dissertation Universität Hannover IBNM–Bericht*, in print, 1997.

[28] U. Brink and E. Stein. A–posteriori error estimation in large–strain elasticity using equilibrated local Neumann problems. *Computer Methods in Applied Mechanics and Engineering*, in print, 1998.

[29] E. Stein and S. Ohnimus. Coupled model– and solution–adaptivity in the finite–element method. *Computer Methods in Applied Mechanics and Engineering*, 150:327–350, 1997.

[30] I. Babuška, T. Strouboulis, S.K. Gangaraj, and C.S. Upadhyay. A-posteriori estimation and adaptive control of the pollution-error in the h-version of the finite element method. *International Journal for Numerical Methods in Engineering*, 38:4207–4235, 1995.

THREE–DIMENSIONAL SHELL FORMULATION AND ELEMENTS FOR LARGE DEFORMATIONS

M. BISCHOFF, E. RAMM
Institute of Structural Mechanics,
University of Stuttgart, Germany

Abstract

Well–known finite element concepts like the Assumed Natural Strain (ANS) and the Enhanced Assumed Strain (EAS) techniques are combined to derive efficient and reliable finite elements for a continuum based, three–dimensional shell formulation including thickness stretch. The formulation is capable to handle large strain and rotation problems. Furthermore, any three–dimensional constitutive law, either linear–elastic, hyperelastic, elastic–plastic or others, can be implemented without modifications.

For the derivation of the non–linear EAS formulation, an alternative way to the one proposed by Simo and Armero [19] is chosen. In context of the question for a sensible choice of trial functions the relation of hybrid stress and EAS elements is investigated. A modification and extension of the thickness strains is necessary to eliminate "thickness locking" problems resulting from the three–dimensional stress and strain state.

1 Scope of Study

The well-known phenomenon "locking" of displacement based finite elements for thin–walled beams, plates and shells is caused by an unbalance of the trial functions. This unbalance, described in innumerable papers, can be cured by either reduction or enhancement of the degrees of freedom to tare the element–trials; or to use a more vivid explanation: Each static variable has to find its kinematic counterpart in an energy expression. From the numerous remedies the following two have been proven to be in particular successful, namely

- the Assumed Natural Strain (ANS) Method, where special interpolations for selected strain components are chosen to eliminate their parasitic influence (Hughes, Tezduyar [11], Dvorkin, Bathe [10]). The procedure which is extremely simple and efficient is in the meantime a classical component of many element formulations and computer codes.
- the Enhanced Assumed Strain (EAS) Method, where extra incompatible strain components are added to the conventional displacement dependent compatible strains. The

H. A. Mang and F. G. Rammerstorfer (eds.),
IUTAM Symposium on Discretization Methods in Structural Mechanics, 27–34.

method has been originally proposed by Simo and Rifai [18] and later on extended to large displacements by Simo, Armero [19]. One objective of the present study is a modified extension of the EAS method to geometrical non–linearities. Instead of a multiplicative decomposition of the deformation gradient as proposed by Simo, Armero [19] the Green–Lagrange strain tensor is additively decomposed into the compatible and the enhanced parts (Büchter, Ramm [8], Braun et al. [7]).

The basis for the presented elements is a combination of the ANS method for transverse shear and EAS extensions for inplane strains, as proposed in Andelfinger, Ramm [1] for a geometrically linear, 5–parameter, four–node shell element. In the present study this concept is extended to geometrical non–linearities within the framework of a three–dimensional, so–called 7–parameter shell formulation, suitable for the treatment of arbitrarily large displacements and rotations. In addition to the four–node element a new nine–node quadrilateral is presented.

Due to the three–dimensional features of the theory, some arising problems have to be solved. In particular an extra artificial stiffening effect for curved structures [14], termed "curvature thickness locking", has to be eliminated.

2 Non–Linear EAS Formulation

The variational formulation of the EAS method starts from the three field Hu–Washizu functional. The decisive step is the *reparametrization* of the Green–Lagrange strain tensor: $\boldsymbol{E} = \boldsymbol{E}^u + \tilde{\boldsymbol{E}}$. This means that the displacement dependent compatible strains $\boldsymbol{E}^u$ are enriched by a set of additional independent enhanced assumed strains $\tilde{\boldsymbol{E}}$. This leads to a modified three field variational principle.

$$\tilde{\Pi}(\boldsymbol{u}, \tilde{\boldsymbol{E}}, \boldsymbol{S}) = \int_B W_s\left(\boldsymbol{E}^u + \tilde{\boldsymbol{E}}\right) dV - \int_B \boldsymbol{S} : \tilde{\boldsymbol{E}}\, dV - W^{ext}(\boldsymbol{u}) \tag{1}$$

2.1 DISCRETIZATION AND LINEARIZATION

Geometry and displacements are discretized by an isoparametric interpolation with nodal values $\boldsymbol{x}_e$ and $\boldsymbol{d}$ for coordinates and displacements, respectively. Enhanced strains and independent stresses are interpolated by functions $\boldsymbol{M}$ and $\boldsymbol{P}$ based on the parameters $\boldsymbol{\alpha}$ and $\boldsymbol{\beta}$, respectively. As usual h denotes the approximated fields.

$$\boldsymbol{x} \approx \boldsymbol{x}_h = \boldsymbol{N} \cdot \boldsymbol{x}_e; \;\; \boldsymbol{u} \approx \boldsymbol{u}_h = \boldsymbol{N} \cdot \boldsymbol{d}; \;\; \tilde{\boldsymbol{E}} \approx \tilde{\boldsymbol{E}}_h = \boldsymbol{M} \cdot \boldsymbol{\alpha}; \;\; \boldsymbol{S} \approx \boldsymbol{S}_h = \boldsymbol{P} \cdot \boldsymbol{\beta} \tag{2}$$

In order to reduce the three field functional $\tilde{\Pi}_h(\boldsymbol{u}_h, \tilde{\boldsymbol{E}}_h, \boldsymbol{S}_h)$ to a two field functional $\tilde{\tilde{\Pi}}_h(\boldsymbol{u}_h, \tilde{\boldsymbol{E}}_h)$ Simo and Rifai [18] suggested to choose the interpolation of $\boldsymbol{S}_h$ orthogonal to $\tilde{\boldsymbol{E}}_h$. Introduction of the above equations into the energy formulation and linearization with respect to **d** and **α** leads to an element matrix of the following form.

$$\begin{bmatrix} \boldsymbol{K}_{e+u} + \boldsymbol{K}_g & \boldsymbol{L}^T \\ \boldsymbol{L} & \boldsymbol{D} \end{bmatrix} \cdot \begin{bmatrix} \Delta \boldsymbol{d} \\ \Delta \boldsymbol{\alpha} \end{bmatrix} = \begin{bmatrix} \boldsymbol{P}^* \\ \boldsymbol{0} \end{bmatrix} - \begin{bmatrix} \boldsymbol{R} \\ \tilde{\boldsymbol{R}} \end{bmatrix} \tag{3}$$

Here, P^* denotes the element load vector, R and $\tilde{R}$ are the residual forces. For details concerning the process of discretization and linearization the reader is referred to Bischoff, Ramm [5]. As shown by Braess [6] the format of equation (3) is a strong indication for an element model having the features of a minimum rather than that of a stationary principle.

Since the "internal" strain parameters do not have to be compatible across the element boundaries they can be eliminated on the element level, leading to a modified stiffness expression for the EAS element

$$\left(K_{e+u} + K_g - L^T \cdot D^{-1} \cdot L\right) \cdot \Delta d = P^* - R + L^T \cdot D^{-1} \cdot \tilde{R} \tag{4}$$

Equation (4) is the basis for typical pathfollowing solution schemes, e. g. arc–length methods in connection with Newton–Raphson iterations. Compared to the formulation of Simo, Armero [19] two advantages can be realized: First, only one additional expression K_g in the tangential stiffness matrix has to be computed, compared to four in Simo, Armero [19], Second, since non–linear terms do not result from the EAS extension, a 2x2 Gauss quadrature is sufficient for the bilinear element (3x3 for the nine node element), contrary to the original EAS formulation.

3 Three–Dimensional Shell Formulation

The use of arbitrary three–dimensional constitutive equations without any modification was the driving force behind the higher order so–called 7–parameter formulation proposed in Büchter, Ramm [8] and further elaborated in Büchter et al. [9]. This "extension into the three dimensions" covers also large strain problems and – most important – still keeps the basic and efficient shell assumption, namely the explicit integration across the thickness leading to stress resultants. The higher order kinematics is described by a linear variation of the thickness stretch (see section 3.5) which in turn counterbalances the linearly varying corresponding thickness stress.

The basic idea to include a thickness stretch into a shell formulation has already been followed by others, although in a different way; see e.g. Reissner [15].

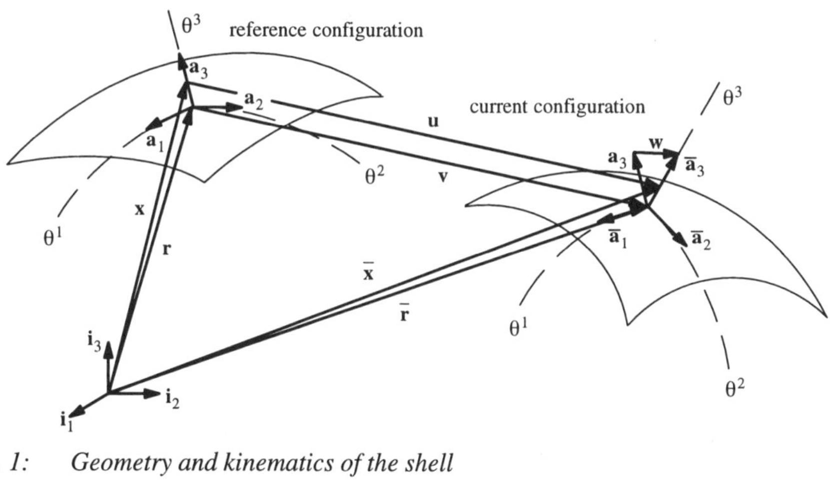

Fig. 1: Geometry and kinematics of the shell

3.1 6–PARAMETER DISPLACEMENT FORMULATION AS BASIS

The geometry and kinematic description of the shell body is illustrated in Fig. 1. Values referring to the deformed configuration are denoted by a bar. The midsurface of the shell is described by position vectors $\boldsymbol{r}$. The covariant base vectors $\boldsymbol{a}_\alpha = \boldsymbol{r}_{,\alpha}$ define the metric of the shell at a given point of its midsurface, $\boldsymbol{a}_3$ is the *director* and defines the thickness direction of the shell. The position vector of an arbitrary point of the shell body and its metric vectors are given by

$$\boldsymbol{x} = \boldsymbol{r} + \theta^3 \boldsymbol{a}_3 \ ; \qquad \boldsymbol{g}_\alpha = \boldsymbol{x}_{,\alpha} = \boldsymbol{a}_\alpha + \theta^3 \boldsymbol{a}_{3,\alpha} \ ; \qquad \boldsymbol{g}_3 = \boldsymbol{a}_3 \tag{5}$$

The displacement $\boldsymbol{u}$ of a point of the shell body is composed of the displacement $\boldsymbol{v}$ of the corresponding point on the midsurface and a difference vector $\boldsymbol{w}$.

$$\boldsymbol{u} = \bar{\boldsymbol{x}} - \boldsymbol{x} = \boldsymbol{v} + \theta^3 \boldsymbol{w} \ ; \qquad \boldsymbol{v} = \bar{\boldsymbol{r}} - \boldsymbol{r} \ ; \qquad \boldsymbol{w} = \bar{\boldsymbol{a}}_3 - \boldsymbol{a}_3 \tag{6}$$

Different from a conventional shell formulation with five independent degrees of freedom, in general three displacements and two rotations, here the difference vector for the update of the director introduces a sixth displacement degree of freedom representing a thickness stretch of the shell, which reflects almost the same parametrization used for a brick element with two nodes on the upper and lower surface.

3.2 TREATMENT OF MEMBRANE STRAINS

In order to avoid both in–plane shear and membrane locking, the membrane strains are enriched by means of the EAS method; here the essential point is a sensible choice of the extra enhanced strain modes. These have to be linearly independent from the compatible strains obtained by differentiating the displacement field.

The subsequent discussion of enhanced strain interpolations refers to the components of the linearized Green Lagrange strain tensor for the in–plane strains of the shell,

$$E_{\alpha\beta} = \frac{1}{2}\left(u_{\alpha,\beta} + u_{\beta,\alpha}\right) = E^c_{\alpha\beta} + \theta^3 E^l_{\alpha\beta} \ ; \qquad \alpha,\beta \in \{1,2\} \tag{7}$$

where $E^c_{\alpha\beta}$ and $E^l_{\alpha\beta}$ are the membrane and the bending parts of the strains.

3.2.1 *Four Node Element*

For a bilinear interpolation of the displacements a maximum of 7 bilinear enhanced strain components may be supplemented for a complete polynomial

$$\begin{pmatrix} \tilde{E}^c_{11} \\ \tilde{E}^c_{22} \\ \tilde{E}^c_{12} \end{pmatrix} \in span \left[\begin{pmatrix} \xi \\ 0 \\ 0 \end{pmatrix} \begin{pmatrix} 0 \\ \eta \\ 0 \end{pmatrix} \begin{pmatrix} 0 \\ 0 \\ \xi \end{pmatrix} \begin{pmatrix} 0 \\ 0 \\ \eta \end{pmatrix} \begin{pmatrix} \xi\eta \\ 0 \\ 0 \end{pmatrix} \begin{pmatrix} 0 \\ \xi\eta \\ 0 \end{pmatrix} \begin{pmatrix} 0 \\ 0 \\ \xi\eta \end{pmatrix} \right] \tag{8}$$

Recently Andelfinger and Ramm [1] have pointed to the equivalence of hybrid stress elements based on the Hellinger–Reissner (HR) formulation and EAS elements, related to the Hu–Washizu variational principle. It has been stated that this equivalence holds if the trials for stresses in the HR elements are complementary to those for strains in the EAS approach.

As an example of the equivalence the EAS–7 membrane element and the hybrid–mixed element of Pian, Sumihara [12] is mentioned (see also Atluri, Punch [2])

A:	**Pian–Sumihara:**	**B:**	**EAS–7:**
	Interpolations of stresses		Interpolations of strains

$$S^{11} = \beta^1 + \beta^4\eta \qquad \tilde{E}_{11} = \alpha_1\xi + \alpha_5\xi\eta$$
$$S^{22} = \beta^2 + \beta^5\xi \qquad \tilde{E}_{22} = \alpha_2\eta + \alpha_6\xi\eta \tag{9}$$
$$S^{12} = \beta^3 \qquad \tilde{E}_{12} = \alpha_3\xi + \alpha_4\eta + \alpha_7\xi\eta$$

$\beta^1,...,\beta^5$ and $\alpha_1,...,\alpha_7$ are the unknown parameters in the stress and strain trial, respectively. It is apparent that no polynomial term in the stress interpolation in **A** shows up in its corresponding strain interpolation in **B**. Stress and strain interpolation of **A** and **B** "sum up" to a complete bilinear polynomial. For this element the equivalence of both formulations has been verified mathematically by Yeo and Lee [20]. In fact it can be shown analytically for elements of arbitrary interpolation order, as will be shown in a subsequent paper by the authors (see also Braess [6]).

3.2.2 *Nine Node Element*

The membrane strains are modified via the EAS method. The objectives of this procedure are to obtain an element formulation that is less sensitive to mesh distortions, variationally sound and satisfies the patchtest. A nine node assumed stress element has been presented by Sansour [17]. Following the concept discussed in section 3.2.1 (eqs. (9)), the optimal choice for the corresponding enhanced strain interpolation is

$$\tilde{E}_{11} \in span\ (1 - 3\xi^2, \eta - 3\xi^2\eta, \eta^2 - 3\xi^2\eta^2)$$
$$\tilde{E}_{22} \in span\ (1 - 3\eta^2, \xi - 3\xi\eta^2, \xi^2 - 3\xi^2\eta^2) \tag{10}$$
$$\tilde{E}_{12} \in span\ (1 - 3\xi^2, 1 - 3\eta^2, \eta - 3\xi^2\eta, \xi - 3\xi\eta^2, 1 - 3(\xi^2 + \eta^2) + 9\xi^2\eta^2)$$

The advantage of this method is twofold: The patchtest for constant *and linear* strains is satisfied and the distortion sensitivity of the element is further reduced, even for the case of nearly incompressible materials.

3.3 TREATMENT OF BENDING STRAINS

The bending strains may be enhanced in the same way as the membrane strains. However, compared to the additional numerical expense, the minor improvement of results does in general not justify this extension.

3.4 TREATMENT OF TRANSVERSE SHEAR STRAINS

The meanwhile "classical" ANS approach to eliminate shear locking effects due to Dvorkin and Bathe [10] is also used herein. Thus the transverse shear strains of the displacement model are calculated at specially chosen sampling points and then interpolated within the element by linear–constant trial functions. For the nine node element, the ANS method is used as well to modify the transverse shear strain interpolation. The corresponding trial functions and the coordinates of the sampling points can be found for example in Pinsky, Jang [13].

3.5 EXTENSION OF TRANSVERSE NORMAL STRAINS

The 6–parameter formulation has a severe deficiency in bending dominated cases with a non–vanishing Poisson's ratio ν. Due to the ν–coupling artificial stresses in thickness direction occur ("Poisson Thickness Locking"). Due to the fact that the origin of the stiffening lies in the formulation of the shell–theory rather than its finite element implementation, the resulting error does not diminish with mesh refinement (Braun et al. [7]).

A possibility to remove the effect without increasing the number of degrees of freedom on the structural level has been proposed by Büchter, Ramm [8]. The idea is to enrich the strain field of the element by an additional linear component $\tilde{E}^{l}_{33}$ of the thickness strain. This extra strain is introduced in the sense of a hybrid–mixed formulation, by the EAS method described in section 2; for details see Büchter et al. [9] and Braun et al. [7].

It has recently been shown that the corresponding (1,1,2)–plate theory, (linear variation for two inplane displacements, quadratic for the out–of–plane displacement), is the lowest one in a hierarchical series which is asymptotically correct *without* any modification of the constitutive equations, i.e. also without shear correction factors. Again this formulation might only be asymptotically better for certain boundary conditions. For a detailed discussion see Rössle et al. [16].

3.6 INTERPOLATION OF TRANSVERSE NORMAL STRAINS

One further problem still has to be overcome if coarse meshes and strong initial curvatures occur. This phenomenon termed "Curvature Thickness Locking" appears in case of bending of initially curved structures. Due to the fact that the difference displacements are interpolated from the nodes, artificial strains and stresses in thickness direction lead to an unwelcome stiffening of the elements.

The key feature of these parasitic strains is the fact that they are zero at the nodes. Therefore, it suggests itself that their interpolation should be based on the *nodal* values instead of an evaluation at the *Gauss quadrature points*.

$$\hat{E}_{33}(\xi,\eta) = \sum_{K=1}^{NN} \tilde{N}^K(\xi,\eta) \cdot E_{33}(\xi^K,\eta^K) \tag{11}$$

This can be regarded as an ANS method with the nodes as sampling points. For the interpolation functions $\tilde{N}^K$ the bilinear functions used for the displacements are the natural choice. The method has been presented in Bischoff, Ramm [4] (see also Bischoff, Ramm [5]), independently Betsch et al. [3] developed the same idea. A similar effect can be observed for the nine node element, even though not so dominant. An interpolation according to (11) with quadratic interpolation functions noticeably improves the element behavior.

4 Numerical Example: Hyperelastic Cylindrical Shell

To demonstrate the properties of the proposed shell elements the geometrically and material non–linear analysis of a cylindrical shell subjected to a line load is presented. The example has already been investigated by Büchter et al. [9], using displacement elements.

Due to the relatively small radius–to–thickness ratio it is definitely a large strain problem. The material law applied is a compressible Neo–Hooke formulation, suitable to describe the behavior of rubber–like materials.

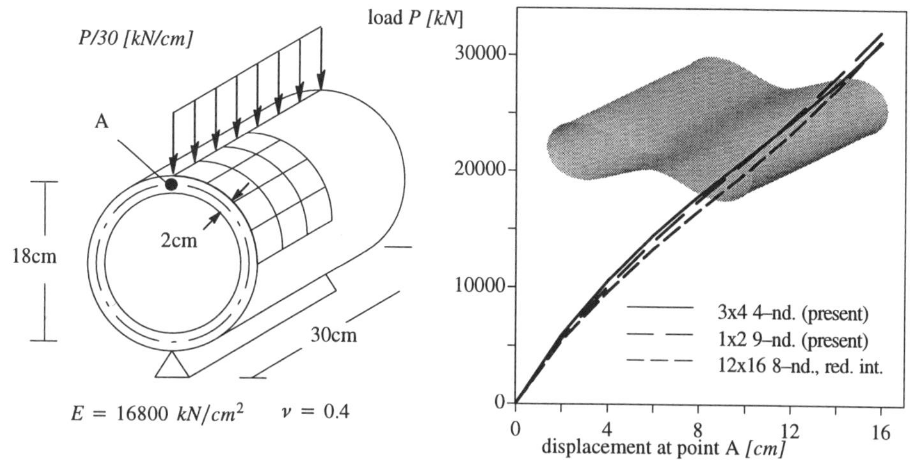

Fig. 2: Hyperelastic cylindrical shell

Making use of the symmetry, one eighth of the cylinder has been discretized for the calculation. It can be seen, that the elements are capable to represent the structural behavior in the presence of large strains. The results obtained with coarse meshes show only slight deviations to the reference solution, calculated with a fine mesh and 8–node reduced integrated elements (2x2 Gauss points).

Conclusions

It has been shown that a non–linear version of the Enhanced Assumed Strain Method alternative to the one proposed by Simo and Armero [19] is possible, namely the additive decomposition of the Green–Lagrange strain tensor. For the investigated example the results proved to be practically identical to those obtained by the decomposition of the deformation gradient.

The presented elements are suitable to handle large strain problems and formulated in a way that allows the use of non–modified three–dimensional constitutive laws. Due to the applied modifications the elements are free of locking effects and the sensitivity to mesh distortions is remarkably reduced. The extension into the three–dimensional stress and strain state usually causes a special kind of locking, termed "curvature thickness locking". This extra locking could also be overcome by an elegant and easy application of the ANS method.

Acknowledgement

The present study is supported by a grant of the German National Science Foundation (DFG, project Ra 218/7). This support is gratefully acknowledged.

References

[1] Andelfinger, U., Ramm, E. (1993) EAS–Elements for two–dimensional, three–dimensional, Plate and Shell structures and their Equivalence to HR–Elements, *International Journal for Numerical Methods in Engineering* **36**, 1311–1337.

[2] Atluri, S.N., Punch, E. (1984) Developmenet and Testing of Stable, Invariant, Isoparametric Curvilinear 2– and 3–D Hybrid Stress Elements, *Comp. Meth. in Applied Mechanics and Engineering* **47**, 331–356.

[3] Betsch, P., Gruttmann, F., Stein, E. (1996) A 4–Node Finite Shell Element for the Implementation of General Hyperelastic 3D–Elasticity at Finite Strains, *Computer Methods in Applied Mechanics and Engineering* **130**, 57–79.

[4] Bischoff, M., Ramm, E.. (1995) EAS Methods for Higher Order Nonlinear Shell Formulations, *3rd. U.S. National Congress on Computational Mechanics (USNCCM)* at Dallas, Texas (No proceedings).

[5] Bischoff, M., Ramm, E. (1997) Shear Deformable Shell Elements for Large Strains and Rotations, Accepted for publication in *International Journal for Numerical Methods in Engineering*.

[6] Braess, D. (1997) Enhanced Assumed Strain Elements and Locking in Membrane Problems, submitted to *Numerical Methods in Applied Mechanics and Engineering.*

[7] Braun, M., Bischoff, M., Ramm, E. (1994) Nonlinear Shell Formulations for Complete Three–Dimensional Constitutive Laws, Including Composites and Laminates, *Computational Mechanics* **15**, 1–18.

[8] Büchter, N. Ramm, E. (1992) 3D–Extension of Nonlinear Shell Equations Based on the Enhanced Assumed Strain Concept, *Computational Methods in Applied Sciences*, Ch. Hirsch (ed.), Elsevier Science Publishers B.V., 55–62.

[9] Büchter, N. Ramm, E. Roehl, D. (1994) Three–Dimensional Extension of Nonlinear Shell Formulation based on the Enhanced Assumed Strain Concept, *International Journal for Numerical Methods in Engineering* **37**, 2551–2568, 1994.

[10] Dvorkin, E.N., Bathe, K.J. (1984) A Continuum Mechanics Based Four–Node Shell Element for General Nonlinear Analysis, *Engineering Computations* **1**, 77–88.

[11] Hughes, T.J.R., Tezduyar, T.E. (1981) Finite Elements Based Upon Mindlin Plate Theory with Particular Reference to the Four–Node Isoparametric Element, *Journal of Applied Mechanics* **48**, 587–596.

[12] Pian, T.H.H., Sumihara, K. (1984) A Rational Approach for Assumed Stress Finite Elements, *International Journal for Numerical Methods in Engineering* **20**, 1685–1695.

[13] Pinsky, P.M., Jang, J. (1987) A C^0–Elastoplastic Shell Element Based on Assumed Covariant Strain Interpolations, *Proceedings of the International Conference NUMETA 1987*, (G.N. Pande, J. Middleton eds.), Swansea.

[14] Ramm, E., Bischoff, M. Braun, M. (1994) Higher Order Nonlinear Shell Formulations – A Step Back into Three Dimensions, In: '*From Finite Elements to the Troll Platform*' (ed. K. Bell), Dptmt. of Structural Engineering, Norwegian Institute of Technology, Trondheim, Norway, 65–88.

[15] Reissner, E. (1989) On Finite Axi–Symmetrical Deformations of Thin Elastic Shells of Revolution, *Computational Mechanics* **4**, 387–400.

[16] Rössle, A. Bischoff, M., Ramm, E., Wendland, W. (1997) Mathematical Foundation of the (1,1,2) Hierarchical Model for Plate Bending, *(Preprint) University of Stuttgart, Germany.*

[17] Sansour, C., Bufler, H. (1992) An Exact Finite Rotation Shell Theory , its Mixed Variational Formulation, and its Finite Element Implementation, *Int. Journal for Num. Meth. in Engineering* **29**, 73–115.

[18] Simo, J.C., Rifai, S. (1990) A Class of Mixed Assumed Strain Methods and the Method of Incompatible Modes, *International Journal for Numerical Methods in Engineering* **29**, 1595–1638.

[19] Simo, J.C., Armero, F. (1992) Geometrically Non–Linear Enhanced Strain Mixed Methods and the Method of Incompatible Modes, *International Journal for Numerical Methods in Engineering* **33**, 1413–1449.

[20] Yeo, S.T., Lee, B.C. (1996) Equivalence Between Enhanced Assumed Strain Method and Assumed Stress Hybrid Method Based on the Hellinger–Reissner Principle, *International Journal for Numerical Methods in Engineering* **39**, 3083–3099.

AN ADAPTIVE FINITE ELEMENT METHOD FOR LARGE DEFORMATION FRICTIONAL CONTACT PROBLEMS

O. SCHERF, P. WRIGGERS
Institut für Mechanik, Technische Hochschule Darmstadt
Hochschulstr. 1, 64289 Darmstadt, Germany

1. Introduction

Frictional contact problems in engineering are often associated with very complex geometries and nonlinear material behaviour. For such problems only numerical methods yield the required solution and especially the finite element method has proved to be a flexible numerical tool in this context. Since numerical methods yield approximate solutions it is necessary to control the errors inherited in the method. During the last years adaptive finite element techniques have been developed, providing automatically a numerical model which is accurate and reliable. Here we introduce an adaptive method for frictional contact problems which ensures successive improvement of the numerical solution via an iterative solution procedure to refine the finite element mesh.

2. Formulation of the Contact Problem

Let us consider the bounded domains of two bodies, $\mathcal{B}^s$ and $\mathcal{B}^m$. The boundary $\partial\mathcal{B}^s$ of body $\mathcal{B}^s$ can in general be split into three parts: $\partial\mathcal{B}^s_\sigma$ with prescribed surface loads, $\partial\mathcal{B}^s_u$ with prescribed displacements and $\partial\mathcal{B}^s_c$ where body $\mathcal{B}^s$ comes into contact with body $\mathcal{B}^m$. An analogous notation is used for body $\mathcal{B}^m$.

2.1. CONTACT KINEMATICS

Assume that two bodies come into contact. The associated mathematical formulation can be found in e.g. Laursen, Simo [1], Wriggers, Miehe [2]. It yields the non–penetration condition in terms of the coordinates $\mathbf{x}$ of the current configuration of the bodies:

$$g_N = (\mathbf{x}^s - \mathbf{x}^m(\bar{\xi^\alpha}))\cdot\mathbf{n}_c \geq 0\,. \tag{1}$$

g_N is the gap function. Here, $\xi^\alpha, \alpha = 1,2$, denote convective coordinates used for parametrisation of the surface of body $\mathcal{B}^m$ and $\mathbf{n}_c$ is the surface normal on $\partial\mathcal{B}^m_c$ with respect to the current configuration. The point $\mathbf{x}^m(\bar{\xi^\alpha})$ is found by solving a minimal distance problem which associates to every point $\mathbf{x}^s$ on the slave contact

H. A. Mang and F. G. Rammerstorfer (eds.),
IUTAM Symposium on Discretization Methods in Structural Mechanics, 35–42.

surface $\partial\mathcal{B}_c^s$ a point $\bar{\mathbf{x}}^m = \mathbf{x}^m(\bar{\xi^\alpha})$ on $\partial\mathcal{B}_c^m$ via

$$||\mathbf{x}^s - \bar{\mathbf{x}}^m|| = \min_{\mathbf{x}^m \subseteq \partial\mathcal{B}_c^m} ||\mathbf{x}^s - \mathbf{x}^m||\,, \tag{2}$$

see e.g. Wriggers, Miehe [2]. In view of the penalty formulation, which will be applied to solve the contact problem, we introduce a penetration function as follows:

$$g_N^- = \begin{cases} g_N & \text{for} \qquad g_N < 0 \\ 0 & \text{otherwise}\,. \end{cases} \tag{3}$$

Function (3) indicates the penetration of one body into the other and shows in which parts of $\partial\mathcal{B}_c^s$ the constraint equations preventing penetration have to be activated. Thus (3) can be used to determine the active contact area.

In case of friction the relative tangential velocity $L_v\mathbf{g}_T$ has to be computed. Since the slip between two bodies is related to the change of the solution point $\bar{\xi^\alpha}$ of the minimal distance problem, the relative tangential velocity can be defined as

$$L_v\mathbf{g}_T = \mathbf{v}^s - \bar{\mathbf{v}}^m = \dot{\bar{\xi}}^\alpha \mathbf{a}_\alpha\,.$$

According to Wriggers, Miehe [2] the change of the Gaussian parameter $\bar{\xi^\alpha}$ can be obtained by solving the following system of equations:

$$(\bar{a}_{\alpha\beta} - g_N\bar{b}_{\alpha\beta})\dot{\bar{\xi}}^\beta = (\mathbf{v}^s - \bar{\mathbf{v}}^m)\cdot\bar{\mathbf{a}}_\alpha + g_N\bar{\mathbf{n}}_c\cdot\bar{\mathbf{v}}^m_{,\alpha}\,,$$

where $\bar{a}_{\alpha\beta}$ are the coefficients of the metric tensor and $\bar{b}_{\alpha\beta}$ are the coefficients of the curvature tensor at $\bar{\xi^\alpha}$.

2.2. WEAK FORMULATION OF THE CONTACT PROBLEM

The weak formulation of the contact problem is given in the following: Find the displacement $\mathbf{u} \in \mathcal{V}$ such that

$$\begin{aligned} G(\mathbf{u},\mathbf{v}) \quad &= \int_{\mathcal{B}_0} \mathbf{P} : \operatorname{Grad}\mathbf{v}\,dV - \int_{\mathcal{B}_0} \rho_0\,\hat{\mathbf{b}}\cdot\mathbf{v}\,dV - \int_{\partial\mathcal{B}_{0\sigma}} \hat{\mathbf{T}}\cdot\mathbf{v}\,dA \\ &- \int_{\partial\mathcal{B}^s_{0c}} (-\mathrm{T}_N\bar{\mathbf{n}}_c + \mathbf{T}_T)\cdot(\mathbf{v}^s - \bar{\mathbf{v}}^m)\,dA = 0 \quad \forall\,\mathbf{v}\in\mathcal{V}_0\,. \end{aligned} \tag{4}$$

The subscript '0' indicates, that integration over the bodies and boundaries has to be performed in the reference configuration. Without superscript 's' or 'm', $\mathcal{B}$ denotes the union of slave and master body, e.g. $\mathcal{B}_0 = \mathcal{B}_0^s \cup \mathcal{B}_0^m$. $\mathbf{P}$ is the first Piola–Kirchhoff stress tensor, $\hat{\mathbf{b}}$ and $\hat{\mathbf{T}}$ applied volume and surface loads and $\mathbf{v}$ virtual displacements. The spaces of displacements and virtual displacements are defined as $\mathcal{V} = \{\,\mathbf{u}\in\mathbf{W}^{1,p}\,|\,\mathbf{u} = \hat{\mathbf{u}}\,\}$ and $\mathcal{V}_0 = \{\,\mathbf{u}\in\mathbf{W}^{1,p}\,|\,\mathbf{u} = \mathbf{0}\,\}$, respectively, with given displacements $\hat{\mathbf{u}}$ on $\partial\mathcal{B}_u$. The integer p which determines the Sobolev space $\mathbf{W}^{1,p}$ depends on the chosen constitutive relation, see the next section. T_N

is the contact normal pressure, which is assumed to be negative and $\mathbf{T}_T$ denotes the tangential stress vector on the contact boundary.

2.3. CONSTITUTIVE EQUATIONS

For rubber materials undergoing moderately large deformations, the Neo–Hooke model, see e.g. Ogden [3], relates the second Piola–Kirchhoff stress tensor $\mathbf{S} = \mathbf{F}^{-1}\mathbf{P}$ and the right Cauchy–Green deformation tensor $\mathbf{C} = \mathbf{F}^T\mathbf{F}$, where $\mathbf{F}$ denotes the deformation gradient, via

$$\mathbf{S} = \mu\,(\mathbf{I} - \mathbf{C}^{-1}) + \lambda\,J\,(J-1)\,\mathbf{C}^{-1}\,,$$

with Lamé constants λ and μ. For Neo–Hooke materials, the admissible solution space is given by $\mathcal{V} = \{\,\mathbf{u} \in \mathbf{W}^{1,p}\,|\,\mathbf{u} = \hat{\mathbf{u}}\,\}$, $p > 3$, see e.g. Curnier, He, Telega [4].

Concerning contact in the normal direction, a penalty formulation is used for regularisation of the non–penetration condition (1), see e.g. Kikuchi, Oden [5]. Within this framework (small) penetrations are admissible and the constitutive relation between the contact pressure and the penetration is given by

$$\mathrm{T}_N = \varepsilon_N \mathrm{g}_N^-\,, \tag{5}$$

where $\varepsilon_N > 0$ denotes the penalty parameter.

A frictional interface law can be formulated within the framework of plasticity, see e.g. Wriggers [6]. The key idea is the split of the tangential displacement $\mathbf{g}_T$ into an elastic (stick) part $\mathbf{g}_T^e$ and a plastic (slip) part $\mathbf{g}_T^p$:

$$\mathbf{g}_T = \mathbf{g}_T^e + \mathbf{g}_T^p\,. \tag{6}$$

In this way the stick part still includes elastic micro displacements which apart from a physical interpretation represents a regularisation of Coulomb's frictional law. Hence a very simple model related to a linear spring is assumed to describe the dependence of the tangential stress from the stick displacement:

$$\mathbf{T}_T = \varepsilon_T \mathbf{g}_T^e\,, \tag{7}$$

where $\varepsilon_T > 0$ is a material (or regularisation) parameter. The tangential plastic slip $\mathbf{g}_T^p$ is governed by a constitutive evolution equation. Thus, as in plasticity theory, a plastic slip criterion function for a given contact pressure T_N can be formulated:

$$f = ||\mathbf{T}_T|| - \mu|\mathrm{T}_N| \leq 0\,. \tag{8}$$

μ denotes the frictional coefficient of Coulomb's model. Finally, the constitutive evolution equation for frictional slip in form of a non-associated slip rule completes together with the loading–unloading conditions in Kuhn–Tucker form the formulation of the frictional law:

$$L_v \mathbf{g}_T^p = -\lambda \frac{\partial f}{\partial \mathbf{T}_T}\,, \qquad \lambda \geq 0\,, \qquad f \leq 0\,, \qquad \lambda f = 0\,. \tag{9}$$

The factor of proportionality λ represents the magnitude of plastic slip.

3. Discretisation by Finite Elements

To discretise the contact problem defined above, we divide the domain $\mathcal{B}$ into non–overlapping finite elements T of diameter h_T and introduce a standard finite element space

$$\mathcal{V}_h = \{\mathbf{v} \in \mathcal{V} \mid \mathbf{v}|_T \in \mathbf{P}(T) \ \forall\, T\}\,, \tag{10}$$

where $\mathbf{P}(T)$ is a space of polynomials of degree p_T on T and p_T is a positive integer. Moreover the boundary is assumed to be piecewise affine such that the triangulation covers $\mathcal{B}$ exactly and the type of boundary condition does not change on a side of an element. With this the discrete form of the contact problem reads: Find the displacement $\mathbf{u}^h \in \mathcal{V}_h$ such that

$$\begin{aligned} G(\mathbf{u}^h,\mathbf{v}^h) \;=\; & \int\limits_{\mathcal{B}_0} \mathbf{P}^h : \operatorname{Grad}\mathbf{v}^h\, dV - \int\limits_{\mathcal{B}_0} \rho_0\, \hat{\mathbf{b}}\cdot\mathbf{v}^h\, dV - \int\limits_{\partial\mathcal{B}_{0\sigma}} \hat{\mathbf{T}}\cdot\mathbf{v}^h\, dA \\ & - \int\limits_{\partial\mathcal{B}^s_{0c}} (-\mathrm{T}^h_N \bar{\mathbf{n}}_c + \mathbf{T}^h_T)\cdot(\mathbf{v}^{hs} - \bar{\mathbf{v}}^{hm})\, dA = 0 \ \ \forall\, \mathbf{v}^h \in \mathcal{V}_{h0}\,. \end{aligned} \tag{11}$$

Newton's method is applied to solve this nonlinear equation. For spatial discretisation of the contact boundary node to segment contact elements are used. A detailed treatise of this discretisation method can be found in, e.g., Wriggers [7].

In the numerical model a backward Euler integration scheme is used in combination with a return–mapping algorithm for temporal discretisation, see e.g. Wriggers [6]. First trial stresses $\mathbf{T}^{trial}_{T\,n+1}$ are computed under the assumption of stick behaviour. Then the slip function is evaluated with the trial stresses to determine if they are admissible. If not, they are situated outside the slip surface and a return–mapping onto the frictional cone has to be performed. Due to the simple frictional law, the return mapping algorithm can be stated here in closed form:

$$\begin{aligned} \mathbf{T}^{trial}_{T\,n+1} &= \mathbf{T}_{T\,n} + \varepsilon_T(\mathbf{g}_{T\,n+1} - \mathbf{g}_{T\,n}) \\ \mathbf{T}^{RM}_{T\,n+1} &= \begin{cases} \mathbf{T}^{trial}_{T\,n+1} & \text{for } f \le 0 \\ \mu T_{N\,n+1} \dfrac{\mathbf{T}^{trial}_{T\,n+1}}{\|\mathbf{T}^{trial}_{T\,n+1}\|} & \text{else}\,. \end{cases} \end{aligned} \tag{12}$$

4. Error Indicators

Let $\mathbf{u}$ denote the exact solution of (4) and let $\mathbf{u}^h$ denote the solution of the discrete problem (11). With

$$\mathbf{e} = \mathbf{u} - \mathbf{u}^h \tag{13}$$

we define the error in the displacement field. Here the error due to the penalty regularisation will not be discussed, for details see e.g. Kikuchi, Oden [5].

For linear elastic frictionless contact problems an a posteriori error estimator has been developed by the authors, see Wriggers, Scherf, Carstensen [8]:

$$r = ||\mathbf{e}||^2_{E(\mathcal{B}_0)} + ||\mathbf{e}||^2_{E(\partial\mathcal{B}_{0c})} \leq C \sum_T E_T \tag{14}$$

$$\begin{aligned} E_T = \; & h_T^2 \, || \operatorname{Div} \mathbf{P}^h + \rho_0 \hat{\mathbf{b}} \, ||^2_{L_2(T)} \; + \; \frac{1}{2} \, h_T \, || \, [\mathbf{T}^h] \, ||^2_{L_2(\partial T \notin \mathcal{B})} \\ & + h_T \, ||\hat{\mathbf{T}} - \mathbf{T}^h||^2_{L_2(\partial T \cap \partial \mathcal{B}_\sigma)} + h_T \, ||\varepsilon_N \, \mathrm{g}_N^-(\mathbf{u}^h)\mathbf{n}_c - \mathbf{T}^h||^2_{L_2(\partial T \cap \partial \mathcal{B}_c)} \end{aligned} \tag{15}$$

Here $||\mathbf{e}||^2_{E(\mathcal{B}_0)} = \int_{\mathcal{B}_0} \mathbf{E}(\mathbf{u} - \mathbf{u}^h) : \boldsymbol{\mathcal{C}} : \mathbf{E}(\mathbf{u} - \mathbf{u}^h) \, dV$ denotes the energy norm of the error, with $\mathbf{E}$ as strain measure for geometrically linear problems and $\boldsymbol{\mathcal{C}}$ as Hooke's elasticity tensor. $||\mathbf{e}||^2_{E(\partial\mathcal{B}_{0c})} = \int_{\partial\mathcal{B}^s_{0c}} \varepsilon_N |\mathrm{g}_N^-(\mathbf{u}) - \mathrm{g}_N^-(\mathbf{u}^h)|^2 \, dA$ represents the energy norm of the error in the contact interface. Inequality (14) yields an upper bound for the error measure which is bounded by the deviation of the discrete solution from equilibrium and the element size. C denotes an interpolation constant. The first and the third term of the right hand side of (15) contribute to the error bound if local equilibrium and the traction boundary condition $\mathbf{T}^h = \hat{\mathbf{T}}$ on $\partial\mathcal{B}_\sigma$ are violated. Furthermore, the equilibrium condition requires that the jumps of the discrete tractions over element edges are zero, $[\mathbf{T}^h] = \mathbf{0}$, which is associated with the second term. The fourth term requires, that the contact pressure $\mathbf{T}_N = \varepsilon_N \mathrm{g}_N^- \mathbf{n}_c$, due to the material law (5), is in equilibrium with the discrete finite element stress $\mathbf{T}^h$ on the contact boundary.

In order to apply the error indicators (15) to frictional contact problems, we use a heuristic extension. Therefore the term representing the contact normal stress $\mathbf{T}_N = \varepsilon_N \mathrm{g}_N^- \mathbf{n}_c$ appearing in the contact residual is changed to include also the tangential contributions $\mathbf{T}_T^{RM}$ arising from the return mapping procedure. From a physical viewpoint equations (12) define a material law for the frictional stress. The new contact contribution to the residuals can therefore also be interpreted as the difference between the contact stress vector given by the constitutive equation in the contact interface and the discrete stress vector $\mathbf{T}^h$, given by the finite element solution. This yields the modified contact residual

$$|| \, \varepsilon_N \, \mathrm{g}_N^-(\mathbf{u}^h)\mathbf{n}_c + \mathbf{t}_T^{RM} - \mathbf{t}^h \, ||^2_{L_2(\partial T \cap \partial \mathcal{B}_c)} \, . \tag{16}$$

The presented error indicators can be used to calculate the error distribution due to the spatial discretisation error. Since frictional processes are path, and with this time dependent, an estimation for the magnitude of the time step used to integrate the frictional constitutive equations (6) to (9) is still missing. However, in this study we will neglect this fact and prescribe the time steps a priori.

In case of large deformations, error indicators for geometrically linear problems can be successfully applied within a load increment of the solution procedure. A mathematical discussion of this topic is given in e.g. Rheinboldt [9]. Assumptions are, that the curvature of the nonlinear load–displacement curve is small and that only regular points are considered. Furthermore the incremental equations of the large deformation problem must show the same structure as the equations of the linear problem. Then the computed quantities for an actual load level can be used to compute the error indicators (15), (16), which yields the spatial error distribution within the considered load step.

5. Adaptive Finite Element Method

The error indicators (15), (16) are computed for a given load level from the finite element solution. Depending on the error indicator distribution a new partially refined mesh is automatically generated. Therefore elements with a high error value are divided into elements with smaller h_T. The condition

$$E_T \leq \frac{TOL}{N}, \tag{17}$$

where N denotes the number of elements in the mesh, serves as refinement criterion. The adaptive process stops, when the global error is below the user given tolerance TOL:

$$\sum_T E_T \leq TOL\,. \tag{18}$$

For further details about the refinement strategy see e.g. Johnson [10], Wriggers, Scherf [11]. During refinement a transfer of frictional history data and displacements from the previous mesh to the new generated mesh is performed via an interpolation technique. The overall algorithm of our h–adaptive method for large deformation frictional contact problems is stated in the following:

1 Set initial values: $l = 0$, $i = 0$, $t_0 = 0$, Δt
2 Generation of start mesh: $\mathcal{T}_i = \{T\}$
3 Increase load increment: $t_{l+1} = t_l + \Delta t$
   3.1 Newton iteration to solve nonlinear problem
   3.2 Mesh optimisation
   - Compute $E_T \; \forall \, T$, using (15), (16)
   - Check stopping criterion: IF $\sum_T E_T < TOL \Longrightarrow$ GOTO 3
   - Check refinement criterion: IF $E_T > TOL\,/\,N \Longrightarrow$ refine T
   - Generate new mesh $\mathcal{T}_{i+1}$
   - Interpolate displacements and history variables on $\mathcal{T}_{i+1}$
   - GOTO 3.1

6. Numerical Example

We consider an example of a rubber sealing in which an elastic block (material data $\lambda_b = 100$, $\mu_b = 10$) is pressed inbetween two almost rigid jaws (material data

$\lambda_j = 1000$, $\mu_j = 500$). The frictional coefficient has the value $\mu = 0.2$. The initial mesh is depicted in Fig. 1. Due to the symmetry of the problem, only half of the structure is discretised.

The results of the adaptive computation are compared with a reference solution, which was obtained by using a uniformly fine mesh with 15687 nodes and a load increment $\Delta t = 0.005$. The load steps and refinements of the adaptive computation are given in the following table:

t	0.4	0.5	0.6	0.7	0.8	0.85	0.9
Refinement					1		2
t	0.95	1.00	1.02	1.04	1.06	1.08	1.10
Refinement		3					

The tolerance for the adaptive algorithm is $TOL = 3000$. The refined meshes and the associated number of nodes are given in Figs. 2 to 4.

The distribution of the tangential stress in the contact interface, depicted in Fig. 5, shows a good agreement of the adaptive and the reference solution. The comparison of the displacement history at the top edge node of the sealing, Fig. 6, shows more deviations from the reference solution, which are due to the coarse mesh and the large load increments at the beginning. A mesh refinement step ($t = 0.8, 0.9, 1.0$) is connected with a jump of the adaptive solution closer to the reference solution. Then, for the following load steps, the deviation of the adaptive solution from the reference solution is approximately constant. Finally, after the last refinement step, both, the reference and the adaptive mesh, yield the same displacements, despite the relatively few number of nodes and large load steps used for the adaptive computation. This indicates, that the adaptive algorithm is efficient and reliable.

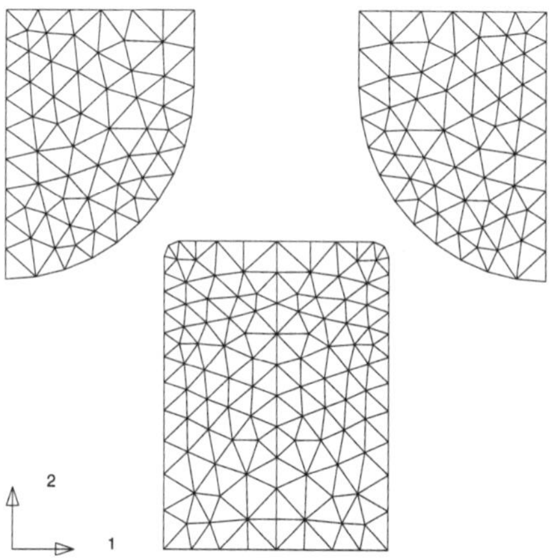

Figure 1. Initial mesh, $t = 0$, 181 nodes

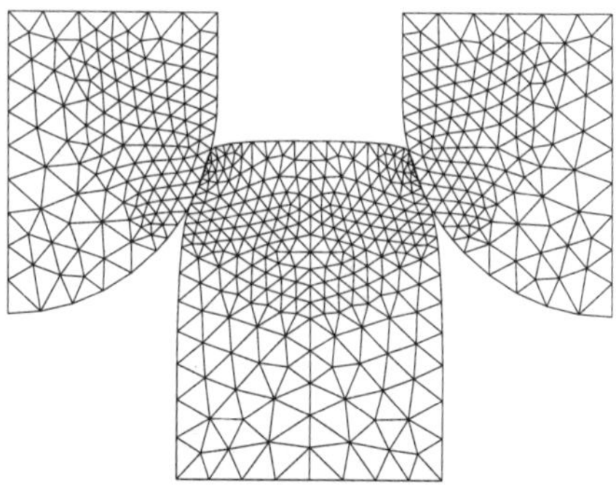

Figure 2. 2nd mesh $t = 0.8$, 293 nodes

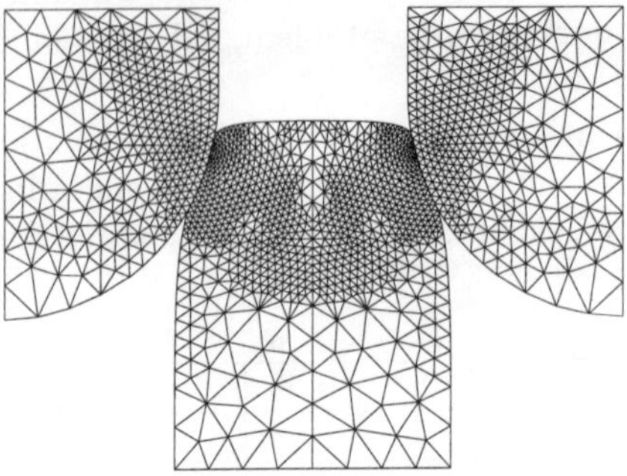

Figure 3. 3rd mesh, $t = 0.9$, 816 nodes

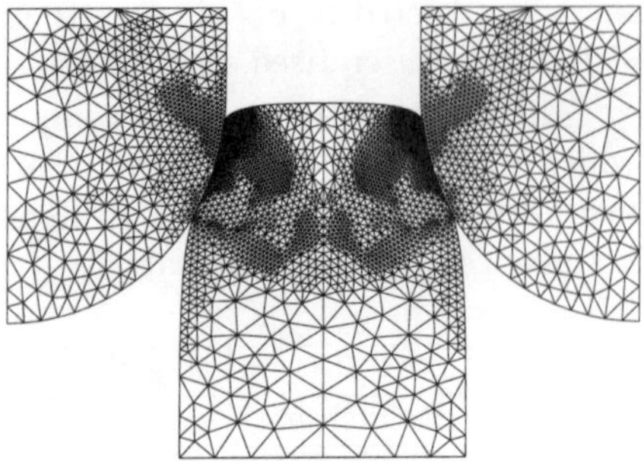

Figure 4. 4th mesh, $t = 1.0$, 1767 nodes

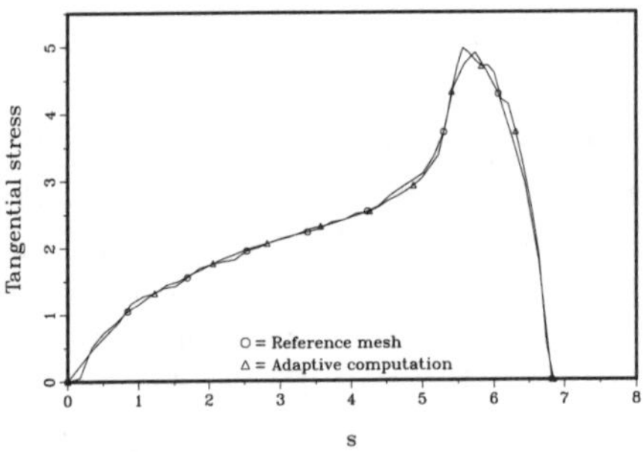

Figure 5. $\|\mathbf{T}_T\|$, $t = 1.1$

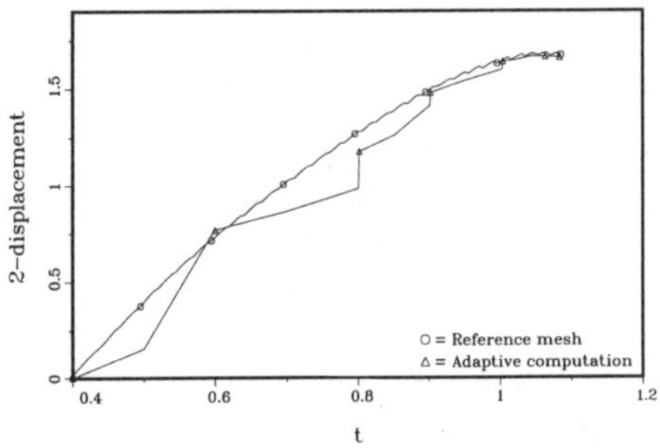

Figure 6. u_2–displacement

7. References

[1] Laursen, T. A., Simo, J. C., A Continuum–based Finite Element Formulation for the Implicit Solution of Multibody, Large Deformation Frictional Contact Problems, Int. J. Num. Meth. Engng., **36**, 3451–3485, 1993

[2] Wriggers, P., Miehe, C., Contact Constraints within Coupled Thermomechanical Analysis, Comp. Meth. Appl. Mech. Engrg., **113**, 301–319, 1994

[3] Ogden, R. W., *Non–linear Elastic Deformations*, Ellis Horwood, Chichester, 1984

[4] Curnier, A., He, Q. C., Telega, J. J., Formulation of Unilateral Contact between Two Elastic Bodies Undergoing Finite Deformations, C. R. Acad. Sci. Paris, **314**, 1–6, 1992

[5] Kikuchi, N., and Oden, J. T., *Contact Problems in Elasticity*, SIAM, Philadelphia, 1988

[6] Wriggers, P., On Consistent Tangent Matrices for Frictional Contact Problems, in: *Proc. NUMETA 87 Conf.*, eds. Middleton et al., Nijhoff Publishers, Dorbrecht, 1987

[7] Wriggers, P., Finite Element Algorithms for Contact Problems, Arch. Comp. Meth. Engng., **4**, 1–49, 1995

[8] Wriggers, P., Scherf, O., Carstensen, C., Adaptive Techniques for the Contact of Elastic Bodies, in: *Recent Developments in Finite Element Analysis*, eds. Hughes et al., CIMNE, Barcelona, 1994

[9] Rheinboldt, W. C., Error Estimates for Nonlinear Finite Element Computations, Comp. Struct., **20**, 91–98, 1985

[10] Johnson, C., Adaptive Finite Element Methods for the Obstacle Problem, Technical Report, Chalmers University of Technology, Göteborg, 1991

[11] Wriggers, P., Scherf, O., An Adaptive Finite Element Technique for Nonlinear Contact Problems, in: *Contact Mechanics, Proc. 2nd Contact Mech. Int. Sym. 1994*, eds. Raous et al., Plenum Publishing, London, 1995

ADAPTIVE METHODS FOR NONLINEAR FE ANALYSIS OF THIN–WALLED STRUCTURES

F. CIRAK, E. RAMM
Institute of Structural Mechanics,
University of Stuttgart, Germany

Abstract

The classical energy norm based adaptive strategies for linear problems are extended to mesh adaption with respect to locally or globally defined variables. The present error estimation techniques are applied to geometrically nonlinear problems and to the deformation theory of plasticity considering the linearized problem. The time discretized flow theory of plasticity leads to the stationary deformation theory type problems within each time increment and the error estimators can be applied in an incremental sense.

1. Introduction

Different error estimators and indicators have been developed for linear elastic problems so far. The most frequently used estimators are based on either residuals as initially introduced by Babuska and Rheinboldt [1] or postprocessed stresses as proposed by Zienkiewicz and Zhu [16]. These error estimators and their related adaptive algorithms are formulated through estimates in energy norms. In the mean time also error estimators and adaptive algorithms are available for local variables [6],[3],[4]. The methods rely on duality arguments or in other notions on the reciprocal theorem and are related to the influence line and surface concept of classical structural analysis. Based on the latest developments we introduce an error estimator to control different globally or locally defined norms including single variables. The error estimator is evaluated solving the set of equations for an additional right hand side and applying the classical error estimators (based on residuals or postprocessing) two times.

According to Rheinboldt [8], Pousin et.al. [7] and Stein et.al. [11] the introduced error estimators can simply be extended to the geometrically nonlinear regime if they are applied to the linearized problem at the regular points of the loading path. The linearized problem is singular at the bifurcation and limit points and the situation is very complex. An a–posteriori error estimator for singular points is proposed by Verfürth [15] based on earlier works on a–priori estimates. For plasticity not only the nonlinear material behavior but also the path dependency must be taken into account. A first step for the classical flow theory is therefore to neglect the path dependency so that basically a deformation theory type of plasticity is investigated.

H. A. Mang and F. G. Rammerstorfer (eds.),
IUTAM Symposium on Discretization Methods in Structural Mechanics, 43–51.

2. Error estimation for linear problems

We briefly introduce the equilibrium equations in strong and weak form and define our notation. The equilibrium in strong form and the boundary conditions are given by

$$\begin{aligned} &\operatorname{div} \boldsymbol{\sigma}(\mathbf{u}) + \mathbf{p} = \mathbf{0} \qquad \text{on } \Omega \\ &\boldsymbol{\sigma}(\mathbf{u})\tilde{\mathbf{n}} = \hat{\mathbf{f}} \quad \text{on } \Gamma_N \qquad \mathbf{u} = \mathbf{0} \quad \text{on } \Gamma_D \end{aligned} \tag{1}$$

where $\boldsymbol{\sigma}$ are the stresses, $\mathbf{u}$ are the displacements, $\mathbf{p}$ are body forces, $\hat{\mathbf{f}}$ are the tractions prescribed on the force boundary Γ_N, $\tilde{\mathbf{n}}$ is the unit normal vector to Γ_N and Γ_D is the displacement boundary. The weak form or the principle of virtual work can be obtained through multiplication of the equilibrium equations with a test function $\mathbf{v}$ and a subsequent integration by parts.

$$\mathrm{B}(\mathbf{u}, \mathbf{v}) = (\mathbf{p}, \mathbf{v}) + (\hat{\mathbf{f}}, \mathbf{v})_{\Gamma_N} \tag{2}$$

For shells the bilinear operator B(,) includes the virtual work of the membrane forces, bending moments and shear forces.

2. 1. BRIEF REVIEW OF ENERGY NORM ERROR ESTIMATORS

The approximation of the continuous displacements and test functions in the virtual work equation (2) by the finite element shape functions leads to discretization errors. The energy norm is a natural choice to measure the errors $\mathbf{e} = \mathbf{u} - \mathbf{u}^h$.

$$\| \mathbf{e} \|_e^2 = \mathrm{B}(\mathbf{e}, \mathbf{e}) = \int_\Omega (\boldsymbol{\sigma}(\mathbf{u}) - \boldsymbol{\sigma}(\mathbf{u}^h)) : (\boldsymbol{\epsilon}(\mathbf{u}) - \boldsymbol{\epsilon}(\mathbf{u}^h)) dx = \int_\Omega \boldsymbol{\sigma}(\mathbf{e}) : \boldsymbol{\epsilon}(\mathbf{e}) dx \tag{3}$$

Integration by parts of equation (3) yields the external work of the element internal residuals and jump terms across the element boundaries.

$$\begin{aligned} \int_\Omega \boldsymbol{\sigma}(\mathbf{e}) : \boldsymbol{\epsilon}(\mathbf{e})\, dx = \sum_{K=1}^{NEL} \Bigg\{ \Bigg[&\int_{\Omega_K} (\operatorname{div} \boldsymbol{\sigma}(\mathbf{u}^h) + \mathbf{p})\mathbf{e}\, dx \Bigg] \\ &+ \int_{\Gamma_K \subseteq \Gamma_N} (\hat{\mathbf{f}} - \boldsymbol{\sigma}(\mathbf{u}^h)\tilde{\mathbf{n}})\mathbf{e}\, dx - \int_{\Gamma_K \nsubseteq \Gamma} \boldsymbol{\sigma}(\mathbf{u}^h)\tilde{\mathbf{n}}\mathbf{e}\, dx \Bigg\} \end{aligned} \tag{4}$$

Here, NEL is the total number of elements, Ω_K is the domain and Γ_K are the edges of a finite element K. The first integral represents the element internal residuals $\mathbf{R}$ and the second and third terms the edge jumps $\mathbf{J}$ of the finite element solution.

$$\mathbf{R} = \operatorname{div} \boldsymbol{\sigma}(\mathbf{u}^h) + \mathbf{p} \quad \text{on } \Omega_K \qquad \mathbf{J} = \begin{cases} \boldsymbol{\sigma}(\mathbf{u}^h)\tilde{\mathbf{n}} - \hat{\mathbf{f}} & \text{on } \Gamma_K \subseteq \Gamma_N \\ \frac{1}{2}\big((\boldsymbol{\sigma}(\mathbf{u}^h)\tilde{\mathbf{n}})|_{K*} + (\boldsymbol{\sigma}(\mathbf{u}^h)\tilde{\mathbf{n}})|_K\big) & \text{on } \Gamma_K \nsubseteq \Gamma \end{cases} \tag{5}$$

K and K * are two elements sharing a common edge. The discretization errors can be computed applying the residuals $\mathbf{R}$ and $\mathbf{J}$ as loading to the structure as indicated by equation (4). The new problem has of course the same complexity as the original problem eq. (2). For energy norm error estimation it is common to solve this equation as a sequence of local problems including only a patch of one or a few elements loaded by the residuals. An explicit residual based energy norm estimator can be derived following the recipes given by Johnson et. al. [6].

$$\| \mathbf{e} \|_e \leq \sum_{K=1}^{NEL} \left\{ \sqrt{\frac{h_K^2}{24Ep}} \| \mathbf{R} \|_{\Omega_K} + \sqrt{\frac{h_K}{24Ep}} \| \mathbf{J} \|_{\Gamma_K} \right\} \tag{6}$$

$\| \cdot \|_{\Omega_K}$ is the L_2–norm on Ω_K. A different type of error estimators was proposed by Zienkiewicz and Zhu [16]. It is well known, that stresses at different points on a finite element mesh show different rates of convergence. With this information we can construct a stress field $\boldsymbol{\sigma}^*$ over the whole domain which has a higher accuracy than the finite element stresses $\boldsymbol{\sigma}^h$. The energy norm error can be computed by the difference $\boldsymbol{\sigma}^* - \boldsymbol{\sigma}^h$.

2. 2. ERROR ESTIMATION FOR LOCAL VARIABLES

The initially proposed energy norm estimators are only appropriate for error control in a global sense. However for practical applications, like sizing of structural members, the variables of interest are mean or maximum values of stresses or displacements at some particular sections. The errors in local variables can be estimated using the reciprocal theorem or in other notions duality arguments as used in the past for a–priori analysis ("Aubin–Nitsche trick" [12]) or for computation of stress intensity factors [2]. In connection with adaptivity they have been introduced in a general framework by Johnson and co–workers [6]. Lately the procedure was extended for error estimation and mesh refinement for local quantities [3], [4]. It is remarkable that the same approach was already described by Tottenham [14] at the beginning of the seventies. In the following we introduce the error estimator and the related refinement indicator for displacements and subsequently for stresses and integral variables.

As for energy norm error estimators, the starting point is the representation of the discretization errors depending on element internal residuals $\mathbf{R}$ and jump terms $\mathbf{J}$ (eqs. (4) and (5)):

$$\begin{gathered} \sum_{K=1}^{NEL} \{ \text{div } \boldsymbol{\sigma}(\mathbf{e}) + \mathbf{R} = \mathbf{0} \text{ on } \Omega_K \} \\ \sum_{K=1}^{NEL} \{ \boldsymbol{\sigma}(\mathbf{e})\tilde{\mathbf{n}} = \mathbf{J} \text{ on } \Gamma_K \nsubseteq \Gamma_D, \quad \mathbf{e} = \mathbf{0} \text{ on } \Gamma_K \subseteq \Gamma_D \} \end{gathered} \tag{7}$$

In order to estimate the error of a specific displacement in the component i, e.g. at point $\mathbf{x} = \overline{\mathbf{x}}$, we additionally consider the following problem, also named the dual problem

$$\begin{gathered} \text{div } \boldsymbol{\sigma}(\mathbf{G}) + \boldsymbol{\delta}_i(\overline{\mathbf{x}}) = \mathbf{0} \text{ on } \Omega \\ \boldsymbol{\sigma}(\mathbf{G})\tilde{\mathbf{n}} = \mathbf{0} \text{ on } \Gamma_N \qquad \mathbf{u} = \mathbf{0} \text{ on } \Gamma_D \end{gathered} \tag{8}$$

$\boldsymbol{\delta}_i$ is the load vector for the dual problem and its i–th component is a Dirac Delta (a single unit load) e.g. i=2

$$\boldsymbol{\delta}_2(\overline{\mathbf{x}}) = 0 \cdot \mathbf{e}_1 + \delta(\overline{\mathbf{x}}) \cdot \mathbf{e}_2 + 0 \cdot \mathbf{e}_3 \tag{9}$$

where $\mathbf{e}_1, \mathbf{e}_2$ and $\mathbf{e}_3$ are the base vectors. The solution of the dual problem is the Green's function $\mathbf{G}$. Applying the reciprocal theorem of Betti and Rayleigh to equation (7) and

(8), we obtain the following relation for the error, which depends on the residuals $\mathbf{R}$ and $\mathbf{J}$ and the Green's function $\mathbf{G}$.

$$(\mathbf{e}, \boldsymbol{\delta}_i(\overline{\mathbf{x}})) = \sum_{K=1}^{NEL} \left\{ (\mathbf{R}, \mathbf{G})_{\Omega_K} + (\mathbf{J}, \mathbf{G})_{\Gamma_K} \right\} = \sum_{K=1}^{NEL} B(\mathbf{e}, \mathbf{G})_{\Omega_K} \tag{10}$$

The term on the left hand side is the i–th component of the displacement error e_i at point $\overline{\mathbf{x}}$. Further, using the Galerkin orthogonality and applying the Cauchy–Schwarz inequality gives an upper bound for the local error

$$(e_i(\overline{\mathbf{x}}))^2 \leq \sum_{K=1}^{NEL} B(\mathbf{e}, \mathbf{e})_{\Omega_K} \; B(\mathbf{G}-\mathbf{G}^h, \mathbf{G}-\mathbf{G}^h)_{\Omega_K} \tag{11}$$

$\mathbf{G}^h$ is a finite element approximation to the Green's function computed on the same mesh as $\mathbf{u}^h$. This means that we can apply the residual or postprocessing based energy norm estimators introduced in the previous section to primal and dual problems and compute an estimate for the local error. This multiplicative procedure can be interpreted as follows: The second term representing the errors of the dual problem serves as weighting function ("influence line") and filters out the influence of the overall energy norm errors over the displacement error of interest. For second order differential equations and in two or three dimensions the internal energy of the structure loaded by point loads is infinite. The inequality (11) is still true but has no practical importance. Therefore some sort of regularization must be applied in order to bound the energy [5].

Utilizing the classical influence line/surface concept the errors in stresses or stress resultants can be estimated in a straight forward manner. In order to compute the errors for a stress variable at point $\overline{\mathbf{x}}$ a discontinuity on the related displacement variable must be applied in the dual problem

$$\begin{gathered} \operatorname{div} \boldsymbol{\sigma}(\mathbf{z}) + \frac{\partial}{\partial x_i} \boldsymbol{\delta}_j(\overline{\mathbf{x}}) = \mathbf{0} \quad \text{on } \Omega \\ \boldsymbol{\sigma}(\mathbf{z})\tilde{\mathbf{n}} = \mathbf{0} \quad \text{on } \Gamma_N \qquad \mathbf{z} = \mathbf{0} \quad \text{on } \Gamma_D \end{gathered} \tag{12}$$

If we apply now the reciprocal theorem and subsequently the divergence theorem, the error in the displacement gradient is

$$(\mathbf{e}, \frac{\partial}{\partial x_i} \boldsymbol{\delta}_j(\overline{\mathbf{x}})) = -\frac{\partial e_j}{\partial x_i} = B(\mathbf{e}, \mathbf{z}-\mathbf{z}^h) \tag{13}$$

Afterwards the errors in strains and stresses can be computed. In two or three dimensions it is not possible to apply a discontinuity to an individual displacement. A simple strategy in the finite element method is to apply two loads with opposite directions at two neighboring

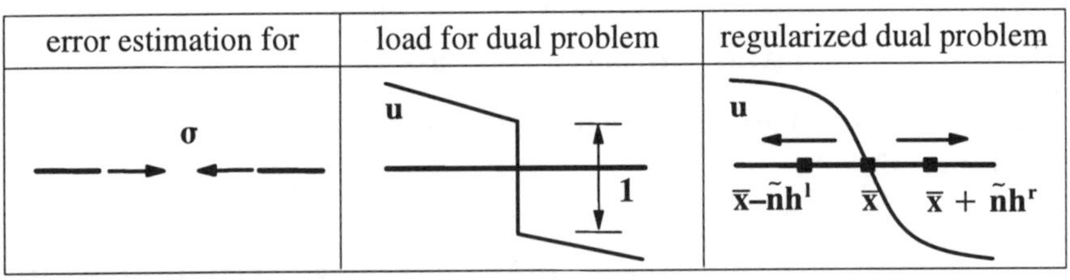

Fig. 1: Loading for dual problem

nodes to point $\bar{\mathbf{x}}$ (Fig. 1). The described concept for local variables can be easily extended by mechanical reasoning to integral variables like boundary reactions [5].

3. Error estimation for geometrical nonlinear problems

Especially thin–walled structures can exhibit very large deformations and the nonlinear dependence of the strains on the displacements must be considered. The equilibrium equations in the reference configuration for the nonlinear problem are

$$\begin{aligned} \operatorname{div}\, \mathbf{t}(\mathbf{u}) + \mathbf{p} &= \mathbf{0} \qquad \text{on } \Omega \\ \mathbf{t}(\mathbf{u})\tilde{\mathbf{n}} = \hat{\mathbf{f}} \quad \text{on } \Gamma_N \qquad & \mathbf{u} = \mathbf{0} \quad \text{on } \Gamma_D \end{aligned} \tag{14}$$

where $\mathbf{t}$ are the first Piola–Kirchhoff stresses, $\mathbf{p}$ are body forces, $\hat{\mathbf{f}}$ are the tractions applied on the force boundary Γ_N. In the following we use for the boundary value problem a formal operator equation (assuming for simplicity only displacement boundary conditions).

$$\mathbf{G}(\mathbf{u}, \lambda) = 0 \tag{15}$$

λ is a scalar load parameter and defined with respect to a given reference load $\mathbf{p}$ and $\hat{\mathbf{f}}$. The finite element approximation $\mathbf{u}^h$ can be computed using the weak form for a fixed load parameter λ^e. For estimating the discretization errors the linearized problem at the finite element solution $\mathbf{u}^h$ is considered. The formal argumentation to this idea is given in the following [8],[11]. We consider the Taylor series for the exact equilibrium point $(\mathbf{u}^e, \lambda^e)$ expanded at point $(\mathbf{u}^h, \lambda^e)$

$$\begin{aligned} \mathbf{G}(\mathbf{u}^e, \lambda^e) &= \mathbf{G}(\mathbf{u}^h, \lambda^e) + \\ &+ \mathbf{G}'(\mathbf{u}^h, \lambda^e)(\mathbf{u}^e - \mathbf{u}^h) + \ldots = \mathbf{0} \end{aligned} \tag{16}$$

The same procedure for the linearized problem with the solution $\mathbf{w}$ yields (Figure 2):

$$\mathbf{L}(\mathbf{u}^h, \lambda^e)[\mathbf{w}] = \mathbf{G}(\mathbf{u}^h, \lambda^e) + \mathbf{G}'(\mathbf{u}^h, \lambda^e)(\mathbf{w} - \mathbf{u}^h) = \mathbf{0} \tag{17}$$

The residuals $\mathbf{G}$ and the tangent stiffness $\mathbf{G}'$ are computed at $(\mathbf{u}^h, \lambda^e)$.The difference of both equations leads after some algebra to

$$(\mathbf{w} - \mathbf{u}^h) = (\mathbf{u}^e - \mathbf{u}^h)\left\{1 + \frac{1}{2}(\mathbf{G}')^{-1}\mathbf{G}''(\mathbf{u}^e - \mathbf{u}^h) + \ldots\right\} \tag{18}$$

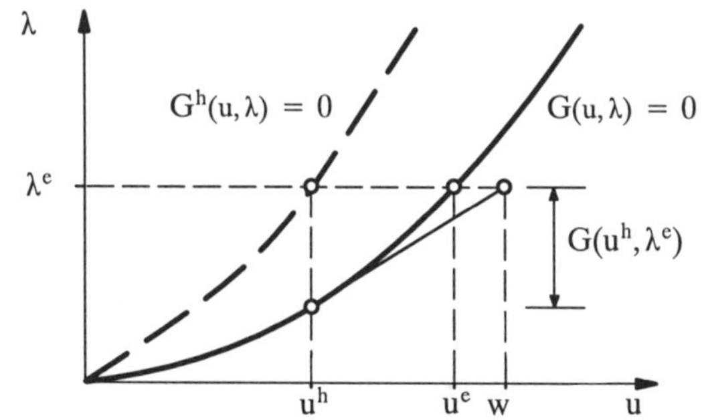

Fig. 2: One dimensional example for linearization

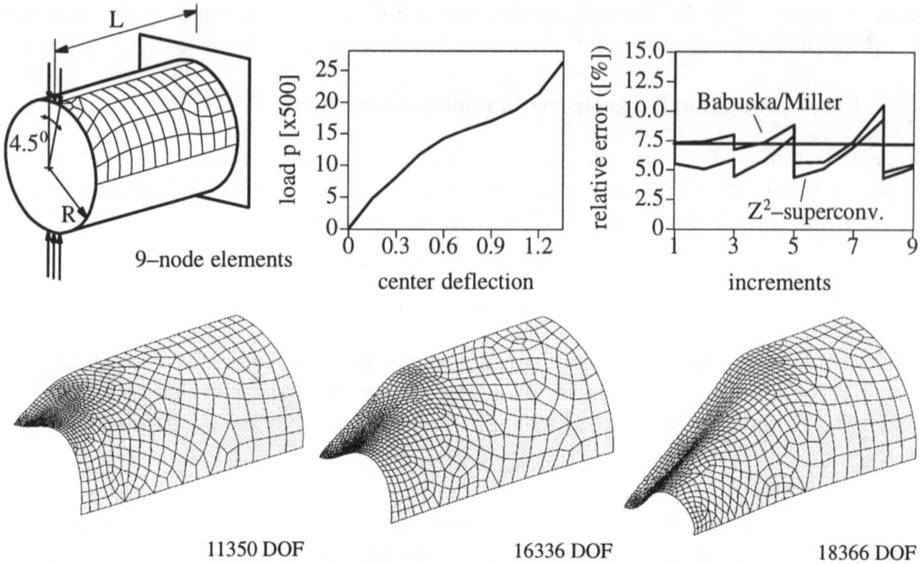

adaptively generated meshes at different loads (deformations on real scale)

Fig. 3: *Adaptive analysis of open clamped cylinder*
($R=1.016$, $L=3.048$, $t=0.03$, $E=2.0685\cdot 10^7$, $\nu=0.3$)

Provided that the finite element solutions $\mathbf{u}^h$ is a sufficient approximation for the exact solution $\mathbf{u}^e$, the higher order terms in the series can be neglected and the discretization error of the nonlinear problem can be estimated

$$\| \mathbf{w}-\mathbf{u}^h \| \approx \| \mathbf{u}^e-\mathbf{u}^h \| \text{ for } \| \mathbf{u}^e-\mathbf{u}^h \| \rightarrow 0 \tag{19}$$

We conclude that the error estimators of the linear theory can be applied at regular points, where $(\mathbf{G}')^{-1}$ is bounded for the linearized problem. The residuals or stresses of the linearized problem can be obtained from the residuals or stresses of the finite element solution and a further perturbed state. The perturbation vector $\mathbf{w}*$ is the solution of the linearized finite element problem and can be computed by

$$\mathbf{Kw}* = \epsilon\mathbf{P} \tag{20}$$

where $\mathbf{K}$ is the stiffness matrix, $\mathbf{P}$ is the reference load vector and ϵ is a small parameter. The difference of the residuals or stresses computed at $\mathbf{u}^h$ and $\mathbf{w} + \mathbf{u}^h$ is the solution for the linearized problem and is used for error estimation. For the relative error the small parameter ϵ is of minor influence to the result.

3. 1. EXAMPLE

A clamped open cylinder is loaded by two distributed loads with opposite directions at the free edge (Figure 3). For small deformations the critical area is around the loaded region. For increasing loads the deformations propagate through the cylinder. The computation is performed with 7.5 % relative error tolerance. The error is estimated for the linearized problem through the residual as well as the superconvergent patch recovery based estimator. Both estimators show a similar tendency for the present example.

4. Error estimation for inelastic materials

As a first step an error estimator for the deformation theory of plasticity based on the reciprocal theorem is considered. In contrast to the incremental Prandtl–Reuss theory the nonlinear constitutive equations are given in total form. Using the principle of virtual work, the nonlinear constitutive relations and the linear strain displacement relationship, the computation of a finite element approximation is a straight forward procedure. The discretization errors in the stresses can be computed similar to the linear case (eqs. (4), (7)) by a nonlinear problem

$$\begin{gathered}\sum_{K=1}^{NEL}\left\{\operatorname{div}\left\{\boldsymbol{\sigma}(\mathbf{u}^h)-\boldsymbol{\sigma}(\mathbf{u})\right\}+\mathbf{R}=\mathbf{0}\ \text{on}\ \Omega_K\right\}\\ \sum_{K=1}^{NEL}\left\{\left\{\boldsymbol{\sigma}(\mathbf{u}^h)-\boldsymbol{\sigma}(\mathbf{u})\right\}\tilde{\mathbf{n}}=\mathbf{J}\ \text{on}\ \Gamma_K\not\subseteq\Gamma_D,\quad \mathbf{e}=\mathbf{0}\ \text{on}\ \Gamma_K\subseteq\Gamma_D\right\}\end{gathered}\tag{21}$$

For error estimation the nonlinear stress strain relationship is expanded in a Taylor series at $\mathbf{u}^h$.

$$\boldsymbol{\sigma}(\mathbf{u})=\boldsymbol{\sigma}(\mathbf{u}^h)+\frac{\partial\boldsymbol{\sigma}}{\partial\boldsymbol{\epsilon}}:\boldsymbol{\epsilon}(\mathbf{u}-\mathbf{u}^h)+\frac{1}{2}\left(\frac{\partial^2\boldsymbol{\sigma}}{\partial\boldsymbol{\epsilon}^2}:\boldsymbol{\epsilon}(\mathbf{u}-\mathbf{u}^h)\right):\boldsymbol{\epsilon}(\mathbf{u}-\mathbf{u}^h)+\ldots.\tag{22}$$

We consider again the linearized problem for error estimation.

$$\boldsymbol{\sigma}(\mathbf{w})=\boldsymbol{\sigma}(\mathbf{u}^h)+\mathbf{A}:\boldsymbol{\epsilon}(\mathbf{w}-\mathbf{u}^h)\tag{23}$$

The tangential material tensor $\mathbf{A}$ is the usual elasticity tensor $\mathbf{C}$ for elastic states and the elastoplastic tangent modulus $\mathbf{C}^{ep}$ for plastic states without making a distinction between loading and unloading as in the flow theory.

$$\mathbf{C}^{ep}=\mathbf{C}-\frac{\mathbf{C}:\frac{\partial F}{\partial\boldsymbol{\sigma}}\frac{\partial F}{\partial\boldsymbol{\sigma}}:\mathbf{C}}{\frac{\partial F}{\partial\boldsymbol{\sigma}}:\mathbf{C}:\frac{\partial F}{\partial\boldsymbol{\sigma}}+H}\tag{24}$$

whereby F is the yield criterion and H is a parameter describing the hardening. Combining (21) and (23) the discretization errors can be computed by

$$\begin{gathered}\sum_{K=1}^{NEL}\left\{\operatorname{div}\left\{\mathbf{A}:\left(\boldsymbol{\epsilon}(\mathbf{w})-\boldsymbol{\epsilon}(\mathbf{u}^h)\right)\right\}+\mathbf{R}=\mathbf{0}\ \text{on}\ \Omega_K\right\}\\ \sum_{K=1}^{NEL}\left\{\mathbf{A}:\left\{\boldsymbol{\epsilon}(\mathbf{w})-\boldsymbol{\sigma}(\mathbf{u}^h)\right\}\tilde{\mathbf{n}}=\mathbf{J}\ \text{on}\ \Gamma_K\not\subseteq\Gamma_D,\quad \mathbf{w}-\mathbf{u}^h=\mathbf{0}\ \text{on}\ \Gamma_K\subseteq\Gamma_D\right\}\end{gathered}\tag{25}$$

In the same manner as in elasticity, for error estimation of a specific displacement $\mathbf{w}$ in the component i at the point $\mathbf{x}=\overline{\mathbf{x}}$ we consider the dual problem

$$\begin{gathered}\operatorname{div}\left(\mathbf{A}:\boldsymbol{\epsilon}(\mathbf{G})\right)+\boldsymbol{\delta}_i(\overline{\mathbf{x}})=\mathbf{0}\ \text{on}\ \Omega\\ \mathbf{A}:\boldsymbol{\epsilon}(\mathbf{G})\tilde{\mathbf{n}}=\mathbf{0}\quad\text{on}\ \Gamma_N\qquad \mathbf{u}=\mathbf{0}\quad\text{on}\ \Gamma_D\end{gathered}\tag{26}$$

Both equations for the displacement errors and the Green's function are linear and the reciprocal theorem can be applied.

$$w_i(\overline{\mathbf{x}})-u_i^h(\overline{\mathbf{x}})=\sum_{K=1}^{NEL}\left\{(\mathbf{R},\mathbf{G})_{\Omega_K}+(\mathbf{J},\mathbf{G})_{\Gamma_K}\right\}\tag{27}$$

The external work equals to the internal work of the related stresses and strains.

$$u_i^h(\overline{\mathbf{x}})-w_i(\overline{\mathbf{x}})=\sum_{K=1}^{NEL}B\left(\boldsymbol{\sigma}(\mathbf{w})-\boldsymbol{\sigma}(\mathbf{u}^h),\boldsymbol{\epsilon}(\mathbf{G})\right)_{\Omega_K}\tag{28}$$

Furthermore, due to the Galerkin orthogonality, the finite element approximation to the Green's function $\mathbf{G}^h$ can be inserted.

$$u_i^h(\bar{\mathbf{x}}) - w_i(\bar{\mathbf{x}}) = \sum_{K=1}^{NEL} B(\boldsymbol{\sigma}(\mathbf{w}) - \boldsymbol{\sigma}(\mathbf{u}^h), \boldsymbol{\epsilon}(\mathbf{G}) - \boldsymbol{\epsilon}(\mathbf{G}^h))_{\Omega_K} \tag{29}$$

The exact stresses and strains in the bilinear form can be replaced by smoothed values and the errors can be simply estimated.

$$u_i^h(\bar{\mathbf{x}}) - w_i(\bar{\mathbf{x}}) \approx \sum_{K=1}^{NEL} B(\boldsymbol{\sigma}^*(\mathbf{w}) - \boldsymbol{\sigma}(\mathbf{u}^h), \boldsymbol{\epsilon}^*(\mathbf{G}) - \boldsymbol{\epsilon}(\mathbf{G}^h))_K \tag{30}$$

The important characteristic of plasticity, namely the different behavior for loading and unloading, is not reflected by the deformation theory. For the instationary flow theory the constitutive equations are given in a incremental form. Introducing a time incrementation to the problem the time derivatives are replaced by finite increments. The procedures for the deformation theory can also be used for flow theory if applied in the incremental sense. The total error is computed by the sum of the contributions of each time increment.

4. 1. EXAMPLE

A perfect plastic perforated tension strip with plane strain conditions is investigated. Due to the symmetry a quarter of the structure is discretized with bilinear quadrilateral elements. The total load of 375 and 450 are applied in one step, in order to simulate the deformation theory by a classical algorithm for J_2 plasticity. We control the errors of displacements at different finite element nodes in the elastic and plastic region indicated by arrows (Figure 4). The loading of the dual problem consists of a single nodal force. The estimated error η^{est} shows for fine meshes a very good agreement with the *quasi exact* errors η^{ex}.

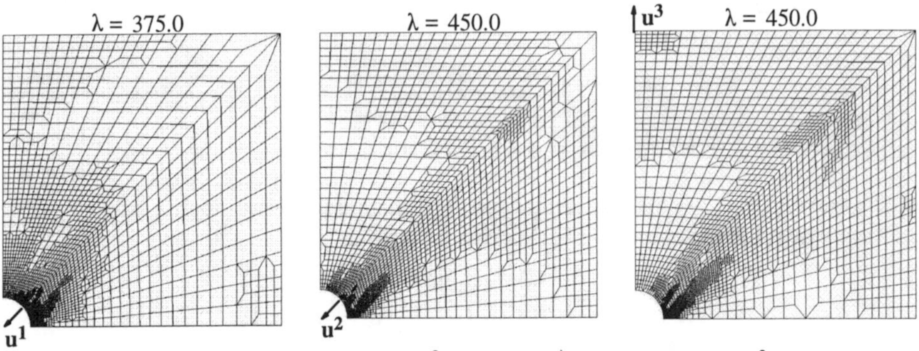

NDOF	u^1	η^{est}	η^*
360	4.21·10e–3	3.50	5.70
1222	4.41·10e–3	0.60	0.90
2426	4.43·10e–3	0.33	0.45
7995	4.45·10e–3	0.10	

NDOF	u^2	η^{est}	η^*
360	7.90·10e–3	9.19	15.8
1333	8.88·10e–3	3.31	3.04
2397	8.99·10e–3	2.05	1.80
8918	9.15·10e–3		

NDOF	u^3	η^{est}	η^*
360	2.37·10e–2	2.62	4.20
744	2.42·10e–2	1.72	2.07
1224	2.44·10e–2	1.08	0.82
2159	2.46·10e–2	0.65	0.40
17038	2.47·10e–2		

Fig. 4: Perforated tension strip (L=10, R=1, E=206900, ν=0.29, σ_y=450.0)

Conclusions

In this paper the relationship of the error estimators based on duality arguments to the influence line/surface concept frequently applied in structural mechanics is explained. The concepts of error estimation and adaptivity are extended into the nonlinear range, including geometrical and material nonlinearities. For regular points on the load path this is a straight forward extension if the techniques are applied for the linearized problem. Deformation theory in plasticity is again a nonlinear problem. The procedures can also be used for flow theory if applied in the incremental sense.

Acknowledgement

This work is part of the DFG research project Ra 218/11–1 "Algorithms, Adaptive methods, Elastoplasticity". The support is gratefully acknowledged.

References

[1] Babuska, I., Rheinboldt, W.C. (1978) Error Estimates for Adaptive Finite Element Computations, *SIAM J. Numer. Anal.* **15**, 736–754.

[2] Babuska, I., Miller, A. (1984) The Post–Processing Approach in the Finite Element Method–Part 2: The calculation of Stress Intensity Factors, *Int. j. numer. methods eng.* **20**, 1111–1129.

[3] Babuska I., Strouboulis, T., Upadhyay, C.S., Gangaraj, S.K. (1995) A posteriori estimation and adaptive control of the pollution error in the h–version of the finite element method, *Int. j. numer. methods eng.* **38**, 4207–4235.

[4] Becker, R., Rannacher, R. (1996) A Feed–Back Approach to Error Control in Finite Element Methods: Basic Analysis and Examples, *EAST–WEST J. Numer. Math.* **4**, 237–264.

[5] Cirak, F., Ramm, E. (1997) A–posteriori error estimation and adaptivity for linear elasticity using the reciprocal theorem, Accepted for publication in *Comput. Methods Appl. Mech. Eng.*

[6] Johnson, C., Hansbo, P. (1992) Adaptive Finite Element Methods in Computational Mechanics, *Comput. Methods. Appl. Mech. Eng.* **82**, 143–181.

[7] Pousin, J., Rappaz, J. (1994) Consistency, Stability, A Priori and A Posteriori Errors for Petrov–Galerkin Methods Applied to Nonlinear Problems, *Numer. Math.* **69**, 213–231.

[8] Rheinboldt, W. C. (1985) Error Estimates for Nonlinear Finite Element Computations, *Computers&Structures* **20**, 91–98.

[9] Ramm, E., Cirak, F. (1997) Adaptivity for Nonlinear Thin–walled Structures, in D.R.J. Owen, E. Onate, E. Hinton (eds.), *Computational Plasticity*, CIMNE, Barcelona, pp. 145–163.

[10] Schrefler, B.A., Zavarise, G., Gori, R. (1993) A Fast Error Check for Structural Analysis Using the Virtual Force Principle, *Int. j. numer. methods eng.*, **36**, 3223–3237.

[11] Stein, E., Seifert, B., Ohnimus, S., Carstensen, C. (1994) Adaptive Finite Element Analysis of Geometrically Non–Linear Plates and Shells, Especially Buckling, *Int. j. numer. methods eng.* **37**, 2631–2655.

[12] Strang, G., Fix, G.J. (1973) *An Analysis of the Finite Element Method*, Prentice Hall, Englewood Cliffs, N.J.

[13] Suttmeier, F. T. (1997) Adaptive Finite Element Approximations of Problems in Elasto–Plasticity Theory. Ph.–D Dissertation, Institut für Angewandte Mathematik, Universität Heidelberg.

[14] Tottenham, H. (1970) Basic Principles, in H. Tottenham, C. Brebbia (eds.) *Finite Element Techniques in Structural Mechanics*, Southampton University Press.

[15] Verfürth, R. (1996) *A Review of A Posteriori Error Estimation and Adaptive Mesh–Refinement Techniques*, John Wiley & Sons, Chichester, B.G. Teubner, Stuttgart.

[16] Zienkiewicz, O.C., Zhu, J.Z. (1987) A simple error estimator and adaptive procedure for practical engineering analysis, *Int. j. numer. methods eng.* **24**, 337–357.

EFFICIENT DISCRETISATION METHODS FOR THE NUMERICAL SIMULATION IN GEOMECHANICS

G. BEER

Institute for Structural Analysis

Technical University Graz, Austria

1 Introduction

The computer simulation of problems in geomechanics places particular demands on a numerical model. On the one hand, the ground, which for all practical purposes can be assumed to be infinite, has to be modelled. On the other hand, the fact that the material soil or rock is also heterogeneous, jointed and exhibits nonlinear behaviour has to be considered.

The very popular Finite Element Method (FEM) has no problem in dealing with heterogeneous, jointed and nonlinear material behaviour but has difficulty modelling infinite domains and for three-dimensional problems may lead to a significant discretisation effort.

The Boundary Element Method (BEM) on the other hand leads to small meshes and hence a small discretisation effort and can deal easily with an infinite domain because fundamental solutions are used instead of locally defined shape functions. However, most implementations of the BEM have difficulty in dealing with heterogeneous, jointed and inelastic material. Recent developments, however, have overcome some of these difficulties and it will be shown in this paper that the modified multi-region boundary element method can better deal with joint behaviour than the FEM.

The purpose of the paper is to critically examine the applicability and efficiency of available discretisation methods to the simulation of problems in geomechanics. Emphasis will be placed on problems involving underground excavations and the modelling of joints and faults. The main numerical methods which will be examined are the boundary element method and the finite element method.

2. Modelling of joints and faults

The modelling of discontinuous displacement behaviour is very important in the numerical simulation of problems in geomechanics. Slip and separation for example can occur in foundation, excavation, earthquake and reservoir modelling. Discontinuous behaviour along joints can be modelled with either the FEM or the multi-region BEM.

H. A. Mang and F. G. Rammerstorfer (eds.),
IUTAM Symposium on Discretization Methods in Structural Mechanics, 53–60.

2.1. FINITE ELEMENT METHOD

For the modelling of such behaviour with the FEM, special joint elements have been developed many decades ago by Goodman [1] and improved upon by several authors [2,3,4]. The approach is basically a penalty method and springs of high stiffness are used at the nodes or the Gauss integration points to prevent interpenetration and slip prior to failure. At failure the stiffnesses are reduced to residual values to allow separation and slip to occur. Figure 1 shows an example of a two-dimensional joint element.

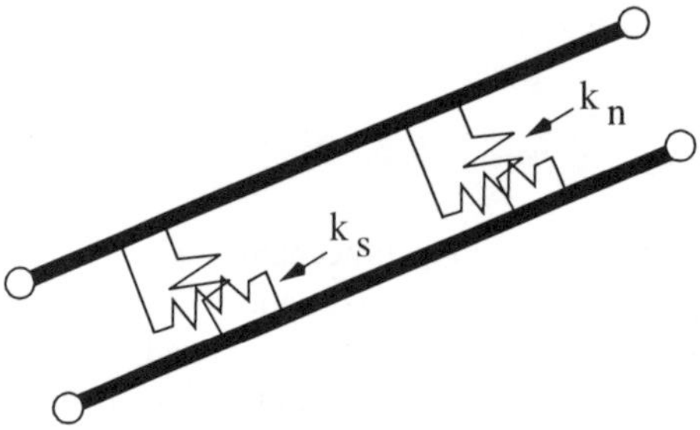

Figure 1. Two-dimensional joint finite element

The theoretical background of this element has been described in detail in [2].
There are at least 6 parameters necessary to describe the element behaviour:

- The initial value of normal stiffness k_n and shear stiffness k_s
- The residual values of normal and shear stiffness which are applied after failure occurs
- The cohesion and the angle of friction.

In addition there may be a dilation angle required for the modelling of dilatant joint behaviour. The normal stiffness and the shear stiffness are parameters which are not easily determined from field values. They can either be interpreted as penalty parameters which are specified an arbitrary high value to prevent interpenetration and slip prior to failure or as parameters describing the elastic joint behaviour, for example the compression or deformation of asperities.
Among other workers Hohberg [5] has found at a very early stage that these elements are very sensitive to the values of stiffness specified and sometimes exhibit oscillatory behaviour of the contact stresses inside the joint element especially when high values of stiffnesses are used. Hohberg suggested that the situation could be improved if a Lobatto instead of a Gauss integration was used. Day[6] investigated the problem with oscillatory behaviour further and found that its occurrence not only depended on the value of the normal and shear stiffness but also on the size of the solid elements adjacent to the joint element. The test example used by Day to demonstrate the problems with oscillatory behaviour is that of an elastic block on a rigid foundation. Figure 2 shows the dimensions of the problem and the loading. The block is loaded by a normal stress of 200 MPa and a shear stress at the top

which is increased until toppling occurs.

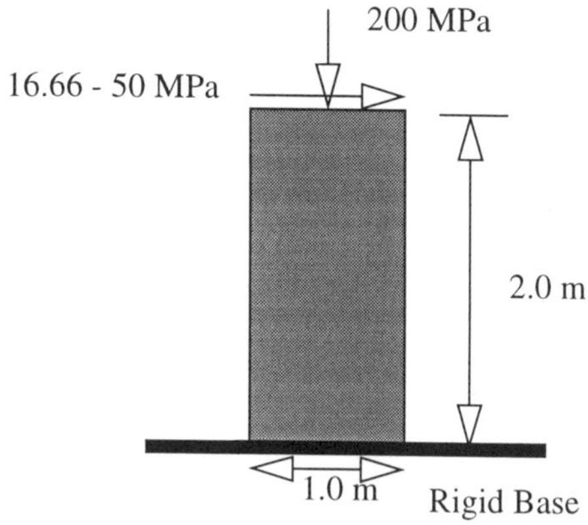

Figure 2. Test example to show oscillatory behaviour of joint finite elements

The theory predicts that tension will start to occur at the left edge of the interface when the shear stress reaches 16.66 MPa. At 50 MPa toppling will occur. The theoretical shear stress distribution for this problem is not known and the solution will be somewhere between the solution for a infinitely long rigid block on flexible foundations (constant distribution) and for an infinitely long flexible block on a rigid foundation (parabolic distribution).

Figure 3 shows the distribution of shear stress at the interface for an applied shear stress of 16.66 MPa depending on the ratio of joint stiffness to elastic modulus of the block (E) used. It can be clearly seen that for the mesh used the stresses at the interface have oscillatory behaviour at ratios k/E of larger than 100 (where k is both k_s and k_n). Figure 4 shows,however, that if the size of the element adjacent to the joint is reduced to 1/4 of its original size then this oscillatory behaviour disappears again, even if a stiffness value of 1000 is used.

2.2. BOUNDARY ELEMENT METHOD

There are a number of ways in which discontinuous behaviour can be modelled with the boundary element method. One way is to use so called Displacement Discontinuity Elements as explained for example by Beer and Watson [7]. Here the joint surfaces are represented by one boundary element with two surfaces which are coincidental. Another way of simulating joint behaviour is with the multi-region boundary element method together with a special algorithm which allows to deal with discontinuous behaviour at the interface between regions. If one is not interested in the ‚elastic‘ behaviour of the joint prior to failure one can use a rigid-plastic model whereby no relative movement is allowed to occur in the joint plane prior to failure. The rigid-plastic boundary element model is explained in detail in [8] and here only the main features are discussed. The BEM is ideally suited for modelling joint behaviour because the primary variables are tractions and not nodal point forces as they are used in the FEM.This fact makes it much easier to check the Mohr-Coulomb failure conditions at the interface nodes.

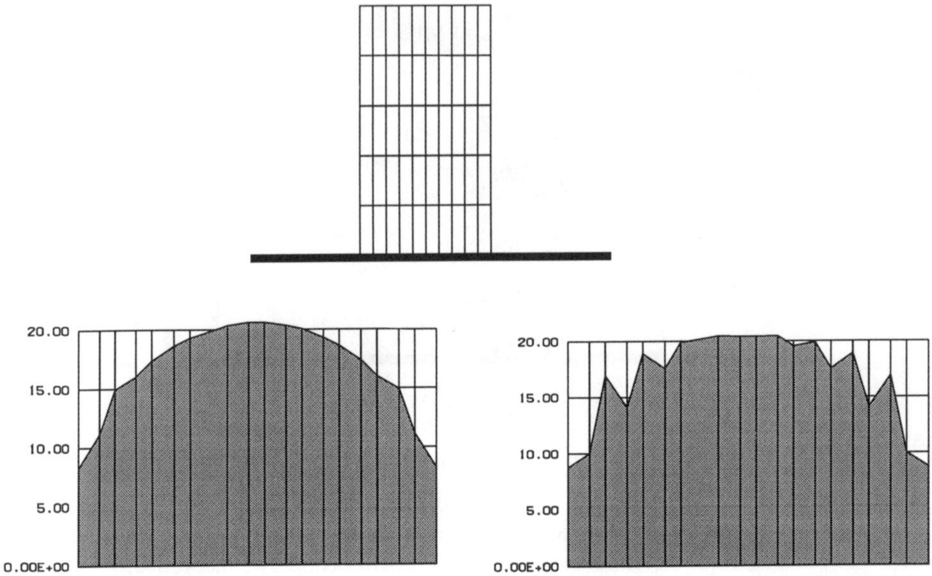

k/E = 100. k/E = 1000.

Figure 3. Distribution of shear stress at the interface for different joint stiffness

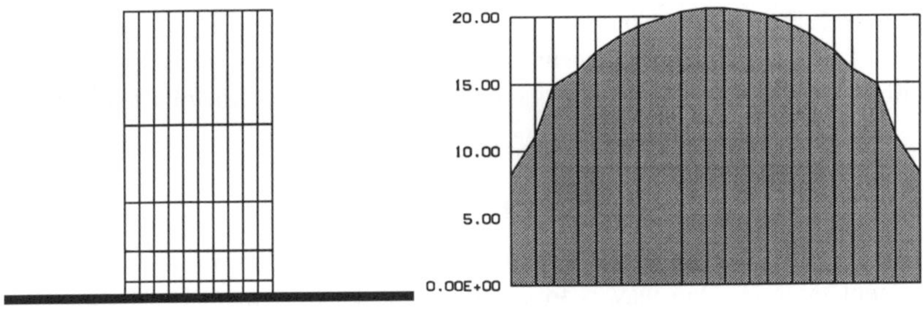

k/E = 1000.

Figure 4. Distribution of shear stress on interface for refined mesh

With the multi-region rigid-plastic joint model one can compute the tractions acting upon region I at the interface i as:

$$\{t^i\}^I = \{t^{i0}\}^I + [K^i]^I \{u^i\}^I \tag{1}$$

In Equation (1) the vector $\{t^{i0}\}^I$ contains the interface tractions when the interface dis-

placements $\{u^i\}^I$ are zero and $[K^i]^I$ is a „Stiffness Matrix“ which relates the **tractions** to the displacements at the interface nodes. To be able to check the Mohr-Coulomb and no-tension conditions on the interface and to be able to disconnect nodes to allow slip and separation to take place at the interface the tractions and displacements have allready been transformed into a local (joint) coordinate system.
Using Equation (1) the problem can be reduced to one of assembling the system of equations similar to the FEM where „Stiffness“ and „Force“ terms of each region are added. If no failure on the interface occurs then for each node there is one set of displacements only. If a failure in shear or tension occurs at a node then there are two different values of displacement at a node depending whether it belongs to region I or II. This will change the way the system of equations is assembled. Details are explained in [8].
The method just described has been applied to the test problem of Fig. 2 and the results of the analysis are shown in Fig. 5.

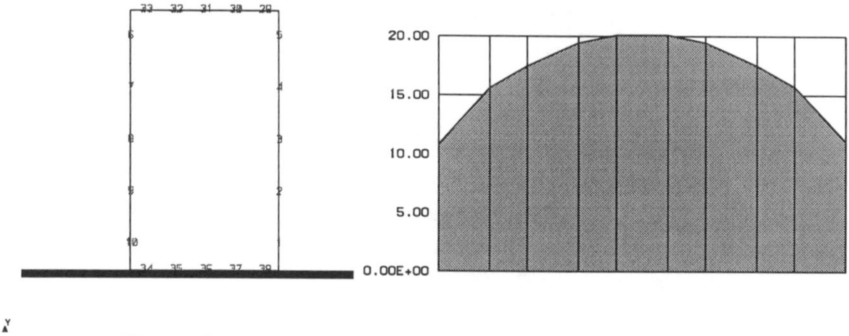

Figure 5. Boundary element mesh and shear stress distribution

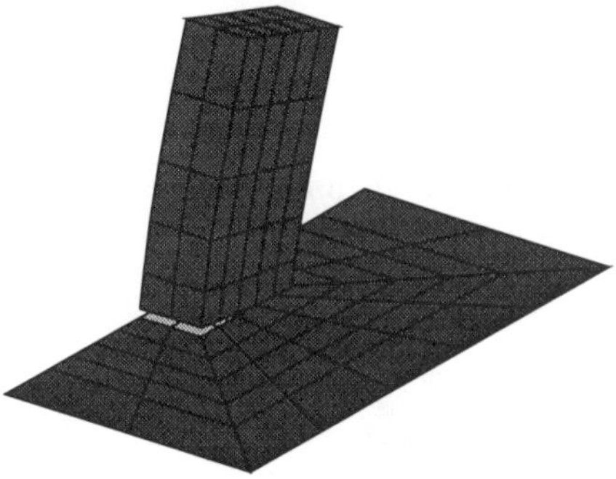

Figure 6. Displaced shape after 3 iterations for a shear stress of 50 MPa (half mesh shown)

In contrast to the FEM discretisation there are no elements inside the block, it is not necessary to specify a joint stiffness and the results of the analysis do not depend on the discretisation. Furthermore there are fewer iterations required to model the toppling of the block. To demonstrate that the procedure also works well in 3-D the problem was solved as a 3-D block resting on an elastic half-space. In Fig. 6 the opening of the interface can be seen.

3. Analysis of stresses in a fault zone

To demonstrate the efficiency of finite and boundary element discretisations for modelling faults a practical example is presented here. The example is taken from an actual tunnel site in Austria where an accident occurred when the tunnel was driven trough a soft zone of ground just before it reached harder rock. The purpose of the analysis was to confirm with a numerical simulation the proposition put forward by the experts investigating the accident that the virgin stress in the disturbed zone had been significantly altered by the difference in stiffness of the two zones and the presence of faults. This modification in the stress was thought to be caused by arching as the soft zone „creeps" under selfweight (Fig. 7).

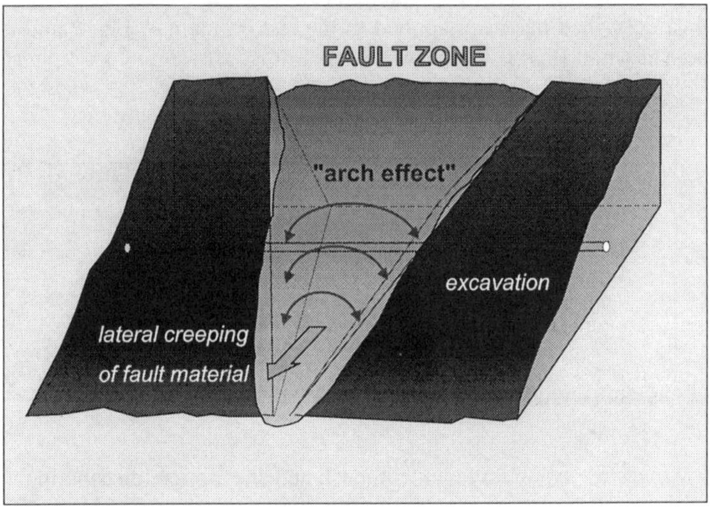

Figure 7. Schematic of the arching effect being modelled

For the analysis two different discretisations were compared. One is a multi-region boundary element discretisation with a rigid-plastic joint model as discussed previously the other one is a finite element discretisation with penalty-type joint elements (three-dimensional equivalents of the joint element depicted in Fig. 1).

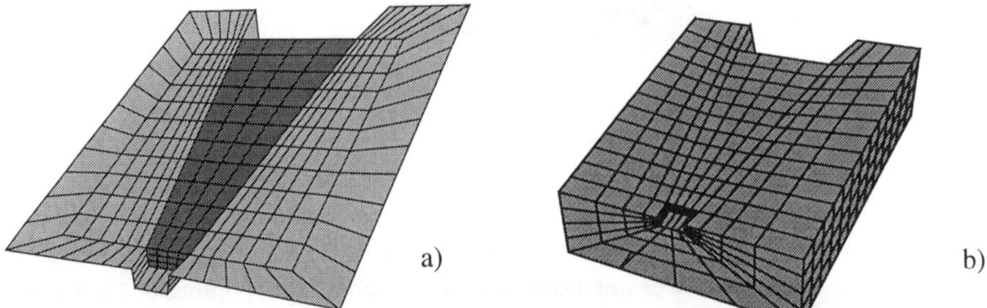

Figure 8. a) Boundary element and b) finite element discretisation

Fig. 8 a shows the boundary element mesh consisting of a finite and a semi-infinite region made up of 8 node boundary elements and 6 node infinite boundary elements whereas Fig 8 b shows a mesh with 20 node solid elements and 16 node joint elements. The base of the FEM mesh was fixed in the three coordinate directions.

The material properties were chosen as follows:

- Undisturbed zone : E = 150 000 MPa, $\nu = 0.3$
- Disturbed zone : E = 50 000 MPa, $\nu = 0.3$, $\gamma = 3\ 000\ \text{kg/m}^3$
- Fault: Angle of friction= 15°, Cohesion= 0.5 Mpa

To simulate the arching effect gravity was applied to the finite region or the finite elements of the disturbed zone, respectively. Figure 9 shows some results of the BEM analysis after convergence had been achieved after only 3 iterations. The slip in the fault plane and the arching of the stresses can be clearly seen. Further results are presented in [9].

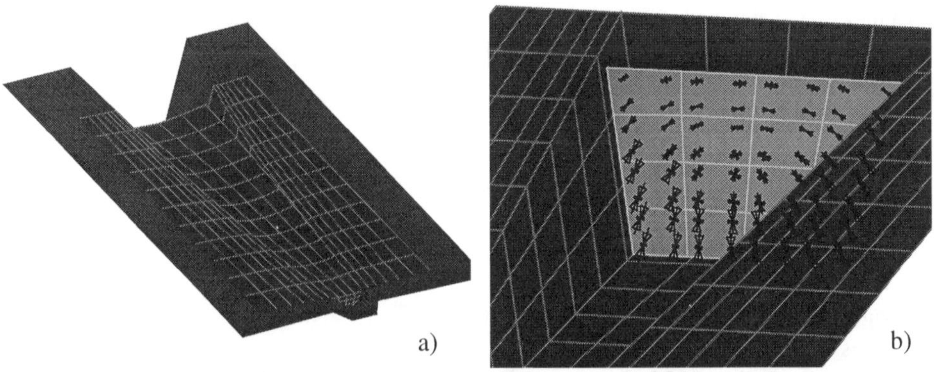

Figure 9. a) Displaced shape and b) stress vectors plotted on a result plane inside region 2

Both FEM and BEM discretisations gave similar results. The computing effort however was significantly different. Table 1 shows the size of the meshes in number of elements and degrees of freedom and the run times needed to obtain the solution. It can be seen that the FEM requires about 10 times the number of degrees of freedom as the BEM.

TABLE 1. Comparison boundary and finite element discretisations

Item	BEM	FEM
Number of Elements	376	932
Number of Unknowns	972	11 899
Run time initial solution	214 sec	3900 sec
Run time one iteration	15 sec	40 sec

The values for the run time given for each iteration step includes the updating of joint conditions on the interface for the BEM but does not include the updating of the joint stiffness for the FEM (i.e. a modified Newton Raphson procedure with constant stiffness is assumed). Each time the joint stiffness values are updated for the FEM a run time equal to the initial solution would be required. The run times are for a Frontal solver. Significant reduction in run times can be expected if one uses a conjugate gradient solver. However the task of keeping and updating Gauss point stresses of finite elements remains.

4. Summary and Conclusions

An attempt has been made to look at the efficiency of different discretisation methods in geomechanics. Because of space limitations only one aspect, namely, the modelling of joints and faults could be examined. It was shown that for problems where the nonlinear behaviour is restricted to the fault plane the multi-region rigid-plastic BEM is more efficient than the FEM with joint elements.

Examples have also been presented in [10] where a coupled boundary and finite element discretisation is more efficient than the FEM for some problems.

5. References

1 Goodman R.E, Taylor R.L and Brekke A.M. (1973) A model for the mechanics of jointed rock *ASCE J.* **94** SM10, 833-847.

2 Beer G. (1985) An isoparametric joint/interface element for FE analysis. *Int. J. Num. Meth. Eng.* **21**, 585-600.

3 Ghabhoussi J., Wilson E.L., Isenberg J. (1973) Finite element for rock joints and interfaces *ASCE J.* **99** SM10, 833-847.

4 Desai C.S., Zaman M.M., Lightner J.G. Siriwardane H.J. (1984) Thin-layer element for interfaces and joints. *Int. J. Num. Meth. Geomech.* **8**, 19-43.

5 Hohberg J. (1988) Trennflaechenformulierungen fuer die statische und dynamische Berechnung von Bogenstaumauern. *Institut fuer Baustatik und Konstruktion, ETH Zuerrich.*

6 Day, R.A. and Potts, D.M. (1994) Zero thickness interface elements - numerical stability and Application. *Int. J. Num. Meth. Geomech.* **18**, 689-708.

7 Beer G. and Watson J.O. (1992) *Introduction to Finite and Boundary Elements for Engineers*, Wiley, Chichester.

8 Beer G. (1995) Rock Joints - BEM Computations, *Mechanics of Matrial Interfaces,* Selvadurai and Boulon (eds), Elsevier B.V.

9 Beer G. (1996) Numerische Simulation im Tunnelbau. *Baustatik Baupraxis - Tagungsheft BB6*. Hochschule fuer Architektur und Bauwesen Weimar.

10 Beer G. and Sigl O. (1997) BEFE helps design new CERN facility in Geneva. *Benchmark,* NAFEMS, UK, January 1998.

TRAVERSING LIMIT POINTS IN NONLINEAR STRUCTURAL ANALYSIS USING EFFICIENT ITERATIVE EQUATION SOLVERS

K. SCHWEIZERHOF AND TH. ROTTNER
Institute for Mechanics,
University of Karlsruhe, 76128 Karlsruhe, Germany

AND

G. ALEFELD AND I. LENHARDT
Institute for Applied Mathematics,
University of Karlsruhe, 76128 Karlsruhe, Germany

1. Introduction

The traversal of limit points and the detection of bifurcation points in structural analysis using direct solvers is a well-known standard technique in Finite Element simulations, for a summary see e.g. [2], [11], [13] and [15]. Iterative solvers have only rarely been used, see e.g. [9] and [8], and were rather limited to small size beam problems and did not show considerable efficiency. For very large well conditioned structures, however, iterative solvers are known as a very promising tool, in particular as parallelization of these iterative algorithms is very efficient.

The main obstacle for using iterative solvers is the ill conditioning of the stiffness matrices in structural mechanics. This is firstly a result of the structural model e.g. shells where the membrane stiffness terms are considerably larger than the bending stiffness terms, or in continuum analysis when the structure becomes rather thin at some locations and the stiffness terms in thickness direction for this region become considerably larger than the other coefficients. Secondly the stiffness matrices become singular or nearly singular when e.g. so-called snap-through points are passed in the calculation of a fully nonlinear load deflection-path of some shell-type structures.

The experience with applications of iterative solvers for such problems has shown, see [1], [12] and [7], that they can very well be used for such an

H. A. Mang and F. G. Rammerstorfer (eds.),
IUTAM Symposium on Discretization Methods in Structural Mechanics, 61–70.

analysis and the performance can be even improved compared to standard direct equation solvers even on serial machines. The solvers investigated are various preconditioned Conjugate Gradient and Lanczos-type solvers with compact storage, which are compared to fast sparse direct solvers on some numerical examples. Also the parallelization concept of the iterative solvers is described. The performance on a selected example is demonstrated.

2. Nonlinear analyses using iterative solvers

In nonlinear Finite Element Analysis usually Newton's method is applied in a stepwise fashion to compute a load deformation path. Then in every step for each Newton iteration several systems of linear equations have to be solved. These systems of equations are sparse, often large and the coefficient matrix – in structural mechanics the stiffness matrix – can be ill conditioned or even indefinite in the case of bifurcation and limit points. Often Newton's method is combined with arc-length procedures to enable the solution procedure to pass limit-points. To compute a new load level then an additional equation containing the solution variables is introduced – mostly also linearized within the Newton process –, which leads to an unsymmetric system of linear equations,

$$\begin{bmatrix} \mathbf{K} & -\mathbf{p} \\ \mathbf{v}^T & \alpha \end{bmatrix} \begin{bmatrix} \Delta\mathbf{x} \\ \Delta\lambda \end{bmatrix} = - \begin{bmatrix} \Delta\mathbf{r} \\ f \end{bmatrix} \tag{1}$$

with the symmetric tangent stiffness $\mathbf{K}$, the unknown increments of displacements $\Delta\mathbf{x}$, the out-of-balance forces $\Delta\mathbf{r}$, the load vector $\mathbf{p}$, the new increment of the load level $\Delta\lambda$ and $\mathbf{v}$, the scalars α and f defining the arc-length constraint (see e.g. [2], [11] or [13]). $\Delta\mathbf{x}$ and $\Delta\lambda$ are then combined as solution variables. In general this unsymmetric linear system of equations is solved by using Gaussian block elimination:

$$\Delta\mathbf{u}^I = \mathbf{K}^{-1}\mathbf{p} \quad \text{and} \quad \Delta\mathbf{u}^{II} = -\mathbf{K}^{-1}\Delta\mathbf{r} \tag{2}$$

$$\Delta\lambda = -\frac{f + \mathbf{v}^T\Delta\mathbf{u}^{II}}{\alpha + \mathbf{v}^T\Delta\mathbf{u}^I} \quad \text{and} \quad \Delta\mathbf{x} = \Delta\mathbf{u}^{II} + \Delta\lambda\Delta\mathbf{u}^I. \tag{3}$$

Thus, two symmetric linear systems of equations (2) have to be solved and then the unknown increments of displacements $\Delta\mathbf{x}$ and the increment of the new load level $\Delta\lambda$ can be computed using equation (3).

Another possibility to solve iteratively equation (1) is to use an iterative scheme for unsymmetric systems of equations like QMR or GMRES. But then per iteration two matrix-vector products have to be computed and the computational effort is at least the same as using equations (2) and (3) separately. A more efficient method is proposed in [12] by the authors but will not be discussed here due to the limited space available.

The efficiency of various known solution strategies is strongly problem dependent (see e.g. [1]) and therefore it is difficult to give general statements. Thus a special example containing the mentioned features is chosen and discussed in detail. The presented results are exactly only valid for this example and the hardware used. Nevertheless they can be used for some more general judgment. It has to be noted, that the iterative solvers have been applied with success to other strongly nonlinear problems, for example snap-back problems or problems with sharp edges in the load deformation curve.

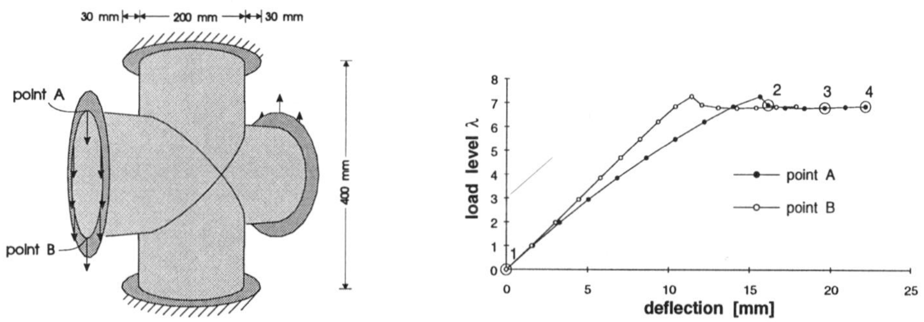

Figure 1. Flange – system and load deflection curve

The example is a cross pipe with stiffeners at the upper and lower, left and right edge of the shells. The so-called flange is hinged at the top and the bottom and subjected to some shear load (see figure 1). The material behavior is assumed to be linear elastic with $E = 2.1 \cdot 10^5 \frac{N}{mm^2}$ and $\nu = 0.3$ (steel). The shell thickness is 3 mm, at the stiffeners 6 mm. The FE-mesh contains 22 678 nodes and 22 400 4-noded bilinear shell elements [4] which leads to 112 500 equations.

In the load deflection curve (figure 1) a snap-through behavior can be seen, after reaching a maximal load value the deformation increases with decreasing load. The maximal condition number is about 10^8. In figure 2 the deformed shapes of the flange at various states are shown. Because of the snap-through behavior Newton's method is used in combination with arc-length procedures (see for example [2], [13]). Then per Newton iteration *two* linear systems of equations have to be solved with the same coefficient matrix, which is an advantage of the direct solver over the iterative one, as can be seen later.

For the control of the arc-length method the well known stiffness parameters are used and the definiteness of the stiffness matrix is monitored. The application of the preconditioned Lanczos algorithm has shown that

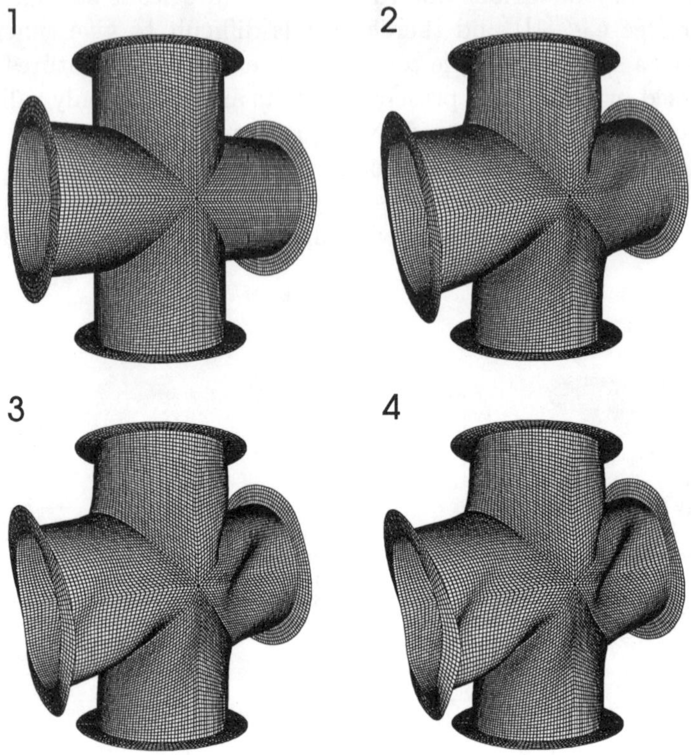

Figure 2. Flange – deformed shapes at various states, see curve in figure 1

the indefiniteness of the stiffness matrix is preserved in the tridiagonal matrix in the examples analyzed.

2.1. PRECONDITIONING STRATEGIES FOR THE SOLUTION OF LINEAR SYSTEMS OF EQUATIONS

In this section the iterative solvers used are described shortly. They are compared to a direct sparse solver from the SMPAK-library [14]. For positive definite (linear) problems a standard CG Method is used, whereas for indefinite problems a preconditioned Lanczos method [10] is taken. In combination with indefinite preconditioning matrices, necessary in some cases, the symmetric QMR method [3] is used.

The preconditioning strategies are:

- Jacobi,
- SSOR with overrelaxation parameter $\omega = 1.0$,
- incomplete LDL^T factorization on the sparsity pattern of the stiffness

matrix (MPILU),
- a block version of MPILU; here some degrees of freedom may be separated (Block-MPILU) and
- incomplete LDL^T factorization allowing fill-in of the first level; fill-in caused by fill-in is not taken into account (FLILU).

2.2. NUMERICAL COMPARISON OF VARIOUS SOLVERS

In figure 3 the total computation times and the memory requirements of direct and iterative solvers on an IBM RS6000 workstation are shown. The direct solver (noted SMPAK) is efficient in computation time but has fairly large memory requirements.

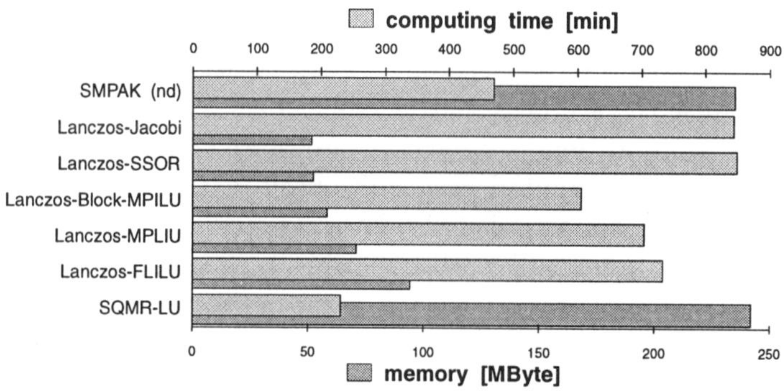

Figure 3. Computation time and memory requirements for nonlinear flange problem

All iterative solvers need significantly less memory, but the computing time increases. This is mainly an effect concerning the application of the arc-length procedure and the necessity to solve two linear systems of equations per Newton iteration. The most efficient iterative solver is Lanczos-Block-MPILU concerning memory and CPU-time as a whole, where all degrees of freedom are decoupled in the preconditioning matrix.

The most efficient solution concerning computation time – though requiring large memory – is achieved by a combination of direct and iterative schemes (SQMR-LU). For this, one complete LDL^T factorization is computed in every Newton step. This factored matrix is then reused as a preconditioner in the following Newton iterations. In case of indefinite stiffness matrices this preconditioner may be indefinite and therefore the symmetric QMR method is used for the iterative solutions.

During the nonlinear analysis the condition numbers of the stiffness matrices increase from values about 10^5 to 10^8 near to singular points. As a consequence the iteration numbers needed for the iterative solution of each corresponding linear system of equations increase (using for example MPILU preconditioning) from about 450 to 900.

2.3. PRECONDITIONING STRATEGIES IN NONLINEAR COMPUTATIONS

In this section the question of updating the preconditioning matrix during the nonlinear computation is investigated. Three variants of updating periods are compared:

(a) For each new tangent stiffness matrix a new preconditioning matrix is constructed.
(b) In every new Newton step a new preconditioner is constructed.
(c) Always the first computed preconditioning matrix is used.

If simple preconditioning strategies like Jacobi and SSOR preconditioning are used, there is no reason not to built a new preconditioner for every new stiffness matrix. For SSOR with the Eisenstat-trick the latter is necessary, however, because the stiffness matrix has to be scaled explicitly.

For other preconditioners which require more computational effort the reuse of an already computed preconditioning matrix seems to be a good choice. If the stiffness matrix is only changing slightly this works very well. Some alternatives are possible, either using a preconditioning matrix for k linear systems of equations with k fixed or like (b) computing a new preconditioning matrix once in every load step (in the equilibrium state) and reuse this for all iterations. The latter is a more general strategy and has proven to be very efficient for most problems analyzed.

If the system is only mildly nonlinear it is even possible to use a constant preconditioning matrix for the computation of the whole load deflection path with e.g. the preconditioning matrix corresponding to the linear problem (c).

In the authors experience (a) is the best choice in combination with Jacobi and SSOR preconditioning, whereas for the variants of incomplete factorization (b) should be preferred. In some cases (a) does not lead to convergence of Newton's method if preconditioners constructed from stiffness matrices related to non-equilibrium states are taken. Method (c) is preferably used with very good and computationally "expensive" preconditioners as e.g. complete LDL^T factorization. In cases of stiffening structures (c) however does not lead to an efficient solution.

In figure 4 for the flange example the CPU times needed for the three variants (a) to (c) are shown in combination with MPILU preconditioning. Here variant (b) is only slightly more efficient then (a). Variant (c) works well until the snap-through. Then the deviation of the preconditioner from the actual stiffness matrix is too large, and the preconditioning is less efficient.

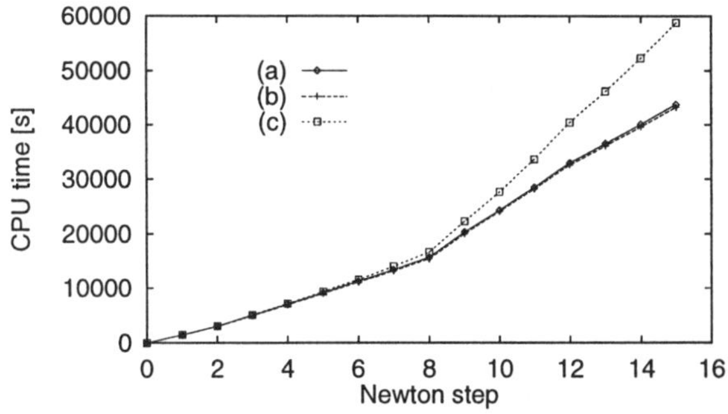

Figure 4. Computation time for different preconditioning matrix update intervals

2.4. PERMUTATION STRATEGIES FOR EFFICIENT ITERATIVE SOLUTIONS OF LINEAR EQUATIONS

When different variants of incomplete factorization are used as preconditioners, it is possible to influence the convergence behavior of the iterative method by permuting the stiffness matrix. In combination with direct solution methods this is a standard fill-in reducing technique. For band/profile solvers well known band and profile minimizing strategies are standard, whereas minimum degree and nested dissection strategies are applied in the context of sparse solvers. A permuted stiffness matrix has also an effect on the incomplete factorization; it is desired then but not guaranteed that the number of neglected terms is smaller or the neglected terms are of minor importance and the computed factor is of higher quality. Then the number of iterations per linear system can be reduced.

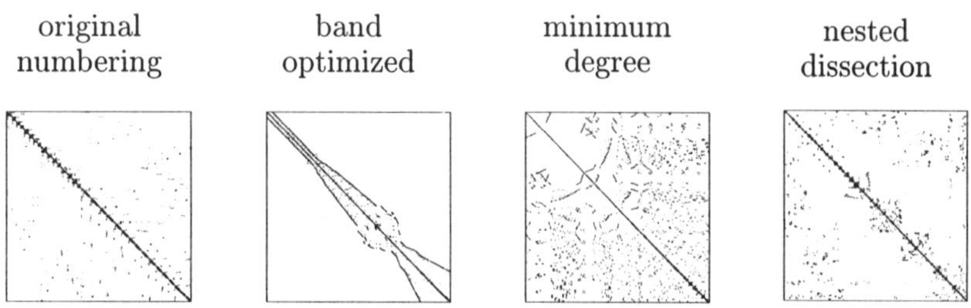

Figure 5. Matrix structures for various permutations

In table 1 for the flange example the iteration numbers and the solution times using various permutations are shown for the solution of the linear problem. The corresponding matrix structures are depicted in figure 5. As the Finite Element mesh is generated and thus not regularly numbered; therefore the band optimizer reduces the bandwidth significantly. Then

TABLE 1. Iteration numbers and solution times for the linear problem (CG-MPILU)

permutation	# of iterations	solution time [s]
original numbering	593	252.4
band optimized	494	212.9
minimum degree	705	306.5
nested dissection	660	280.6

less fill-in will be discarded and the number of iterations decreases considerably. Minimum degree and nested dissection strategies (see e.g. [5]) have no positive effects. Here also the number of fill-in terms is reduced but in this case they are of high importance and as a result the quality of the preconditioner decreases.

Experiences with other problems show also, that in general for carefully numbered meshes no permutation can be recommended, for automatically numbered or adaptive meshes however a band or profile minimization is advisable.

3. Parallel Implementation

In the parallel implementation the SPMD concept on distributed memory machines is taken. Therefore in a sequential pre-processing phase a domain decomposition is performed; the grid is divided into nonoverlapping domains by domain decomposition methods as given in [6] (for the decomposition of the flange problem see figure 6).

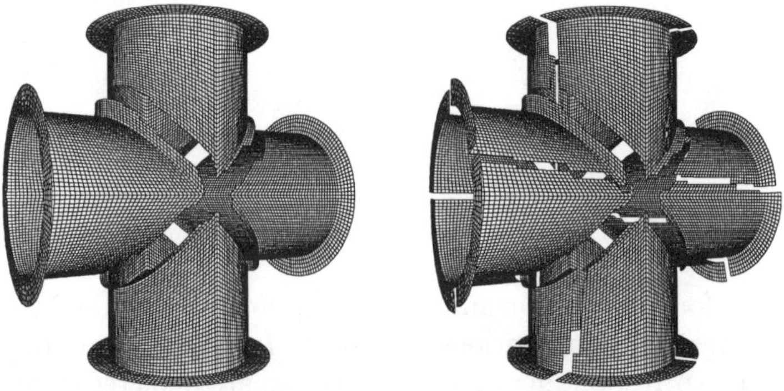

Figure 6. Domain decomposition for 4 and 16 processors

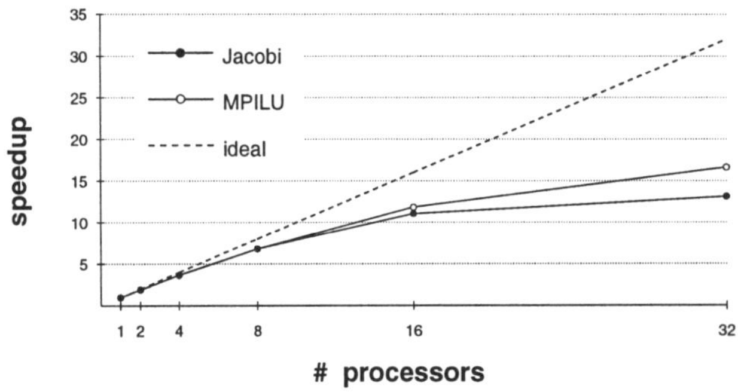

Figure 7. Speedup for the flange problem on a IBM SP2 parallel computer

The different processors then work independently on their partial grid. The preconditioning matrices are computed locally on the partial matrices resulting from those local grids. Here, however, the local entries of the partial stiffness matrices corresponding to the inner boundaries are globalized to ensure positive definiteness of the matrices on the various processors (in case of a positive definite stiffness matrix). A more detailed description is given in [12].

In figure 7 the achieved speedup for the flange problem is shown. For up to 16 processors the speedup is satisfactory. Using more processors does not lead to a further significant speedup. This is certainly a function of the problem size: by using 32 processors, each processor works on a local system of only 3 500 equations. The communication tasks dominate then the computation task with the consequence of decreasing speedup. It is expected to achieve a better speedup also with a larger number of processors when the problems are bigger.

4. Concluding Remarks

The above investigations – not only for the presented example but also for a wide class of bad-conditioned problems in structural mechanics – have shown, that iterative solvers perform in a robust manner. In combination with arc-length procedures, however, iterative solvers are in general not as efficient as direct sparse solvers with respect to CPU-time but have the advantage of fairly low memory requirements, as is well known. In addition iterative solvers are easy to parallelize. The fastest solution of nonlinear problems is achieved by the combination of iterative and direct schemes. Not updating the preconditioning matrix in every Newton iteration but only once in every load step leads to a more efficient iterative solution.Also performing some profile/band minimization speeds up the iterative solution of equations with factorization included for the preconditioner if the FE mesh is not regularly numbered.

A further speedup of the iterative solution method can be achieved by applying a modified Lanczos algorithm on the unsymmetric system of equations (1) as proposed in [12].

The parallel implementation of the iterative solvers leads to very good speedups not only for the presented example.

References

1. Alefeld, G., Lenhardt, I., Rottner, Th., and Schweizerhof, K.: Strukturmechanische Berechnungen auf Hochleistungsrechnern – iterative und direkte Verfahren im Vergleich. In R. Krahl G. Wolf, editor, *Statustagung des BMBF, HPSC 97, Stand und Perspektiven des Parallelen Höchstleistungsrechnens und seiner Anwendungen 25.-26. Februar 1997* (1997), 107–122.
2. Crisfield, M.A.: *Non-linear Finite Element Analysis of Solids and Structures*, John Wiley & Sons, 1991.
3. Freund, R.W., and Nachtigal, N.M.: Software for simplified Lanczos and QMR algorithms, *Appl. Numer. Math.* **19** (1995), 319–341.
4. Gebhardt, H.: *Finite Element Konzepte für schubelastische Schalen*, PhD thesis, Institut für Baustatik, Universtität Karlsruhe, 1990.
5. George, A., and Liu, J.W.: *Computer Solution of Large Sparse Positive Definite Systems*, Prentice Hall, 1981.
6. Karypis, G., and Kumar, V.: A fast and high quality multilevel scheme for partitioning irregular graphs, Technical report, TR 95-35, Department of Computer Science, University of Minnesota, July 1995.
7. Lenhardt, I., and Rottner, Th.: Lösung nichtlinearer Gleichungssysteme aus der Strukturmechanik unter Einsatz einer parallelen Version des Lanczos-Algorithmus. In G. Wolf R. Krahl, editor, *Statustagung des BMBF, HPSC 95, Stand und Perspektiven des Parallelen Höchstleistungsrechnens und seiner Anwendungen, 11.-14. September 1995* (1996), 127–138.
8. Papadrakakis, M.: A truncated Newton-Lanczos method for overcoming limit and bifurcation points, *Int. J. Numer. Methods Eng.* **29** (1990), 1065–1077.
9. Papadrakakis, M., and Gantes, C.J.: Preconditioned conjugate- and secant-Newton methods for nonlinear problems, *Int. J. Numer. Methods Eng.* **28** (1989), 1299–1316.
10. Papadrakakis, M., and Smerou, S.: A new implementation of the Lanczos method in linear problems, *Int. J. Numer. Methods Eng.* **29** (1990), 141–159.
11. Ramm, E.: Strategies for tracing the nonlinear response near limit points. In E. Stein K.J. Bathe and W. Wunderlich, editors, *Europe-US-Workshop on nonlinear FE Analysis in Structural Mechanics*, Springer (1980), 63–89.
12. Rottner, Th., Lenhardt, I., Alefeld, G., and Schweizerhof, K.: Nonlinear structural finite element analysis using the preconditioned Lanczos method on serial and parallel computers, *BIT* **37** (1997), 759–769.
13. Schweizerhof, K.: Quasi-Newton Verfahren und Kurvenverfolgungsalgorithmen für die Lösung nichtlinearer Gleichungssysteme in der Strukturmechanik. Universität Karlsruhe, 1989.
14. SMPAK: Scientific Computing Associates, 265 Church Street, New Haven, CT 06510-7010, USA.
15. Wagner, W.: Zur Behandlung von Stabilitätsproblemen der Elastostatik mit der Methode der Finiten Elemente, Forschungs- und Seminarberichte aus dem Bereich der Mechanik, Nr. F91/1, Universität Hannover, 1991.

COMPUTATION OF GLOBAL EQUILIBRIUM PATHS

An Iteration-Free Approach

ZS. GÁSPÁR AND G. DOMOKOS
Technical University of Budapest
H-1521 Budapest, Hungary

1. Introduction

Some interesting aspects of numerical computation of bar structures subject to quasi-static load are dealt with. Beyond civil engineering applications, elastic beams are used in a variety of fields as models (e.g. for DNA molecules), so the ideas presented here are addressed to all scientists applying beams as models.

The mathematical formulation of the problem is usually a Boundary Value Problem (BVP) associated with a system of Ordinary Differential Equations (ODEs) . In almost all practical cases these ODEs are non-integrable so one applies discretization methods. This level of discretization is *not* addressed in this paper. Even after discretizing the beam, the BVP remains a nonlinear problem which is customarily approached by an incremental-iterative algorithm (cf. Riks (1979), Hermann (1987)). These methods rely on local approximation of the BVPs solution curve (also called equilibrium path) by truncated Taylor series. Most frequently all nonlinear terms of the Taylor series are neglected and the solution curve is extrapolated along the tangent. The error at the extrapolated point is subsequently decreased by an iteration procedure. The disadvantage of incremental-iterative methods is the fact that they rely on extrapolation, thus errors can only be reduced by iterations whichmayor maynot converge. Moreover, these algorithms fall into ths category of continuation methods, i.e. they are only capable of following a path, a point of which was previously identified by some other computation method or consideration.

The approach which we outline in this paper is rather different. Instead of extrapolation it relies on interpolation, eliminating the cumbersome and computation-intensive iterative phase. Moreover, our approach (which is based on the so-called PL algorithm presented by Allgower and

H. A. Mang and F. G. Rammerstorfer (eds.),
IUTAM Symposium on Discretization Methods in Structural Mechanics, 71–78.

Georg (1990) and which we call the Simplex Method, not to be confused with the simplex algorithm in linear programming) offers two alternatives: it can be either applied as an iteration-free continuation technique, or, it can be used as a global, "scanning" algorithm, identifying all equilibria in a given parameter-domain, regardless whether these branches are connected to each other or not. We will describe the mathematical basis of the Simplex Method in section 2.

Naturally one has to pay toll for these advantages. The Simplex Method has its sensitive and weak sides: the continuation code can be halted by internal singularities and has to be re-started. In the case of complicated frame structures the method becomes very sensitive and high digital accuracy is required. The "scanning" version is more robust in the sense that the computer program is virtually never halted, however, besides producing an approximation for all branches in the given domain, it often delivers spurious solutions which are not easy to filter. Although the scanning version is very computation-intensive, it was parallelized very efficiently, so powerful parallel machines or networks of workstations can be used for simultaneous computation. The advantages and disadvantages of the method will be summarized in section 3.

The Simplex Method can be applied to a relatively large variety of problems. In section 4 we will present a few, including the application of the scanning version to the classical Euler buckling problem, path continuation in the case of a simple frame, investigation of elasto-plastic frames and nonlinear contact problems. In section 5 we draw conclusions and indicate further development directions and possible applications.

2. Mathematical Background

The mathematical backgound of the Simplex Method can be readily explained on a simple example: the uniform, incompressible cantilever beam with bending stiffness EI is loaded by a compressive axial force P (see figure 1).

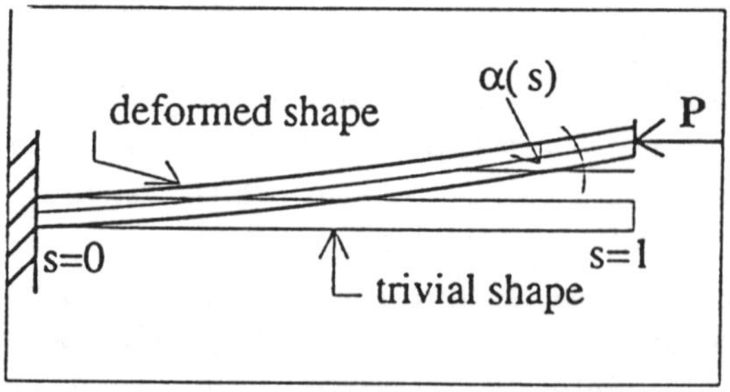

Figure 1: The cantilever beam

The slope α of the beam is described as a function of the arclength $s \in [0, 1]$; the equilibrium shape is described by Euler's classical differential equation

$$EI\alpha'' + P \sin \alpha = 0, \qquad (1)$$

supplemented by the boundary conditions

$$\begin{array}{lrcl} a/ & \alpha(0) & = & 0 \\ b/ & \alpha'(1) & = & 0 \end{array} \qquad (2)$$

expressing zero slope and zero curvature at the left and right end, respectively. For the time being, we ignore the far-end condition (2/b) and concentrate on (2/a). This condition eliminates one of the "variable" initial conditions as a constant, so all trajectories which *might* meet the boundary conditions can be uniquely represented in the $[\alpha'(0), P]$ plane. Thus we managed to project the infinite-dimensional space of all geometrically possible configurations to a 2 dimensional space in such a way that the latter space is in a one-to-one correspondence with the set of IVPs. The relevant BVP solutions can be regarded as a subset of these IVPs. The scalars $\alpha'(0)$ and P are the *global coordinates*, or, *variables* for this BVP, the plane (space) spanned by them will be called the *global representation space* (GRS) of the BVP. Since the BVP contains one parameter (P), the solutions will typically appear as one-dimensional manifolds, i.e. lines. Algebraically these lines can be expressed as the solutions of the nonlinear equation (corresponding to the far-end condition (2/b)), since the value $\alpha'(1)$ can be expressed as a *function* of the *variables* defined previously:

$$\alpha'(1) = f_1(\alpha'(0), P) = 0. \qquad (3)$$

The far-end value of $\alpha'(1)$ is uniquely determined by our chosen variables $\alpha'(0)$ and P since the investigated ODE satisfies the conditions of Peano's Uniqueness Theorem [Bieberbach, 1923]. Our approach guarantees that whenever two solution lines intersect, the equilibria corresponding to those lines also coincide. We call such a diagram *topologically correct.*

On this simple example we introduced the basic concepts of our method. In the case of more complicated problems the GRS has more dimensions, however, it still remains a finite-dimensional space. If the investigated GRS is n-dimensional, then the analogous equation system to (3) contains $n-1$ equations, determining - as in the case of the cantilever - 1-dimensional solution sets, i.e. lines. The assembly of those lines is called the global equilibrium path, or, the global bifurcation diagram. The global bifurcation diagram of the cantilever beam is illustrated (for a finite domain of the GRS) in Figure 2:

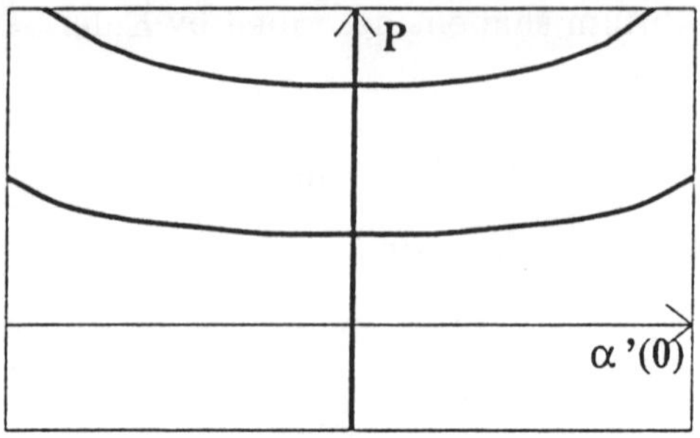

Figure 2: The bifurcation diagram of the cantilever beam

The computation of the diagram relies on the discretization of the GRS. The most natural way to discretize is to choose a simplectic grid. An n-dimensional simplex is defined by $n+1$ points. (For example, a two-dimensional simplex is a triangle.) There are many ways to construct a simplectic grid. We choose the following method: in the first step we construct an orthogonal (cubic) grid and then, in the second step we subdivide each cube into $n!$ simplices. Our algorithm (which is an adaptation of the so-called "PL" algorithm described by Allgower and Georg (1990)) computes all $(n-1)$ *function* values at the $n+1$ vertices of the simplex and interpolates all functions linearly inside the simplex by an $(n-1)$-dimensional hyperplane. The intersection of $n-1$ hyperplanes of dimension $n-1$ is typically a straight line. This line has typically either zero or two intersection points with the investigated simplex. In the former case there is no solution, in the latter case we obtained a piece of the polygonal approximation of the equilibrium path. The individual steps of this computation, involving mainly linear algebra techniques, are described by Gáspár et al.(1997).

3. Advantages and disadvantages

The Simplex Method has two versions: (A) the path continuation version and (B) the scanning version. Version A can be compared with other path continuation codes. The main advantage of the Simplex Method is that it does not involve iterative steps and the choice of variables is relatively free: one can incorporate unknown support reactions, unknown displacements or unknown locations of extremal displacements among the variables. It is possible to change the number of variables during the computation; a step required by switching to a different mathematical model. We will illustrate this feature in the next section on the analysis of elasto-plastic frames. The computation effort grows approximately linearly with the dimension of the

GRS (number of variables), which is in fair accordance with other methods. Computation times are hard to compare presently since the Simplex Method's stepsize has not been optimized yet. The application of version A raises two difficulties: - the code in the present form can not handle bifurcation points (it selects one of the external branches effectively at random), - if the linearized solution lies very close to the vertex of a simplex, numerical instabilities can arise, the program might be halted. We hope to eliminate this difficulties by a simple subroutine.

Version B has (as far as we know) no competition among other methods: the computation of *all* equilibria in a given domain without any prior information is the most general question that can be asked. Version B produces all solutions, regardless whether they are connected to the trivial one or not. Bifurcation points are linearized, however, all branches are computed. This version has two main drawbacks:

- it is very computation-intensive, the computation effort grows exponentially with the number of variables. This problem can be reduced by the application of parallel computers. The parallel code has been developed under the PVM system (Parallel Virtual Machine), has been successfully tested and can be implemented on any UNIX environment (cf. Gáspár et al. (1997)).

- it produces (besides approximations of the true solutions) spurious solutions. There is no general algorithm at hand for the removal of spurious solutions, however, in all problems computed thus far the number of spurious solutions could be radically reduced to a non-disturbing level. For more details see Gáspár et al. (1997).

4. Applications

We will review here a few applications with reference to the original source.

The first example is the application of version B to the classical Euler buckling problem. The ODE (1) is supplemented now either by the boundary conditions

$$\begin{aligned} a/ \quad \alpha'(0) &= 0 \\ b/ \quad \alpha'(1) &= 0 \end{aligned} \tag{4}$$

expressing zero curvature at both ends (hinge and roller support) or by

$$\begin{aligned} a/ \quad \alpha(0) &= 0 \\ b/ \quad \alpha(1) &= 0 \end{aligned} \tag{5}$$

expressing zero slope at both ends (clamped support). In the first case the GRS is spanned by the horizontal (H) and vertical force (V) and slope (α) at $s = 0$, in the second case H, V and the moment M at $s = 0$. The surprisingly complex bifurcation diagrams are shown in figure 3, for more details see Domokos (1994).

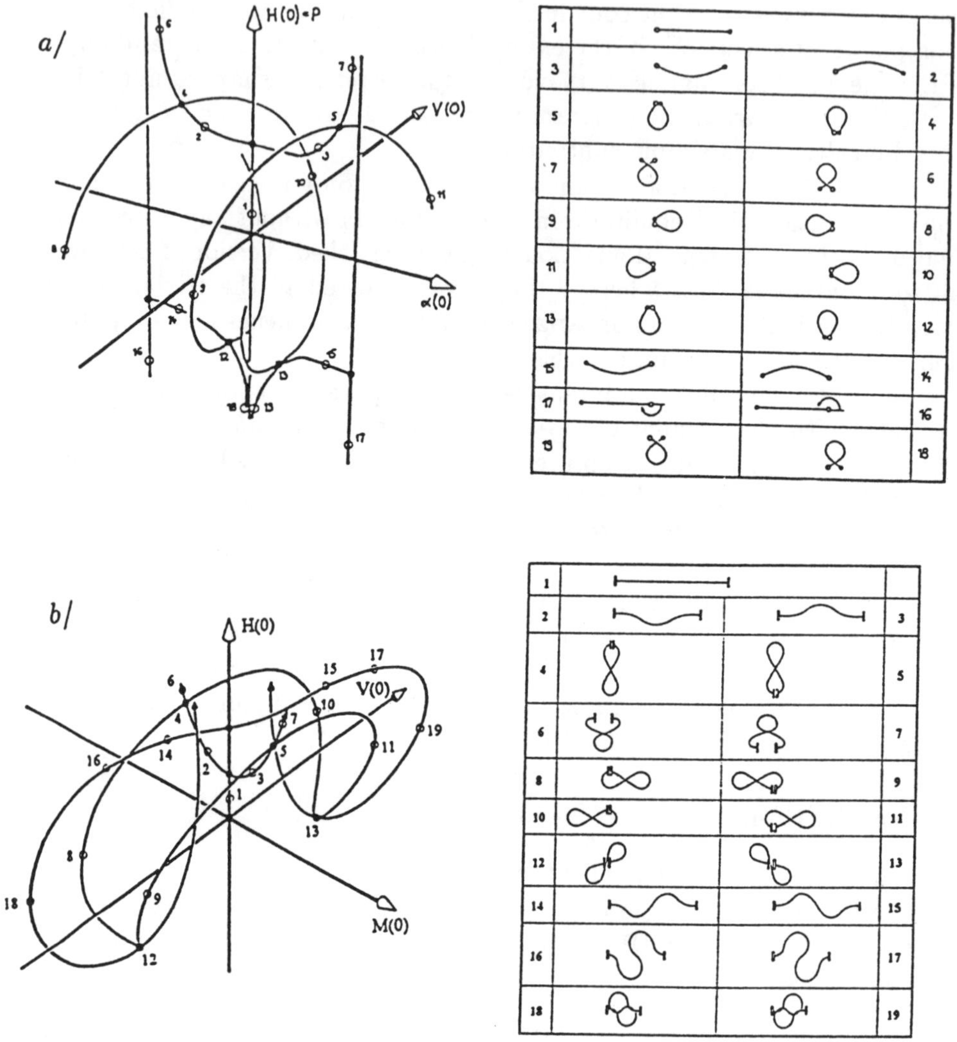

Figure 3: The global bifurcation diagrams for the Euler buckling problem with (a) hinge and roller supports and (b) with clamped supports.

The second example is closer to classical civil engineering applications; we applied version A to an elastic two-storey frame. The GRS is 7 dimensional and figure 4 shows the deformation of the structure along the equilibrium path. For more details see Domokos and Gáspár (1995).

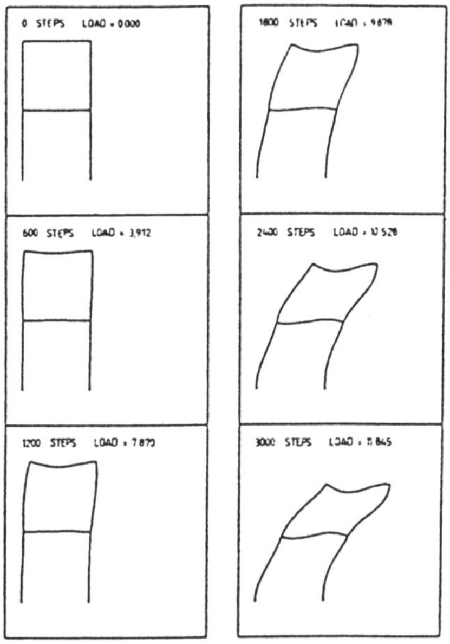

Figure 4: Deformation of the frame along the equilibrium path.

A similar structure is analysed by Gáspár and Nédli (1997), though with elasto-plastic theory, resulting a change in the number of variables whenever a plastic hinge appears or disappears.

The last example is constrained Euler buckling: the classical Euler beam is constrained by two rigid walls, at equal distance h to both sides from the bar axis. When the maximum of the lateral displacement in the unconstrained problem reaches h, contact is established and the GRS has to be modified. At larger axial load P the point contact evolves into line contact, resulting again a modification of the GRS. This apparently simple mechanical problem was investigated by Domokos et al. (1997). All numerical computation was carried out with version A of the Simplex Method. Figure 5 shows one of the startingly complex bifurcation diagrams obtained by this computation. The numerical results have been confirmed both by exact analysis and physical experiments.

5. Summary

In this short paper we introduced the Simplex Method, an algorithm suitable for the global numerical computation of bar structures. After describing the mathematical background we discussed the advantages and disadvantages for both (path continuation and scanning) version of the algorithm and reviewed a collection of examples.

Besides the illustrated ones, the Simplex Method is capable of handling a large variety of structural problems, including active static control and the kinematic description of single degree of freedom mechanisms; these directions still have to be explored. The scanning version is presently applied to a 5-dimensional problem; with increasing computer capacity we hope that our method will be applied to a widening range of problems.

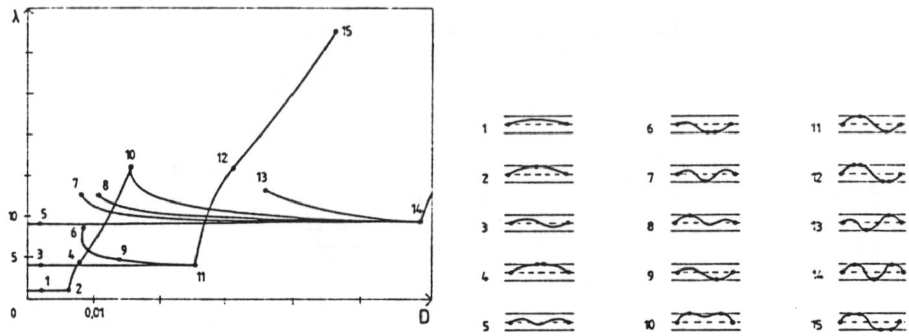

Figure 5: The bifurcation diagrams for the constrained Euler problem.

Acknowledgements

The support of OTKA grants F021307, T015851 and MKM grant 815 is gratefully acknowledged.

References

Allgower, E.L. and Georg, K. (1990):*Numerical continuation methods : an introduction* Springer, Berlin

Bieberbach, L. (1923):*Differentialgleichungen* Springer, Berlin

Domokos, G., (1994):Global Description of Elastic Bars *ZAMM* **74** (4) T289-T291

Domokos, G. and Gáspár Zs. (1995): A Global, direct algorithm for path-following and active static control of elastic bar structures *J. of Structures and Machines* **23** (4) pp 549-571

Domokos, G., Holmes, P. and Royce, B. (1997): Constrained Euler Buckling *J. of Nonlinear Sci.***7** pp 281-314

Gáspár, Zs., Domokos, G. and Szeberényi, I. (1997): A parallel algorithm for the global computation of elastic bar structures *CAMES* **4** pp 55-68.

Gáspár, Zs. and Nédli, P. (1997): Global numerical analysis of elasto-plastic frames *CAMES* [in press]

Hermann, M. (1987): Shooting methods for nonlinear TPBVPs, *11th Int. Congress on Applications of Mathematics in Engineering* , Berichte 4, Weimar, Germany

Riks, E. (1979): An incremental approach to the solution of snapping and buckling problems *Int. J. Solids Struct.* **15** pp 529-551

NEW THIN PLATE AND SHELL TRIANGLES WITH TRANSLATIONAL DEGREES OF FREEDOM ONLY

E. OÑATE and F. ZARATE
E.T.S. Ingenieros de Caminos, Canales y Puertos
Universitat Politècnica de Catalunya
08034 Barcelona, Spain

1. Introduction

The need for efficient plate and shell elements is critical for solving large scale industrial problems such as the analysis of civil engineering shell structures, vehicle crash-worthiness situations and sheet stamping processes. The derivation of simple triangles capable of accurately representing the shell deformation under complex loading conditions is still nowadays the objective of intensive research.

Several authors have tried to derive finite elements with displacements as the only nodal variables. So far the methods proposed limit their applicability to triangular element shapes only. Barnes (1977) proposed a method for deriving a three node triangle with three nodal degrees of freedom (d.o.f.) based on the computation of the curvatures in terms of the normal rotations at the mid-side points determined from the nodal deflections of adjacent elements. This method has also been exploited by Hampshire *et al.* (1992) assuming that the elements are hinged together at their common boundaries, the bending stiffness being represented by torsional springs resisting the rotations about the hinge lines. Phaal and Calladine (1992, 1992) have proposed a similar class of triangles for plate and shell analysis. Yang *et al.* (1993) derived a family of elements of this type based on the so-called bending energy augmented membrane approach which basically reproduces the hinge bending stiffness procedure of Hampshire *et al.* (1992). Brunet and Sabourin (1994) proposed a different approach to compute the constant curvature field within each element in terms of the six node displacements of a macro-element. The element was successfully applied to nonlinear shell analysis using an explicit dynamic approach. More recently Rio *et al.* (1994) have used the concept of side hinge bending stiffness to derive a thin shell triangle of "translational" kind for explicit dynamic analysis of sheet stamping problems.

Oñate and Cervera (1993) have proposed a general finite volume type procedure for deriving thin plate elements with translational degrees of freedom only and presented a competitive three d.o.f. plate triangle. This element has been

H. A. Mang and F. G. Rammerstorfer (eds.),
IUTAM Symposium on Discretization Methods in Structural Mechanics, 79–89.

successfully extended to shell analysis by Zárate (1996). The nine d.o.f. thin shell element has been further developed to deal with sheet forming problems using a non linear explicit dynamic program [Cendoya (1996) and Oñate *et al.* (1996)].

In this paper the derivation of the rotational d.o.f. free thin plate/shell triangles is described in the framework of a mixed Hu-Washizu formulations where concepts originating from finite element (FE) and finite volume (FV) methods are also used. Two conceptually different triangles emerge from this approach the "basic plate/shell triangle" originating from the so called cell vertex FV schemes, and the "plate/shell patch" where cell centered FV patches surrounding a node are used [Oñate and Zienkiewicz (1994)]. Details of the derivation of both element types are given in the paper together with some examples of application of the new elements to a range of plate and shells problems in structures and sheet stamping problems.

2. Basic theory

Let us consider the plate of Figure 1. We will assume Kirchhoff's thin plate conditions to hold, i.e.,

$$\theta_x = \frac{\partial w}{\partial x} \qquad \text{and} \qquad \theta_y = \frac{\partial w}{\partial y} \tag{1}$$

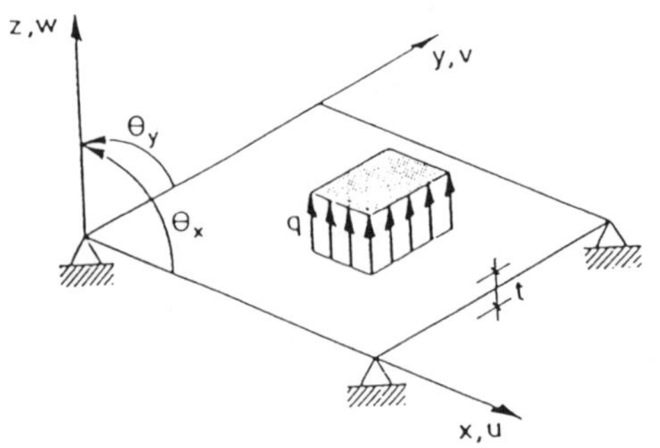

Figure 1. Sign convention for the deflection and the rotations in a plate.

The curvature field and the moment-curvature relationship can be expressed in the usual manner as

$$\boldsymbol{\kappa} = \mathbf{L}w \quad , \quad \mathbf{m} = \mathbf{D}\boldsymbol{\kappa} \tag{2}$$

where

$$\boldsymbol{\kappa} = [\kappa_x, \kappa_y, \kappa_{xy}]^T \quad , \qquad \mathbf{m} = [m_x, m_y, m_{xy}]^T$$
$$\mathbf{L} = \left[-\frac{\partial^2}{\partial x^2}, -\frac{\partial^2}{\partial y^2}, -2\frac{\partial^2}{\partial x \partial y}\right]^T \quad , \quad \mathbf{D} = \frac{Et^3}{12(1-\nu^2)}\begin{bmatrix} 1 & \nu & 0 \\ \nu & 1 & 0 \\ 0 & 0 & \frac{1-\nu}{2} \end{bmatrix} \tag{3}$$

where E and ν are the Young's modulus and Poisson's ratio, respectively.

The set of governing equations will be obtained in integral form from the standard Hu-Washizu functional

$$\Pi = \frac{1}{2}\int_A \boldsymbol{\kappa}^T \mathbf{D}\boldsymbol{\kappa}\, dA + \int_A [\mathbf{L}w - \boldsymbol{\kappa}]^T \mathbf{m}\, dA - \int_A qw\, dA \tag{4}$$

where q is the distributed loading and A is the area of the plate. Variation of Π leads to the following three equations

$$\int_A \delta\boldsymbol{\kappa}^T [\mathbf{D}\boldsymbol{\kappa} - \mathbf{m}]\, dA = 0 \tag{5.a}$$

$$\int_A \delta\mathbf{m}^T [\mathbf{L}w - \boldsymbol{\kappa}]\, dA = 0 \tag{5.b}$$

$$\int_A [\mathbf{L}\delta w]^T \mathbf{m}^T\, dA - \int_A \delta wq\, dA = 0 \tag{5.c}$$

Eqs.(5.a), (5.b) and (5.c) represent the global satisfaction of the constitutive, kinematic and equilibrium equations, respectively. Eqs. (5) are the basis for the FE/FV discretization to be defined next.

3. Finite element/finite volume discretization

Let us consider a discretization of the plate into standard three node triangles. The curvature and the bending moments are fist described by *constant* fields within specific control volumes (CV) as

$$\begin{aligned} \mathbf{m} &= \delta\mathbf{m} = \mathbf{I}_3 \bar{\mathbf{m}}^{(p)} \\ \boldsymbol{\kappa} &= \delta\boldsymbol{\kappa} = \mathbf{I}_3 \bar{\boldsymbol{\kappa}}^{(p)} \end{aligned} \tag{6}$$

where $\bar{(\cdot)}^{(p)}$ denotes values for the pth control volume (CV) and $\mathbf{I}_3$ is the 3×3 unit matrix.

Two modalities of CV are considered: a) a CV formed by a single triangular element (Figure 2a), and b) the CV formed by 1/3 of the areas of the elements surrounding a node (Figure 2b). The two CV's are termed in the FV literature "cell centered" and "cell vertex" schemes, respectively [(Oñate, Cervera and Zienkiewicz (1994)].

The area integrals in eqs.(5) can be expressed as sum of contributions over the different CV's. The terms involving the curvature operator are then simplified taking advantage of the constant moment field within each CV with area A_p as

$$\begin{aligned} \int_{A_p} \delta\mathbf{m}^T \mathbf{L}w\, dA &= \delta\mathbf{m}^T \int_{A_p} \mathbf{L}w\, dA = \delta\mathbf{m}^T \oint_{\Gamma_p} \mathbf{T}\nabla w\, d\Gamma \\ \int_{A_p} [\mathbf{L}\delta w]^T \mathbf{m}\, dA &= \left(\int_{A_p} [\mathbf{L}\delta w]^T\, dA\right)\mathbf{m} = \left(\oint_{\Gamma_p} \mathbf{T}\nabla\delta w\, d\Gamma\right)\mathbf{m} \end{aligned} \tag{7}$$

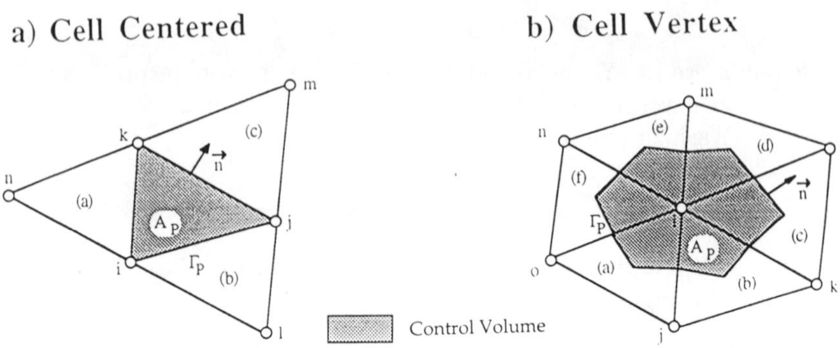

Figure 2. Cell centered and cell vertex control volumes.

where

$$\mathbf{T} = \begin{bmatrix} n_x & 0 & n_y \\ 0 & n_y & n_x \end{bmatrix}^T \quad , \quad \boldsymbol{\nabla} = \begin{Bmatrix} \frac{\partial}{\partial x} \\ \frac{\partial}{\partial y} \end{Bmatrix} \tag{8}$$

and $\mathbf{n} = [n_x, n_y]^T$ is the unit normal vector to the boundary Γ_p surrounding a CV.

The deflection field is finally interpolated within each triangle using standard linear shape functions N_i (Zienkiewicz and Taylor, 1989) as

$$w = \sum_{i=1}^{3} N_i \bar{w}_i = \mathbf{N}^{(e)} \bar{\mathbf{w}}^{(e)} \tag{9}$$

where $\bar{w}_i$ denote nodal deflection values, $\mathbf{N}^{(e)} = [N_1, N_2, N_3]$ and $\bar{\mathbf{w}}^{(e)} = [w_1, w_2, w_3]^T$.

Substituting eqs.(6) and (9) into eqs.(5) and making use of eqs.(7) gives the following linear system of equations, after elliminating the curvature and moment variables from eqs.(5a) and (5b),

$$\mathbf{K}\bar{\mathbf{w}} = \mathbf{f} \tag{10}$$

where the vector $\bar{\mathbf{w}}$ contains the nodal deflections. The stiffness matrix $\mathbf{K}$ is obtained by assembling the contributions from each CV given by

$$\mathbf{K}_p = [\mathbf{B}_p]^T \mathbf{D} \mathbf{B}_p A_p \tag{11}$$

where $\mathbf{B}_p$ is the CV curvature matrix defined below.

The components of the nodal force vector $\mathbf{f}$ are obtained as in standard linear finite element triangles. For a constant distributed force q this gives

$$f_i = \sum_e \frac{qA^{(e)}}{3} \tag{12}$$

where the sum is extended to all triangles sharing the ith node and $A^{(e)}$ is the area of element e.

4. Computation of the CV curvature matrix

The computation of the curvature matrix $\mathbf{B}_p$ is different for the cell vertex and cell centered control volumes.

4.1 CELL CENTERED SCHEME. BPT ELEMENT

Note that eqs.(7) require necessarily the computation of the deflection gradient along the CV sides. This poses a difficulty in cell centered configurations as the term $\boldsymbol{\nabla} w$ is discontinuous at the element edges. A simple method to overcome this problem proposed in Oñate and Cervera (1993) is to compute the deflection gradient at the element sides as the average value of the gradients contributed by the two elements sharing the side. The expression of $\mathbf{B}_p$ in this case can be written as

$$\mathbf{B}_p = \frac{1}{2A_p} \oint_{\Gamma_p} \mathbf{T}[\boldsymbol{\nabla}\mathbf{N}_p + \boldsymbol{\nabla}\mathbf{N}^{(i)}]\, d\Gamma \tag{13}$$

where $\mathbf{N}^{(i)}$, $i = a, b, c$ denote the shape function matrices of the linear triangles surrounding the central element p (see Figure 2a).

Note that $\mathbf{B}_p$, in this case, is a 3×6 matrix involving the deflections at each of the six nodes contributing to a CV and therefore $\mathbf{K}_p$ is a 6×6 matrix.

The application of the boundary conditions is simple. In clamped and symmetry edges the contribution of the zero edge rotations is neglected in the computation of (13). In free and simply supported edges only the contribution from the central triangle p is accounted for. The thin plate element resulting from these approximations is termed BPT (for Basic Plate Triangle). This element was originally developed in Oñate and Cervera (1993) where full details of the formulation can be found.

4.2 CELL VERTEX SCHEME. BPN ELEMENT

The deflection gradient is now continuous along the sides of the control volume and the CV curvature matrix is now simply

$$\mathbf{B}_p = \frac{1}{A_p} \oint_{\Gamma_p} \mathbf{T}\boldsymbol{\nabla}\mathbf{N}\, d\Gamma \tag{14}$$

Matrix $\mathbf{N}$ contains contributions from all the elements forming the patch p contributing to the CV. For instance, for the patch of Figure 2b $\mathbf{N} = [\mathbf{N}^{(a)}, \mathbf{N}^{(b)}, \mathbf{N}^{(c)}, \mathbf{N}^{(d)}, \mathbf{N}^{(e)}, \mathbf{N}^{(f)}]$. An explicit form of $\mathbf{B}_p$ can be easily found taking into account that $\boldsymbol{\nabla}\mathbf{N}^{(i)}$ is constant within each element. For details see Zárate (1996).

The boundary conditions are even simpler than for the BST case. If a rotation is zero along an edge this can be simply imposed by neglecting the contribution of the rotation in the line integral (14).

The mixed character of the formulations allows to impose "a priori" vanishing edge bending moments at free and simply supported sides. This can be simply done in practice by making zero the adequate terms of the constitutive matrix at the boundary nodes where the bending moments are constrained [Zárate (1996)].

The resulting plate element is termed BPN (for Basic Plate Nodal patch).

5. Extension to shells. BST and BSN elements

The extension to the shell case of the BPT and BPN elements previously described is straightforward. The key details are given below.

5.1 BASIC SHELL TRIANGLE (BST) ELEMENT

The BPT element of the previous section is combined with the standard Constant Strain Triangle (CST) to model membrane behaviour. The resulting element is termed BST (for Basic Shell Triangle) (Figure 3). Note that the element curvatures are expressed in local axed within each element. The (constant) curvatures within each CV are expressed in terms of the (local) deflection gradients at the CV edges by means of expressions equivalent to eq.(7). These gradients are in turn expressed in terms of the values at the adjacent elements sharing each side. A transformation of the local bending stiffness matrix to global axes is then performed for assembly with the in-plane contributions in the standard manner [Zárate (1996)].

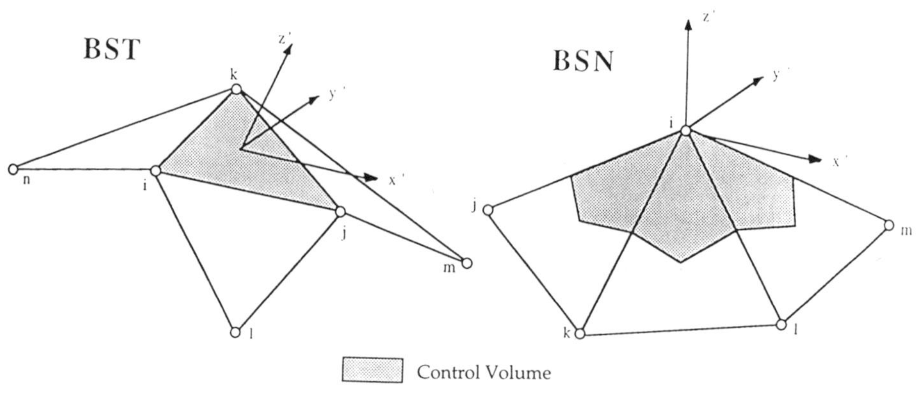

Figure 3. BST and BSN element control volumes for shell analysis.

5.2 BASIC SHELL NODAL PATCH (BSN) ELEMENT

The BPN element of previous section and the CST element can be combined to give the so called BSN element (Figure 3). The constant curvature and bending moment fields are defined now in local nodal axes. The application of eq.(14) requires now a transformation of the rotations from nodal to element axes for each of the elements forming the patch. Details can be found in [Zárate (1996)].

6. Examples

6.1 CLAMPED SQUARE PLATE UNDER UNIFORM LOADING

The square plate of Figure 4 has been analyzed with different uniform meshes of BPT and BPN elements. Table I shows the convergence of the central deflection with the number of degrees of freedom for each mesh. Results obtained with the DKT (Batoz, 1992) and the Morley triangles are also shown for comparison. Excellent performance of both BPT and BPN elements is found with a slight better behaviour of the latter for coarse meshes.

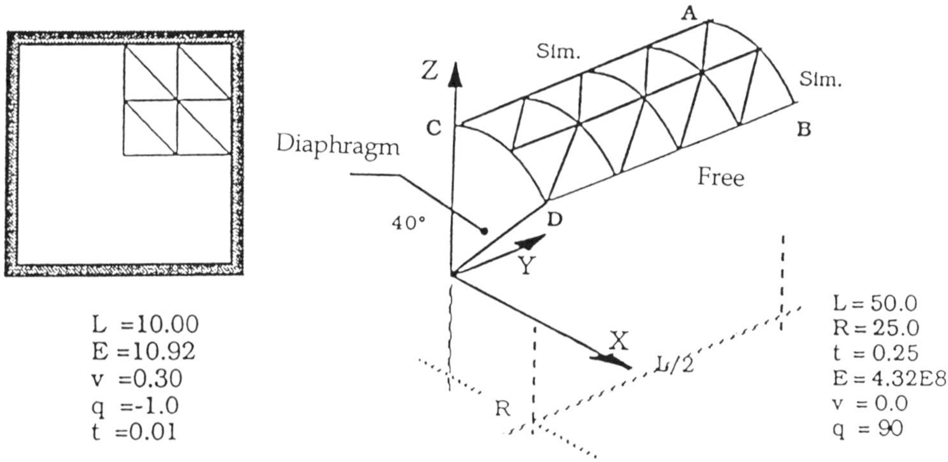

Figure 4. Geometry of clampled square plate and cylindrical shell.

TABLE I.

CLAMPED SQUARE PLATE. UNIFORM LOAD							
Mesh	dof	BPN	BPT	dof	DKT	dof	MORLEY
1 × 1	1	1.2980	2.6042	1	1.0365	2	4.5094
2 × 2	4	1.5819	1.9014	8	1.2144	12	2.6245
4 × 4	16	1.3987	1.4582	40	1.2578	56	1.6634
8 × 8	65	1.3051	1.3144	176	1.2636	240	1.3702
16 × 16	257	1.2756	1.2774	736	1.2649	992	1.2920
32 × 32	1024	1.2678	1.2683	3008	1.2652	4032	1.2720
Exact [Timoshenko]		1.2653	1.2653		1.2653		1.2653

6.2 CYLINDRICAL SHELL UNDER UNIFORM LOADING

The standard Scordelis-Lo cylindrical shell (Scordelis and Lo, 1969) is analyzed with the BST, BSN and DKT elements. Results for the deflection at the edge point B are shown in Table II for different meshes. Excellent results are obtained again with both the BST and BSN elements for relatively coarse meshes.

TABLE II.

CYLINDRICAL SHELL UNDER UNIFORM LOADING						
Mesh	No.Elem	dof	BSN	BST	dof	DKT
4 × 4	32	56	0,2418	0,2080	92	0,2220
8 × 8	128	208	0.2759	0.2537	344	0.2641
16 × 16	512	800	0.2971	0.2832	1326	0.2895
32 × 32	2048	3136	0.3026	0.2942	5216	0.2976
Exact [Timoshenko]			0.3024	0.3024		0.3024

The performance of the BSN element is found to be slightly superior to that of the BST. Both BSN and BST elements show a better accuracy/dof ratio than the DKT in this case.

6.3 SPHERICAL DOME UNDER IMPULSE PRESSURE LOADING

This example illustrates the efficiency of the new elements for non linear dynamic analysis of shells using an explicit time integration scheme. The geometry of the dome and the material properties are shown in Figure 5. The loading is due to a uniform pressure of 600 psi acting impulsively. Numerical results for the evolution of the central deflection obtained with three different meshes of BST elements with 507, 867 and 1324 d.o.f. are shown in Figure 4. Note the accuracy of the response for the coarse mesh. A comparison of the results for the 1324 d.o.f. mesh with those obtained with the standard DKT-15 shell element (Batoz *et al.*, 1980) is shown in Figure 6.

7. Concluding Remarks

A family of new plate and shell elements with translational degrees of freedom only has been derived. The elements show a promising behaviour for analysis of static and dynamic analysis of shell structures. Recent applications of the elements to nonlinear dynamic problems involving frictional contact conditions [Oñate *et al.* (1996a and b), Cendoya (1996) and Cendoya *et al.* (1996)] indicate that the new rotational-dof free triangles are excellent candidates for efficient solution of large scale structural and sheet stamping problems.

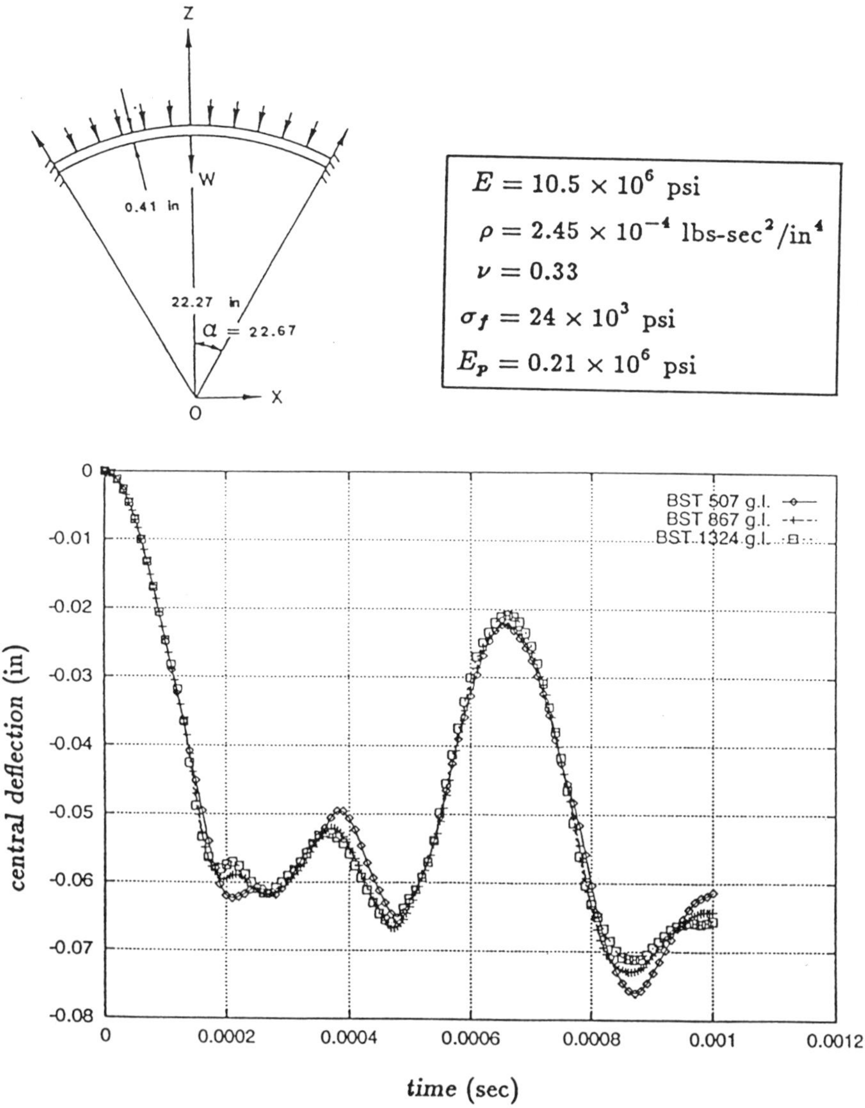

Figure 5. Spherical dome. Time evolution of central deflection.

References

Batoz, J.L., Bathe, K.J and Ho, L.W. (1980), "A study of three-node triangular plate bending elements", *Int. J. Num. Meth. Engn.*, Vol. **15**, 1771–1812.

Batoz, J.L. (1992), *Modèllisation des structures par èlèments fins*, Volume 3 coques, Hermes, Paris.

Barnes, M.R. (1977), "Form finding and analysis of tension space structure by dynamic relaxation", *Ph.D. Thesis*, Dept. of Civil Engineering, The City University, London.

Brunet, M. and Sabourin, F. (1994), "Prediction of necking and wrinkles with a

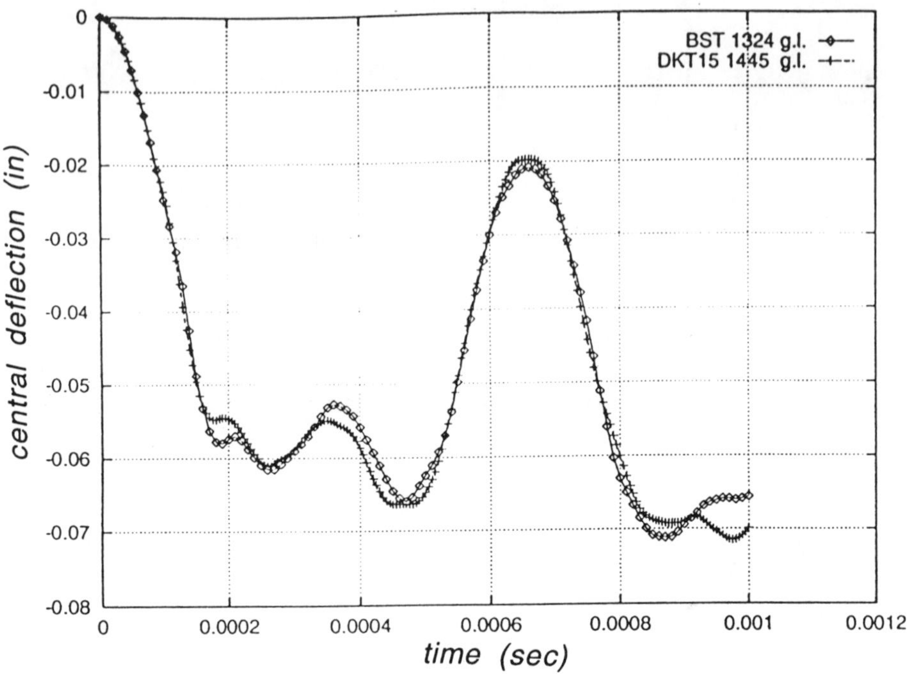

Figure 6. Spherical dome. Comparison of results obtained with BST and DKT15 elements.

simplified shell element in sheet forming", *Int. Conf. of Metal Forming Simulation in Industry*, Vol. II, pp. 27–48, B. Kröplin (Ed.).

Cendoya, P. (1996), "Explicit dynamic analysis of shells using rotational dof-free triangles", *Ph.D. Thesis* (in Spanish), UPC, Barcelona.

Cendoya, P., Oñate, E., Miquel, J. and Zárate, F. (1996), "Nuevos elementos finitos para el análisis dinámico elastoplástico no lineal de láminas", *Actas del III Congreso de Métodos Numéricos en Ingeniería*, Zaragoza, Spain.

Hampshire, J.K., Topping, B.H.V and Chan, H.C. (1992), "Three node triangular elements with one degree of freedom per node", *Eng. Comput.* Vol. **9**, 49–62.

Morley, L.S.D. (1968), "The triangular equilibrium element in the solution of plate bending problems", *Aero Quart*, Vol. **19**, 149–169.

Morley, L.S.D. (1971), "On the constant moment plate bending element", *J. Strain Analysis*, Vol.**6**, 10–14.

Oñate, E. and Cervera, M. (1993), "Derivation of thin plate bending elements with one degree of freedom per node", *Eng. Comput.* Vol. **10**, 543–561.

Oñate, E., Cervera, M. and Zienkiewicz, O.C. (1994), "A finite volume format for structural mechanics", *Int. J. Num. Meth. Eng.*, **37**, 181–201.

Oñate, E., Cendoya, P., Rojek, J. and Miquel, J. (1996), "A simple thin shell triangle with translational degrees of freedom for sheet stamping analysis", at *3rd International Conference on Numerical Simulation of 3D Sheet Forming Processes (NUMISHEET'96)*, Dearbon, Michigan, USA, 29 Sept. - 3 Oct.

Oñate, E., Cendoya, P., Rojek, J. and Miquel, J. (1996), "Non linear explicit dynamic analysis of shell structures using a simple triangle with translational degrees of freedom only", at the *International Conference on Computational Engineering Science (ICES'97)*, San Jose, Costa Rica, May 4–9.

Phaal, R. and Calladine, C.R. (1992), "A simple class of finite elements for plate and shell problems. I: Elements for beams and thin plates", *Int. J. Num. Meth. Engn.*, Vol. **35**, 955–977.

Phaal, R. and Calladine, C.R. (1992), "A simple class of finite elements for plate and shell problems. II: An element for thin shells with only translational degrees of freedom", *Int. J. Num. Meth. Engn.*, Vol. **35**, 979–996.

Rio, G., Tathi, B. and Laurent, H. (1994), "A new efficient finite element model of shell with only three degrees of freedom per node. Applications to industrial deep drawing test", in *Recent Developments in Sheet Metal Forming Technoloy*, Ed. M.J.M. Barata Marques, 18th IDDRG Biennial Congress, Lisbon.

Scordelis, A.C. and Lo, K.S. (1969), "Computer analysis of cylindrical shells" *J. Amer. Concrete Institute*, Vol. **61**, 539–561.

Timoshenko, S.P. and Goodier, J.N. (1968), *Teoría de la elasticidad*, Edic. Urmo.

Yang, D.Y., Jung, D.W., Song, L.S., Yoo, D.J. and Lee, J.H. (1993), *NUMISHEET'93*, Eds. Makinouchi, A., Nakamachi, E., Oñate, E. and Wagoner, R.H., RIKEN, 35–42, Tokyo.

Zárate, F. (1996), "New finite element for plate and shell analysis" (in Spanish), *Ph.D. Thesis*, Univ. Politècnica de Catalunya, Barcelona.

Zienkiewicz, O.C. and Taylor, R.C. (1989), *The finite element method*, 4th Edition, Vol. 1, McGraw Hill.

FINITE ELEMENTS WITH NONLINEAR STATICS AND DYNAMICS

M. A. CRISFIELD[1], G. JELENIĆ[1] and U. GALVANETTO[2]

[1] Department of Aeronautics, Imperial College, London, UK

[2] Dipartimento di Costruzioni e Trasporti, Università di Padova, Padua, Italy

Abstract

The paper briefly considers three separate aspects of the non-linear finite element method, in which particular care must be taken in relation to the adopted discretisation. In relation to statics, we discuss

(i) the objectivity of strain measures in finite element implementations of geometrically exact 3D beam theory

In relation to dynamics, we discuss

(ii) energy and momentum conserving algorithms for continua, beams and shells

(iii) the introduction of physical and numerical damping

1. The objectivity of strain measures in finite element implementations of geometrically exact 3D beam theory

The adopted strain measures (see Simo [1]) are given by

$$\boldsymbol{\gamma} = \boldsymbol{\Lambda}^t \boldsymbol{r}' - \begin{Bmatrix} 1 \\ 0 \\ 0 \end{Bmatrix} \; ; \; \widehat{\boldsymbol{\kappa}} = \boldsymbol{\Lambda}^t \boldsymbol{\Lambda}', \tag{1.1}$$

where $\boldsymbol{\gamma}$ and $\boldsymbol{\kappa}$ are the translational and rotational strains, $\boldsymbol{r}$ is the position vector, $\boldsymbol{\Lambda}$ is the rotation triad, the prime ($'$) denotes differentiation with respect to the axial direction and the superimposed hat ($\widehat{\ }$) denotes the skew-symetric matrix of a respective vector.

These measures are objective in the following sense. Consider a constant translation $\boldsymbol{r}_R$ and rotation $\boldsymbol{\Lambda}_R$ superposed onto the current configuration $\boldsymbol{r}, \boldsymbol{\Lambda}$ so that as a result we have

$$\underline{\boldsymbol{r}} = \boldsymbol{\Lambda}_R (\boldsymbol{r}_R + \boldsymbol{r}) \; ; \; \underline{\boldsymbol{\Lambda}} = \boldsymbol{\Lambda}_R \boldsymbol{\Lambda}. \tag{1.2}$$

It is easily demonstrated that

H. A. Mang and F. G. Rammerstorfer (eds.), IUTAM Symposium on Discretization Methods in Structural Mechanics, 91–98.

$$\gamma(\boldsymbol{r}, \boldsymbol{\Lambda}) = \gamma(\underline{\boldsymbol{r}}, \underline{\boldsymbol{\Lambda}}) \; ; \; \boldsymbol{\kappa}(\boldsymbol{\Lambda}) = \boldsymbol{\kappa}(\underline{\boldsymbol{\Lambda}}). \tag{1.3}$$

Once we introduce the finite element approximations (denoted by a superscript h), it can be shown [2] that the above is true provided we can demonstrate that

$$\underline{\boldsymbol{r}}^h = \boldsymbol{\Lambda}_R \left(\boldsymbol{r}_R + \boldsymbol{r}^h \right) \; ; \; \underline{\boldsymbol{\Lambda}}^h = \boldsymbol{\Lambda}_R \boldsymbol{\Lambda}^h \tag{1.4}$$

as a result of any constant $\boldsymbol{r}_R$ and $\boldsymbol{\Lambda}_R$.

In order to apply (1.1), finite element formulations apply updatings of the form

$$\boldsymbol{\Lambda}^h = \exp \widehat{\boldsymbol{\vartheta}}^h \boldsymbol{\Lambda}_r, \tag{1.5}$$

where $\boldsymbol{\Lambda}_r$ depends on the particular implementation. For some implementations, $\boldsymbol{\Lambda}_r$ is the rotation matrix at the previous iteration [1], for others it is the rotation matrix at the end of the previous increment [3] and for others it is the identity matrix [4]. Each use of $\boldsymbol{\Lambda}_r$ requires a different definition of $\boldsymbol{\vartheta}$ but, in each case, the authors have shown that the formulation is non-objective in that it does not satisfy (1.4) except at specific nodal points or more generally in the limit as the number of nodal variables is increased towards infinity. In some cases, the formulations are also history dependent. The problem can be overcome by stealing an idea from the co-rotational formulation and using an "element triad" for $\boldsymbol{\Lambda}_r$ [2]. Following the co-rotational approach, this triad should rotate with the element.

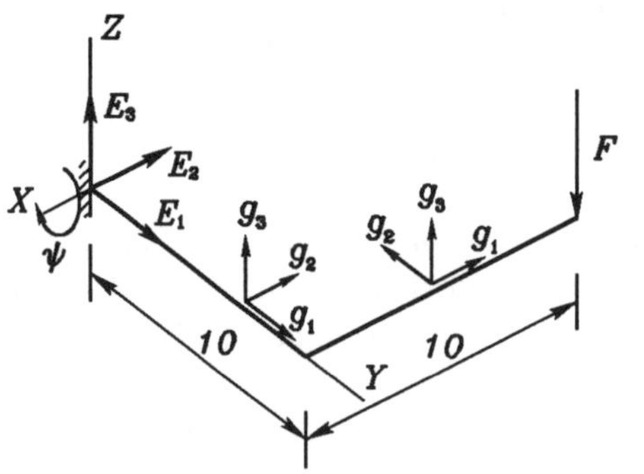

Stiffness properties:
$EA = GA_2 = GA_3 = 10^6$
$GI_t = EI_2 = EI_3 = 10^3$

Finite element mesh:
one quadratic isoparametric element per leg

Concentrated load: $F = 5$

Figure 1. 3D cantilever frame subject to a fixed vertical force at the free end and a prescribed rotation around base vector $\boldsymbol{E}_2$ at the fixed end

To illustrate the problem, we (statically) analyse a rectangular elbow cantilever subject to a concentrated vertical force at its free end and a prescribed rotation at its fixed end (Fig. 1). The frame is composed of two equally long mutually perpendicular legs, the orientation of which is defined by the orthonormal element frames $\boldsymbol{g} = [\boldsymbol{g}_1\, \boldsymbol{g}_2\, \boldsymbol{g}_3]$, where the unit base vector $\boldsymbol{g}_1$ is always directed along the centroidal axis and the base vectors $\boldsymbol{g}_2$ and $\boldsymbol{g}_3$ are directed along the principal axes of inertia. The stiffness properties of legs are given in Fig. 1. Each leg is modelled by a single 3-noded isoparametric element, with quadratic interpolation for the displacement and iterative rotation fields in Algorithm X [1] or the displacement and current local rotation fields in Algorithm PG [2]. The initial orientation of each element frame with respect to the orthonormal global frame $\boldsymbol{E} = [\boldsymbol{E}_1\, \boldsymbol{E}_2\, \boldsymbol{E}_3]$ is shown in Fig. 1. The applied force F is of a fixed magnitude and constant vertical direction. The applied prescribed rotation ψ rotates the whole structure around the unit vector $\boldsymbol{E}_2$ of the global frame. This rotation does not change the direction of the load F.

This problem will be solved for displacements U, V and W along coordinate axes X, Y, Z of the global frame $\boldsymbol{E}$ for different prescribed rotations ψ.

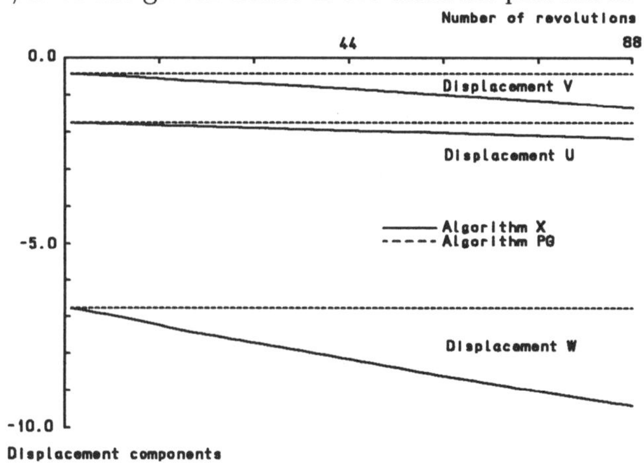

Figure 2. Displacement components at the free end of the cantilever at ends of each revolution of the cantilever

If the frame is left spinning with the prescribed rotation being applied in increments of $\frac{\pi}{2}$ and the displacement components at the free end are monitored at the end of each revolution, it can be observed that the displacements obtained by Algorithm X do not remain constant. The inability of Algorithm X to remain unaffected by a rigid rotation is related to the nonobjectivity of the strain measures in the corresponding finite element implementation [2]. In contrast, Algorithm PG, which was designed to guarantee the objectivity of strain measures, preserves the displacement components at the end of each revolution. The results for both algorithms for first eighty-eight revolutions are collected and graphically presented in Fig. 2.

2. Energy and momentum conserving algorithms

It is now well known that finite element procedures directly based on the trapezoidal Newmark procedure are not stable in a non-linear geometric environment [5,6,7]. One tentative solution involves the addition of numerical damping, using for example the "α method" [8]. However, this technique can only be proved to lead to dissipation in the linear regime. In the non-linear regime, the associated solution procedures can still lead to "energy blow-up" [9,10]. Another possible way of overcoming the problem is to add constraints on both the energy and the momenta via Lagrangian multipliers [11]. However, it is not obvious how one may extend such methods to include time-dependent loads or damping, be it real or numerical.

We will now briefly discuss a range of "conserving algorithms", based on a form of mid-point algorithm which can be easily extended to encompass dissipation. The "conserving mid-point algorithm" was originally described for continua by Simo and Tarnow [5]. It is best explained in relation to a total Lagrangian formulation using the second Piola-Kirchhoff stresses, $\boldsymbol{S}$, and Green strains, $\boldsymbol{E}$. In these circumstances, the "mid-point" dynamic equilibrium equations can be expressed via:

$$\bar{\boldsymbol{g}}_m \equiv \int_{V_0} \left(\frac{\boldsymbol{B}_n + \boldsymbol{B}_{n+1}}{2}\right)^T \frac{\boldsymbol{S}_n + \boldsymbol{S}_{n+1}}{2} dV_0 - \boldsymbol{q}_{em} + \boldsymbol{M}\frac{\dot{\boldsymbol{p}}_{n+1} - \dot{\boldsymbol{p}}_n}{\Delta t} =$$

$$\boldsymbol{q}_{im} - \boldsymbol{q}_{em} + \boldsymbol{q}_{mas,m} = \boldsymbol{0}. \tag{2.1}$$

Here $\boldsymbol{B}_n$ is the conventional non-linear $\boldsymbol{B}$ matrix relating the infinitesimal changes in Green strain to the equivalent changes in nodal variables ($\delta \boldsymbol{p}$) at step n. The change in strain energy over the step can now be expressed as

$$\Delta\phi = \Delta \boldsymbol{p}^T \int_{V_0} \left(\frac{\boldsymbol{B}_n + \boldsymbol{B}_{n+1}}{2}\right)^T \frac{\boldsymbol{S}_n + \boldsymbol{S}_{n+1}}{2} dV_0 = \Delta \boldsymbol{p}^T \boldsymbol{q}_{im},$$

while the change of kinetic energy is given by

$$\Delta K = \frac{1}{2}\dot{\boldsymbol{p}}_{n+1}^T \boldsymbol{M} \dot{\boldsymbol{p}}_{n+1} - \frac{1}{2}\dot{\boldsymbol{p}}_n^T \boldsymbol{M} \dot{\boldsymbol{p}}_n = \frac{1}{2}\left(\dot{\boldsymbol{p}}_{n+1} - \dot{\boldsymbol{p}}_n\right)^T \boldsymbol{M}\left(\dot{\boldsymbol{p}}_{n+1} + \dot{\boldsymbol{p}}_n\right) =$$

$$\frac{1}{\Delta t}\left(\dot{\boldsymbol{p}}_{n+1} - \dot{\boldsymbol{p}}_n\right)^T \boldsymbol{M}\left(\boldsymbol{p}_{n+1} - \boldsymbol{p}_n\right) = \frac{1}{\Delta t}\left(\dot{\boldsymbol{p}}_{n+1} - \dot{\boldsymbol{p}}_n\right)^T \boldsymbol{M}\Delta \boldsymbol{p}.$$

It follows that for fixed applied loads, the total change in energy will be zero once the dynamic residual, $\bar{\boldsymbol{g}}_m$, has been iterated to zero.

For continua, the resulting algorithm not only conserves energy but also the momenta [5]. For beams and shells, the formulation is complicated by the non-additive rotation variables [12]. One possible solution is to use the "co-rotational approach" [12] in which we would replace (2.1) by

$$\bar{\boldsymbol{g}}_m \equiv \left(\frac{\boldsymbol{T}_n + \boldsymbol{T}_{n+1}}{2}\right)^T \frac{\boldsymbol{q}_{li,n} + \boldsymbol{q}_{li,n+1}}{2} - \boldsymbol{q}_{em} + \boldsymbol{q}_{mas,m} = \boldsymbol{0}, \qquad (2.2)$$

where $\boldsymbol{q}_{li,n}$ are now the local internal forces at increment n and $\boldsymbol{T}$ is the equivalent transformation matrix that relates the infinitesimal changes in local nodal variables to the equivalent changes in global nodal variables ($\delta\boldsymbol{p}$). The resulting algorithm is now "nearly" but not "fully" energy conserving because, in contrast to the equivalent situation with the Green strain, we now have an approximation in the relationship

$$\Delta\boldsymbol{p}_l \simeq \frac{1}{2}\left(\boldsymbol{T}_n + \boldsymbol{T}_{n+1}\right)\Delta\boldsymbol{p}.$$

In some circumstances it is possible to derive a "secant transformation matrix" whereby $\Delta\boldsymbol{p}_l = \boldsymbol{T}_{\text{sec}}\Delta\boldsymbol{p}$. In these circumstances, we can replace the internal forces, $\boldsymbol{q}_{im}$, in (2.2) with

$$\boldsymbol{q}_{im} = \boldsymbol{T}_{\text{sec}}^T \frac{\boldsymbol{q}_{li,n} + \boldsymbol{q}_{li,n+1}}{2},$$

and the algorithm becomes fully energy conserving. (There are other issues associated with mass terms [12] that we will not discuss here). The authors and co-workers have derived such fully energy conserving procedures for truss elements [6], 2-dimensional beams [7] and facet shells [10]. However, for 3D co-rotational beams, so far we have only achieved an approximately energy conserving procedure using (2.2) [9].

3. The introduction of damping

Physical damping can be added to the "mid-point formulation" (care must be taken with one's definition of a "mid-point formulation"; here we mean one based on (2.1) or (2.2)) by adding the term

$$\boldsymbol{C}\frac{\dot{\boldsymbol{p}}_{n+1} + \dot{\boldsymbol{p}}_n}{2} \qquad (3.1)$$

to the residual $\bar{\boldsymbol{g}}_m$ of (2.1) or (2.2). It is worth noting that the difficulties associated with the energy blow-up in Newmark's method are not relieved by the introduction of physical damping. This is illustrated in the second example (Fig. 3) which relates to the non-linear analysis of a simple pendulum with large deformations [6] and involves a truss element formulation (with no bending).

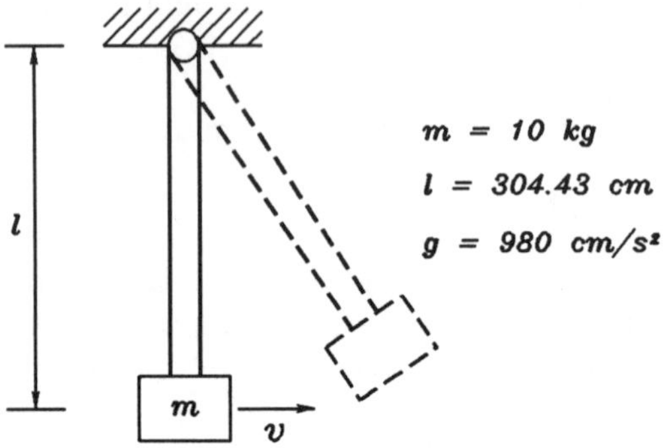

Figure 3. Simple pendulum

The initial velocity $v = 772.5\frac{\mathrm{cm}}{\mathrm{s}}$ is such that the pendulum should rise to the horizontal position before falling. The period is of the order of 4.0 seconds and the adopted time step was given by $\Delta t = 0.1$ seconds. Viscous damping of the form $\boldsymbol{C}\dot{\boldsymbol{d}}$ for the Newmark's method or (3.1) for the mid-point formulation with $\boldsymbol{C} = \boldsymbol{I}$ has been added to the free end of the pendulum.

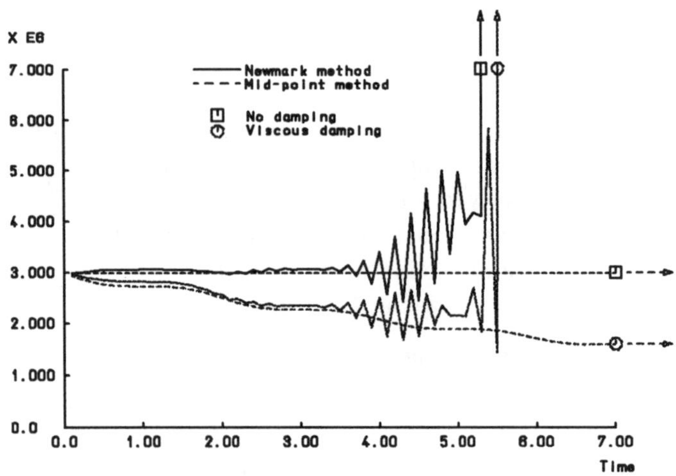

Figure 4. Total energy of the simple pendulum using different integration schemes and different amounts of structural damping

With the mid-point formulation, we obtain a satisfactory (and significant) energy decay (Fig. 4) with the elastic energy decaying by 47 % over a time of seven seconds. In contrast, when the end-point formulation is applied in conjunction with the Newmark's method, we experience a sudden jump in energy although the solution procedure does not fail to converge. Fig. 4 also shows the same and well-known deficiency of the Newmark's method in the absence of physical damping.

Numerical damping can be introduced using a method inspired by a paper by Armero and Petőcz [13] which relates directly to dynamic contact. However, the concept has a much wider potential. In relation to (2.1), we simply modify the static internal forces to give

$$\begin{aligned} \boldsymbol{q}_{im} = & \int_{V_0} \left(\frac{\boldsymbol{B}_n + \boldsymbol{B}_{n+1}}{2} \right)^T \frac{\boldsymbol{S}_n + \boldsymbol{S}_{n+1}}{2} dV_0 + \\ & \xi \int_{V_0} \left(\frac{\boldsymbol{B}_n + \boldsymbol{B}_{n+1}}{2} \right)^T \frac{\boldsymbol{S}_{n+1} - \boldsymbol{S}_n}{2} dV_0. \end{aligned} \tag{3.2}$$

If we now consider the inner product, $\bar{\boldsymbol{g}}_{mc}^T \Delta \boldsymbol{p}$, where, in $\bar{\boldsymbol{g}}_{mc}$, the second term in (3.2) is added to $\bar{\boldsymbol{g}}_m$ of (2.1) and assume fixed applied loads, we have

$$\begin{aligned} \bar{\boldsymbol{g}}_{mc}^T \Delta \boldsymbol{p} = \bar{\boldsymbol{g}}_m^T \Delta \boldsymbol{p} + \xi \Delta \boldsymbol{p}^T \int_{V_0} \left(\frac{\boldsymbol{B}_n + \boldsymbol{B}_{n+1}}{2} \right)^T \frac{\boldsymbol{S}_{n+1} - \boldsymbol{S}_n}{2} dV_0 = \\ \Delta \phi + \Delta K + \frac{1}{2} \xi \Delta \boldsymbol{E}^T \Delta \boldsymbol{S}, \end{aligned}$$

where $\Delta \boldsymbol{E}^T \boldsymbol{S}$ is always positive. It follows that once we have iterated to the position whereby $\bar{\boldsymbol{g}}_{mc} = \boldsymbol{0}$, the change of total energy, $\Delta \phi + \Delta K$, will be negative for any positive constant ξ. This method has been applied by the authors to an approximately energy conserving 3D co-rotational beam formulation whereby the concepts are applied to (2.2) rather than (2.1) [9].

4. Conclusions

The paper has briefly discussed a number of issues associated with the discretisation in both time and space of non-linear structural equilibrium equations. A common theme is that great care must be taken with the detail of the discretisation procedure if key basic properties are to be preserved.

As a first example, it is indicated that while for a 3D beam formulation the underlying strain measures may be objective, this property is not automatically inherited by the discretised formulation.

A second example re-emphasises that while for the key governing equations it may be better to directly consider the differential equations of motion leading to an infinitesimal change in time of the momenta, in a discretised form it is better to start with the algorithmic difference in the momenta over an adopted time step. It is shown that both real and numerical damping can easily be added to the resulting "conserving algorithms" and it is indicated that "conventional" trapezoidal Newmark procedures can encounter difficulties even in the presence of damping.

Acknowledgements

This work has been financially supported by EPSRC.

References

1. Simo, J.C. and Vu Quoc, L.: A three-dimensional finite strain rod model. Part II: Computational aspects, *Comp. Meth. Appl. Mech. Eng.* **58** (1986), 79-116.

2. Jelenić, G. and Crisfield, M.A.: Objectivity of strain measures in finite element implementations of geometrically exact 3D beam theory, submitted to *Comp. Meth. Appl. Mech. Eng.* (1997)

3. Cardona, A. and Géradin, M.: A beam finite element non-linear theory with finite rotations, *Int. J. Num. Meth. Eng.* **26** (1988), 2403-2438.

4.Ibrahimbegović, A., Frey, F. and Kožar, I.: Computational aspects of vector-like parametrization of three-dimensional finite rotations, *Int. J. Num. Meth. Eng.* **38** (1995), 3653-3673.

5. Simo, J.C. and Tarnow, N.: The discrete energy-momentum method. Conserving algorithms for nonlinear elastodynamics, *J. Appl. Math. Phys. (ZAMP)* **43** (1992), 757-792.

6. Crisfield, M.A. and Shi, J.: A co-rotational element/time-integration strategy for non-linear dynamics, *Int. J. for Num. Meth. Eng.* **37** (1994), 1897-1913.

7. Galvanetto, U. and Crisfield, M.A.: An energy-conserving co-rotational procedure for the dynamics of planar beam structures, *Int. J. Num. Meth. Eng.* **39** (1996), 2265-2282.

8. Hilber, H.M., Hughes, T.J.R. and Taylor, R.L.: Improved numerical dissipation for time integration algorithms in structural dynamics, *Earthq. Eng. Struct. Dyn.* **5** (1977), 283-292.

9. Crisfield, M.A., Galvanetto, U. and Jelenić, G.: Dynamics of 3-D co-rotational beams, submitted to *Comput. Mech.* (1997)

10. Zhong, H.-G. and Crisfield, M.A.: An energy-conserving co-rotational procedure for the dynamics of shell structures, submitted to *Eng. Comput.* (1996)

11. Kuhl, D. and Ramm, E.: Constraint energy momentum algorithm and its application to non-linear dynamics of shells, *Comp. Meth. Appl. Mech. Eng.* **136** (1996), 293-315.

12. Crisfield, M.A.: *Non-linear finite element analysis of solids and structures, Vol. 2, Advanced Topics*, John Wiley, Chichester, 1997.

13. Armero, F. and Petőcz, É.: A new class of conserving algorithms for dynamic contact problems, in J.-A. Désidéri, P. Le Tallec, E. Oñate, J. Périaux and E. Stein (eds.), *Proceedings of the second ECCOMAS Conference on Numerical Methods in Engineering*, 9-13 September 1996, Paris, France, John Wiley & Sons, Chichester, pp. 861-867, 1996.

NUMERICAL TREATMENT OF THE DYNAMIC SOIL–STRUCTURE–FLUID INTERACTION OF UNANCHORED LIQUID STORAGE TANKS

Parallel computing of a semi–analytical FE–model

W. WUNDERLICH, M. RAPOLDER
Lehrstuhl für Statik
Technische Universität München
Arcisstr. 21, D–80333 München, Germany
e–mail: ww@statik.bauwesen.tu–muenchen.de

1. Introduction

A realistic analysis of containments, liquid–filled storage tanks, silos, and similar structures with rotational geometry should include the interaction effects between the structure and both the liquid and the ground. This is especially important for dynamic loading such as excitations caused by earthquakes or explosions that may cause damages in various parts of the structure [1].

The coupling of the various effects leads to considerable numerical effort especially as nonlinear dynamical calculations of relatively large systems are necessary. The discretisation with two– and three–dimensional finite elements creates nonlinear systems of algebraic equations with a very high number of degrees of freedom.

The advantage of the rotational geometry of the containments is used in a special semi–analytical approach with Fourier series in the circumferential direction. Although this method substantially reduces the discretization effort, further acceleration of the process is necessary to obtain realistic responses. Therefore, the algorithms of the semi–analytical method are parallelised for application on workstation clusters.

2. Numerical Modelling

In this paper a complete dynamic analysis in the time domain is carried out in which the nonlinear behaviour of the structure, the soil and the interface are taken into account simultaneously. Merely the fluid is described with linear relations and taken as inviscid and incompressible. The soil region is regarded to be composed of two parts: a near field which is fully discretized permitting nonlinearities, e.g. plastification of the soil or uplifting of the tank, and the far field that enables the treatment of radiation of energy to infinity.

The same semi–analytical approach as used in the linear case [2] was extended to the geometrical and physical nonlinear range. To avoid the explicit nonlinear coupling of the

H. A. Mang and F. G. Rammerstorfer (eds.),
IUTAM Symposium on Discretization Methods in Structural Mechanics, 99–106.

various Fourier harmonics, all terms arising from nonlinearities are treated as pseudo–loads. In each iteration step of the nonlinear analysis their current components are evaluated numerically at certain stations, and then the updated distribution is again developed into Fourier series. As a result, all relations are orthogonally decoupled both on the element and the system level [3].

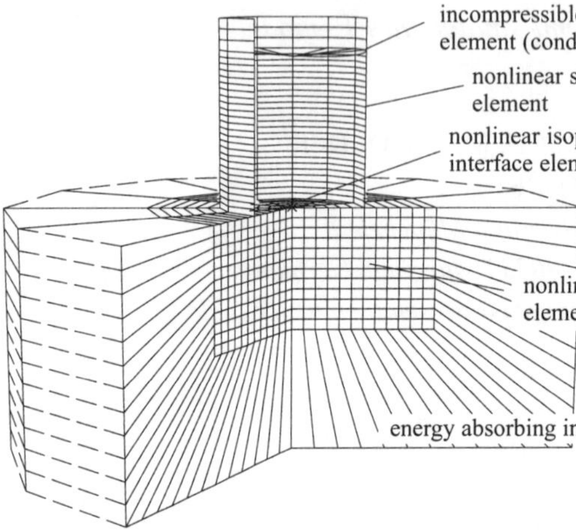

Figure 1. The soil–structure–fluid system

As this specific modelling is described in detail in various former papers, only a short outline is given below accompanied by some references.

2.1. THE INCOMPRESSIBLE FLUID

The interaction between the shell and the fluid is commonly approximated by a 'lumped–mass' model. In the nonlinear range it becomes necessary to employ a finite element discretisation also for the fluid, but it is sufficient to use formulations for an inviscid and incompressible fluid and the equations for an acoustic medium. Its influence on the behaviour of the shell is taken into account by an additional mass matrix [4].

2.2. THE SHELL ELEMENT FORMULATION

The shell equations in incremental form are based on a mixed variational principle of Hellinger–Reissner type. They are valid for large displacements and moderate rotations. The circumferential variation of all geometrical and field variables is approximated by Fourier series. The pseudo–load vectors are functions of the current distribution of the load–dependent tangent modulus throughout the shell, its increment in the actual state and the displacements of the reference state as well as the displacement rates. In addition, the dynamic forces are also treated as inhomogeneous terms [3].

The inclusion of plastic deformations implicitly accounts for the hysteretic damping of the shell material. This in turn requires a constitutive description which realistically predicts the time history of the material's tangent moduli and its finite increments appearing in the pseudo–load vector. This is achieved by a two–surface plasticity model, which accounts for non–proportional loading conditions in multi–axial stress states in an efficient manner [5].

2.3. INTERFACE–ELEMENT FOR SOIL–STRUCTURE–INTERACTION

The boundary nonlinearity is treated through pseudo material nonlinearities of a special interface element with thin finite thickness accounting for the slip and the separation behaviour. The constitutive equations expressed by the local interface stresses define the relevant deformation modes. The shear strains are calculated from the relative in–plane displacements of the element. The tangential shear modulus depends on several factors, such as amplitude of the relative displacements, interface thickness, shear stress, normal stress, interface roughness and the number of cycles of loading.

The element simulates the separation between tank and soil in case that the absolute normal stress becomes positive (tensile stress). In this case neither normal nor shear traction can be transmitted through the interface and the stresses become zero. The element returns to contact for negative normal strains and negative slip functions [6].

2.4. SOIL ELEMENTS FOR THE NEAR AND THE FAR FIELD

To obtain reliable numerical results for soil–structure–interaction problems, all typical characteristics of the soil have to be taken into account. Therefore, linear and nonlinear isoparametric continuum ringelements are implemented in the numerical model. The element model realistically reproduces the deformations and shear failure of soils [7].

The excited tank acts as a transmitter of waves propagating into the unbounded domain of the soil. Modelling the soil–structure–system with finite elements produces an artificial boundary between the discretized near field of the soil and the surrounding region. This outer region is treated as a linear elastic halfspace. The energy radiation is simulated by viscous damping forces at the artificial boundary [7].

3. Solution of the nonlinear equations of motion

3.1. TIME INTEGRATION

Due to the strong nonlinearity and non–smoothness of the discussed problem, the choice of a proper time integration scheme is very important. It is essential that the algorithm is unconditionally stable especially when simulating the highly nonlinear contact process, which may induce also arbitrary high frequencies that have no mechanical background. Moreover, it should eliminate these frequencies because they are artifacts of the discretisation and cannot be described by finite element meshes [8][9].

It is well known that explicit methods have shortcomings in stability and need very short time steps. Therefore implicit methods are widely used. A second order, unconditionally stable scheme (linear case) is the trapezoidal rule. Unfortunately, it may be unstable in the nonlinear case and neglects numerical damping [10] (Figure 2). The stability of Newmark's method is much better for nonlinear calculations and it supplies numerical damping for high oscillations. Although it is only first order accurate, it is very useful for contact problems [9] and for this reason is used here.

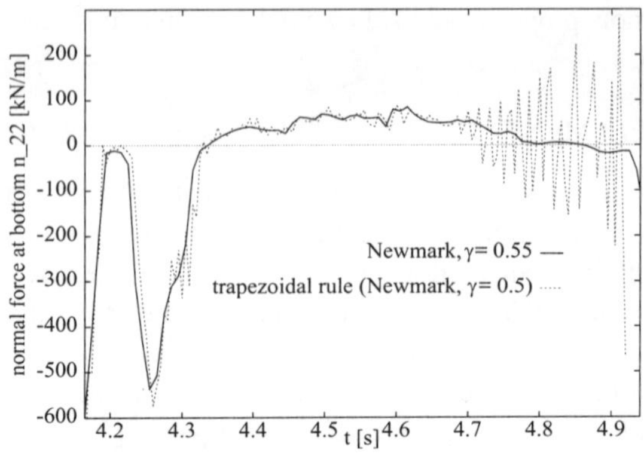

Figure 2. Normal force at the bottom of an uplifting tank for Newmark's time integration scheme

In addition, some newer proposals of integration algorithms were investigated.The HHT–α scheme [11] is second order accurate and allows numerical dissipation but becomes unstable for the discussed problem due to nonlinearities [12]. Further research is necessary with respect to schemes recently developed, e.g. the generalized–α method [13]. Another group of time integration algorithms have been developed for elastic material behavior using energy–conserving methods [10]. Although these algorithm are unconditionally stable in the nonlinear case, the embedded energy conservation is doubtful for material– and numerical damping.

3.2. SOLUTION OF EQUATIONS

The method to obtain the nonlinear equilibrium is based on a gradual iteration where all nonlinear terms are taken into account only through pseudo–loads on the right–hand–side of the system. For this iteration a kind of Quasi –Newton –method (or BFGS–method) is used [6].

4. Parallel Programming of semi–analytical FE–model

The main problem that currently limits the application of FE–programs simulating large nonlinear systems in the time domain is their very long computation time. Although the semi–analytical FE–model substantially reduces the computation time (solution of linear equations of systems with a very small bandwith, dynamic substructure–methods for linear parts), the numerical process may be further accelerated using parallel computing.

4.1. PARALLEL PROGRAMMING MODEL

Detailed analyses showed, that most of the CPU time is needed for calculations on the element–level (stresses, plastic strain, internal forces, nonlinear pseudo–loads). The solution of the linear equations takes only little time due to the small bandwith of the system–matrices.

As the Fourier–harmonics are coupled on the element level (computing nonlinear parts) but decoupled on a less CPU–intense part (solution of equations), it would not make sense to parallelize the harmonics.

In standard FE–analysis the FE–mesh is usually partitioned and then calculated on distributed CPU's, so only the boundary variables have to be transmitted to the neighbours. This approach would create many disadvantages in the context of the semi–analytical method. For example, the iteration–step on the system level and the postprocessing would be very time–consuming. Therefore, a model was developed that consists of two different programs including a master for calculating the small CPU–intense parts on the system level and a slave for calculations on the element level. In this manner more than 85 percent of the total CPU–time is distributed to the slaves. Depending on the given problem, the theoretical speedup may reach 6–7. For small workstation–clusters consisting of no more than 10 computers this is a very satisfactory result.

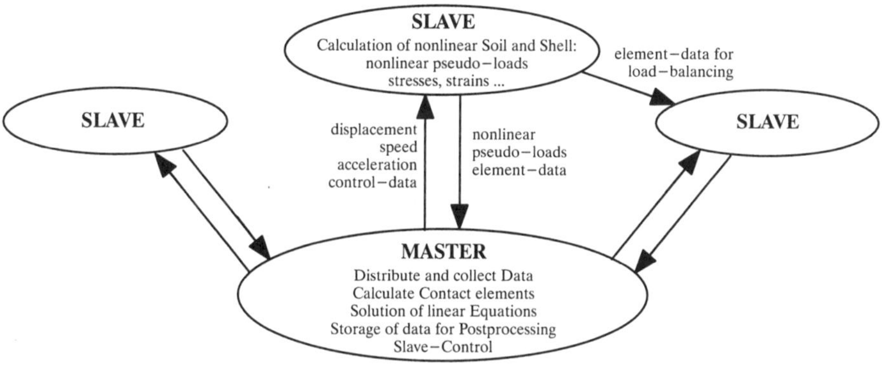

Figure 3. Data–Transfer and Calculation for each iteration step

In each iteration step the master sends the updated matrices for displacements, speeds, accelerations and their increments to the corresponding slaves. While all slaves calculate stesses, plastic strains and internal forces for their assigned shell and soil elements, the master works on contact elements. Every slave that has completed its task, sends its part of the pseudo–load vector to the master which collects this data and also solves the linear equations (Figure 3).

The element data are stored on every slave and sent to the master when the iteration is finished. For postprocessing, the masters stores these element–data.

Communication handling as well as dynamic process control is programmed using the well known internet software–package PVM (Parallel Virtual Machine) that supports programming even on heterogenous workstation–clusters [14].

4.2. DYNAMIC LOAD BALANCING

The aim of dynamic load balancing is the distribution of data–blocks (tasks) to obtain solutions in optimum time. This is necessary, because a heterogeneous network consists of slow and fast computers and moreover, they are influenced by other users with external loads.

One simple but normally quite efficient method is the so called 'pool of tasks' (Task Farming), in which every slave gets one task if its preceding task is complete.

Unfortunately, this method leads to a huge amount of data transfer in the problem considered.

Less Element Data–Transfer is required using the 'greedy method' described in [15], where fixed data–blocks are stored on every slave and locally updated at each iteration step. If load balancing is necessary, some blocks are migrated from slave to slave, which takes only moderate time. The master decides, whether load balancing is necessary and which slave has to move a specified amount of data blocks to another slave.

Moreover, a heuristic approach is used that takes 'historical' behavior into account: Let ΔT_i denote the time–difference between responding back to the master of the first slave and i.th slave ($i = 1..N_{slave}$) and History(i) a parametric vector, then

$$ {}_{n}History(i) = \frac{{}_{n-1}History(i) * weight + \frac{\Delta T_i}{\max(\Delta T)}}{weight + 1} \quad (1) $$

where weight is an empirical determined function of $\max(\Delta T_i)$ and N_{slave}, that could also be evaluated through a neuronal network. If

$$ max\ (\Delta T_i) > \varepsilon_{\Delta T} \ \wedge\ max(History) - min(History) > \varepsilon_{History} \quad (2) $$

then elements from slave with $max_{History}$ to slave with $min_{History}$ are moved, with number of Elements as an empirical function of $N_{elements}$ and (max(History) – min(History)).

With this simple approach to a 'learning machine' unnecessary data–transfer is avoided and a relatively fast load balancing is achieved.

4.3. EXPERIMENT

For measuring the speedup a homogeneous network of HP 9000/712 workstations linked with ethernet was used, and the algebraic system consisted of about 2000 equations (Figure 4).

The program needs some iteration steps to 'learn' the current system–loads. After this short phase, data blocks are moved accordingly. In a network that is not influenced by other users with varying external loads, the program moves data–blocks until a static level is reached. When other users produce external load, the program moves data blocks (accordingly CPU–intensive load) to underloaded machines (Figure 5).

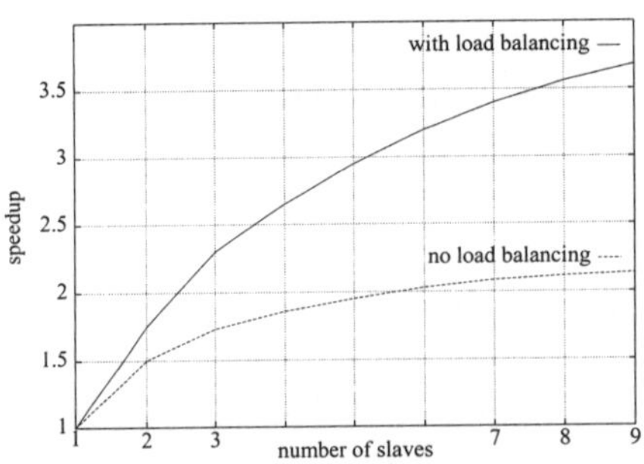

Figure 4. Speedup with load balancing and without load balancing

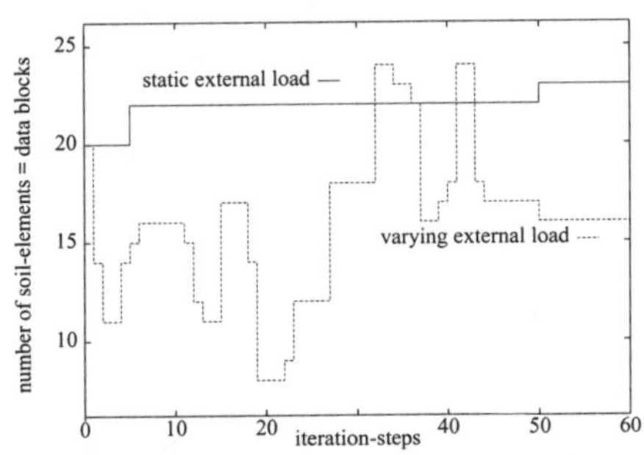

Figure 5. Number of elements evaluated on one slave (of five) with static and dynamic external load

5. Numerical Investigations

5.1 EARTHQUAKE EXCITATION

For detailed investigations of the highly nonlinear behavior of unanchored tanks, it is necessary to have several accelerograms available that match characteristic properties such as acceleration response spectrum, peak ground acceleration (PGA) and a certain Arias–intensity (Energy). Therefore, it is useful to generate synthetic accelerograms from a stationary nonwhite process [16]. In comparison, recorded ones from actual earthquakes may be used, too (Figure 6).

5.2 BEHAVIOUR OF LIQUID FILLED TANKS

Tanks anchored to the ground and nonanchored tanks that do not uplift show nearly linear behavior for small loads [7]. For design purposes, quasi–static or modal analyses may be used. Uplifting tanks exhibit a totally different, strongly nonlinear behavior.

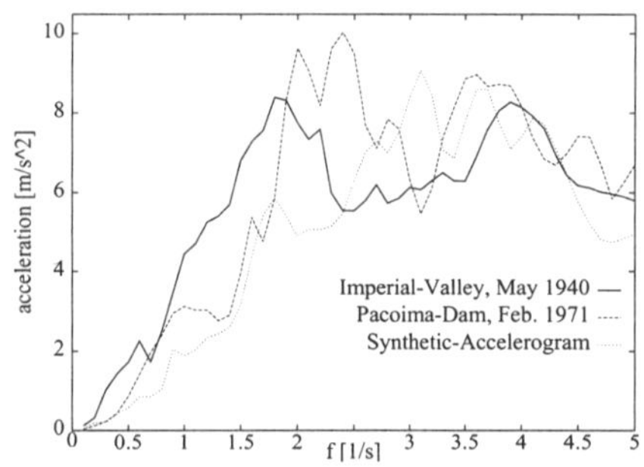

Figure 6. Response–spectra of earthquakes (5 % damping)

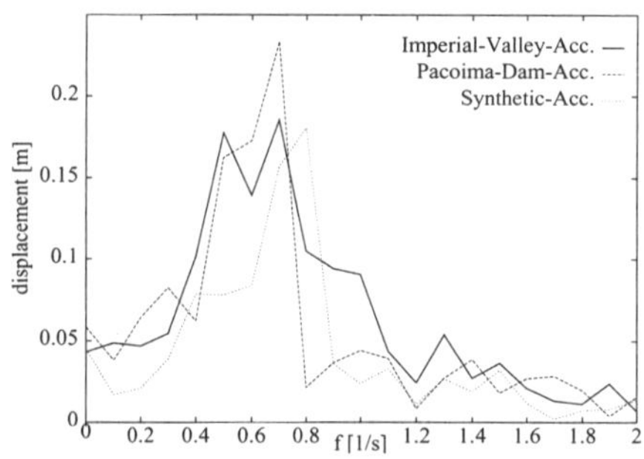

Figure 7. Spectra of displacements of a shell–node at the roofing of an uplifting tank

As the natural–frequencies of the shell–fluid–interaction strongly depend on the uplifting–amplitude of the tank it responds to different frequencies during an earthquake. It is known, that the frequency–content of earthquakes also changes in time. Therefore, a fully nonstationary earthquake model might be necessary, because these results may differ from those of a stationary excitation [17]. But first calculations have shown that the response of actual and synthetic accelerograms are quite similar (Figure 7). It seems that the results depend more on the total frequency–content than on the variation in time. Further investigations on this subject are necessary.

6. Conclusions

The analysis of liquid storage tanks under earthquake excitation has to include all interaction effects to obtain a realistic behaviour of the coupled tank–soil–fluid system. The FE–model employed by the authors takes all these effects into account and has proven to be a sophisticated tool for earthquake simulations of liquid filled tanks. For further reduction of the running time on workstation clusters the semi–analytical method has been adapted to parallel computing. For load–balancing a simple but efficient method was developed.

Due to the strongly nonlinear behaviour of the unanchored tank, especially arising from the contact problem, it is necessary to use a special time–integration scheme. Newmark's algorithm with numerical damping turns out as a stable scheme which is very suitable for contact problems.

The numerical results show that it is inevitable to analyse every parameter of the complex system in detail, because even small changes in the input may influence the results considerably.

References

1. Fischer, F.D., Rammerstorfer, F.G., Scharf, K. (1990) Earthquake resistant design of anchored and unanchored liquid storage tanks under three–dimensional earthquake excitations, in G.I. Schuëller (eds.) *Structural Dynamics – Recent Advances*, Springer, Heidelberg, Germany
2. Wunderlich, W. (1967) On the analysis of shells of revolution by transfer matrices (in German). *Ing.–Arch.* **36**, 262–279
3. Wunderlich, W., Cramer, H., Obrecht, H. (1985) Application of ring elements in the nonlinear analysis of shells of revolution under nonaxisymmetric loading, *Comp. meth. in appl. mech. Eng.* **51**, 259–275
4. Wunderlich, W., Springer H. and Goebel, W. (1989) Discretization and solution techniques for liquid filled shells of revolution under dynamic loading, *Discretization Methods in Structural Mechanics*, H.A. Mang, G. Kuhn (eds.), Springer–Verlag, Heidelberg, pp. 145–155
5. Wunderlich, W., Obrecht, H., Springer H. and Lu Z. (1989) A semi–analytical approch to the nonlinear analysis of shells of revolution, *Analytical and Computational Models for Shells – ASME Winter Annual Meeting 1989*, Noor, A.K., Belytschko, L.T, Simo, J.C. (eds.), ASME, New York, pp. 509–536
6. Wunderlich, W. and Temme, C., (1995) Soil–fluid–structure interaction of uplifting tanks., in N.E. Wiberg, (eds.) *Advances in finite element technology*, CIMNE, Barcelona, pp. 189–211
7. Wunderlich, W., Schäpertöns, B., Temme, C. (1994) Dynamic stability of nonlinear shells of revolution under consideration of the soil–fluid–structure–interaction. *Int. J. for num. meth. in Eng.* **37**, 2679–2697
8. Hughes, T.J.R. (1987) *The Finite Element Method*, Prentice–Hall, Englewood Cliffs, New Jersey, USA
9. Laursen, T.A. and Simo, J.C., (1993) A continuum–based Finite Element formulation for the implicit solution of multibody, large deformation frictional contact problems, *Int. J. for num. meth. in Eng.* **36**, 3451–3485
10. Simo, J.C. and Tarnow, N. (1995) Non–linear dynamics of three–dimensional rods: Exact energy and momentum conserving algorithms, *Int. J. for num. meth. in Eng.* **38**, 1431–1473
11. Hilber, H.M., Hughes, T.J.R. and Taylor, R.L. (1977) Improved numerical dissipation for time integration algorithms in structural dynamics, *Earthquake Engineering and Structural Dynamics* **5**, 283–292
12. Bauchau, O.A. et. al. (1995) Numerical integration of non–linear elastic multi–body systems, *Int. J. for num. meth. in Eng.* **38**, 2727–2751
13. Chung, J. and Hulbert G.M. (1993) A time integration algorithm for structural dynamics with improved numerical dissipation: The generalized–α method, *Journal of Applied Mechanics* **60**, 371–375
14. Dongarra, J.J., Geist, G.A., Manchek, R., and Sunderam, V.S. (1996) Integrated PVM framework supports heterogeneous network computing, in A.Y Zomaya (eds.), *Parallel Computing*, International Thomson Computer Press, London, pp. 435–451
15. Chien, Y.P., Carpenter, F., Ecer, A., and Akay, H.U. (1995) Load–balancing for parallel computation of fluid dynamics problems, *Comp. meth. in appl. mech. Eng.* **120**, 119–130
16. Klein, H. H. (1988) Realistische seismische Lastannahmen für Bauwerke, *Bauingenieur* **63**, 221–228
17. Conte, J.P. and Peng, B. F. (1997) Fully nonstationary analytical earthquake ground–model, *J. of Engineering Mechanics, ASCE* **123**, 15–24

CONTROL OF GLOBAL RELATIVE ERROR OF SPATIAL -TEMPORAL DISCRETIZATION IN DYNAMIC ANALYSIS

CHANG-KOON CHOI and HEUNG-JIN CHUNG
Department of Civil Engineering, Korea Advanced Institute of Science and Technology, Taejon 305-701, Korea

1. Introduction

An adaptive analysis should find optimal discretization at the least cost, such that the local error is uniformly distributed and remains within a given tolerance over the entire spatial/time domain. In this study an adaptive procedure of finite element method to solve dynamic problems is suggested. Both the kinetic and strain energy errors induced by space and time discretization were estimated in a consistent manner and controlled by the simultaneous use of the adaptive mesh generation and the automatic time stepping.

In the time discretization error estimation a successive quadratic function is used for the locally exact value of the acceleration and the corresponding parameters for the function are obtained from accelerations at three time stations at each time stage. The local error can be estimated simply by comparing the solutions obtained by direct time integration method (Newmark method and Wilson method) with the locally exact solutions. Based on this local error estimate an adaptive time stepping technique in global sense is also proposed.

To control the relative error the time step size and the mesh distribution should be modified at the same time based on the relative error estimate and the prescribed relative error tolerance. Relative error can be defined as an error norm of energy divided by either the total energy norm of current time stage or that of the final time stage. In this study it was found that the use of relative error with the total energy norm of the final time stage is more efficient. At this case, however, the energy norm of final time stage is desired to be assumed appropriately.

H. A. Mang and F. G. Rammerstorfer (eds.),
IUTAM Symposium on Discretization Methods in Structural Mechanics, 107–114.

Also an optimal relative ratio of spatial discretization error to temporal discretization error in global sense was discussed. In this study it was found that the best performance can be obtained when the relative values of specified spatial and temporal discretization errors have the same value.

2. Discretization Errors

To solve a dynamic problem by the finite element method, the domain of interest is subdivided first into a number of elements. Then the semidiscrete Galerkin approximation can be used to obtain an integral formulation which is usually referred to as the weak form. The aforementioned discretization of continuous spatial domain during the formulation of finite elements induces the spatial discretization error, $e_S(x,t_n)$ which can be written as follows:

$$e_S(x,t_n) = u(x,t_n) - u^h(x,t_n) \tag{1}$$

where $u(x,t_n)$ is the exact solution of dynamic problem and $u^h(x,t_n)$ the solution to semidiscrete Galerkin approximation. To obtain the transient responses, the semidiscrete Galerkin approximation is solved with some time integration scheme. Generally, when a single-step scheme like the Newmark method is used for the direct time integration, the variation of the acceleration in each time step is assumed to be either constant or linear. This approximation yields a discontinuous distribution for the acceleration in the time domain and induces the temporal discretization error, $e_T(x,t_n)$ as follows:

$$e_T(x,t_n) = u^h(x,t_n) - U_n(x) \tag{2}$$

where $U_n(x^h)$ is the solution to time integration at time $t = t_n$. The total discretization error which contains both the spatial and temporal discretization errors in the finite element solution can be expressed as

$$e(x,t_n) = u(x,t_n) - U_n(x) \tag{3}$$

Then, for any choice of norm [2]

$$\begin{aligned} \|e(x,t_n)\| &= \|u(x,t_n) - U_n(x) + u^h(x,t_n) - u^h(x,t_n)\| \\ &= \|e_S(x,t_n) + e_T(x,t_n)\| \;\le\; \|e_S(x,t_n)\| + \|e_T(x,t_n)\| \end{aligned} \tag{4}$$

Thus the spatial discretization error and time discretization error can be estimated separately and the upper bound of the total discretization error can be evaluated by adding up both errors.

3. Estimate of Time and Spatial Discretization Error

Let us consider a time interval $[t, t+\Delta t]$ and assume that $\tau \in [t, t+\Delta t]$. The time discretization error of acceleration at time τ is

$$\ddot{e}(\tau) = \ddot{u}^n - \ddot{u}^{ex}(\tau) \tag{5}$$

where $\ddot{u}^n$ is the assumed acceleration and $\ddot{u}^{ex}$ is the real acceleration which varies continuously. Suppose that the solutions at time station t are exact. Then, in Newmark method, the time discretization error of velocity and diaplacement solution at time τ and $t+\Delta t$ can be estimated by

$$\dot{e}(t+\Delta t) = \ddot{u}^{nv}\Delta t - \int_t^{t+\Delta t} \ddot{u}^{ex}(\tau')d\tau' \tag{6}$$

$$e(t+\Delta t) = \frac{\ddot{u}^{nd}}{2}\Delta t^2 - \int_t^{t+\Delta t}\int_t^{t+\tau} \ddot{u}^{ex}(\tau')d\tau' d\tau \tag{7}$$

where $\ddot{u}^{nv} = (1-\gamma)\,\ddot{u}_t + \gamma\,\ddot{u}_{t+\Delta t}$ and $\ddot{u}^{nd} = (1-2\beta)\,\ddot{u}_t + 2\beta\,\ddot{u}_{t+\Delta t}$. In Wilson method, the time discretization error of velocity and displacement at time $t+\theta\Delta t$ can also be estimated by

$$\dot{e}(t+\theta\Delta t) = \int_t^{t+\theta\Delta t} \ddot{u}^w(\tau')d\tau' - \int_t^{t+\theta\Delta t} \ddot{u}^{ex}(\tau')d\tau' \tag{8}$$

$$e(t+\theta\Delta t) = \int_t^{t+\theta\Delta t}\int_t^{t+\tau} \ddot{u}^w(\tau')d\tau' d\tau - \int_t^{t+\theta\Delta t}\int_t^{t+\tau} \ddot{u}^{ex}(\tau')d\tau' d\tau \tag{9}$$

where $\ddot{u}^w(\tau') = \ddot{U}_t + \frac{\tau'}{\theta\Delta t}\left(\ddot{U}_{t+\theta\Delta t} + \ddot{U}_t\right)$. Since the exact value of the acceleration $\ddot{u}^{ex}(\tau)$ cannot be obtained in most real problems, it is desirable to approximate the acceleration by a higher order function than the order of the assumed acceleration function in the Newmark method. Zienkiewicz and Xie proposed a local error estimator by comparing the Newmark solution with the exact solutions expanded in the Taylor series[5]. Wiberg and Li derived a formulation for linearly varying third-order derivatives and, based on this, they obtained *a posteriori* error estimates for displacements and velocities[3]. In this study, a quadratic function is used for the approximation for $\ddot{u}^{ex}(\tau)$ and the corresponding parameter for the function is obtained from accelerations at three time stations; $t-\Delta t$, t and $t+\Delta t$. This approximation for the exact acceleration gives a more accurate error estimation than a linear approximation of acceleration since it consists of connecting the successive three points as a group on the curve expressed by a second-degree parabola. The sum of the local error estimates by this method will give a reasonably good estimate for the global error. If a sufficiently small time step size is chosen, the total energy of the error is mainly due to the strain energy of the error at most stages. The spatial discretization error estimator in the displacement can be obtained by comparing the smoothed stress and the finite element stress.

4. Combined Adaptive Procedure

To control the relative error which is defined as an error norm of energy divided by the total energy norm, the time step size and the mesh distribution should be modified at the same time based on the local error estimate and the prescribed error tolerance. On the other hand, nevertheless the user specifies the relative error tolerance η, and the portion of the spatial discretization error out of the total error and that of the temporal discretization error should also be specified. Accordingly, parameters can be defined as follows

$$\varepsilon = \frac{\eta_S}{\eta} \qquad \text{and} \qquad \delta = \frac{\eta_T}{\eta} \tag{10}$$

where η_S and η_T are the prescribed relative energy norm of spatial and temporal discretization errors, respectively, and the sum of these two parameters is 1.0 ($\varepsilon + \delta = 1.0$). Optimal values for this parameters will be discussed in the numerical studies.

Generally the tolerance specified by the user is given as a relative value of the error norm to the energy norm. Thus the energy norm of the final time stage is desired to be assumed appropriately. Assuming that the accumulated energy with time varies linearly, at the certain time stage, $t = t_i \ (\le t_n)$ the anticipated energy norm $E_n^{1/2}$ at the final time stage, $t = t_n$, can be estimated as

$$E_n^{1/2} = \frac{\left(\sum_{j=1}^{i} E_j^{1/2} \Delta t_j\right) + (t_n - t_i) E_i^{1/2}}{t_n} \tag{11}$$

where $E_i^{1/2}$ is the energy norm at time $t = t_i$. In order to perform the computation more efficiently and economically for a given tolerance, a lower error limit and an upper error limit are introduced as

$$\gamma_1 \eta E_n^{1/2} \frac{\Delta t_i}{t_n} \le E(e, \dot{e})^{1/2} \le \gamma_2 \eta E_n^{1/2} \frac{\Delta t_i}{t_n} \tag{12}$$

where $0 \leq \gamma_1 \leq 1$ and $\gamma_2 \geq 1$ are two parameters and η is the relative error tolerance at the final time stage. When Eq. (12) is satisfied, the solution is accepted and the time integration is proceeded to the next step without changing the time step size or mesh distribution. However, if the upper limit is violated, the solution is rejected, the time step size or mesh distribution is updated, and a re-calculation for the current time is performed.

The combined spatial and temporal adaptive procedure proposed in this paper can be summarized as follows :

[STEP 1] Carry out the finite element analysis with the initial or previous mesh and time step until the prescribed termination time is reached.

[STEP 2] Estimate the temporal discretization error and check whether the value of temporal discretization error is within the range of specified error tolerance. If the error is not within that range, change the current time step size and go to [STEP 1]. Otherwise go to the next step.

[STEP 3] Estimate the spatial discretization error and check whether the value of the spatial discretization error is within the range of specified error tolerance. If the error is not within the range, change the mesh distribution and go to [STEP 1]. Otherwise forward one time step and go to [STEP 1]. For practical reasons, when there is no change in the mesh distribution after mesh generation, forward to the next time stage.

5. Numerical Examples

To investigate the performance of the proposed time discretization error estimate, a SDOF model subjected to a sinusoidal loading for which an exact solution can be obtained analytically is considered. The relative error distributions for the energy norm with different integration schemes are shown in Fig. 1 . Comparisons with the real global error are shown in Fig. 2

An one-dimensional elastic bar subjected to a step loading is analyzed to demonstrate the effectiveness of the proposed adaptive analysis procedure (Fig. 3). The accuracy of the constant average acceleration method in the transient analysis of this wave propagation problem depends not only on the temporal discretization but also on the spatial discretization[1]. Therefore, combined effect of both discretization errors should be considered in wave propagation problems.

An one-dimensional elastic bar subjected to a sinusoidal pulse is analyzed to demonstrate the effectiveness of the proposed adaptive analysis procedure. In Fig. 4 the variation of time step sizes and the mesh adaptation is plotted.

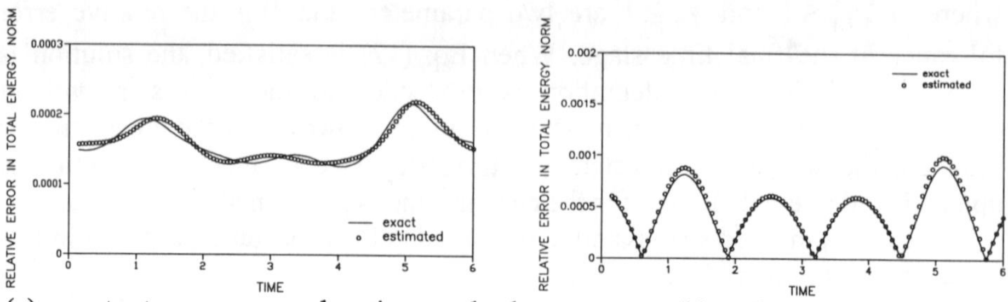

(a) constant average acceleration method (b) Wilson method

Figure 1. Comparisons of relative local error of energy norm, $\Delta t = 0.05$

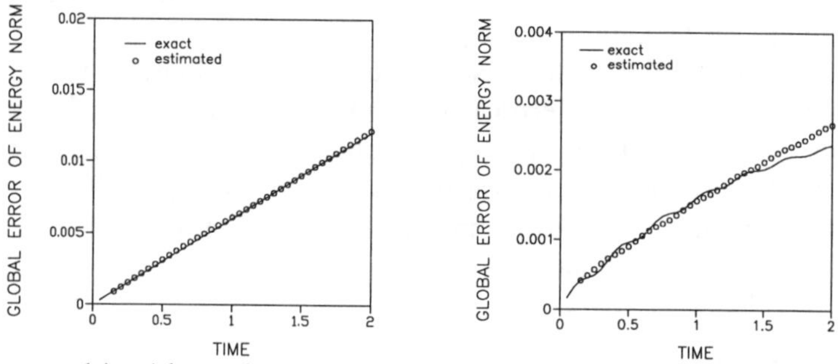

(a) without dmping (b) with damping

Figure 2. Global error for constant average acceleration method, $\Delta t = 0.05$

In order to find the optimal value of ε or δ, a parametric study on the problem used in the previous section is carried out with a various ratios of the spatial and time discretization error tolerance. In Fig. 5 the relationship between the total computing time and ε or δ is shown. In this study it can be observed that the reasonable value of ε or δ is about 0.5 in terms of computing time which is also the value used in the preceding example.

6. Conclusions

In this study the spatial discretization error and the temporal discretization error were estimated consistently and an effective algorithm which controls the errors automatically and simultaneously by the adaptive modification of the mesh distribution and time step size is proposed. In such a way, the best performance attainable by the finite element analysis of dynamic problems can be obtained.

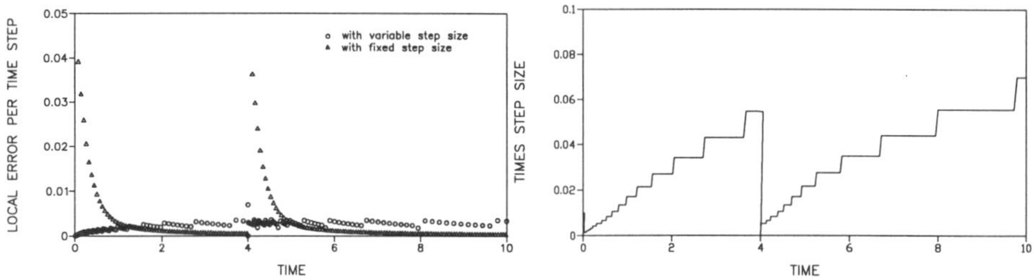

(a) Comparison of local error distribution (b) Variation of time step size

Figure 3. Adaptive analysis of one dimensional elastic bar

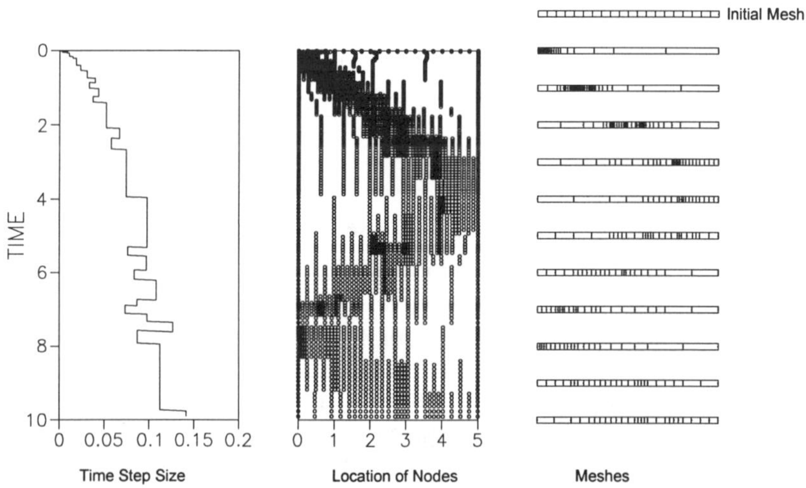

Figure 4. Variation of time step sizes and meshes in the bar

The temporal discretization error can be estimated simply by comparing the solution obtained by the Newmark method with solutions obtained by the locally exact quadratic function. The global and local error estimates and the adaptive control procedure proposed in this study was adopted directly to an adaptive control procedure of the spatial-temporal discretization error. This error estimate converges to the exact error as the size of time step is decreased. Since the error estimate by quadratic function is not affected by the specific time integration method used, the temporal discretization error estimate proposed in this study can be applied to the various single step method such as Wilson method. Based on the parametric studies a reasonable ratio of spatial and time discretization errors, which should be specified by the user in an adaptive analysis of dynamic problem is also proposed.

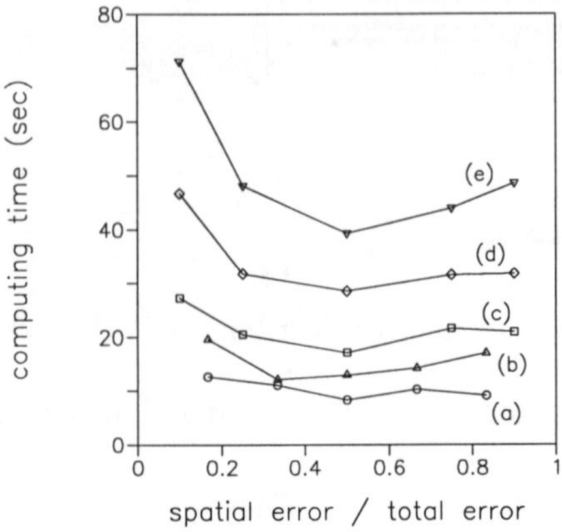

Figure 5. Comparison of computing time : (a) $\eta = 0.12$; (b) $\eta = 0.06$; (c) $\eta = 0.04$; (d) $\eta = 0.02$; (e) $\eta = 0.01$

7. References

1. C. K. Choi and H. J. Chung (1995): An adaptive control of spatial-temporal discretization error in finite element analysis of dynamic problems. *Structural Engineering and Mechanics* **3**.
2. Strang, G. and Fix, G. J. (1973): *An Analysis of the Finite Element Method*, Prentice-Hall, Englewood Cliffs, N. J.
3. Wiberg, N. E. and Li, X. D. (1993): A post-processing technique and an a posteriori error estimate for the Newmark method in dynamic analysis, *Earthquake eng. struct. dyn.*, 22, 465-489.
4. Zeng, L. F. and Wiberg, N. E. (1992): Spatial mesh adaptation in semidiscrete finite element analysis of linear elastodynamic problems , *Comput. mech.*, 9, 315-332.
5. Zienkiewicz, O. C. and Xie, Y. M. (1991): A simple error estimator and adaptive time stepping procedure for dynamic analysis, *Earthquake eng. struct. dyn.*, 20, 871-887.
6. Zienkiewicz, O. C. and Zhu, J. Z. (1987): A simple error estimator and adaptive procedures for practical engineering analysis, *Int. j. numer. methods eng.*, 24, 337-357.

NONLINEAR STRUCTURAL DYNAMIC ANALYSIS BY A DISCONTINUOUS GALERKIN FINITE ELEMENT METHOD

N.–E. WIBERG and **X. D. LI***
Department of Structural Mechanics, Chalmers University of Technology, S−412 96 Göteborg, Sweden
** Also at Division of Solid Mechanics, Department of Mechanical Engineering, Linköping Institute of Technology, S−581 83 Linköping, Sweden*

Abstract

This paper discusses computational aspects of the discontinuous Galerkin (DG) finite element method as applied to nonlinear structural dynamic problems, by which displacements and velocities are approximated as piecewise bilinear functions in space-time and may be discontinuous at the discrete time levels. Both implicit and explicit iterative algorithms for solving the resulting system of coupled equations are derived. An h-adaptive procedure based on the Zienkiewicz-Zhu error estimate using the SPR technique is described. Numerical examples are provided to show the suitability of the DG method for nonlinear structural dynamics.

1. Introduction

Over the past years, application of the discontinuous Galerkin (DG) finite element method to second-order hyperbolic equations has been investigated by several researchers, e.g. Hughes and Hulbert [4–7], Johnson [9] and French [3]. It is shown that the DG method often leads to A-stable, higher-order accurate time-stepping schemes and can introduce numerical dissipation for spurious high modes without change of the form of the original differential equations. Since the DG method is implemented over each space-time slab with the time step size as thickness and the continuity of displacements and velocities is enforced only in some weak form, in principle, different meshes can be used on different slabs to follow characteristics of the exact solution and, also, different time step sizes can

H. A. Mang and F. G. Rammerstorfer (eds.),
IUTAM Symposium on Discretization Methods in Structural Mechanics, 115–122.

be used in different domains to apply the so-called subcycling technique. This facility is very useful for the development of efficient and reliable adaptive finite element procedures for nonelliptic and nonlinear problems which, according to Babuska [1], is one of the two most outstanding problems in finite element analysis in 1990s.

In order to make the DG method economically competitive with those existing time integration schemes, the development of an efficient solution procedure is desired, because the DG method has inherent computational complexity. For example, when using displacements and velocities as independent unknowns and approximating them with piecewise bilinear function in space-time, a system of coupled equations which is four times larger than the system of semidiscrete structural dynamic equations is generated. In [11–13], an implicit iterative solution algorithm which involves only one factorization for each fixed time step size and a few iterations at each time step is developed for linear problems. The algorithm is unconditionally stable, of third-order accuracy and capable of filtering out the effects of spurious high modes. An h–adaptive procedure using the Zienkiewicz-Zhu error estimate in space and the jumps of displacements and velocities in the total energy norm as an error indicator in time is also presented.

In this paper we report some results of our effort to apply the DG method to nonlinear structural dynamics. Attention is paid to materially nonlinear problems in the context of classical incremental elastoplasticity theory for metals, though other nonlinear problems can be analyzed by use of the same lines of reasoning. An outline of the paper follows. In Section 2 we define the weak form of DG finite element formulation. In Section 3, we present two implementation algorithms, implicit and explicit, for solving the resulted system of equations. We also describe the properties regarding spectral radius, numerical dissipation and dispersion and the error measure used for adaptive analysis. Section 4 contains three different examples to demonstrate the performance of the DG method. Finally, conclusions are drawn in Section 5.

2. Weak Form of DG Method

As long as a space-time slab is discretized using conventional finite elements in space and one-layer uniform rectangular elements in time, the properties regarding spatial discretization will be the same as those based on semidiscretization and, consequently, we can start from the ordinary differential equations of the following two-field form

$$\boldsymbol{M}\dot{\boldsymbol{v}} + \boldsymbol{P}(\boldsymbol{d}) = \boldsymbol{F}(t) \tag{1}$$

$$\dot{\boldsymbol{d}} - \boldsymbol{v} = \boldsymbol{0} \tag{2}$$

where $\boldsymbol{d}$ and $\boldsymbol{v}$ are the vectors of nodal displacements and velocities, respectively, $\boldsymbol{M}$ is the mass matrix, $\boldsymbol{F}$ is the vector of applied external forces and $\boldsymbol{P}$ is the vector of internal resisting forces given as

$$\boldsymbol{P} = \int_{\Omega} \boldsymbol{B}^{\mathrm{T}} \boldsymbol{\sigma} \, d\Omega \tag{3}$$

We introduce the notations

$$\boldsymbol{w}_n^+ = \lim_{\epsilon \to 0^+} \boldsymbol{w}(t_n + \epsilon),\ \boldsymbol{w}_n^- = \lim_{\epsilon \to 0^-} \boldsymbol{w}(t_n - \epsilon)\ and\ [\boldsymbol{w}_n] = \boldsymbol{w}_n^+ - \boldsymbol{w}_n^- \tag{4a,b,c}$$

where $[\boldsymbol{w}_n]$ represents the jump of $\boldsymbol{w}_n$ at t_n.

The local weak form of DG finite element in a typical time step $I_n = (t_n, t_{n+1})$ can be expressed as

$$\int_{I_n} \boldsymbol{w}_v(\boldsymbol{M}\dot{\boldsymbol{v}} + \boldsymbol{P} - \boldsymbol{F})dt + \int_{I_n} \boldsymbol{w}_d\boldsymbol{K}(\dot{\boldsymbol{d}} - \boldsymbol{v})dt + \boldsymbol{w}_d(t_n)\boldsymbol{K}[\boldsymbol{d}(t_n)] + \boldsymbol{w}_v(t_n)\boldsymbol{M}[\boldsymbol{v}(t_n)] = \boldsymbol{0} \tag{5}$$

We notice that the terms in equation (5) have been arranged so as to correspond to energy expressions and displacements and velocities are required only to satisfy some weak continuity conditions through strain and kinetic energy inner products, respectively.

3. Implementation and Adaptivity

Let $\boldsymbol{d}^-$, $\boldsymbol{v}^-$ denote the vectors of displacements and velocities at t_n^-,which are known from either the previous step's calculation or, if $n = 1$, the initial data. And let $\boldsymbol{d}_1, \boldsymbol{v}_1$ and $\boldsymbol{d}_2$, $\boldsymbol{v}_2$ denote the vectors of displacements and velocities at t_n^+ and t_{n+1}^-, respectively, which are the four unknown vectors involved in this step's calculations. Thus, the vectors of displacements and velocities at an arbitrary time can be expressed as

$$\boldsymbol{d} = N_1\boldsymbol{d}_1 + N_2\boldsymbol{d}_2,\ \boldsymbol{v} = N_1\boldsymbol{v}_1 + N_2\boldsymbol{v}_2 \tag{6a,b}$$

respectively, in which

$$N_1 = \frac{t_{n+1} - t}{\Delta t},\ N_2 = \frac{t - t_n}{\Delta t} \tag{7a,b}$$

are shape functions in time. Substituting equations (6a,b) into equation (5), we obtain

$$\begin{cases} \boldsymbol{d}_1 = \boldsymbol{d}^- + \frac{1}{6}\Delta t\,(\boldsymbol{v}_1 - \boldsymbol{v}_2) \\ \boldsymbol{d}_2 = \boldsymbol{d}^- + \frac{1}{2}\Delta t\,(\boldsymbol{v}_1 + \boldsymbol{v}_2) \\ \boldsymbol{M}\boldsymbol{v}_1 + (\boldsymbol{P}_1 - \boldsymbol{P}_2) - (\boldsymbol{F}_1 - \boldsymbol{F}_2) - \boldsymbol{M}\boldsymbol{v}^- = \boldsymbol{0} \\ \boldsymbol{M}\boldsymbol{v}_2 + (\boldsymbol{P}_1 + \boldsymbol{P}_2) - (\boldsymbol{F}_1 + \boldsymbol{F}_2) - \boldsymbol{M}\boldsymbol{v}^- = \boldsymbol{0} \end{cases} \tag{8a–d}$$

in which

$$\boldsymbol{P}_1 = \int_{I_n} N_1\boldsymbol{P}dt,\ \boldsymbol{P}_2 = \int_{I_n} N_2\boldsymbol{P}dt,\ \boldsymbol{F}_1 = \int_{I_n} N_1\boldsymbol{F}dt,\ \boldsymbol{F}_2 = \int_{I_n} N_2\boldsymbol{F}dt \tag{9}$$

We mention here that, instead of $\boldsymbol{d}_1$ and $\boldsymbol{v}_1$, the jumps of displacements $[\boldsymbol{d}]$ and velocities $[\boldsymbol{v}]$ can also be used as unknowns and handled as a kind of hierarchical functions. In this way, the DG method discussed here may be regarded as a special case of hierarchical formulation in time as presented by Wiberg and Möller [18] for first-order equations.

3.1 SOLUTION ALGORITHMS

We apply the Newton-Raphson process to equations (8c–d) in terms of $\boldsymbol{v}_1$ and $\boldsymbol{v}_2$, which leads to

$$\begin{Bmatrix} \boldsymbol{r}_1^i \\ \boldsymbol{r}_2^i \end{Bmatrix} + \boldsymbol{J} \cdot \begin{Bmatrix} \Delta \boldsymbol{v}_1^i \\ \Delta \boldsymbol{v}_2^i \end{Bmatrix} = \boldsymbol{0} \tag{10}$$

in which $\Delta \boldsymbol{v}_1^i$ and $\Delta \boldsymbol{v}_2^i$ are the iterative corrections to $\boldsymbol{v}_1^i$ and $\boldsymbol{v}_2^i$, respectively, $\boldsymbol{r}_1^i$ and $\boldsymbol{r}_2^i$ are the residuals when $\boldsymbol{v}_1^i$, $\boldsymbol{v}_2^i$ and $\boldsymbol{d}_1^i$, $\boldsymbol{d}_2^i$ are substituted into equations (8c–d), respectively, and $\boldsymbol{J}$ is the Jacobian matrix. 'Exact' calculation for $\boldsymbol{J}$ will be too expansive and, thus, some modifications are made, which leads to different solution algorithms.

Implicit. An implicit iterative algorithm may be obtained by recasting equation (10) as the following form

$$\begin{bmatrix} \boldsymbol{M} + \frac{1}{6}\Delta t^2 \boldsymbol{K} & \frac{2}{3}\boldsymbol{M} \\ \frac{1}{3}\Delta t^2 \boldsymbol{K} & \boldsymbol{M} + \frac{1}{6}\Delta t^2 \boldsymbol{K} \end{bmatrix} \begin{bmatrix} \Delta \boldsymbol{v}_1^i \\ \Delta \boldsymbol{v}_2^i \end{bmatrix} = - \begin{bmatrix} \frac{2}{3}\boldsymbol{r}_2^i + \boldsymbol{r}_1^i \\ \boldsymbol{r}_2^i \end{bmatrix} \tag{11}$$

This algorithm is related the one we used for solving linear problems in [11–13]. It involves only one factorization and is unconditionally stable.

Explicit. If keeping only the contributions of masses in the $\boldsymbol{J}$ matrix, we get an explicit algorithm of the form

$$\begin{bmatrix} \boldsymbol{M} & \boldsymbol{0} \\ \boldsymbol{0} & \boldsymbol{M} \end{bmatrix} \begin{bmatrix} \Delta \boldsymbol{v}_1^i \\ \Delta \boldsymbol{v}_2^i \end{bmatrix} = - \begin{bmatrix} \boldsymbol{r}_1^i \\ \boldsymbol{r}_2^i \end{bmatrix} \tag{12}$$

However, this algorithm is only conditionally stable.

In Figures 1(a)–1(c), the spectral radius, the algorithmic damping ratio and the relative period error of the DG method are presented and compared with those commonly used time-stepping schemes. The results demonstrate the superior numerical dissipation and dispersion properties of the DG method.

3.2 ADAPTIVITY

For linear dynamic problems, we have used the Zienkiewicz-Zhu error estimate [19,20] in the following total energy norm

$$\|\bar{\boldsymbol{e}}_s\|_E = \left[\int_\Omega (\boldsymbol{\sigma}^* - \boldsymbol{\sigma}^h)^T \boldsymbol{D}^{-1} (\boldsymbol{\sigma}^* - \boldsymbol{\sigma}^h) d\Omega + \int_\Omega (\boldsymbol{v}^* - \boldsymbol{v}^h)^T \rho (\boldsymbol{v}^* - \boldsymbol{v}^h) d\Omega \right]^{1/2} \tag{13}$$

In above, the term of kinetic energy may normally be discarded, as the velocities have often a higher-order convergence rate than the stresses.

For problems of plasticity, the Zienkiewicz-Zhu error estimate is still useful. But so far the applications listed in literature are few [2, 14, 15]. In this paper, we shall test the L_2 norm of total strain errors

$$\|\bar{e}_s\|_\epsilon = \left[\int_\Omega (\epsilon^* - \epsilon^h)^T(\epsilon^* - \epsilon^h) d\Omega \right]^{1/2} \tag{14}$$

In time, we have used the jumps of displacements $[\boldsymbol{d}_n]$ and velocities $[\boldsymbol{v}_n]$ in the total energy norm as a local error indicator for the implicit algorithm. When an explicit algorithm is used, the time step size is primarily determined by the critical stability property. It may be possible to use different step sizes in different spatial domains. However, this could lead to lack of convergence as shown by Klisinki and Moström [10].

Based on the error estimate, we update the current spatial mesh (and the time step size) according to the modification strategy we described previously in [13, 16].

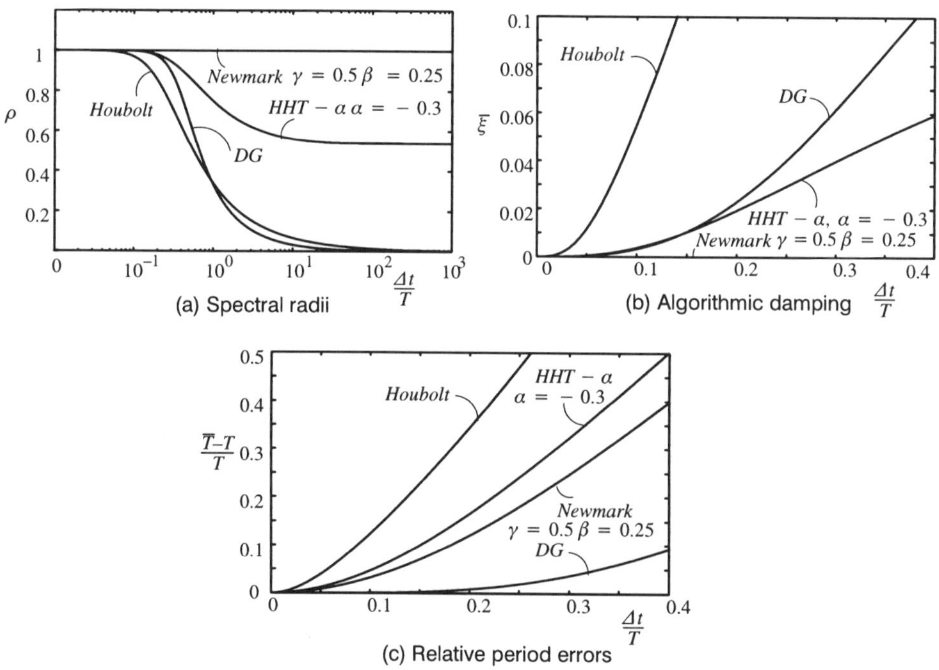

Figure 1. *Properties of DG, Newmark, HHT–α and Houblot methods.*

4. Numerical Examples

Example 1: A SDOF model. We consider a modified Duffing equation of the form

$$\ddot{d} + d^3 = -\cos(t) + \cos^3(t) \tag{15}$$

with $d_0 = 1$ and $v_0 = 0$. The exact solution is thus $d = \cos(t)$. The explicit DG method is tested and compared with the conventional central difference method. Plots of the history of displacements and the rate of convergence are given in Figure 2. It shows that the DG method considered here is of third-order accuracy.

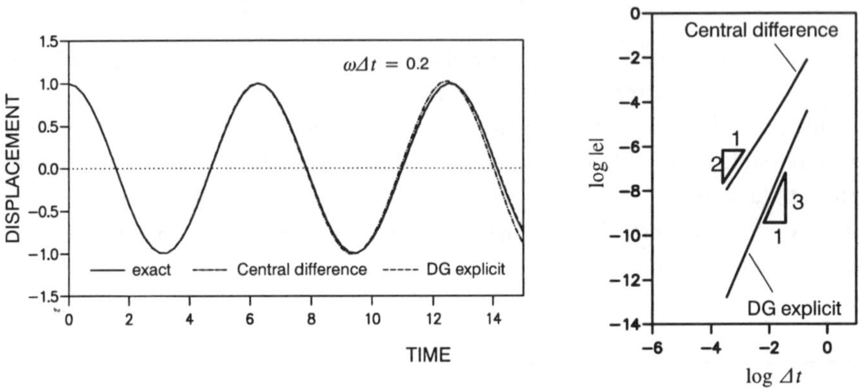

Figure 2. *History of displacements and rate of convergence at t=10s for the nonlinear SDOF model.*

Example 2: Uniformly Loaded Beam. Application of the DG method to two-dimensional elastoplastic dynamic problems is demonstrated by this example. The geometry of the problem and the finite element mesh used are illustrated in Figure 3. The history of the vertical displacement at point A is plotted in Figure 4.

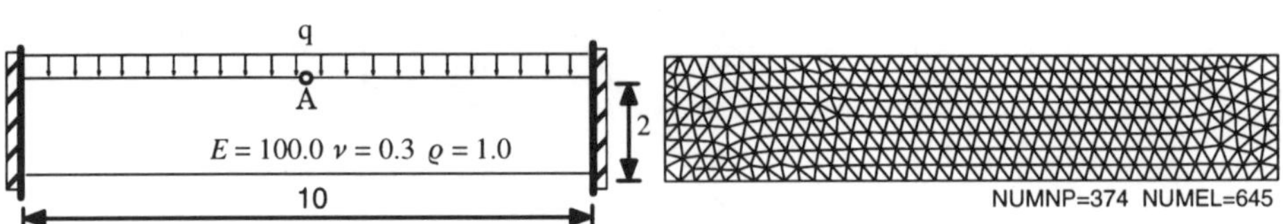

Finite element mesh

Figure 3. *Dynamic elastoplastic bending of a uniformly loaded rectangular beam*

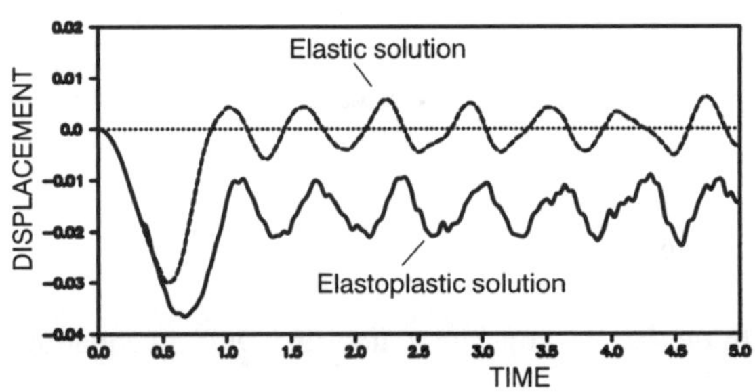

Figure 4. *Displacement response at point A for the uniformly loaded beam example*

Example 3: Waves in Elastoplastic Bar, Adaptivity. Finally, we consider the problem of wave propagation and reflection in a bar of bilinear material. As shown in Figure 5, the bar travels freely to left and impinges upon a fixed rigid wall. Then, an elastic and a plastic shock front propagate from the impact at different speeds and meet each other when the elastic front reflects from the free end. An adaptive analysis is performed and meshes predicted for some specific time stations are plotted in Figures 6. We can see the moving of the plastic wave front from the figure.

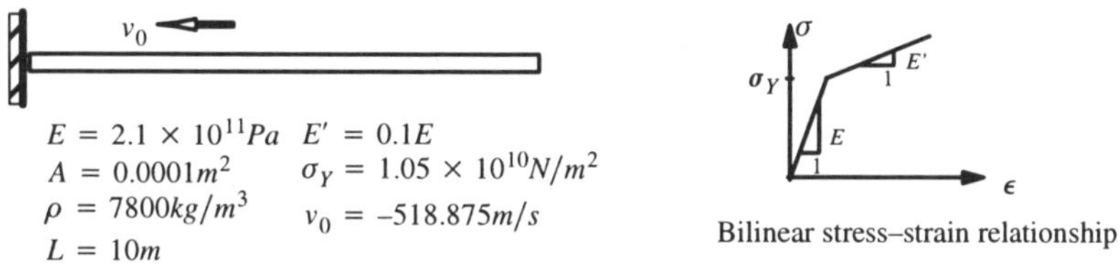

Figure 5. *Impact of a finite bar of bilinear material*

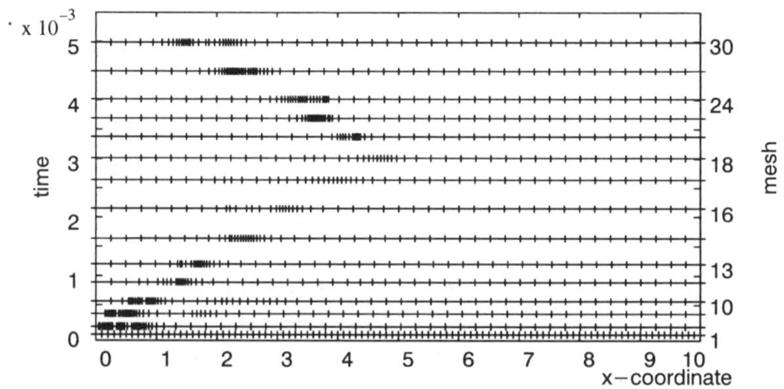

Figure 6. *Meshes used in the adaptive analysis for the impact problem of a bar of bilinear material.*

5. Concluding Remarks

Computational aspects of the DG finite element method that approximates both displacements and velocities as piecewise linear functions in space-time and permits them being discontinuous at the discrete time levels have been studied. Two iterative solution procedures, an implicit one and an explicit one, have been presented. An *h*-adaptive procedure based on the Zienkiewicz–Zhu error estimate has been tested for one-dimensional elasto-plastic problems. Based on our numerical experiments, we believe that the DG method is well suited to both linear and nonlinear dynamic analysis. Applications of the DG method to other types of nonlinear dynamic problems such as contact-impact and cracking and, especially, the related space-time adaptive procedures will be studied in the future.

6. References

1. Babuska, I.: Courant element: before and after, in Krizek, M., Neittaanmäki, P. and Stenberg, R. (eds.), *Finite Element Methods, fifty years of the Courant element*, 37–52, (1993).
2. Belytschko, T. and Tabbara, M.: H-adaptive finite element methods for dynamic problems, with emphasis on localization, *Int. j. numer. methods eng.*, **36**, 4245–4265, (1993).
3. French, D.A.: A space–time finite element method for the wave equation, *Comput. Methods Appl. Mech. Engrg.* **107**, 145–157, (1993).
4. Hughes, T.J.R. and Hulbert, G.: Space–time finite element methods for elastodynamics: Formulations and error estimates, *Comput. Meth. Appl. Mech. Engrg.* **66**, 339–363, (1988).
5. Hulbert G. and Hughes, T.J.R.: Space–time finite element methods for second order hyperbolic equations, *Comput. Methods Appl. Mech. Engrg.* **84**, 327–348, (1990).
6. Hulbert, G.: *Space–time finite element methods for second-order hyperbolic equations*, Ph.D. Thesis, Department of Mechanical Engineering, Stanford University, 1989.
7. Hulbert, G.: Time finite element methods for structural dynamics, *Int. j. numer. methods eng.* **33**, 307–331, (1992).
9. Johnson, C: Discontinuous Galerkin finite element methods for second order hyperbolic problems, *Comput. Methods Appl. Mech. Engrg.* **107**, 117–129, (1993).
10. Klisinki, M. and Moström, A.: On stability of multi-time step integration procedures, (submitted to *J. of Engineering Mechanics*) (1995).
11. Li, X.D.: *Adaptive finite element procedures in structural dynamics,* Ph.D. Thesis, Publ. 96:2, Dept. of Structural Mechanics, Chalmers University of Technology, Göteborg, Sweden, 1996.
12. Li, X.D. and Wiberg, N.–E.: Structural dynamic analysis by a time–discontinuous Galerkin finite element method, *Int. j. numer. methods eng.*, **39**, 2131–2152, (1996).
13. Li, X.D. and Wiberg, N.-E.: Implementation and adaptivity of a space-time finite element method in structural dynamics, (1997), (to appear in *Comput. Meth. Appl. Mech. Engrg.*).
14. Peric, D., Yu, J. and Owen, D.R.J.: On error estimates and adaptivity in elasto-plastic solids: application to the numerical simulation of strain localization in classical and Cosserat continua, *Int. j. numer. methods eng.*, **37**, 1351–1379, (1994).
15. D.R.J. Owen, E. Oñate and E. Hinton (Eds.): *Computational Plasticity, Fundamentals and Applications*, CIMNE, Barcelona, 1997.
16. Wiberg, N.–E. and Li, X.D.: A postprocessed error estimate and an adaptive procedure for the semidiscrete finite element method in dynamic analysis, *Int. j. numer. methods eng.*, **37**, 3583 –3603, (1994).
18. Wiberg, N.–E. and Möller, P.: Formulation and solution of hierarchical finite element equations, *Int. j. numer. methods eng.*, **37**, 1213–1233, (1988).
19. Zienkiewicz, O.C. and Zhu, J.Z.: A simple error estimator and adaptive procedure for practical engineering analysis, *Int j. numer. methods eng.*, **24**, 333–357, (1987).
20. Zienkiewicz, O.C. and Zhu, J.Z.: The superconvergent patch recovery and *a posteriori* error estimates, Part 1: The recovery technique; Part 2: Error estimates and adaptivity, *Int. j. numer. methods eng.*, **33**, 1331– 1382, (1992).

SEISMIC RESPONSE OF CONCRETE GRAVITY DAMS – A CONTINUUM DAMAGE MECHANICS APPROACH

S. VALLIAPPAN

Professor

M. YAZDCHI

Research Associate

School of Civil and Environmental Engineering,

The University of New South Wales, Sydney 2052, Australia

Abstract– In this paper, the seismic response of concrete gravity dams is presented using the concept of Continuum Damage Mechanics (CDM) and adopting the hybrid Finite Element–Boundary Element technique. The finite element is used for discretization of the near field and the boundary element method is employed to model the semi–infinite far field. The proposed method is shown to be computationally efficient and consistent in its treatment of damage evolution in gravity dams.

1. Introduction

Earthquake resistant design of dams is particularly important in view of catastrophic consequences such as loss of life and property, if the dam fails. The numerical simulation of concrete dams subjected to earthquakes involves such complexities as nonlinear structural behaviour as a result of cracking, effect of reservoir–dam–foundation interaction and realistic representation of input excitation.

For a comprehensive analysis, simple but realistic material laws for mass concrete, which can be easily implemented, are required. Such material laws have to take into consideration conditions which typically occur under seismic action. In problems where tensile cracking plays a significant role on the behaviour of the structure, the normal approach is to adopt fracture mechanics (discrete or smeared crack modelling) and plasticity theory combined with tensile strength criteria. Among the few researchers who investigated the response of dams under seismic action using a fracture mechanics model, Chapuis et al. (1985) developed a finite element approach based on LEFM combined with a discrete crack model to analyze the crack propagation of Pine Flat dam. Ayari and Saouma (1990) presented a LEFM approach for the structural response of dams under seismic action. Among the smeared crack procedures, the work reported by Bhattacharjee and Leger (1993) employs nonlinear fracture mechanics to study seismic cracking of concrete dams. But none of these nonlinear analyses have included the effect of dam–foundation interaction with proper

H. A. Mang and F. G. Rammerstorfer (eds.),
IUTAM Symposium on Discretization Methods in Structural Mechanics, 123–130.

modelling of semi–infinite medium. In spite of the success of these approaches in solving numerous problems which take into account the stress singularity at the crack tip and the inherent nature of concrete cracking, their usefulness is limited due to the factors such as the need for the use of quite an arbitrary shear retention factor to ensure some shear resistance along the crack, and the difficulties of defining stress paths following the opening and closing of cracks under cyclic loading conditions. Therefore, the application of fracture mechanics is limited to problems where only a few well defined fractures are encountered. For large scale problems such as concrete gravity dams where extensive micro–cracking may develop, it would be inefficient, especially in dynamic analysis.

Some of these limitations could be overcome if a single constitutive model could be used that simulates the nonlinear behaviour of concrete. When the crack profiles are not known a *priori,* the continuum damage mechanics approaches are computationally very attractive.

It is essential that the numerical simulation of a structure subjected to earthquakes should include the structure–foundation interaction effects. The FEM has shortcomings for wave propagation problems due to wave reflections at the boundaries of FE meshes when the semi–infinite domain of the foundation is also included. The BEM on the other hand, is capable of taking into account wave radiation to infinity and is therefore suitable to model the foundation medium. A significant limitation of this approach is, however, that only linear elastic homogeneous domain can be treated efficiently. In view of these problems, a combined BEM and FEM approach may be advantageous.

It is the purpose of this paper to model the nonlinear behaviour of concrete dams, including the interaction of dam–foundation with hybrid FE–BE method within the framework of continuum damage mechanics. The structure and a small portion of the supporting foundation (near field) is discretizied by finite elements, while the remaining foundation medium (far field) is represented by boundary elements. The selected time marching algorithm is based on its ability to dissipate high frequency noises resulting from sudden changes in the stiffness due to closing/opening of the cracks. For this purpose the HHT$-\alpha$ algorithm of Hilber, Hughes and Taylor (1977) was found to be particularly efficient.

2. Equations of Dynamic Equilibrium for Damaged Materials

Figure 1 shows a concrete gravity dam along with the applied loads. Two regions (dam and foundation) are to be analyzed by FEM and BEM respectively. Hence two sets of equations have to formulated and combined together.

2.1. FINITE ELEMENT FORMULATION

The dynamic equilibrium equations under seismic excitation including the damage can be written as

$$[M]\{\ddot{U}\} + [C]\{\dot{U}\} + \{P^*(\{U\})\} = \{F_{st}\} + \{F_{eq}\} = \{R^{ext}\} \qquad (1)$$

where $[M]$ is the mass matrix of the system, $[C]$ is the damping matrix. $\{P^*(\{U\})\}$ represents the vector of restoring forces, $\{F_{st}\}$ is the static (preseismic) load vector including self–weight of dam and hydrostatic pressure, and $\{F_{eq}\}$ is the earthquake load vector due to uniform free–field ground acceleration in both horizontal and vertical directions. Also $\{U\}$ is the vector of nodal displacements relative to the free–field input ground motions, and a superdot indicates the time derivative.

The hydrodynamic pressure due to dam–reservoir interaction, is estimated using the Westergaard added mass technique. This gives a fair approximation of hydrodynamic pressure for many practical problems.

2.2. BOUNDARY ELEMENT FORMULATION

At time t, the displacement at point Q can be obtained using dynamic version of Betti's reciprocal theorem (Banerjee et al., 1992), in an integral form as

$$\begin{aligned} c_{ij}(Q)u_i(Q,t) &= \int_{\Gamma}\int_0^t [U_{ij}(P,t;Q,\tau)s_i(P,\tau) - S_{ij}(P,t;Q,\tau)u_i(P,\tau)]d\tau d\Gamma(P) \\ &+ \varrho\int_V\int_o^t U_{ij}(P,t;Q,\tau)b_i(P)d\tau dV(P) \\ &+ \varrho\int_V\left[\bar{\dot{u}}_{i0}(P,0)U_{ij}(P,t;Q,0) + \bar{u}_{i0}(P,0)\dot{U}_{ij}(P,t;Q,0)\right]dV(P) \end{aligned} \quad (2)$$

The coefficient tensor c_{ij} is discontinuity or jump term and its value is equal to δ_{ij} for Q within the domain and $0.5\delta_{ij}$ for Q in the smooth boundary, where δ_{ij} is Kronecker delta. For the general case, it is obtained using the rigid body motion concept. u_i and s_i are the displacement and traction vectors respectively, P is field point whereas Q is source point. b_i is the body force vector per unit mass. Also U_{ij} and S_{ij} are the so–called Stokes' fundamental solutions. The third term in the above equation represents the effect of initial conditions of $\bar{u}_{i0}$ and $\bar{\dot{u}}_{i0}$ in the formulation. Numerical implementation of the above boundary integral equation requires approximations in both temporal and spatial variations of the field variables. The time domain is divided into equal intervals Δt. Both traction and displacement variables are assumed to be linear within that time interval. With respect to space, both traction and displacement fields are assumed to vary quadratically within each element. Finally, by evaluation of the integral terms over each boundary segment (numerically) and over each time interval (analytically) and by collocation at each boundary node and at all time steps $t_N = N\Delta t$ the following discretizied equation analogous to Eq. (1) is obtained

$$\sum_{n=1}^{N}\left(\left[U_1^{N-n+1} + U_2^{N-n}\right]\left\{s^{B(n)}\right\} - \left[S_1^{N-n+1} + S_2^{N-n}\right]\left\{u^{B(n)}\right\}\right) = \{0\} \quad (3)$$

These system of equations can be written in compact form as

$$\left[U^1\right]\left\{s^{B(N)}\right\} = \left[S^1\right]\left\{u^{B(N)}\right\} + \left\{r^{N-1}\right\} \quad (4)$$

where $u^{B(N)}$ and $s^{B(N)}$ are displacement and traction vectors, respectively, at the boundary nodes at time $N\Delta t$, and the vector $\{r^{N-1}\}$ physically represents the influence of all previous time steps $(n < N)$ on the current time node, N. Also U^n and S^n are so–called influence matrices of the n–th time step.

2.3. FEM–BEM COUPLING PROCEDURE

The following formulations are restricted to subdivision of bodies into only one BE and one FE subregion, according to Figure 1,

Finite element system: Separating the nodal points at the interface of FE–BE (subscript i) from the other nodal points of FE system (subscript o) and modifying Eq. (1) according to α–method, equation of motion for the N–th time step is expressed

$$[M]^F\{\ddot{U}\}^{F(N)} + (1+\alpha)[C]^{F(N)}\{\dot{U}\}^{F(N)} - \alpha[C]^{F(N-1)}\{\dot{U}\}^{F(N-1)} + (1+\alpha)[K]^{F(N)}\{U\}^{F(N)} - \alpha[K]^{F(N-1)}\{U\}^{F(N-1)} = (1+\alpha)\{P\}^{F(N)} - \alpha\{P\}^{F(N-1)} \quad (5)$$

where all matrices and vectors are partitioned according to interface and outer nodal degrees of freedoms. In the above equation, $[K]$ is the stiffness matrix of the system.

Boundary element system: To distinguish between the nodal points on the outer boundaries ($\Gamma_o{}^B$, subscript o) and on the interface ($\Gamma_i{}^B$, subscript i), the system of equations, encompassing the subregion Ω^B, have to be rearranged. Also, to obtain consistency between the FE and BE formulations, the interface tractions $s_i{}^B$ have to be transformed to resultant nodal forces $P_i{}^B$, by a transformation matrix.

In the case of a mixed boundary value problem, e.g. when $s_{o1}^B = \bar{s}_1^B$ is known along $\Gamma_{o1}{}^B$ and $u_{o2}^B = \bar{u}_2^B$ is prescribed along $\Gamma_{o2}{}^B$, separation of boundary values and its reordering with respect to knowns and unknowns and transforming of boundary tractions to equivalent loads yields

$$\begin{Bmatrix} \bar{s}_1^{B(N)} \\ \bar{u}_2^{B(N)} \\ P_i^{B(N)} \end{Bmatrix} = \begin{bmatrix} Q_{11} & Q_{12} & Q_{1i} \\ Q_{21} & Q_{22} & Q_{2i} \\ Q_{i1} & Q_{i2} & Q_{ii} \end{bmatrix} \begin{Bmatrix} \bar{u}_1^{B(N)} \\ \bar{s}_2^{B(N)} \\ u_i^{B(N)} \end{Bmatrix} + \begin{Bmatrix} R_1^{N-1} \\ R_2^{N-1} \\ R_i^{N-1} \end{Bmatrix} \quad (6)$$

In order to generate the dynamic interaction between the elastic structure Ω^F and the half–plane Ω^B, the systems (5) and (6) have to be combined by imposing the compatibility and equilibrium conditions at the interface. First, the equilibrium condition is used to reduce the number of unknowns by eliminating the unknown nodal forces along Γ_i. By separating $\{P_i{}^{F(N)}\}$ from the FE system (5), and $\{P_i{}^{B(N)}\}$ from the BE relation (6), this interface condition can be satisfied. Finally by incorporating the compatibility condition, and after some manipulations the stiffness matrix and load vector in Eq. (5) are modified to include the interface compatibility as given below:

$$[K^B] = [Q_{ii}] - \begin{bmatrix} Q_{i1} & Q_{i2} \end{bmatrix} \begin{bmatrix} Q_{11} & Q_{12} \\ Q_{21} & Q_{22} \end{bmatrix}^{-1} \begin{Bmatrix} Q_{1i} \\ Q_{2i} \end{Bmatrix}$$

$$\{P^{B(N)}\} = -\{R_i^{N-1}\} - \begin{bmatrix} Q_{i1} & Q_{i2} \end{bmatrix} \begin{bmatrix} Q_{11} & Q_{12} \\ Q_{21} & Q_{22} \end{bmatrix}^{-1} \begin{Bmatrix} \bar{s}_1^{B(N)} - R_1^{N-1} \\ \bar{u}_2^{B(N)} - R_2^{N-1} \end{Bmatrix} \quad (7)$$

$$\{P^{B(N-1)}\} = -\{R_i^{N-2}\} - \begin{bmatrix} Q_{i1} & Q_{i2} \end{bmatrix} \begin{bmatrix} Q_{11} & Q_{12} \\ Q_{21} & Q_{22} \end{bmatrix}^{-1} \begin{Bmatrix} \bar{s}_1^{B(N-1)} - R_1^{N-2} \\ \bar{u}_2^{B(N-1)} - R_2^{N-2} \end{Bmatrix}$$

The solution of this equation is obtained using the modified Newton–Raphson approach at each time step.

3. Damage Model and Constitutive Law for Anisotropic Damage State

The constitutive relationship in the principal anisotropic coordinate system is presented as

$$\{\tilde{\sigma}\} = \left[\tilde{E}^{*}\right]\{\tilde{\varepsilon}\} \tag{8}$$

The relationship between Cauchy and net stresses can be written as (Valliappan et al., 1990)

$$\{\sigma^{*}\} = [\Phi^{*}]\{\sigma\} \tag{9}$$

The detailed formulation for constitutive law can be found in Yazdchi et al. (1996).

4. Material Model and Damage Evolution

Based on the hypothesis of strain energy equivalence, the isotropic damage parameters can be defined in terms of Young's modulus (Valliappan et al., 1990) as

$$D = 1 - \sqrt{\frac{E^{*}}{E}} \tag{10}$$

Hence, from Figure 2, the proper definition for damage variable under uniaxial conditions is

$$D = 1 - \sqrt{\left(\frac{\varepsilon_0}{\varepsilon}\right)\left[2e^{-a(\varepsilon-\varepsilon_0)} - e^{-2a(\varepsilon-\varepsilon_0)}\right]} \tag{11}$$

where the constant a defined as (Ghrib and Tinawi, 1995)

$$a = \frac{3}{\varepsilon_0\left[\frac{2EG_f}{l_{ch}f'^{2}_{t}} - 1\right]} \geq 0 \tag{12}$$

where the characteristic length l_{ch} is a material property and is equal to the length of the element normal to the cracks. The parameter l_{ch} plays the role of relating the specific energy per unit volume (area under stress–strain curve) and the fracture energy per unit area, G_f. Earthquake loading being cyclic, when the sign of the stresses change from tension to compression (unloading), cracks are closed. In the unloading regime, concrete can not recover all of its deformation. According to Figure 2, some part of the maximum strain, is inelastic and permanent. By introducing a fraction δ of the maximum principal strain, the total strain is divided into two parts: recoverable elastic ε^{e} and inelastic strain ε^{in}. The value of δ=0.2 provides predictions that reflect all essential experimental features (Dahlblom and Ottosen ,1990). When the compressive stress develops, the material gains its original stiffness. Reloading of the crack follows the unloading path until the principal strain is greater than ε_{max}. Also when the strain is less than ε^{in} it is supposed that the crack is closed. Then the unloading–reloading modulus is equal to (Figure 2)

$$E_{unl} = E\frac{(1 - D)^2}{(1 - \delta)} \tag{13}$$

5. Discussion of Results

Because of the computational limitations, a small dam similar to Koyna dam is chosen for the analysis. The dam is 15.0 m high, with a width of 10.0 m at the base and 2.0 m at the top. The dam and its foundation are discretizied with four node isoparametric finite elements together with quadratic boundary elements. The material properties of the dam are as follows: E=30.0×10^6 kPa, ν=0.20, ϱ=2600.0 kg/m^3, f_t'=1.5×10^3 kPa, G_f=150.0 N/m, whereas stiffness of foundation varies from 0.5 to 4 times of dam's stiffness. (cases (a) to (d) respectively, in Figures 3 and 4).

For including the effect of high strain rate due to earthquake, the tensile strength and the fracture energy are increased by 20% in the computations. The viscous damping has been used in the finite element model with damping ratio of ς=0.05. Final damage patterns are shown in Figure 4. It can be seen that when the stiffness of the foundation is very high compared to that of the dam, a stress concentration induces more crack propagation at the base of the dam. During the upstream movement of the dam, damage is localized in a zone that almost bridges across the 'neck' of the dam in downstream (Figure 4). These damage patterns can be followed from Figure 3 that shows the history of global damage index.

6. Conclusion

A computational method is presented for the nonlinear seismic response of concrete gravity dams using continuum damage mechanics. The methodology has shown that the local failure can be efficiently modelled by continuum damage mechanics. Because concrete is an elastic–strain softening brittle material under seismic action, the material model presented here is appropriate. The model is capable of describing the rate–dependent nonlinear behaviour of the brittle materials. Damage accumulation is reflected by the continuous degradation of material stiffness during the loading process. Moreover, this model allows for stiffness recovery for seismic analysis of structures. To include the effect of the dam–foundation interaction the hybrid FE–BE method has been employed, and the results show the importance of including foundation in the calculations.

7. References

Ayari, M.L. and Saouma, V.E. (1990) A Fracture Mechanics Based Seismic Analysis of Concrete Gravity Dams Using Discrete Cracks, *Engineering Fracture Mechanics* 35, 587–598.

Banerjee, P.K., Israil, A.S.M. and Wang. H.C. (1992) Time Domain Formulations of BEM for Two–Dimensional, Axissymmetric and Three–Dimensional Transient Elastodynamics, in P.K. Banerjee (eds.), *Developments in Boundary Element Methods,* Elsevier Applied Science 7, 115–154.

Bhattacharjee, S.S. and Leger, P. (1993) Seismic Cracking and Energy Dissipation in concrete Gravity Dams, *Earthquake Engineering and Structural dynamics* 22, 991–1007.

Chapuis, J., Rebora, B. and Zimmermann, T. (1985) Numerical Approach of Crack Propagation Analysis in Gravity Dams During Earthquakes, *Proceedings of Fifteenth ICOLD Congress, Q. 57, R. 26,* Lausanne, 451–474.

Dahlblom, O. and Ottosen, N.S. (1990) Smeared Crack Analysis Using Generalized Fictitious Model, *Journal of Engineering mechanics, ASCE* 116, 55–76.

Ghrib, F. and Tinawi, R. (1995) An Application of Damage Mechanics for Seismic Analysis of Concrete Gravity Dams, *Earthquake Engineering and Structural Dynamics* 24, 157–173.

Valliappan, S., Murti V. and Zhang, W. (1990) Finite Element Analysis of Anisotropic Damage Mechanics Problems, *Engineering Fracture Mechanics* 35(6), 1061–1071.

Yazdchi, M., Valliappan, S. and Zhang, W. (1996) A Continuum Model for Dynamic Damage Evolution of Anisotropic Brittle Materials, *International Journal of Numerical Methods in Engineering* 39, 1555–1583.

Hilber, H.M., Hughes, T.J.R. and Taylor, R.L. (1977) Improved Numerical Dissipation for Time Integration Algorithms in structural Dynamics, *Earthquake Engineering and Structural Dynamics* 5, 283–292.

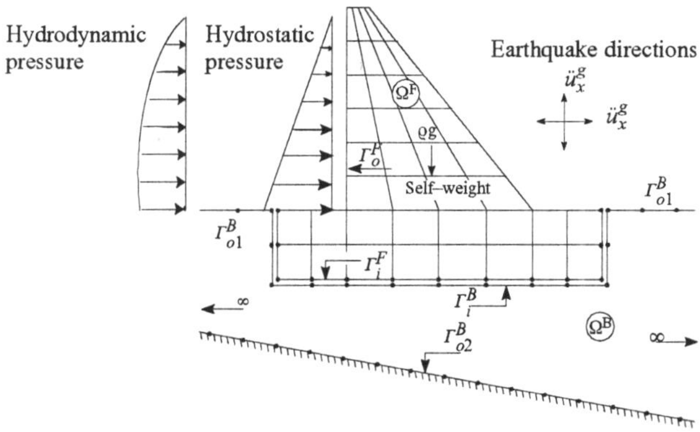

Figure 1. Concrete gravity dam–foundation system

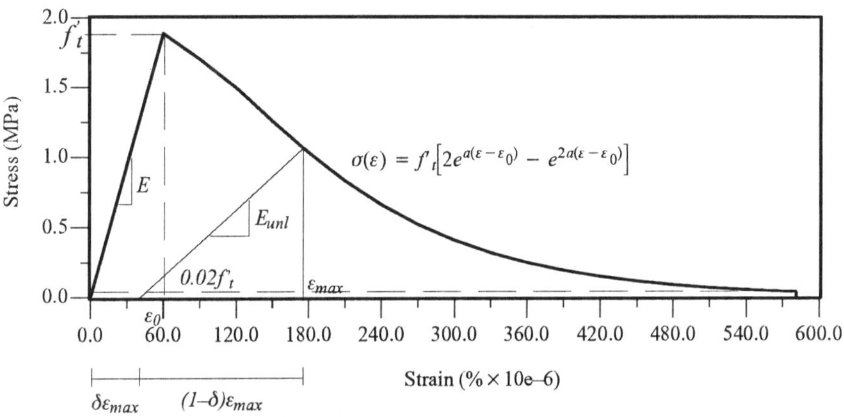

Figure 2. Stress–strain curve for mass concrete with exponential strain softening model and closing/ opening criterion

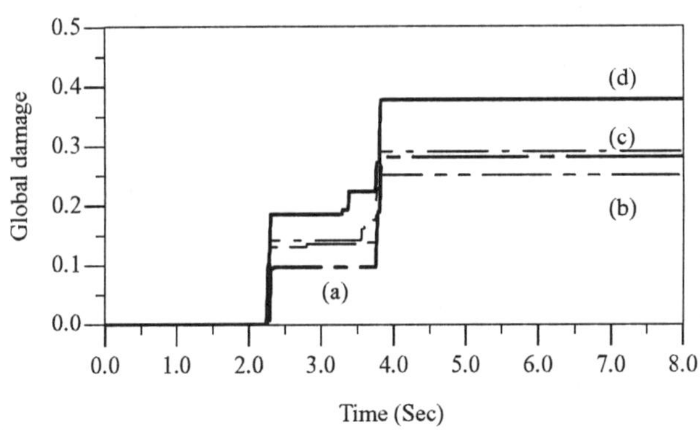

Figure 3. Global damage index for different stiffnesses of the foundation

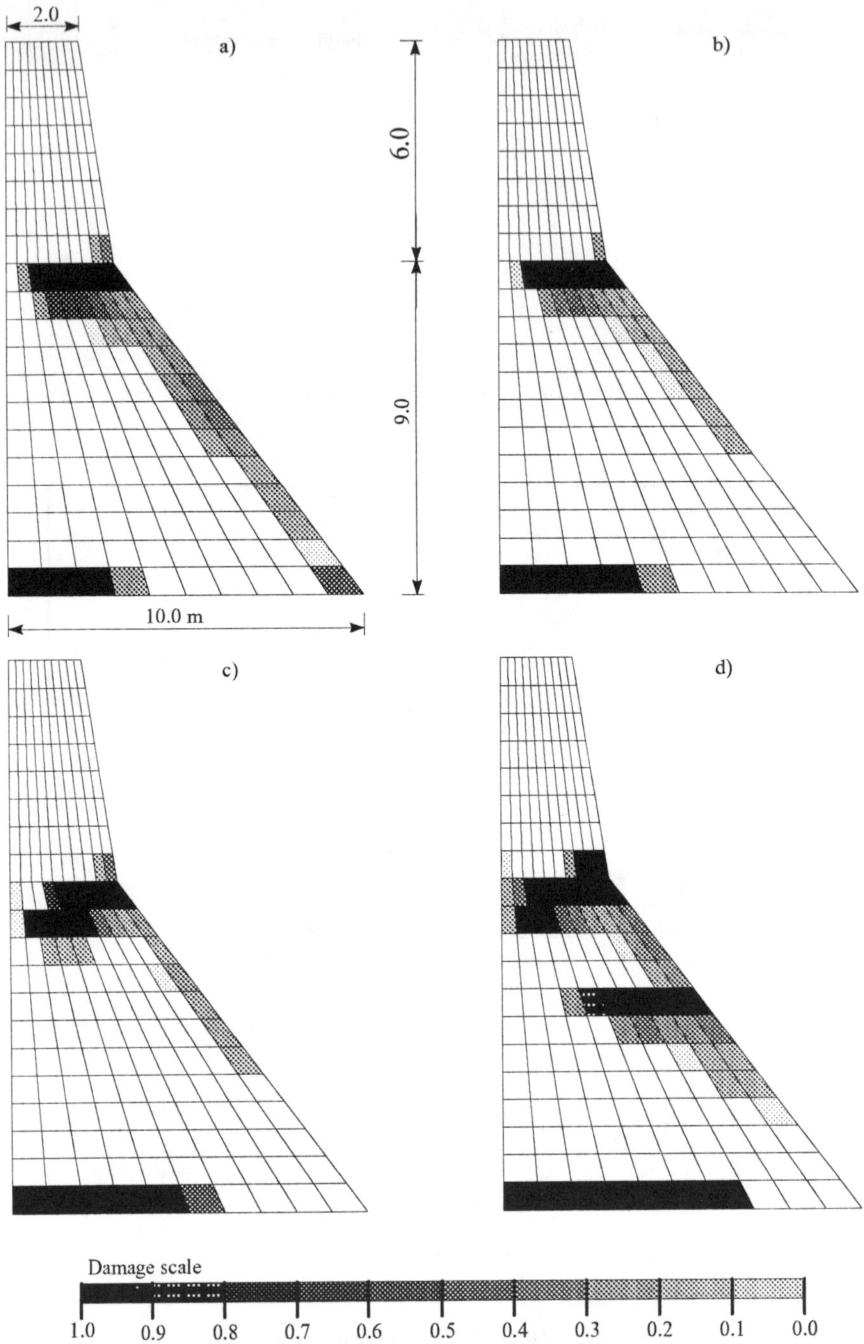

Figure 4. Final damage paterrn of the dam under different foundation rigidity

ON NON-LINEAR STATIC AND DYNAMIC THIN SHELL ANALYSIS

FRANO B. DAMJANIĆ †
Institute of Structural and Earthquake Engineering,
University of Ljubljana, Jamova 2, 1000 Ljubljana, Slovenia
&
Department of Civil Engineering, University of Split,
M.Hrvatske 15, 21000 Split, Croatia

Abstract: Some aspects of continuum formulation for the geometrically non-linear dynamics of elastic thin shells and its temporal discretisation are very briefly presented (At the moment, an elasto-plastic constitutive model based on the von Mises yield criterion and isotropic hardening is incorporated in the static analysis only). The dynamic model follows the principles of the first-order shear deformation large rotation shell theories. Temporal discretisation of the weak form of the equations of motion is performed using an implicit numerical time-integration scheme. In fact, an algorithm, which may be regarded as a special form of the mid-point rule, is employed. It preserves the fundamental constants of the autonomous motion: the total linear and angular momentum as well as, for a Hamiltonian case, the total energy. The performance of the model is illustrated with selected results of numerical examples.

1. Introduction

Various finite element models for simulation of the behaviour of flexible shells are now established and efficient solutions can be obtained using a variety of appropriate elements and formulations. Here, we consider non-linear statics and dynamics of thin shells. The shells under consideration may be composed of a generally elastic material with trough thickness variable mass density and degree of orthotropy. It is common practice to represent the mechanical behaviour of a composite laminate on a macroscopic level. In this case the shell structure is treated as a composition of perfectly bonded layers and each layer is treated as a homogeneous and anisotropic continuum. At small strains the behaviour of a particular layer can be assumed as purely elastic and described with an orthotropic material law (e.g. Basar (1993); Brank, Perić and Damjanić (1995)). As a constitutive law the rate-independent elasto-plastic relation with the non-linear isotropic hardening and von Mises yield condition is adopted for static shell analysis exclusively. More details related to the static elasto-plastic formulation can be found in Brank, Perić, Damjanić (1997). Here, due to limited space, our main aim is to present a numerically stable finite element formulation for the description of

H. A. Mang and F. G. Rammerstorfer (eds.),
IUTAM Symposium on Discretization Methods in Structural Mechanics, 131–138.

long-term response of shells, which may be undergoing large deformations and rotations, and large overall motion. The formulation has a lot in common with the works of Simo, Rifai and Fox (1992) and Simo and Tarnow (1994). The differences are related to the derivation of the continuum dynamic formulation and the implementation procedure. The dynamic shell model is developed from three-dimensional continuum theory by employing standard assumptions on the distribution of the displacement field in the shell body (e.g. Büchter and Ramm (1992). A single director model for thin shells, which is equivalent to the two-dimensional Cosserat continuum presented by Simo and Fox (1989), is obtained by approximating the terms describing the shell geometry. A rotational vector based on large rotation formulation is used to describe the motion of the shell director field. The starting point for the numerical formulation is the continuum weak form of the initial boundary value problem. Time and spatial discretisation of the weak form are subsequently applied. The implicit numerical time-integration scheme follows the works of Simo and Tarnow (1992, 1994) and Simo, Tarnow and Doblare (1995), who constructed the so-called energy-momentum conserving algorithms for various mechanical systems, i.e. solids, shells and rods. The algorithm, adopted in this work, may be regarded as a special form of the mid-point rule. It is of second-order accurate and characterized by exact conservation of the total energy for Hamiltonian shell systems. It has been shown by many authors (see for example Simo and Tarnow (1992)), that the energy conserving time-integration schemes demonstrate unconditional numerical stability. This is in contrast to conventional time integration schemes of the Newmark family, which may exhibit a loss of stability due to the sudden increase in the higher frequency strain energy components. The energy conserving algorithms are therefore suitable for the analysis of long-term response. If the motion of a shell is driven only by specified initial conditions, and no displacement and rotation boundary conditions are prescribed, the adopted algorithm also conserves both the total linear and the total angular momentum of the shell. The spatial discretisation is based on the four noded isoparametric elements. To avoid locking, a so-called assumed natural approach is adopted. Particular attention is devoted to the consistent linearisation of the time discretised weak form of the equations of motion, in order to achieve a quadratic rate of asymptotic convergence typical for the Newton-Raphson based solution procedure. In the following sections the attention is restricted to some key issues of the formulation.

2. Concepts of the Continuum Formulation

The classical one-director shell theory is completely defined in terms of the two objects: (i) a surface M, parametrised by curvilinear coordinates ξ^1, ξ^2, and (ii) a vector field, called shell director field, defined on M. Let the time interval of the interest be $[0,T]$. The position vectors of a material point at the initial time t=0, and at time t>0 are defined as

$$\mathbf{X} = \mathbf{R} + \xi\mathbf{A}_3 , \qquad \text{and} \qquad \mathbf{x} = \mathrm{r} + \xi\mathbf{a}_3 \tag{1}$$

respectively, where $\xi \in [-h/2, h/2]$ and h is the constant initial shell thickness. The shell director $\mathbf{A}_3$ coincides with the normal to the surface M at the point $\mathbf{R} \in M$. It follows from eq. (1) that the displacement vector is a linear function of ξ coordinate: $\mathbf{u} = (\mathbf{r} - \mathbf{R}) + (\mathbf{a}_3 - \mathbf{A}_3)$. To avoid ill conditioned equations in the thin shell, an assumption of the inextensibility of the shell director is further introduced, $\| \mathbf{a}_3 \| = \| \mathbf{A}_3 \| = 1$, which allows introduction of the rotation tensor $\mathbf{Q}$. Relation $\mathbf{a}_3 = \mathbf{Q}\,\mathbf{A}_3$ enables a two-parametric representation of $\mathbf{Q}$. Straining (eq.(2)) is described by the Green-Lagrange strain tensor, which is evaluated with respect to the thickness coordinate ξ, ignoring functions of higher order than one:

$$\mathbf{E} = (\alpha_{ij} + \xi\beta_{ij})\,\mathbf{A}^i \otimes \mathbf{A}^j. \tag{2}$$

$\mathbf{A}_i$ represents generally non-orthogonal covariant base vectors, spanned by the tangents to the curves ξ^α and ξ, while $\mathbf{A}^i \cdot \mathbf{A}_j = \delta^i_j$.

2.1. ELASTO-PLASTIC STATICS

The above briefly described formulation is extended to include elasto-plastic material modelling for static analysis. A finite element formulation based on the 4-node isoparametric element is used. A consistent linearisation of the shell kinematics and elasto-plastic material model is performed in order to achieve an efficient algorithm with a quadratic rate of asymptotic convergence typical for the Newton-Raphson solution procedure. Detailed formulation can be found in Brank, Perić, Damjanić (1997) and the selected numerical results of a static analysis (Damjanić *et al.*, 1997) of an elasto-plastic pinched cylinder are exclusively presented here to illustrate the model. The problem details (1/8 of the cylinder) and the load-displacement curve compared with the results of Simo, Kennedy (1992) are shown in Fig.1. Deformed meshes under increasing static load are illustrated in Fig.2.

2.2. ELASTO-DYNAMICS

The dynamic formulation necessitates the definition of the kinetic and the potential energy. Then the weak form of the canonical Hamilton's equations may be constructed, which further leads to the weak form of the momentum equations. Finally, the weak form of the initial boundary value problem for non-linear elastic shells (together with the initial conditions in eq.(4)) can be expressed as

$$\mathbf{G}_{dyn} = \mathbf{G}_{ine} + \mathbf{G}_{int} - \mathbf{G}_{ext} = 0 \tag{3}$$

$$(\mathrm{r}, \mathbf{a}_3)\big|_{t=0} = (\mathbf{R}, \mathbf{A}_3), \quad (\mathbf{p}, \pi)\big|_{t=0} = (\mathbf{P}, \Pi) \tag{4}$$

where explicit expressions for inertial, internal and external parts are defined as follows

$$\mathbf{G}_{ine} = \int_M [\mathbf{p}' \cdot \delta\mathbf{v} + \pi' \cdot \delta\mathbf{a}_3]\, dM, \qquad \mathbf{G}_{int} = \int_M (n^{ij}\,\delta\alpha_{ij} + m^{ij}\,\delta\beta_{ij})\, dM \tag{5}$$

$$\mathbf{G}_{ext} = \int_M (\mathbf{q} \cdot \delta\mathbf{x} + 1 \cdot \delta\mathbf{a}_3)\, dM + \int_{\partial M} (\mathbf{N}^s \cdot \delta\mathbf{x} + \mathrm{M}^s \cdot \delta\mathbf{a}_3)\, ds. \tag{6}$$

Here, $\mathbf{p} = A\,\mathbf{r}'$ and $\pi = I\,\mathbf{a}'_3$, with A and I defining the 0th and 2nd moment of mass density about the surface M. The n^{ij} and m^{ij} define the 0th and 1st order second Piola-

Kirchhoff stress resultants and **q**, **1**, $\mathbf{N}^s$ and $\mathbf{M}^s$ represent external force resultants related to the surface M and its boundary ∂M.

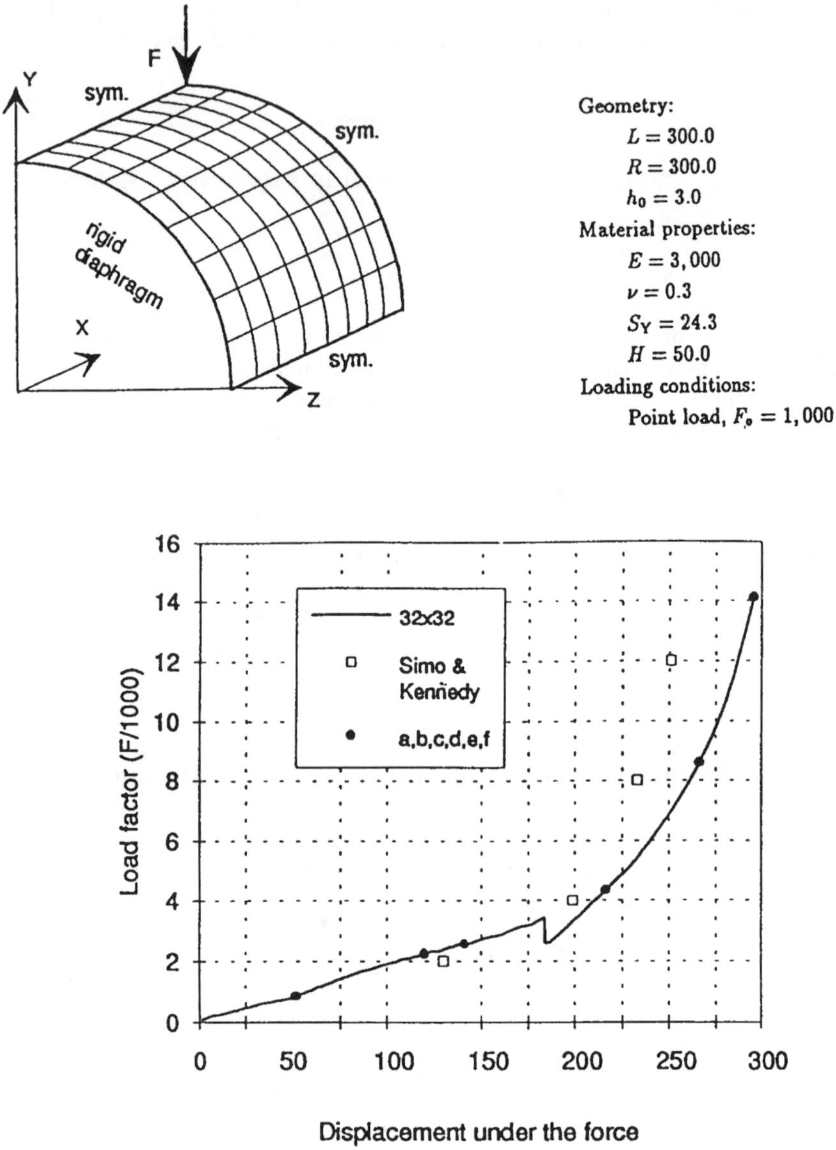

Figure 1. Problem details and the load-displacement curve for the elasto-plastic pinched cylinder under static loading conditions

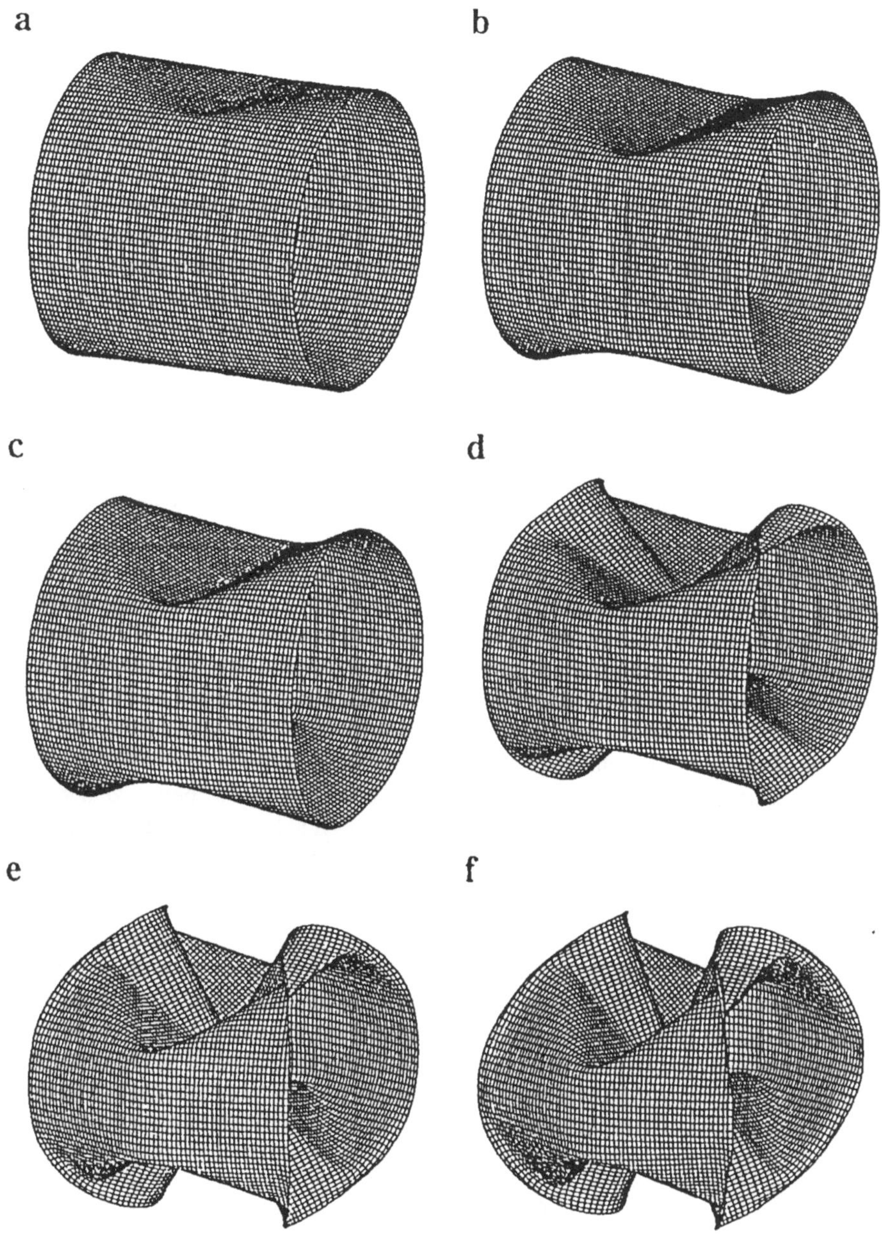

Figure 2. Deformed meshes under increasing static load for the elasto-plastic pinched cylinder (load factors (a-f) are indicated in Figure 1)

3. Concepts of the Dynamic Numerical Procedure

The numerical formulation is based on time and spatial discretisation of the weak form of the initial boundary value problem. Let a time interval of interest be partitioned into N subintervals $_{n=0}\cup^{N-1} [t_n, t_{n+1}]$. A solution is obtained at discrete points by an algorithmic approximation. Considering a typical time subinterval, the algorithmic problem is: Assuming that initial conditions at t_n are given, define algorithmic approximations of the configuration and the velocities at t_{n+1}. As already mentioned, a special form of the mid-point rule is employed in this work. The structure of the algorithm is based on a two-step procedure. First, the weak form of the initial boundary value problem is evaluated at the middle of a time subinterval, with the stress resultants left undefined

$$[\mathbf{G}_{ine} + \mathbf{G}_{int} - \mathbf{G}_{ext}] \big|_{tn+1/2} = 0. \tag{7}$$

Second, the algorithmic stress resultants are defined in such a way that the total energy is conserved for the Hamiltonian case. If a stored energy function is a quadratic form of the Green-Lagrange strain tensor, the energy is conserved with the average of strains at time t_n and t_{n+1}. The agorithmic components of the second Piola-Kirchhoff stress tensor are therefore

$$S_{ij} = C^{ijkl} \cdot \tfrac{1}{2} [(E_{kl})_n + (E_{kl})_{n+1}] \quad ; \quad i,j \neq 3,3, \;\; k,l \neq 3,3 \tag{8}$$

with C^{ijkl} being the components of the fourth order constitutive tensor.
The time discretisation is followed by spatial discretisation, which is based on the four node isoparametric finite element. Locking is avoided by the assumed natural strain approach for the transverse shear part (e.g. Brank, Perić and Damjanić (1995)). The displacement-based formulation is used for the membrane and bending parts. Appropriate linearisation of the eq.(7) is performed in order to achieve a quadratic rate of asymptotic convergence of a Newton-Raphson based solution procedure. The tangent operator is obtained by the directional derivative of the fully discrete weak form in the direction $(\Delta r_{n+1}, \Delta \mathbf{a}_{3n+1})$.

4. Dynamic Numerical Example

The motion of a short cylinder under impulsive point loads is presented. This example was already solved by Simo and Tarnow (1994) in order to compare the performance of different numerical time integration schemes. The problem details and loading conditions are given in Figs.3 and 4. At the reference positions the initial conditions are prescribed to be zero. The cylinder is subjected to short impulsive external loads up to 1 second as shown in Fig. 4. The cylinder motion is computed and traced during a time period of 25 seconds divided into 1250 time steps of Δt=0.02 sec. The complex motion of the cylinder, from two different views, is shown in Fig.5, where deformed shapes are drawn for every 2 seconds. The values of three vector components of total linear and total angular momentum remain constant during the free motion. As expected, the total input energy (see Fig. 6) of the undamped system is also preserved. However, periodic oscillations of internal and kinetic energy can be seen.

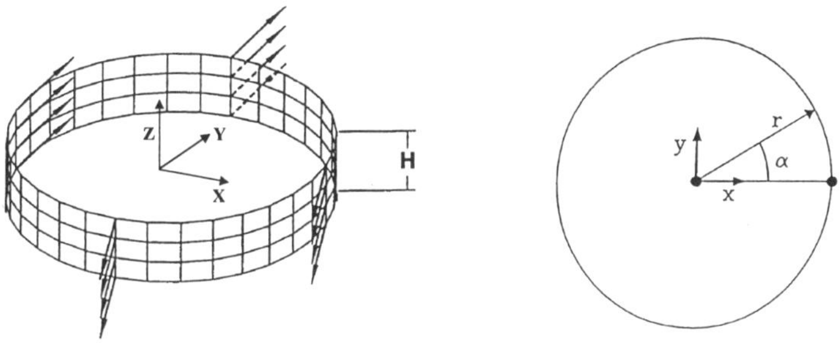

$E = 2 \times 10^8$ N/cm^2
$\nu = 0.5$
$\rho_o = 1.0$ kg/cm^3

$r = 7.5$ cm
$H = 3.0$ cm
$t = 0.02$ cm

Figure 3. Problem details and loading conditions of the short cylinder.

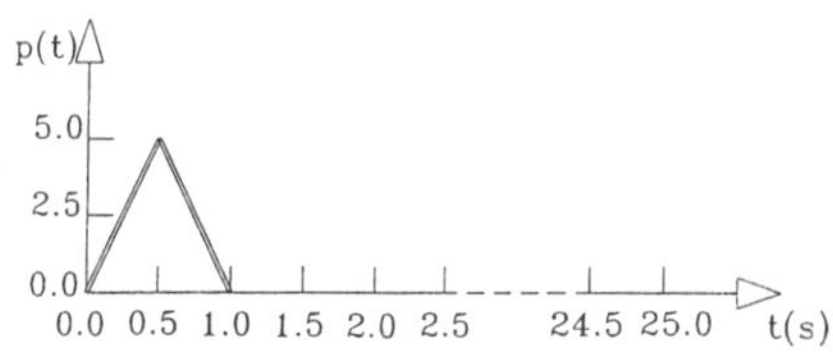

Angle α	0	$\pi/2$	π	$3/2\pi$
Nodal loads	$[0, -1, -1]^T p(t)$	$[1, 1, 1]^T p(t)$	$[1, 1, 1]^T p(t)$	$[0, -1, -1]^T p(t)$

Figure 4. Time history of external loads

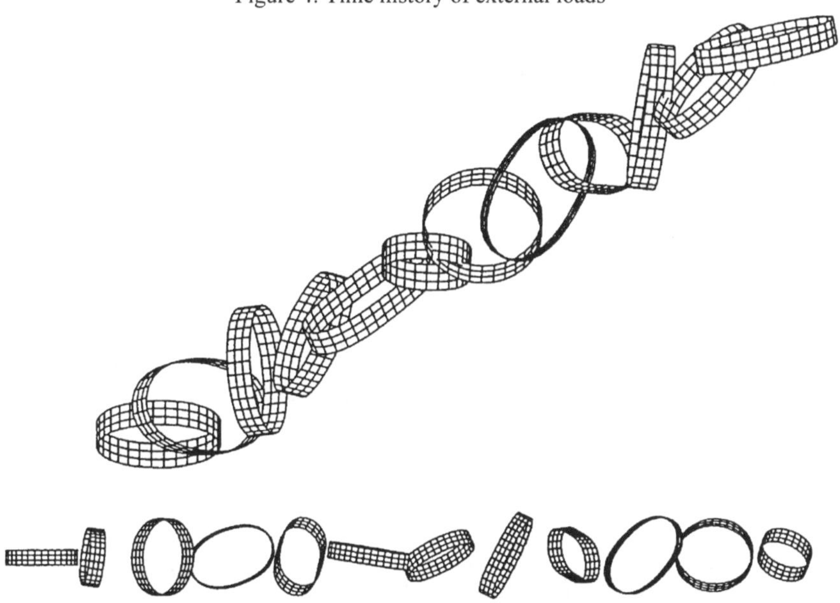

Figure 5. Sequences of deformed finite element mesh for the time interval [0,25 sec] (meshes are drawn for every 2 seconds).

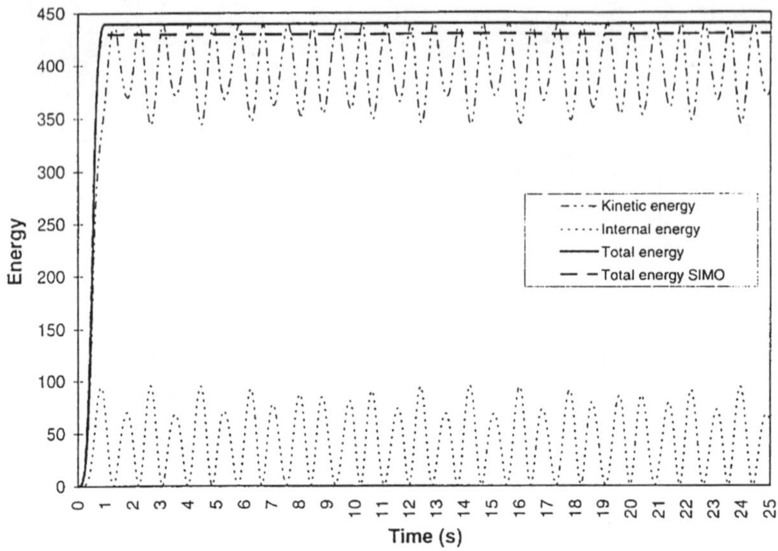

Figure 6. Time history of the kinetic, internal and total energy.

5. References

Basar, Y. (1993) Finite-Rotation Theories for Composite Laminates, *Acta Mechanica* **98**,159-176.

Brank, B., Perić, Đ. and Damjanić, F.B. (1995) On Implementation of a Nonlinear Four Node Shell Finite Element for Thin Multilayered Elastic Shells, *J. Computational Mechanics* **16**, 341-359

Brank, B., Perić, Đ. and Damjanić, F.B (1997) On large deformations of thin elasto-plastic shells: Implementation of a finite rotation model for quadrilateral shell element, *Int. J. Num. Meth. Engng.* **40**, 689-726.

Büchter, N. and Ramm, E. (1992) Shell Theory Versus Degeneration - A Comparison in Large Rotation Finite Element Analysis, *Int. J. Num. Meth. Engng.* **34**, 39-59.

Damjanić, F.B., Brank, B. and Perić, Đ. (1995) On Elasto-Plastic Buckling Analysis of Cylindrical Shells, in S.N. Atluri *et al.* (eds.), *Computational Mechanics '95*, Springer, pp. 1565-1570.

Damjanić, F.B., Brank, B. and Perić, Đ. (1997) Elasto-plasticity in the non-linear thin shell modelling, in D.R.J. Owen *et al.*, *Proc. 5th COMPLAS*, Barcelona, Spain, pp.1890-1897.

Simo, J.C. and Fox, D.D. (1989) On a Stress Resultant Geometrically Exact Shell Model. Part I: Formulation and Optimal Parametrization, *Comp. Meth. Appl. Mech. Engng.* **72**, 267-304.

Simo, J.C. and Kennedy, J.G. (1992) On a stress resultant geometrically exact shell model. Part V: Nonlinear plasticity: formulation and integration algorithms, *Comp. Meth. Appl. Mech. Engng.*, **96,** 133-171.

Simo, J.C. and Tarnow, N. (1992) The Discrete Energy-Momentum Method: Conserving Algorithms for Nonlinear Elastodynamics, *Z. Agnew. Math. Phys.,* **43**, 757-792.

Simo, J.C., Rifai, M.S. and Fox, D.D. (1992) On Stress Resultant Geometrically Exact Shell Model. Part VI: Conserving Algorithms for Nonlinear Dynamics, *Int. J. Num. Meth. Engng.,* **34**, 117-164.

Simo, J.C. and Tarnow, N. (1994) A New Energy and Momentum Conserving Algorithm for the Nonlinear Dynamics of Shells, *Int. J. Num. Meth. Engng.*, **37**, 2527-2549.

Simo, J.C., Tarnow, N and Doblare, M. (1995) Nonlinear Dynamic of Three-Dimensional Rods: Exact Energy and Momentum Conserving Algorithm, *Int. J. Num. Meth. Engng.* **38**, 1431-1474.

NUMERICALLY EFFICIENT FORMULATION OF THE EQUATIONS OF MOTION OF TETHERED SATELLITE SYSTEMS

W. POTH, M. SCHAGERL, A. STEINDL, W. STEINER AND H. TROGER
Vienna University of Technology
A–1040 Vienna, Austria

Abstract. Simulation of tethered satellite systems requires the numerical solution of a coupled system of nonlinear partial and ordinary differential equations.

Due to the technical design of such systems the set of governing equations of motion results in a stiff system of differential equations. In order to avoid long integration times, contrary to the formulation usually used in the literature, an alternative description of the deformation of the tether is given. In the new variables after discretization of the continuous tether in space (for example by Finite Elements or Finite Differences) a system of ordinary differential equations is obtained in which the fast motion is separated from the slow motion. For the integration of such systems specialized software is available reducing integration time considerably.

1. Introduction

The main object of this paper is to present an alternative description of the deformation of the tether of a tethered satellite system which most naturally and almost exclusively in the literature is given by its displacement components in properly moving reference frames (Kohler et al, 1978), (Misra and Modi, 1986), (Vu-Quoc and Simo, 1987), (Liangdong and Bainum, 1989), (Beletskii and Levin, 1993), and (Steiner et al, 1995a). Although such a formulation is quite well suited for theoretical investigations, if it comes to numerical simulation, as preliminary numerical experiences indicate, apparently a different formulation will be more advantageous. The reason for this is that the motion of a tethered satellite system as it is well known both from flight experience and from simulations may be separated into several different motions. One is a very fast oscillation in axial direction of the tether and the other two are a comparably slow oscillation in the direction transversal to the tangent to the tether and further the overall motion of the whole

H. A. Mang and F. G. Rammerstorfer (eds.),
IUTAM Symposium on Discretization Methods in Structural Mechanics, 139–146.

system performs also a relatively slow oscillation. These are consequences of the large axial stiffness of tethers and hence the problem resultes in a stiff system of differential equations (Hairer and Wanner, 1991).

To integrate a stiff system of differential equations numerically, implicit integration schemes must be used. A strong improvement of the numerical integration still can be achieved if a formulation is given, in which the fast motion and the slow motions are separated. This can be done by representing the deformation of the tether not by its displacement components in a cartesian frame as it is given in (Kohler et al, 1978), (Vu-Quoc and Simo, 1987), (Beletskii and Levin, 1993), and (Steiner et al, 1995a), but by introduction of a special set of variables indicated already in (Beletskii and Levin, 1993) for tethered systems and in (Hairer and Wanner, 1991) for the oscillation of a beam and worked out for the tether motion in (Kuhn, 1995). Here as variables the orientation of the tangent vector to the deformed tether and the elongation of the tether are used.

2. Equations of motion

The tether is modelled as a massive continuous flexible string without bending and torsional stiffnesses (Kohler et al, 1978), (Beletskii and Levin, 1993). In the numerical simulation we use both inextensible and axially visco-elastic tethers (Steiner et al, 1995a). The satellites are modelled both as rigid bodies or as point masses depending on which motions we want to investigate. For example it is shown in (Steiner et al, 1995a) that for long tethers modelling of the satellites as rigid bodies has very little influence on the tether motion. For the modelling of

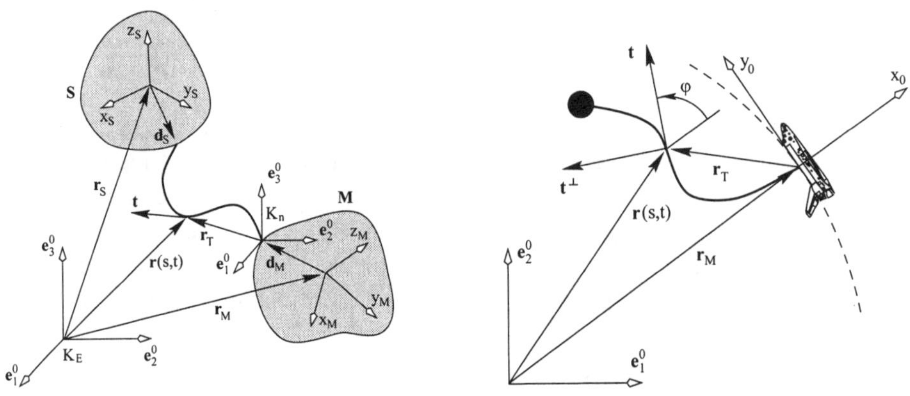

Figure 1. Left: Tether motion described in the local nonrotating floating frame K_n. The position r_T of the tether may be given either by its components x, y, z, or by the elongation and the orientation of the tangent vector t relative to K_n. *Right:* Description of the planar tether motion by the angle φ of the tangent vector t and the elongation η in the rotating orbital frame x_0, y_0.

deployment of an inextensible tether the dynamics of the reeling drum is included.

2.1. CABLE DISPLACEMENT DESCRIBED BY THE COMPONENTS OF THE POSITION VECTOR

In this section we only collect some equations derived in (Steiner et al, 1995a) by means of the variational principle of Hamilton Ostrogradski. Their numerical treatment is given by a discretization with Finite Elements and a time integration with the implicit Newmark algorithm as proposed in (Vu-Quoc and Simo, 1987).

A geometrically exact large displacement but small strain description of the deformation is given. As constitutive relation for the tether the linear Kelvin-Voigt law of viscoelasticity $N = EA(\varepsilon+\alpha\dot{\varepsilon})$ is stipulated, where N is the axial force in the tether, α a dissipation constant, EA the stiffness of the tether and ε the strain defined below. $(\,)^{\cdot}$ denotes the derivative with respect to time t.

In the left frame of Fig. 1 the choice of the coordinate system (Steiner et al, 1995b) is explained where a nonrotating frame K_n moving with the main satellite $\boldsymbol{M}$ is attached to one end point of the tether. Further the decomposition

$$\boldsymbol{r}(s,t) = \boldsymbol{r}_M(t) + \boldsymbol{d}_M(t) + \boldsymbol{r}_T(s,t), \tag{1}$$

of the position vector $\boldsymbol{r}$ is easy to see. In the formulation of the problem the unstrained arclength s along the tether instead of the strained arclength $\overline{s}$ will be used. Introducing the elongation $\eta = d\overline{s}/ds$ in the relation for the tangential vector

$$\boldsymbol{t} = \frac{\partial \boldsymbol{r}}{\partial \overline{s}} = \frac{\partial \boldsymbol{r}}{\partial s}\frac{ds}{d\overline{s}} = \boldsymbol{r}'\frac{1}{\eta} = \boldsymbol{r}'_T\frac{1}{\eta}\,, \tag{2}$$

yields for the strain ε

$$\varepsilon = \frac{d\overline{s} - ds}{ds} = \frac{d\overline{s}}{ds} - 1 = \eta - 1 = \|\boldsymbol{r}'_T\| - 1. \tag{3}$$

Here $\boldsymbol{t}\cdot\boldsymbol{t} = 1$ is used. $(\,)'$ denotes the derivative with respect to s.

The field equation describing the motion of the tether is given by

$$-\mu\ddot{\boldsymbol{r}} + (N\boldsymbol{t})' - \mu\nabla\Phi(\boldsymbol{r}) + \boldsymbol{f} = 0, \tag{4}$$

where $\boldsymbol{t}$ is the tangent vector to the line of centroids. The linear and angular momentum equations describing the motions of the satellites supply the boundary conditions. For satellite M the linear momentum equation reads

$$\begin{aligned} m_M^0\ddot{\boldsymbol{r}}_M + \mu s_M(\ddot{\boldsymbol{r}}_M + \ddot{\boldsymbol{d}}_M) = -\mu\dot{s}_M^2\boldsymbol{r}'(s_M) + N(s_M)\boldsymbol{t}(s_M) \\ +\boldsymbol{f}_M - m_M^0\nabla\Phi(\boldsymbol{r}_M) - \mu s_M\nabla\Phi(\boldsymbol{r}_M + \boldsymbol{d}_M) \end{aligned} \tag{5}$$

and the angular momentum equation is given by

$$\begin{aligned} \mathbf{I}_M^0\dot{\boldsymbol{\Omega}}_M + \boldsymbol{\Omega}_M \times \mathbf{I}_M^0\boldsymbol{\Omega}_M - N(s_M)\mathbf{D}_M \times \mathbf{B}_{EM}^T\boldsymbol{t}(s_M) \\ +\mathbf{D}_M \times \mathbf{B}_{EM}^T\left[\mu\dot{s}_M^2\boldsymbol{r}'(s_M) + \mu s_M(\ddot{\boldsymbol{r}}_M + \ddot{\boldsymbol{d}}_M)\right] = \mathbf{M}_M\,. \end{aligned} \tag{6}$$

Two similar equations follow for the satellite S. The variables s_M and s_S denote the arc lengths at the satellites. The mass and the tensor of inertia of satellite M without the stored part of tether are denoted by m_M^0 and $\mathbf{I}_M^0$. The parameter μ is the tether mass per unit unstrained tether length. $\boldsymbol{f}$, Φ and $\boldsymbol{M}_M$ denote the external forces, the gravitational potential of the Earth and the external moments, respectively.

To avoid time dependent elements for the numerical treatment by means of Finite Elements the deployed length of the tether is transformed to the interval [0,1] by introducing a new spatial coordinate (Kohler et al, 1978).

2.2. CABLE DISPLACEMENT DESCRIBED BY THE TANGENT VECTOR AND THE ELONGATION

Now we derive the equations of motion representing the position vector $\boldsymbol{r}$ by the tangential vector $\boldsymbol{t}$ and the elongation η. Such variables are used in (Hairer and Wanner, 1991), (Beletskii and Levin, 1993) and (Kuhn, 1995). The equations of motion are derived for a system of point masses as end bodies and for the general case of variable tether length that is without prescribing the length change at the satellites. With $\boldsymbol{r}_M(t)$ giving the motion of the origin of K_n and $\boldsymbol{r}_T(s,t)$ the position of the tether in K_n (Fig. 1, left) we have ($\boldsymbol{d}_M = \boldsymbol{0}$)

$$\boldsymbol{r}(s,t) = \boldsymbol{r}_M(t) + \boldsymbol{r}_T(s,t), \tag{7}$$

which gives the relationship between the representation in the inertial frame K_E and K_n.

Integration of (2) with respect to the unstrained arclength s yields

$$\boldsymbol{r}_T(s,t) = \int_0^s \eta(\sigma,t)\boldsymbol{t}(\sigma,t)\,d\sigma\,. \tag{8}$$

The inextensional case is given by $\varepsilon = 0$ or $\eta = 1$.

We follow the derivation given in (Beletskii and Levin, 1993) resulting in a set of equations named *Minakov's equations*. We take the derivative of (4) with respect to s to obtain

$$\ddot{\boldsymbol{r}}' = \frac{1}{\mu}(N\boldsymbol{t})'' - (\nabla\Phi(\boldsymbol{r}))' + \frac{1}{\mu}(\boldsymbol{f})' \tag{9}$$

Substituting from (2) for $\boldsymbol{r}'$ into (9) yields

$$(\eta\boldsymbol{t})^{\cdot\cdot} = \frac{1}{\mu}(N\boldsymbol{t})'' + \boldsymbol{P}' \tag{10}$$

where as abbreviation $\boldsymbol{P} = \boldsymbol{f}/\mu - \nabla\Phi(\boldsymbol{r})$ is introduced. Performing the differentiation in (10) results in

$$\ddot{\eta}\boldsymbol{t} + 2\dot{\eta}\dot{\boldsymbol{t}} + \eta\ddot{\boldsymbol{t}} = \frac{1}{\mu}(N''\boldsymbol{t} + 2N'\boldsymbol{t}' + N\boldsymbol{t}'') + \boldsymbol{P}' \tag{11}$$

Taking the inner product of (11) with $\boldsymbol{t}$ and taking into account the relations $\boldsymbol{t}\cdot\dot{\boldsymbol{t}} = 0$, $\boldsymbol{t}\cdot\boldsymbol{t}' = 0$, $\boldsymbol{t}\cdot\boldsymbol{t}'' = -\boldsymbol{t}'^2$ and $\boldsymbol{t}\cdot\ddot{\boldsymbol{t}} = -\dot{\boldsymbol{t}}^2$ yields the fast equation for the elongation η

$$\ddot{\eta} - \eta\dot{t}^2 = \frac{1}{\mu}\left(N'' - Nt'^2\right) + \boldsymbol{P}'\cdot\boldsymbol{t}. \tag{12}$$

Eqn. (12) allows to eliminate the second derivative $\ddot{\eta}$ from (11) to obtain

$$\eta\ddot{\boldsymbol{t}} + 2\dot{\eta}\dot{\boldsymbol{t}} + \eta\dot{\boldsymbol{t}}^2\boldsymbol{t} = \frac{N}{\mu}(\boldsymbol{t}'' + \boldsymbol{t}'^2\boldsymbol{t}) + \left(\frac{N'}{\eta} + \left(\frac{N}{\eta}\right)'\right)\boldsymbol{t}' + \boldsymbol{P}' - (\boldsymbol{P}'\cdot\boldsymbol{t})\boldsymbol{t}. \tag{13}$$

which gives the required second equation. Eqs. (12) and (13) are equivalent to the original system (11) and the constraint $\boldsymbol{t}\cdot\boldsymbol{t} = 1$. In fact we have one scalar equation (12) and one vector equation (13) which have to be solved. The limit to the inextensible tether can be performed if the axial force N is taken as variable instead of η letting EA tend to infinity.

2.2.1. *Boundary conditions*

We follow the exposition given in (Beletskii and Levin, 1993). From (Fig. 1, left) we have $\boldsymbol{r}_M = \boldsymbol{r}(s_M, t)$. Differentiating with respect to time yields

$$\dot{\boldsymbol{r}}_M = \boldsymbol{r}'(s_M, t)\dot{s}_M + \dot{\boldsymbol{r}}(s_M, t)$$

and

$$\ddot{\boldsymbol{r}}_M = \boldsymbol{r}''\dot{s}_M^2 + 2\dot{\boldsymbol{r}}'\dot{s}_M + \boldsymbol{r}'\ddot{s}_M + \ddot{\boldsymbol{r}}. \tag{14}$$

Introducing the tangent vector $\eta\boldsymbol{t} = \boldsymbol{r}'$ from (2) into (14) we obtain

$$\ddot{\boldsymbol{r}}(s_M, t) = \ddot{\boldsymbol{r}}_M - \boldsymbol{w}_M, \tag{15}$$

where

$$\boldsymbol{w}_M = (\eta'\dot{s}_M^2 + 2\dot{\eta}\dot{s}_M + \eta\ddot{s}_M)\boldsymbol{t} + \eta\dot{s}_M^2\boldsymbol{t}' + 2\eta\dot{s}_M\dot{\boldsymbol{t}}.$$

In this expression we obtain different signs compared to the analogous expression Equ. (2.73) in (Beletskii and Levin, 1993). Now we insert from (4) for $\ddot{\boldsymbol{r}}(s_M, t)$ and for $\ddot{\boldsymbol{r}}_M$ from the equation corresponding to (5) to obtain

$$\frac{1}{\mu}(N\boldsymbol{t})'_{s=s_M} + \mathbf{P}_M = \frac{1}{m_M}\left[\left(N - \mu\eta\dot{s}_M^2\right)_{s=s_M}\boldsymbol{t}_M + \mathbf{Q}_M\right] - \boldsymbol{w}_M \tag{16}$$

where

$$\mathbf{P}_M = \left(\nabla\Phi(\boldsymbol{r}) + \frac{1}{\mu}\boldsymbol{f}\right)_{s=s_M}, \qquad \mathbf{Q}_M = m_M \nabla\Phi(\boldsymbol{r}_M) + \boldsymbol{f}_M.$$

Here $m_M = m_M^0 + \mu s_M$ is the mass of the satellite M. Next we take the inner product of (16) with $\boldsymbol{t}_M$ to obtain

$$N'(s_M) = \mu\left(\frac{\tilde{N}}{m_M} + \mathbf{T}_M \cdot \boldsymbol{t}_M\right) - \mu \boldsymbol{w}_M \cdot \boldsymbol{t}_M \tag{17}$$

where

$$\tilde{N} = N(s_M) - \mu\eta\dot{s}_M^2\,, \qquad \mathbf{T}_M = \frac{\mathbf{Q}_M}{m_M} - \mathbf{P}_M.$$

Performing the differentiation on the left hand side of (16) and substracting (17) after multiplication by $\boldsymbol{t}_M$ we obtain the condition for $\boldsymbol{t}'_M$ in the form

$$\tilde{N}\boldsymbol{t}'_M = \mu\left(\mathbf{T}_M - \boldsymbol{t}_M(\mathbf{T}_M \cdot \boldsymbol{t}_M) - 2\eta\dot{s}_M\dot{\boldsymbol{t}}_M\right). \tag{18}$$

Equations (17) and (18) are the required boundary conditions for (12) and (13). In addition $\boldsymbol{r}_M$ must be either prescribed or calculated from (5). Similar expressions are obtained for satellite S.

The spatial discretization of the field equation is performed by Finite Differences (Kuhn, 1995). The resulting system of ordinary differential equations is integrated by the program LSODI (Hindmarsh, 1983).

3. Results

3.1. FREE DEPLOYMENT OF A SUBSATELLITE

The free deployment of a subsatellite from a main satellite, which is moving on a Keplerian circular orbit with data corresponding to the project Tros-Rapunzel, is calculated following (Schagerl, 1997). The tether is assumed to be inextensible, reeled on a drum of $2R = 1$ m diameter and to perform only a planar motion in the orbital plane. We use the formulation given in section 2.2. The moment of inertia of the drum is taken into account during the deployment. The representation in Fig. 2 (left) is in the orbital frame (x_0, y_0) as shown in Fig. 1 (right). In the left part of Fig. 2 several configurations of the tether and the subsatellite are depicted for the data $m_S = 6$ t and the tether mass $\mu = 3.5$ kg/km, after the subsatellite is released from its initial position which is in the local vertical position 25 m below the spacestation. In the right part of Fig. 2 the length of the tether unreeled from the drum is depicted against deployment time. Obviously at the beginning the process of free deployment is very slow, since the initial gravity difference is very small. Hence it takes a long time until the satellite moves away from the space station, but then the deployed length increases rapidly.

3.2. COMPARISION OF THE NUMERICAL EFFICIENCY

Fig. 3 (left) shows the time evolution of the motion of a subsatellite relative to the orbital frame, which itself is moving on a circular Keplerian orbit (Fig. 1, right). The two different ways of representation and hence calculation are designated by $x - y$, according to section 2.1, and $\eta - \boldsymbol{t}$, according to section 2.2, respectively. For a period of about 4000 sec no differences in accuracy can be recognized. In Fig. 3 (right) the corresponding computational times are depicted. In both cases a maximum in computational time for intermediate values of the axial stiffness

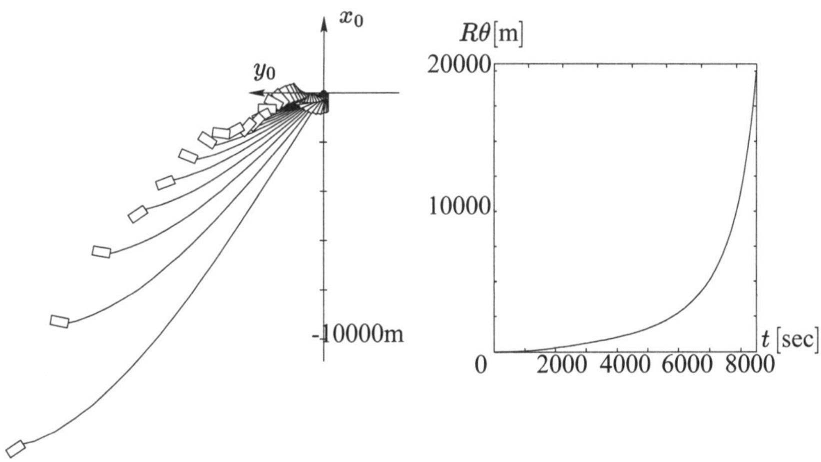

Figure 2. Planar motion of free deployment of a subsatellite ($\Delta t = 1.5$ min) from a main satellite moving on a Keplerian circle with orbital angular velocity $\omega = 1.1288\ 10^{-3}$ rad/sec (left). In the right part the deployed length $R\theta$ at the drum as a function of time is shown. In the initial position the subsatellite is located 25 m below the spacestation in its radial equilibrium.

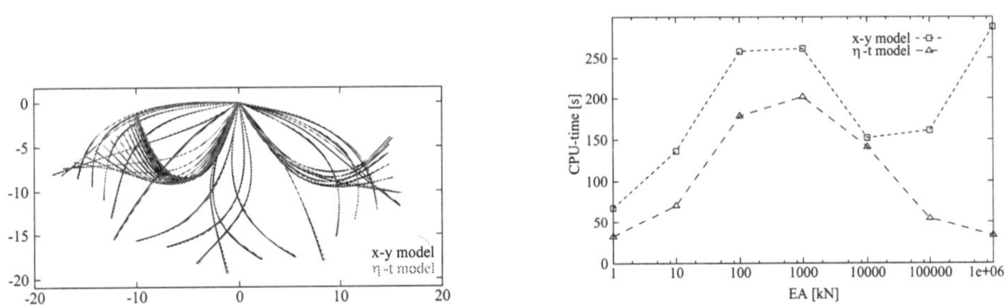

Figure 3. Left: Simulations for the motion of a subsatellite relative to a space station calculated with the two different formulations for a tether stiffness $EA = 10^5$ kN. Upper frame: $x - y$ formulation according to section 2.1, lower frame $\eta - \boldsymbol{t}$ formulation according to section 2.2. *Right:* Comparison of CPU-time for the simulations according to the two different formulations. After passing an initial maximum, increasing the stiffness of the tether results in an increase for the $x - y$ formulation, but in a decrease for the $\eta - \boldsymbol{t}$ formulation.

occurs. For the $x-y$ model this maximum is local because for increasing stiffness after a slight decrease the computation time again increases strongly. For the $\eta-t$ model on the other hand for increasing stiffness the computation time decreases.

Conclusions

For numerical simulations the formulation given in the form of the $\eta - t$ representation can be considered as a serious alternative to the formulation used so far in the literature. For very stiff or inextensible tethers most probably it is the only one which allows in a meaningful way to perform simulations over a long time interval. Moreover the practically important process of free deployment could be easily included.

ACKNOWLEDGEMENT

This research project has been supported by the Austrian Science Foundation (FWF), under the project P10705-MAT and by ESA under project DATES.

References

BELETSKII, V. V., LEVIN, E. M.: Dynamics of Space Tether Systems. Advances of the Astronautical Sciences **83**, 1993.

HAIRER, E., WANNER, G.: Solving Ordinary Differential Equations, II. Stiff and Differential-Algebraic Problems, Springer Verlag, Berlin, Heidelberg 1991. Advances of the Astronautical Sciences **83**, 1993.

HINDMARSH, A. C.: 1983. ODEPACK, Asystematic collection of ODE solvers in scientific computing, R.S. Stepleman et al (eds.) North-Holland, Amsterdam, 55–64.

KOHLER, P., MAAG, W., WEHRLI, R., WEBER, R., BRAUCHLI, H.: Dynamics of a System of two Satellites Connected by a Deployable and Extensible Tether of Finite Mass. ESTEC Contract No. 2992/76/NL/AK(SC), 1978.

KUHN, A.: Numerische Behandlung von "Tethered Satellite Systems" unter besonderer Berücksichtigung längssteifer Verbindungsseile. Dissertation TU-Wien, 1995.

LIANGDONG, L., BAINUM, P. M.: Effect of Tether Flexibility on the Tethered Shuttle SubsatelliX-Cte Stability and Control. J. of Guidance, Control and Dynamics **12** (1989), 866–873.

MISRA, A. K., MODI, V. J.: A survey on the dynamics and control of tethered satellite systems AAS 86-246, 667–719

SCHAGERL, M.: Dynamik von Seilen unter holonomen Zwängen, ZAMM 77(1997) S1,291-292

STEINER, W., STEINDL, A., AND TROGER, H.: 1995. "Dynamics of a Space Tethered Satellite System with Two Rigid Endbodies". In *Proceedings of the Fourth International Conference on Tethers in Space.* Science and Technology Corporation, Hampton, VA, 1367–1379.

STEINER, W., ZEMANN, J. V., STEINDL, A., TROGER, H.: Numerical Study of Large Amplitude Oscillations of a Two Satellites Continuous Tether System with Varying Length. Acta Astronautica **35** (1995), 607–621.

VU-QUOC, L., SIMO, J. C.: Dynamics of Earth-Orbiting Flexible Satellites with Multibody Components. J. Guidance, Control and Dynamics **10** (1987), 549–558.

A FIELD BOUNDARY ELEMENT FORMULATION FOR AXISYMMETRIC FINITE STRAIN ELASTOPLASTICITY[1]

G. KUHN AND O. KÖHLER
Universität Erlangen-Nürnberg
Egerlandstr. 5, D-91058 Erlangen, Germany

1. Introduction

The boundary element method (BEM) is now well established in linear elasticity, where this approach leads to a reduction of the problem by one dimension. When dealing with nonlinear problems an additional domain integral is required in order to represent the nonlinear behaviour. Due to this additional integral the method is then called field boundary element method (FBEM).

Although many geometrically linear problems with inelastic material behaviour have been solved by FBEM, only few have been published for elastoplasticity at finite strains (see e.g. [1, 2, 3, 4] and the references given there). In contrast to most of these approaches, which make use of constitutive equations in rate-form, the FBEM formulation proposed by Foerster [3, 4] is based on the concept of a hyperelastic response relative to an intermediate configuration. This leads to a time-independent formulation on a global level, whereas the evolution of the intermediate configuration is carried out on a nodal level. Because of the Total-Lagrange approach the system matrices need to be calculated only once. The application of this formulation to axisymmetric problems is presented in this paper.

2. Governing Equations

The deformation of a body from the initial to the current configuration can be described by an invertible mapping $x = \phi(X, t)$, relating the coordinates X of a material point in the initial configuration to the coordinates x in the current configuration at a certain time t. According to the rotational symmetry of the considered problems, cylindrical coordinate systems $\{x^a\}$ and $\{X^A\}$ will be used in the following, that is,

[1]Dedicated to the 60th birthday of Prof. Dr. tech. Dr. h.c. F. Ziegler, TU-Wien

H. A. Mang and F. G. Rammerstorfer (eds.),
IUTAM Symposium on Discretization Methods in Structural Mechanics, 147–154.

$$\begin{array}{llll} x^1 = r\,, & x^2 = \varphi\,, & x^3 = z\,; \\ X^1 = R\,, & X^2 = \Phi\,, & X^3 = Z\,. \end{array} \tag{1}$$

The components of the deformation gradient $\boldsymbol{F}$ are given by

$$F^a{}_A(X) = \frac{\partial \phi^a}{\partial X^A}(X)\ . \tag{2}$$

It should be mentioned, that this coordinate expression does not involve any covariant derivatives, because ϕ is not a vector, but a mapping of points. The deformation gradient can be used to express the equation of equilibrium in the initial configuration

$$\boldsymbol{Sh}^{-1}\cdot \mathrm{DIV}(\boldsymbol{F}\cdot\boldsymbol{S}) + \rho\boldsymbol{B} = 0 \tag{3}$$

in terms of the 2nd Piola-Kirchhoff stress tensor $\boldsymbol{S}$ and the volume force density $\rho\boldsymbol{B}$ both defined in the initial configuration. The Shifter $\boldsymbol{Sh}$ is a two-point tensor, that transports vectors emanating from X to vectors emanating from x. The right Cauchy-Green tensor $\boldsymbol{C}$ and the Green strain tensor $\boldsymbol{E}$ are defined by

$$\boldsymbol{C} = \boldsymbol{F}^{T}\cdot \boldsymbol{F} \quad , \quad \boldsymbol{E} = \tfrac{1}{2}[\boldsymbol{C} - \boldsymbol{G}]\ , \tag{4}$$

where $\boldsymbol{G}$ denotes the metric tensor in the initial configuration.

Using the material derivative of the displacement $\boldsymbol{H}$, which with respect to cylindrical coordinates is defined as

$$H^A{}_B(X) = \frac{\partial U^A}{\partial X^B}(X) \tag{5}$$

with the displacement vector $\boldsymbol{U}(X)$ in axisymmetric problems given by

$$U^A(X) = \begin{bmatrix} \phi^r(X) - X^R \\ \phi^\varphi(X) - X^\Phi \\ \phi^z(X) - X^Z \end{bmatrix} = \begin{bmatrix} r - R \\ \varphi - \Phi \\ z - Z \end{bmatrix}\ , \tag{6}$$

leads to the alternative expression for the deformation gradient

$$\boldsymbol{F} = \boldsymbol{Sh}\,(\boldsymbol{G} + \boldsymbol{H})\ . \tag{7}$$

Using this relation, the equation of equilibrium (3) takes the form

$$\mathrm{DIV}[(\boldsymbol{G}+\boldsymbol{H})\cdot\boldsymbol{S}] + \boldsymbol{Sh}^{-1}\cdot \nabla\boldsymbol{Sh} : [(\boldsymbol{G}+\boldsymbol{H})\cdot\boldsymbol{S}] + \rho\boldsymbol{B} = 0 \tag{8}$$

and the Green strain tensor (4b) can be written as

$$\boldsymbol{E} = \tfrac{1}{2}\left[\boldsymbol{H} + \boldsymbol{H}^{T}\right] + \tfrac{1}{2}\boldsymbol{H}^{T}\cdot \boldsymbol{H}\ . \tag{9}$$

The second term in the sum of eqn (8) only occurs in torsion problems, as $\nabla \boldsymbol{Sh}$ is non-zero in this case (see [5] for the covariant derivative of two-point tensors).

Finally, a hyperelastic constitutive law is used to determine the 2nd Piola-Kirchhoff stress tensor

$$\boldsymbol{S} = 2\rho \frac{\partial}{\partial \boldsymbol{C}} \psi(\boldsymbol{C}, \boldsymbol{C}^p, \boldsymbol{\Lambda}) \tag{10}$$

where the free energy ψ depends on the right Cauchy-Green tensor $\boldsymbol{C}$, the intermediate configuration represented by the plastic right Cauchy-Green tensor

$$\boldsymbol{C}^p = \boldsymbol{F}^{p^T} \cdot \boldsymbol{F} \tag{11}$$

and further internal variables $\boldsymbol{\Lambda}$ (see e.g. Simo [6, 7]). The intermediate configuration is introduced by a local multiplicative decomposition

$$\boldsymbol{F} = \boldsymbol{F}^e \cdot \boldsymbol{F}^p \tag{12}$$

of the deformation gradient into an elastic and a plastic part. Since this decomposition is not unique, the plastic right Cauchy-Green tensor $\boldsymbol{C}^p$ is used instead to describe the intermediate configuration. The constitutive equation (10) is accomplished by a set of evolution equations

$$\dot{\boldsymbol{C}}^p = \mathcal{F}(\boldsymbol{C}, \boldsymbol{C}^p, \boldsymbol{\Lambda}, \dot{\boldsymbol{C}}) \quad , \quad \dot{\boldsymbol{\Lambda}} = \mathcal{G}(\boldsymbol{C}, \boldsymbol{C}^p, \boldsymbol{\Lambda}, \dot{\boldsymbol{C}}) \tag{13}$$

for the plastic right Cauchy-Green tensor and the internal variables.

The algorithmic integration of the evolution equations and the computation of the consistent material tangent moduli depend on the explicit structure of eqn (13).

3. Field Boundary Element Formulation

To derive a FBEM formulation the governing eqns (8), (9) and a modified form of eqn (10)

$$\boldsymbol{S} \;=\; \boldsymbol{\mathcal{C}}_0 : \boldsymbol{E} \;+\; \boldsymbol{S}^n \tag{14}$$

are used. This formal split of the constitutive equation into a linear and a nonlinear part will be equivalent to the original hyperelastic law, if a constant elasticity tensor

$$\boldsymbol{\mathcal{C}}_0 \;=\; 4\rho\, \partial^2\psi/\partial \boldsymbol{C}^2 \Big|_{C=C^p=G} \tag{15}$$

is introduced and the nonlinear stress tensor is defined by

$$\boldsymbol{S}^n = 2\rho \frac{\partial \psi}{\partial \boldsymbol{C}} - \boldsymbol{\mathcal{C}}_0 : \boldsymbol{E} \; . \tag{16}$$

The integral equations are derived by starting from the weak form

$$0 = \int_{\mathcal{B}} \tilde{\boldsymbol{U}} \cdot \left\{ \mathrm{DIV}[(\boldsymbol{G}+\boldsymbol{H})\cdot\boldsymbol{S}] + \boldsymbol{Sh}^{-1} \cdot \nabla \boldsymbol{Sh} : [(\boldsymbol{G}+\boldsymbol{H})\cdot\boldsymbol{S}] + \rho\boldsymbol{B} \right\} \, dV \tag{17}$$

of the equation of equilibrium (8) in the initial configuration. Choosing Kelvin's linearelastic fundamental solution as test function $\tilde{\boldsymbol{U}}$ the derivation of the boundary integral equation can be performed in a classical way for the linear part of the governing equations. A standard limiting process $\xi \to \partial\mathcal{B}$ leads to the generalized boundary integral equation

$$\begin{aligned} \mathbf{C}(\xi)\cdot\boldsymbol{U}(\xi) &= \int_{\partial\mathcal{B}} \left[\overset{*}{\boldsymbol{U}}(X,\xi)\cdot\boldsymbol{T}(X) - \overset{*}{\boldsymbol{T}}(X,\xi)\cdot\boldsymbol{U}(X) \right] dA \\ &- \int_{\mathcal{B}} \nabla\overset{*}{\boldsymbol{U}}(X,\xi) : \boldsymbol{Nl}_1(X) \; dV + \int_{\mathcal{B}} \overset{*}{\boldsymbol{U}}(X,\xi)\cdot\boldsymbol{Nl}_2(X) \; dV \\ &+ \int_{\mathcal{B}} \overset{*}{\boldsymbol{U}}(X,\xi)\cdot\boldsymbol{B}(X)\rho(X) \; dV \end{aligned} \tag{18}$$

with two domain integrals accounting for the nonlinearities of the problem. The nonlinear expressions $\boldsymbol{Nl}_1$ and $\boldsymbol{Nl}_2$ are given by

$$\boldsymbol{Nl}_1 = (\boldsymbol{G}+\boldsymbol{H})\cdot\boldsymbol{S} - \boldsymbol{\mathcal{C}}_0 : \boldsymbol{H} \quad \text{and} \quad \boldsymbol{Nl}_2 = \boldsymbol{Sh}^{-1}\cdot\nabla\boldsymbol{Sh} : \left[(\boldsymbol{G}+\boldsymbol{H})\cdot\boldsymbol{S}\right]. \tag{19}$$

In polar coordinates the boundary values $\boldsymbol{U}(X)$, $\boldsymbol{T}(X)$ and the nonlinear terms $\boldsymbol{Nl}_1(X)$, $\boldsymbol{Nl}_2(X)$ do not depend on the angle Φ, so that the known kernel functions in eqn (18) can be integrated analytically with respect to the circumferential coordinate. Thus the three-dimensional problem is reduced to a two-dimensional one on a generator plane of the axisymmetric body. Due to the structure of the nonlinear terms $\boldsymbol{Nl}_1$, $\boldsymbol{Nl}_2$ and the axisymmetric kernel functions $\overset{\circ}{\boldsymbol{U}}$, $\nabla\overset{\circ}{\boldsymbol{U}}$, the two domain integrals can be combined so that the generalized boundary integral equation for axisymmetric problems finally reads

$$\begin{aligned} \mathbf{C}(\xi)\cdot\boldsymbol{U}(\xi) &= \int_{\partial\mathcal{M}} R \left[\overset{\circ}{\boldsymbol{U}}(X,\xi)\cdot\boldsymbol{T}(X) - \overset{\circ}{\boldsymbol{T}}(X,\xi)\cdot\boldsymbol{U}(X) \right] dS \\ &+ \int_{\mathcal{M}} R\overset{\circ}{\boldsymbol{U}}(X,\xi)\cdot\boldsymbol{B}(X)\rho(X) \; dA \\ &- \int_{\mathcal{M}} R\nabla\overset{\circ}{\boldsymbol{U}}(X,\xi) : \boldsymbol{Nl}(X) \; dA \, . \end{aligned} \tag{20}$$

As the nonlinear term $\boldsymbol{Nl}$ in the domain integral of eqn (20) depends on $\boldsymbol{H}$ directly and indirectly via the constitutive equation, a representation formula for the displacement gradient is needed. For source points ξ lying inside the domain $\mathcal{M}$ this can easily be achieved by differentiation of the axisymmetric generalized Somigliana equation with respect to the source point coordinates ξ. This leads to the integral formula for interior points

$$\begin{aligned}\boldsymbol{H}(\xi) = \nabla_{\xi}\boldsymbol{U}(\xi) = &\int_{\partial\mathcal{M}} R\left[\nabla_{\xi}\mathring{\boldsymbol{U}}(X,\xi)\cdot\boldsymbol{T}(X) - \nabla_{\xi}\mathring{\boldsymbol{T}}(X,\xi)\cdot\boldsymbol{U}(X)\right]\,dS \\ &+ \int_{\mathcal{M}} R\,\nabla_{\xi}\mathring{\boldsymbol{U}}(X,\xi)\cdot\boldsymbol{B}(X)\rho(X)\;dA \\ &- \int_{\mathcal{M}^r} R\,\nabla_{\xi}\nabla\mathring{\boldsymbol{U}}(X,\xi) : \boldsymbol{Nl}(X)\;dA \\ &- \nabla_{\xi}\int_{\mathcal{M}^s} R\,\nabla\mathring{\boldsymbol{U}}(X,\xi) : \boldsymbol{Nl}(X)\;dA\end{aligned} \tag{21}$$

where the domain integral is split into a regular part $\mathcal{M}^r$ and a singular part $\mathcal{M}^s$ containing the source point ξ. Source points on the boundary $\partial\mathcal{M}$ can be avoided by using discontinuous domain elements. The differentiation of the singular domain integral requires special care because some components of the kernel $\nabla\mathring{\boldsymbol{U}}$ are strongly singular. For such integrals arising in elastoplastic 3D problems a regularization technique based on Gauss theorem has been published by Dallner and Kuhn [8]. Applying an analogous method to the singular domain integral of eqn (21) results in the following regularized formulation. Due to the different behaviour of the covariant derivative of the axisymmetric kernel with respect to the field point coordinates R, Φ and Z, three expressions are given (no sum over R, Φ, Z). For the kernel components $\mathring{U}^{\bar{A}}{}_{B|R}$ the regularized expression reads

$$\begin{aligned}\frac{\partial}{\partial\xi^D}&\int_{\mathcal{M}^s} R\,\mathring{U}^{\bar{A}}{}_{B|R}(X,\xi)\;Nl^{BR}(X)\;dA \\ &= \int_{\mathcal{M}^s}\Big\{R\frac{\partial}{\partial\xi^D}\mathring{U}^{\bar{A}}{}_{B|R}(X,\xi)\left[Nl^{BR}(X) - Nl^{BR}(\xi)\right] \\ &\qquad - \frac{\partial}{\partial\xi^D}\mathring{U}^{\bar{A}}{}_{B}(X,\xi)\,Nl^{BR}(\xi)\Big\}\;dA \\ &\quad + Nl^{BR}(\xi)\int_{\partial\mathcal{M}^s} R\,\frac{\partial}{\partial\xi^D}\mathring{U}^{\bar{A}}{}_{B}(X,\xi)N_R\;dS\,.\end{aligned} \tag{22}$$

Because the covariant differentiation of the kernel function $\mathring{\boldsymbol{U}}$ with respect to Φ does not increase the singularity, no regularization is required in this case. The resulting domain integral is

$$\begin{aligned}\frac{\partial}{\partial\xi^D}&\int_{\mathcal{M}^s} R\,\mathring{U}^{\bar{A}}{}_{B|\Phi}(X,\xi)\;Nl^{B\Phi}(X)\;dA \\ &= \int_{\mathcal{M}^s} R\,\frac{\partial}{\partial\xi^D}\mathring{U}^{\bar{A}}{}_{B|\Phi}(X,\xi)\;Nl^{B\Phi}(X)\;dA\,.\end{aligned} \tag{23}$$

Finally the derivative with respect to Z leads to the regularized integral

$$\frac{\partial}{\partial \xi^D} \int_{\mathcal{M}^s} R\, \mathring{U}^{\bar{A}}{}_{B|Z}(X,\xi)\; Nl^{BZ}(X)\; dA$$
$$= \int_{\mathcal{M}^s} R\, \frac{\partial}{\partial \xi^D} \mathring{U}^{\bar{A}}{}_{B|Z}(X,\xi) \left[Nl^{BZ}(X) - Nl^{BZ}(\xi) \right] dA \tag{24}$$
$$+ Nl^{BZ}(\xi) \int_{\partial\mathcal{M}^s} R\, \frac{\partial}{\partial \xi^D} \mathring{U}^{\bar{A}}{}_{B}(X,\xi) N_Z\; dS\;.$$

There are two additional situations that require special attention when integrating axisymmetric kernel functions. When the source point ξ falls on the axis of rotation ($\bar{R} = 0$) separate limiting forms of the kernel functions are necessary. These are obtained by substituting the complete elliptic integrals E and K contained in the axisymmetric kernels by a series expansions and performing a limiting process afterwards. The second case concerns the accurate numerical integration of weakly singular boundary and domain integrals. This requires the kernel functions to be rearranged to a form suitable for logarithmically weighted gaussian integration [9].

4. Discretization and Iterative Solution

For the discretization of eqns (20) and (21) the usual standard procedures are used. Applying the collocation technique to the discretized integral equations provides the matrix equations

$$\mathbf{A}\mathbf{u} - \mathbf{B}\mathbf{t} = \mathbf{b} - \mathbf{E}\mathbf{n} \qquad \rightarrow \qquad \tilde{\mathbf{A}}\mathbf{x} = \tilde{\mathbf{B}}\mathbf{y} + \mathbf{b} - \mathbf{E}\mathbf{n} \tag{25}$$

and

$$\mathbf{h} = \tilde{\mathbf{G}}\mathbf{y} + \tilde{\mathbf{H}}\mathbf{x} + \tilde{\mathbf{b}} - \mathbf{E}'\mathbf{n}\;. \tag{26}$$

The vectors $\mathbf{u}$, $\mathbf{t}$, $\mathbf{y}$ and $\mathbf{x}$ contain the values of the boundary displacements and tractions and the known and unknown boundary values, respectively, for the nodes of the boundary elements. The vectors $\mathbf{h}$ and $\mathbf{n}$ contain the values of $\boldsymbol{H}$ and $\boldsymbol{Nl}$ for the nodes of the internal cells. Substituting the unknown boundary values $\mathbf{x}$ in eqn (26) by eqn (25) leads to a nonlinear set of equations

$$0 = \mathbf{f} = \mathbf{h} + \tilde{\mathbf{E}}\mathbf{n} - \overset{(l)}{\mathbf{h}}(\lambda) \tag{27}$$

with a residual vector $\mathbf{f}$ and the abbreviations

$$\tilde{\mathbf{E}} = \tilde{\mathbf{H}}\tilde{\mathbf{A}}^{-1}\,\mathbf{E} + \mathbf{E}'\;;\; \overset{(l)}{\mathbf{h}}(\lambda) = \left[\tilde{\mathbf{G}} + \tilde{\mathbf{H}}\tilde{\mathbf{A}}^{-1}\,\tilde{\mathbf{B}}\right]\, \overset{(l)}{\mathbf{y}}(\lambda) + \tilde{\mathbf{H}}\tilde{\mathbf{A}}^{-1}\,\mathbf{b}(\lambda) + \tilde{\mathbf{b}}. \tag{28}$$

The matrix $\tilde{\mathbf{E}}$ is constant and $\mathbf{h}^{(l)}(\lambda)$ depends on the loading factor λ and can be calculated directly from the prescribed boundary values and volume forces once the system matrices have been computed.

The path dependence of inelastic problems requires an incremental solution strategy for eqn (27) taking into account the integration algorithm of the evolution equation. In the typical $(k+1)^{\text{th}}$ load increment the vector of basic unknowns $\mathbf{h}^{k+1}$ has to be determined as the solution of

$$\mathbf{f}^{k+1} = \mathbf{h}^{k+1} + \tilde{\mathbf{E}}\mathbf{n}(\mathbf{h}^{k+1}, \mathbf{s}(\mathbf{c}^{k+1}, \mathbf{c}^{p\,k+1}, \mathbf{\Lambda}^{k+1})) - \overset{(l)}{\mathbf{h}}{}^{k+1} = 0 \;, \tag{29}$$

in which $\mathbf{c}^{k+1}$ is a function of $\mathbf{h}^{k+1}$. The stresses $\mathbf{s}$ depend on $\mathbf{c}^{k+1}$ and on the intermediate configuration characterized by $\mathbf{c}^{p\,k+1}$ and $\mathbf{\Lambda}^{k+1}$ which result from a local time integration algorithm of the evolution equations at each internal node. The integration starts from the known nodal values $(\boldsymbol{C}, \boldsymbol{C}^p, \boldsymbol{\Lambda})^k$ at the end of the kth load increment to avoid influences of the iteration path. The vector $\mathbf{h}^{(l)k+1}$ represents the new loading in the $(k+1)^{\text{th}}$ load increment.

The application of advanced iteration schemes using gradients (e.g. Newton-Raphson algorithm) requires the consistent linearization of the nonlinear problem with respect to the basic unknowns summarized in $\mathbf{h}$. The Jacobian matrix is given by

$$\mathbf{J} = \frac{\mathrm{d}\mathbf{f}}{\mathrm{d}\mathbf{h}} = \mathbf{I} + \tilde{\mathbf{E}}\frac{\mathrm{d}\mathbf{n}}{\mathrm{d}\mathbf{h}} \;. \tag{30}$$

Using the Newton-Raphson algorithm to solve the nonlinear set of equations (29) involves the solution of a large linear system of equations with $\mathbf{J}$ being the matrix of coefficients. Even for moderate-sized problems this task is by far the most time consuming part of the calculation and for large problems the solution by direct methods is prohibitive. The implementation of iterative solvers to reduce the computational time is currently under investigation. Another possibility of avoiding the decomposition of the Jacobian matrix is to solve the nonlinear set of equations with a fix point iteration in combination with a line-search algorithm. As this method may fail to converge when the current configuration is significantly different from the initial one, an update of the reference configuration to the last converged configuration becomes necessary. But for large problems, the recomputation of the system matrices is less time consuming than the decomposition of the Jacobian matrix in a Newton-Raphson scheme.

5. Implementation and Numerical Results

The FBEM formulation presented has been implemented in a boundary element code, including for example the constitutive model proposed by Simo [6, 7], which is based on a Neo-Hookean hyperelastic relation, the von Mises yield criterion and the principle of maximum plastic dissipation. Using this material law with isotropic exponential hardening, the elastoplastic elongation of the axisymmetric, notched rod shown in fig. 1 was computed in 30 load steps. The material parameters are: shear modulus $\mu = 80194\,\text{MPa}$, bulk modulus $K = 164206\,\text{MPa}$, linear hardening $h = 300\,\text{MPa}$, yield stress $\kappa_0 = 200\,\text{MPa}$, yield stress $\kappa_\infty = 600\,\text{MPa}$, saturation coefficient $\delta = 16$. The problem was discretized with 60 quadratic boundary elements and 495 linear domain cells resulting in 2720 degrees of freedom.

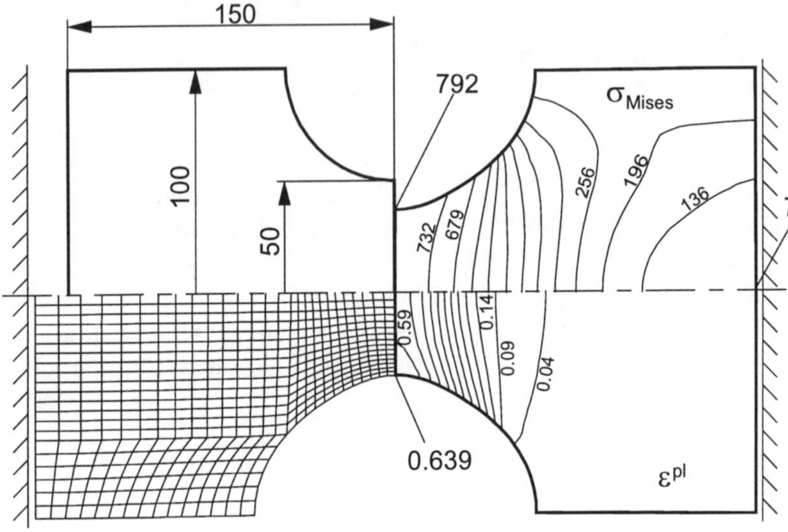

Figure 1. 10% elongation of an axisymmetric rod with a circular notch: model geometry, deformed mesh, contours of equivalent stress (von Mises) and equivalent plastic strain

References

1. Mukherjee, S. & Chandra, A.: Nonlinear formulations in solid mechanics, in *Boundary Element Methods in Mech.* (ed.: D.E. Beskos), 286–331, Elsevier Science Publ. 1987.
2. Okada, H. & Atluri, S.N.: Recent developments in the field-boundary element method for finite/small strain elastoplasticity, *Int. J. Solids Structures* **31**, 1737–1775 (1994).
3. Foerster, A. & Kuhn, G.: A Field Boundary Element Formulation for Material Non-linear Problems at Finite Strains, *Int. J. Solids Structures* **31**, 1777–1792 (1994).
4. Foerster, A.: *Eine Boundary-Element-Formulierung für geometrisch und physikalisch nichtlineare Probleme der Festkörpermechanik*, VDI-Fortschrittsberichte Reihe 18: Mechanik/Bruchmechanik Nr. 140, VDI Verlag, Düsseldorf 1993.
5. Marsden, J.E. & Hughes, T.J.R.: *Mathematical Foundations of Elasticity*, Prentice-Hall, Englewood Cliffs 1983.
6. Simo J.C.: A framework for finite strain elastoplasticity based on maximum plastic dissipation and the multiplicative decomposition: part I. continuum formulation, *Comp. Meth. Appl. Mech. Eng.* **66**, 199–219 (1988).
7. Simo J.C.: A framework for finite strain elastoplasticity based on maximum plastic dissipation and the multiplicative decomposition: part II. computational aspects, *Comp. Meth. Appl. Mech. Eng.* **68**, 1–31 (1988).
8. Dallner, R. & Kuhn, G.: Efficient evaluation of volume integrals in the boundary element method, *Comp. Meth. Appl. Mech. Eng.* **109**, 95–109 (1993).
9. Białecki, R.A., Herding, U., Köhler, O. & Kuhn, G.: Weakly singular 2D quadratures for some fundamental solutions, *Eng. Anal. Boundary Elem.* **18**, 333–336 (1996).

ON THE COMPUTATION OF DERIVATIVES UP TO THE BOUNDARY AND RECOVERY TECHNIQUES IN BEM

W.L. WENDLAND
Mathematisches Institut A, Universität Stuttgart, D–70569 Stuttgart, Germany

H. SCHULZ
Mathematisches Institut A, Universität Stuttgart, D–70569 Stuttgart, Germany

AND

CH. SCHWAB
Seminar f. Angewandte Mathematik, ETH Zürich, CH–8092 Zürich, Switzerland

Abstract. In this paper we present a bootstrapping algorithm for the computation of arbitrary derivatives of the Cauchy data and for the extraction of potentials near and on the boundary for general elliptic boundary value problems of second order if corresponding boundary integral equations are used. The method employs the derivatives of the Green representation formula in terms of Cauchy singular or weakly singular integrals and their composition with derivatives. It allows the robust and superconvergent numerical evaluation of potentials and their derivatives up to and on the boundary. Here we specify this method to the equations of elasticity.

1. Introduction

The computation of potentials, displacements, stresses and their derivatives from their potential representations on and near to the boundary is one of the difficult tasks in boundary element methods. It is, however, well known, that on either side of $\partial\Omega$ the potentials are smooth (even analytic) up to the boundary. So, in principle, there should be a "nonsingular" method for their accurate evaluation. To present such a method, proposed first in [8], with particular attention to the equations of elasticity, is the topic of

H. A. Mang and F. G. Rammerstorfer (eds.),
IUTAM Symposium on Discretization Methods in Structural Mechanics, 155–164.

the present paper. The approach is based upon the following key ideas: If the field point $x \in \Omega$ is sufficiently close to $\partial\Omega$, the potential $U(x)$ may be expanded in a Taylor series about the boundary point x_0 closest to x: $|x - x_0| = \text{dist}(x, \partial\Omega)$. By the smoothness of the potential, this Taylor expansion has good accuracy and the precision increases as $|x - x_0| \to 0$. This is paid for, however, with the requirement to know higher order normal derivatives of U at x_0; however, using the differential equation, these normal derivatives can be recursively calculated based only on the tangential derivatives of the solution at x_0. These, in turn, can be obtained numerically in a stable and efficient fashion, as shown below. This was numerically tested for the Laplacián in [6], where also criterions are determined whether the use of the traditional Green representation formula or a Taylor expansion is preferable.

In [6, 8], also a recovery algorithm is presented for the pointwise extraction of the solution and their derivatives on the boundary itself. Here we will give corresponding numerical results for the case of linear elasticity.

We consider a bounded two- or three-dimensional domain Ω with sufficiently smooth boundary $\Gamma := \partial\Omega$ and the formally positive elliptic system of second order

$$-\mathcal{P}U = -\sum_{k,\ell=1}^{n} \frac{\partial}{\partial x_\ell}\left(a_{k\ell}\frac{\partial}{\partial x_k}U\right) + cU = 0 \quad \text{in } \Omega \subset \mathbb{R}^n, \quad n = 2, 3 \tag{1}$$

with Cauchy data

$$U_{|\Gamma} =: \phi, \qquad \sum_{k,\ell=1}^{n} n_\ell a_{k\ell}\frac{\partial U}{\partial x_k} = TU_{|\Gamma} =: \psi \quad \text{on the boundary } \Gamma. \tag{2}$$

In traditional mixed boundary value problems ϕ is given on one part Γ_D of the boundary and ψ on the complementary part $\Gamma_N = \Gamma\backslash\Gamma_D$. For simplicity we assume here that ϕ is given on the whole boundary $\Gamma = \Gamma_D$ (Dirichlet problem). By using a fundamental solution $E(x, y)$ of $\mathcal{P}$, the solution U can be represented by

$$U(x) = (V\psi)(x) - (W\phi)(x) \quad \text{for } x \in \Omega \tag{3}$$

with a *single layer potential*

$$(V\psi)(x) = \int_\Gamma E(x, y)\psi(y)\, d\Gamma_y,$$

and a *double layer potential*

$$(W\phi)(x) = \int_\Gamma \left(T_{(y)}E(x, y)\right)^\top \phi(y)\, d\Gamma_y.$$

For $x \in \Gamma$ one obtains with the so-called jump relations the integral equation of the first kind

$$(V\psi)(x) = (\frac{1}{2}I + K)\phi(x) \tag{4}$$

where for the Dirichlet problem $F(x) := (\frac{1}{2}I + K)\phi(x)$ is given and

$$(K\phi)(x) = \int\limits_{\Gamma\setminus\{x\}} \left(T_{(y)}E(x,y)\right)^\top \phi(y)\, d\Gamma_y$$

is the direct Cauchy principal value double layer integral operator. In the case of linear elasticity we have

$$-\mathcal{P}U := -\mu\Delta U - (\lambda + \mu)\mathrm{grad}\,\mathrm{div}\, U, \tag{5}$$

where $E(x,y)$ is in the three-dimensional case defined by the Kelvin fundamental solution

$$E_{jk}(x,y) = \frac{\lambda + 3\mu}{8\pi\mu(\lambda + 2\mu)} \left\{ \frac{\delta_{jk}}{|x-y|} + \frac{\lambda + \mu}{\lambda + 3\mu} \frac{(x_j - y_j)(x_k - y_k)}{|x-y|^3} \right\},$$

$j,k = \{1,2,3\}$, where δ_{jk} denotes the Kronecker symbol and λ and μ are the Lamé constants, $T_{(y)}$ is then the boundary stress operator.

After solving the boundary integral equation (4) by a Galerkin or collocation method, the traditional evaluation of derivatives of the solution $U(x)$ in Ω uses the representation formula

$$\frac{\partial}{\partial x_m} U_h(x) = \frac{\partial}{\partial x_m} \int\limits_\Gamma E(x,y)\psi_h(y)\, d\Gamma_y - \frac{\partial}{\partial x_m} \int\limits_\Gamma \left(T_{(y)}E(x,y)\right)^\top \phi_h(y)\, d\Gamma_y$$

or corresponding finite difference formulae. For x near to the boundary this approach leads to a tremendous loss of accuracy, for details see e.g. [8]. An alternative is the following algorithm which will be presented in detail in the next two sections.

1. Rewrite the original system of differential equations with respect to tangential and normal coordinates at a point $x_0 \in \Gamma$.
2. At $x_0 \in \Gamma$, compute the tangential derivatives $\partial_s^\alpha U$ of order $|\alpha| \le L$ (where $s = (s_1, s_2)$ are intrinsic coordinates of Γ, see the Appendix ahead) on Γ with any chosen L.
3. Use Cauchy's algorithm for the computation of the normal derivatives $\left(\frac{\partial}{\partial n}\right)^\ell U$ on Γ, where n is the exterior unit normal vector at x_0.
4. Evaluate the solution or its derivatives at some point x near to the boundary by using a Taylor formula about some boundary point $x_0 \in \Gamma$ which, for convenience, here is chosen satisfying $|x - x_0| = \mathrm{dist}(x, \Gamma)$.

2. Computation of Derivatives on the Boundary

For the following considerations we need the representation of the differential operator $\mathcal{P}$ in a tubular neighbourhood of the boundary in the form

$$\mathcal{P}U = \sum_{k,\ell=1}^{n} \frac{\partial}{\partial x_\ell}\left(a_{k\ell}\frac{\partial}{\partial x_k}U\right) - cU = \left\{\mathcal{P}_2 + \mathcal{P}_1\frac{\partial}{\partial n} + \mathcal{P}_0\left(\frac{\partial}{\partial n}\right)^2\right\}U \quad (6)$$

with tangential operators $\mathcal{P}_m$ of orders m. These operators can be expressed in terms of the curvature, the components of the Riemann-tensor and corresponding derivatives of the coefficients a_{jk}, see [2, Chap. III] or the appendix of [8]. For the case of linear elasticity see the appendix or [7].

Theorem 1 *[2, 8] Suppose $\mathcal{P}U = 0$, Dirichlet boundary conditions and a decomposition (6) are given. Then all normal derivatives $(\frac{\partial}{\partial n})^\ell U$ and tangential derivatives $\partial_s^\alpha U$ are computable on the boundary Γ.*

Proof. Solving the Dirichlet boundary value problem gives the second Cauchy data $\psi = TU$ on Γ which fulfills the boundary integral equation (4). Using (4) one obtains with $A := V$ and $F := (\frac{1}{2}I + K)\phi$ for arbitrary tangential derivatives $\partial_1^k\partial_2^\ell\psi$ up to the order $k+\ell$, the triangular system

$$\begin{aligned} A\psi &= F \\ A(\partial_1\psi) &= \partial_1 F - A_{(1,0)}\psi \\ A(\partial_2\psi) &= \partial_2 F - A_{(0,1)}\psi \\ &\vdots \\ A\left(\partial_1^k\partial_2^\ell\psi\right) &= \partial_1^k\partial_2^\ell F - \sum_{\substack{i,j=0\\ 0<i+j}}^{k,\ell}\binom{k}{i}\binom{\ell}{j}A_{(i,j)}\left(\partial_1^{k-i}\partial_2^{\ell-j}\psi\right) \end{aligned} \quad (7)$$

where $A_{(0,0)} := A$ and $A_{(i+1,j)} := [\partial_1, A_{(i,j)}]$, $A_{(i,j+1)} := [\partial_2, A_{(i,j)}]$.

The system (7) can be solved if A is invertible, at least numerically. The brackets $[\partial, A]$ denote the commutator $(\partial A - A\partial)$. Then, using

$$TU = \psi = \mathcal{P}_0\frac{\partial U}{\partial n} + \alpha_1\frac{\partial\phi}{\partial s_1} + \alpha_2\frac{\partial\phi}{\partial s_2} \quad \text{with } \alpha_\ell(x) = \sum_{j,k=1}^{3} a_{jk}n_j\frac{\partial\chi_k}{\partial s_\ell}, \quad (8)$$

the first normal derivative of U can be computed. The vector function $\chi = \chi(s_1, s_2)$ denotes a (local) representation of the boundary surface. Using the differential equation $\mathcal{P}U = 0$ the application of Cauchy's algorithm gives

$$\begin{aligned} \frac{\partial^2 U}{\partial n^2} &= -\mathcal{P}_0^{-1}\left\{\mathcal{P}_1\frac{\partial U}{\partial n} + \mathcal{P}_2 U\right\}, \\ \frac{\partial^3 U}{\partial n^3} &= -\mathcal{P}_0^{-1}\left\{\mathcal{P}_1\frac{\partial^2 U}{\partial n^2} + \mathcal{P}_2\frac{\partial U}{\partial n}\right\} - \left(\frac{\partial}{\partial n}\mathcal{P}_0^{-1}\mathcal{P}_1\right)\frac{\partial U}{\partial n} - \left(\frac{\partial}{\partial n}\mathcal{P}_0^{-1}\mathcal{P}_2\right)U \\ &\vdots \end{aligned}$$

for the normal derivatives. Since $\mathcal{P}_m$, $m = 0, ..., 2$, are tangential operators, this completes the proof. □

The numerical algorithm works in the same manner. We emphasize that for the computation of the tangential derivative always the same stiffness matrix corresponding to the single layer or the double layer potential can be used which is in a usual method already implemented and computed. The calculation of the first commutators can be done in a straightforward and simple way using the chain rule and kernel properties of the boundary integral operators, see [2, 8].

Using a Galerkin method to solve system (7), one obtains, based on strong ellipticity of the operator A, the error estimate

$$||\partial_s^\varrho \psi - \psi_h^{(\varrho)}||_{H^p(\Gamma)} \leq c(\varrho)\, h^{d-p}||\psi||_{H^{d+|\varrho|}(\Gamma)},$$

provided $-d+2\alpha \leq p \leq \alpha \leq d$, where 2α is the order of the operator A and d is the polynomial degree of the ansatz- and test-space. For more details see [1, 2, 4, 6, 8].

To obtain localized versions of the algorithm, local error estimators etc., the whole procedure can be combined with a window technique using truncation functions similar to [5].

For traditional mixed boundary value problems, the algorithm is also applicable, provided additional regularity at the cross points. For that case, after solving the mixed boundary value problem, the complete Cauchy data ϕ and ψ are given. The tangential derivatives of ϕ can be obtained solving a system similar to system (7), but with $A := \frac{1}{2}I + K$ and $F := V\psi$, respectively.

3. Evaluation of the Solution near to the Boundary

For x near to the boundary the convergence of the traditional evaluation (3) deteriorates significantly. After the computation of the tangential and normal derivatives of the solution on the boundary Γ we are able to evaluate for that case the solution at $x \in \Omega$ via the Taylor formula:

$$U(x) = \phi(x_0) + v\psi(x_0) + \frac{1}{2}v^2\partial_n^2 U(x_0) + \frac{1}{3!}v^3\partial_n^3 U(x_0) + \dots \qquad (9)$$

about a boundary point $x_0 \in \Gamma$ which has to be computed as the point with minimal distance to x. The parameter v is given by $v := |x - x_0|$. Obviously the accuracy of the evaluation (9) strongly depends on v. For the corresponding error estimates and for a comparison of the numerical errors of (3) and (9), see [6].

4. Recovery Techniques for Linear Elasticity

The Galerkin or collocation solutions of the triangular system (7) are possible discontinuous and not suitable for the direct evaluation (9). In this section we present a technique following [6, 8] for the computation of superconvergent point values of the solution and derivatives on the boundary itself.

4.1. THE GENERAL APPROACH

Let $s = (s_1, s_2)$ be the local representation of a boundary point x_0 with respect to an intrinsic coordinate system. Then, using repeated integration by parts, one obtains for any distribution ψ which is smooth at $x_0 \in \Gamma$ the following representation for the tangential derivatives [8]

$$(\partial_s^\gamma \psi)(s) = \sum_{0\leq|\beta|\leq M} \int d_\beta(t-s)(\partial^{\gamma+\beta}\psi)(t) H_k(t-s)\, dt \quad \text{with} \tag{10}$$

$$H_k(t-s) = \frac{(-1)^k \cdot 2}{\pi 4^k\left((k-1)!\right)^2} |t-s|^{2(k-1)} \ln|t-s|,\ M = 2k, \ \text{ for } M \text{ even} \tag{11}$$

and $H_{k-1} = \nabla_t H_k$, $M = 2k-1$, for the remaining case. The functions $d_\beta(t-s)$ are given in terms of derivatives of a sufficiently smooth cut-off function $\zeta(t-s)$ with $\zeta(0) = 1$ and can be obtained with the help of the Leibniz formula for even M from

$$(-1)^k \Delta_t^k \left(\zeta(t-s)(\partial^\gamma\psi)(t)\right) = \sum_{0\leq|\beta|\leq 2k} d_\beta(t-s)(\partial^{\gamma+\beta}\psi)(t) \quad \text{and from}$$

$$(-1)^{k-1}\nabla_t\Delta_t^{k-1}\left(\zeta(t-s)(\partial^\gamma\psi)(t)\right) = \sum_{0\leq|\beta|\leq 2k-1} d_\beta(t-s)(\partial^{\gamma+\beta}\psi)(t)$$

for odd M. The integration in (10) is done with respect to the local coordinate system. Using now on the right-hand side of (10) the Galerkin solution of system (7) instead of the exact derivatives $\partial^{\gamma+\beta}\psi(t)$, one obtains the *superconvergent recovery approximation* $\widehat{\psi}^{(\gamma)}(s)$ for $\partial_s^\gamma\psi$ which fulfills the error estimate [8]

$$\left|(\partial_s^\gamma\psi)(s) - \widehat{\psi}^{(\gamma)}(s)\right| \leq ch^{\ell+m}||\psi||_{H^{\ell+|\gamma|+M+\alpha}(\Gamma)}, \tag{12}$$

provided $M > 1-\alpha$ and $0 \leq \ell \leq \min\{d-\alpha, M-1+\alpha-\varepsilon\}$, $\varepsilon > 0$; $0 \leq m \leq d-\alpha$, $\ell + m \leq 2d - 2\alpha$.

4.2. A TWO-DIMENSIONAL NUMERICAL EXAMPLE

In [6] are numerical experiments concerning the recovery techniques on the boundary and the evaluation of the solution near to the boundary for the Laplace equation with Dirichlet boundary data are given. Here, we present results for the case of linear elasticity and for the pointvalue extraction of the normal stress vector on the boundary using discontinuous trial functions.

We consider the Dirichlet boundary value problem of linear elasticity

$$\mathcal{P}U = 0 \quad \text{in } \Omega, \quad U_\Gamma = \phi \tag{13}$$

with $\mathcal{P}$ defined as in (5) over the elliptical domain

$$\Omega := \left\{x \in \mathbb{R}^2 : \; x_1^2/a^2 + x_2^2/b^2 < 1\right\} \tag{14}$$

with $a = 0.6$, $b = 0.3$. The following algorithm was utilized:

1. Compute a first approximation ψ_h for the unknown Neumann datum by solving equation (4) approximately with a piecewise constant ansatz ($d = 1$) and the Galerkin Method on a *polygonal approximation* Γ_h of Γ.
2. Find an approximation $\psi_h^{(1)}$ for $\partial_s\psi$ by solving the second equation of (7) approximately with the same piecewise constant trial functions and the Galerkin Method on Γ_h.
3. Obtain superconvergent pointvalues $\hat{\psi}_h(x_0)$ via the post-processing formula (10) on the *real boundary* Γ.

For the numerical computations we have chosen the Dirichlet datum in form of the first row of the Kelvin fundamental solution with source point $y = (-0.5, -0.5)^\top \notin \overline{\Omega}$ for the two-dimensional case, so the unknown Neumann datum ψ and the derivatives $\partial_{s_\kappa}\psi$ are explicitly known and it is possible to control the numerical errors of the steps 1.-3. directly. The following tables show the relative L^2 and L^∞ errors for the standard Galerkin solution, the tangential derivative and the recovery solution. For the realization of the Galerkin method the boundary element code of O. Steinbach [9] was used.

N	$\|\|\psi - \psi_h\|\|_{L^2}$	t_1	$\|\|\psi' - \psi_h'\|\|_{L^2}$	$\|\|\psi - \tilde{\psi}_h\|\|_{L^2}$	t_2
16	0.224615	0.16	0.567461	0.125828	0.33
32	0.114549	0.51	0.271384	0.025274	1.12
64	0.057753	1.81	0.138825	0.005386	4.01
128	0.028962	7.25	0.069538	0.001237	15.94
256	0.014492	28.94	0.034765	0.000298	63.45
512	0.007247	115.80	0.017381	7.326e-5	252.58
1024	0.003624	466.36	0.008690	1.817e-5	1016.27

Tab. 1: L^2 error of the Galerkin solution, of the Galerkin solution for the tangential derivative and of the recovered solution

N: Number of boundary elements.

t_1: CPU time in seconds for the computation of ψ_h.

t_2: CPU time in seconds for the whole algorithm, including the evaluation of the recovered solution in $10 \cdot N$ points for the computation of the errors.

N	$\|\|\psi - \psi_h\|\|_{L^\infty}$	α	$\|\|\psi' - \psi_h'\|\|_{L^\infty}$	α	$\|\|\psi - \tilde{\psi}_h\|\|_{L^\infty}$	α
16	0.298329		0.860647		0.197483	
32	0.172736	0.8	0.541414	0.7	0.037380	2.4
64	0.097015	0.8	0.306541	0.8	0.010325	1.9
128	0.048507	1.0	0.156122	1.0	0.002473	2.0
256	0.024229	1.0	0.078192	1.0	0.000592	2.0
512	0.012111	1.0	0.039087	1.0	0.000144	2.0
1024	0.006055	1.0	0.019549	1.0	3.576e-5	2.0

Tab. 2: L^∞ errors of the Galerkin solution, of the recovered solution and corresponding numerical orders of convergence with respect to $\frac{1}{N}$.

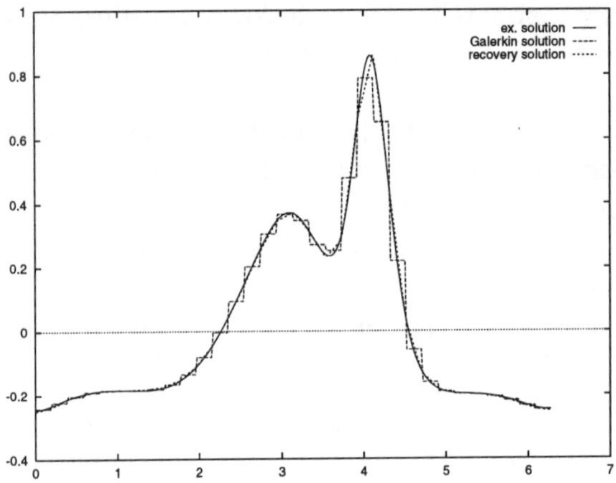

Fig. 1: Second component of the boundary normal stress vector: exact solution, Galerkin solution and recovered solution ($N = 32$)

Appendix

Following the notation in [2, Chap. III] and [8] we define a tubular neighbourhood of Γ via

$$\Phi(s_1, s_2, v) = \chi(s_1, s_2) + vn(s_1, s_2)$$

where $\chi(s_1, s_2)$ is a local parametric representation of Γ and $n(s_1, s_2)$ denotes the exterior unit normal vector to Γ. Then we define the Riemannian tensor $((g_{jk}))_{3\times3}$ as

$$g_{jk} := \sum_{\ell=1}^{3} \Phi_{\ell|j}\Phi_{\ell|k}. \quad \text{and} \quad g := \det((g_{jk}))_{3\times3}.$$

For more details see [2, 3, 8]. Furthermore we define matrices

$$N_i := \frac{1}{2}\left(e_i n^\top + n e_i^\top\right), \quad i = 1, \dots, 3,$$

where e_i is the ith unit vector, and numbers

$$\alpha_{\ell,\tau} := \frac{1}{g}(\Phi_{\ell|\tau} g_{\overline{\tau}\overline{\tau}} - \Phi_{\ell|\overline{\tau}} g_{\tau\overline{\tau}}), \qquad \beta_{jk} := \frac{1}{2}(\alpha_{k\tau}\alpha_{j\kappa} + \alpha_{k\kappa}\alpha_{j\tau})$$

with $\overline{\tau} \equiv \tau + 1$ where $\equiv$ is taken modulo 2 and $\tau \in \{1, 2\}$. Then for the case of linear elasticity one obtains for $\mathcal{P}_m$, $m = 0, \ldots, 2$,

$$\mathcal{P}_0 = \mu((\delta_{jk}))_{3\times 3} + (\lambda + \mu)((n_j n_k))_{3\times 3} \tag{15}$$

$$\begin{aligned} \mathcal{P}_1 &= \sum_{\tau=1}^{2}\sum_{\ell=1}^{3} \alpha_{\ell\tau} \left(\mu \left(n_\ell \frac{\partial}{\partial_{s_\tau}} + n_{\ell|\tau} \right) ((\delta_{jk}))_{3\times 3} \right. \\ &+ \left. (\lambda + \mu) \left(N_\ell \frac{\partial}{\partial_{s_\tau}} + N_{\ell|\tau} \right) \right) + \left(\sqrt{\frac{\gamma}{g}}(vK - H) - \sum_{\kappa=1}^{2} G^{\kappa}_{\kappa 3} \right) \mathcal{P}_0, \end{aligned} \tag{16}$$

$$\begin{aligned} \mathcal{P}_2 &= \sum_{\kappa,\tau=1}^{2} \frac{1}{\sqrt{g}} \frac{\partial}{\partial s_\kappa} \sqrt{g} \left\{ \mu \sum_{i=1}^{3} \alpha_{i\tau}\alpha_{i\kappa}((\delta_{jk}))_{3\times 3} + (\lambda + \mu)((\beta_{jk}))_{3\times 3} \right\} \frac{\partial}{\partial s_\tau} \\ &+ \sum_{\tau=1}^{2} \left(\sum_{i=1}^{3} 2\sqrt{\frac{\gamma}{g}}(vK - H)\alpha_{i\tau} + \sum_{\varrho=1}^{2} g^{\varrho\tau} n_{i|\varrho} \right. \\ &- \left. \sum_{\varrho,\nu=1}^{2} \left(g^{\varrho\tau} G^{\nu}_{\varrho 3} + g^{\nu\varrho} G^{\tau}_{\varrho 3} \right) \phi_{i|\nu} \right) (\mu n_i((\delta_{k\ell}))_{3\times 3} + (\lambda + \mu) N_i) \frac{\partial}{\partial s_\tau}, \end{aligned} \tag{17}$$

where $((g^{\ell r}))_{3\times 3} := ((g_{jk}))^{-1}_{3\times 3}$ and $G^{\tau}_{\varrho k}$, K and H are the Christoffel symbols, the Gaussian curvature and the mean curvature, respectively, see [3].

Proof. The proof is based on the general representation of $\mathcal{P}_0$, $\mathcal{P}_1$ and $\mathcal{P}_2$ for an arbitrary system of second order in [2, Chap. III] and the formula

$$\sum_{j,k=1}^{3} a_{jk} b_{jk} = \mu \sum_{i=1}^{3} b_{ii}((\delta_{kj}))_{3\times 3} + (\lambda + \mu)\frac{1}{2}((b_{k\ell} + b_{\ell k}))_{3\times 3}, \quad b_{jk} \in \mathbb{R}, \tag{18}$$

valid for the system of linear elasticity. The representation for $\mathcal{P}_0$ follows immediately from [2, 8]

$$\mathcal{P}_0 = \sum_{j,k=1}^{3} a_{jk} n_j n_k$$

with $b_{jk} = n_j n_k$. For $\mathcal{P}_1$ we obtain from [2, 8]

$$\begin{aligned}\mathcal{P}_1 &= \sum_{\tau=1}^{2}\left(\sum_{j,k=1}^{3} a_{jk}\sum_{\nu=1}^{2}\Phi_{j|\nu}n_k g^{\nu\tau}\right)\frac{\partial}{\partial s_\tau} \\ &\quad + \left(2\sqrt{\frac{\gamma}{g}}(vK-H)-\sum_{\kappa=1}^{2}G^{\kappa}_{\kappa 3}\right)\sum_{j,k=1}^{3} n_j a_{jk} n_k \\ &\quad - \sum_{j,k=1}^{3}\sum_{\varrho,\tau,\nu=1}^{2} a_{jk} y_{j|\varrho} L^{\nu}_{\tau}\chi_{k|\nu} g^{\varrho\tau},\end{aligned}$$

using Eq. (18), the representation of $\mathcal{P}_0$ and $n_{|\lambda} = \sum_{\nu=1}^{2} L^{\nu}_{\tau}\chi_{|\nu}$ (see e.g. [3, Chap. III 11.7]) and a Computer algebra procedure [7] one finds representation (16). Formula (17) follows by the same manner; for more details see [7]. □

References

1. D. N. Arnold and W. L. Wendland: *On the asymptotic convergence of collocation methods.* Math. Comp. **41** (1983) 349-381.
2. G.C. Hsiao, W.L. Wendland. *Variational Methods for Boundary Integral Equations.* In preparation, Springer-Verlag Berlin.
3. D. Laugwitz: Differentialgeometrie. B.G. Teubner, Stuttgart 1960.
4. S. Prößdorf and B. Silbermann. *Numerical Analysis for Integral and Related Operator Equations.* Birkhäuser–Verlag, Basel, 1991.
5. H. Schulz. *Über lokale und globale Fehlerabschätzungen für adaptive Randelementmethoden.* Doctoral Thesis, University of Stuttgart, 1997.
6. H. Schulz, Ch. Schwab, W.L. Wendland. *The computation of potentials near and on the boundary by an extraction technique for boundary element methods.* Comput. Methods Appl. Mech. Engrg. Elsevier 1997, to appear.
7. H. Schulz, Ch. Schwab, W.L. Wendland. *On the computation of derivatives up to the boundary and recovery techniques in BEM for linear elasticity.* In preparation.
8. Ch. Schwab, W.L. Wendland. *On the Extraction technique in Boundary Integral Equations.* Math. Comp., to appear.
9. O. Steinbach. *Galerkin- und Kollokations-Diskretisierungen für Randintegralgleichungen in 2D (Dokumentation).* Preprint No. 96-5, Mathematics Institute A, Stuttgart University (1996).

PARALLEL COMPUTING WITH FREE MESH METHOD: VIRTUALLY MESHLESS FEM

G. Yagawa, T.Yamada and T.Furukawa
School of Engineering, University of Tokyo
7-3-1 Hongo, Bunkyo-ku, Tokyo, Japan 113

1. Introduction

Recent growth of the computing technology and the popularization of the computer have enabled the finite element method (FEM) to become more widely used to analyze complicated phenomena in heat conduction, structural mechanics, fluid dynamics etc. To simulate such problems accurately, large scale finite element analysis must be carried out. However, problems to be solved have become large in scale, thus difficult to work on a single processor with respect to computation speed and required memory size. This gives rise to parallel computation and, up to now, a number of papers have been reported[1-4] on the parallelization of FEM.

When dealing with more than one million elements on a parallel environment, the mesh generation process is however not easy due to the fact that it has to be conducted prior to the actual analysis, and is not suitable for parallel processing. Therefore various meshless methods[5-8], which do not need element-node connectivity information *a priori*, have received considerable attention. The diffuse element method (DEM)[5] and the element free Galerkin method (EFGM)[6] are some of them where the moving least square method is utilized. The reproducing kernel particle method (RKPM)[7] is another meshless scheme, which is based on particle method and wavelets. However, these meshless methods are based on integration region such as background-cell in EFGM, so that each calculation cannot be done node by node.

On the other hand, the authors have proposed another meshless finite element method, named Free Mesh Method (FMM)[8]. In this method, contrary to the other meshless methods, the global stiffness matrix can be assembled node by node independently, and

H. A. Mang and F. G. Rammerstorfer (eds.),
IUTAM Symposium on Discretization Methods in Structural Mechanics, 165–172.

domain decomposition for parallel computation can be performed with respect to only nodes. The purpose of this paper is to demonstrate techniques for the parallel computation of large scale analysis with FMM and its pre- and post-processing.

In the following section, the fundamental formulation of FMM is presented, and the third section deals with the pre- and post-processing for large scale analysis. Parallelization algorithm of FMM is presented in the fourth section. A large scale problem is solved on parallel processors in the fifth section, and some concluding remarks are given in the final section.

2. Free Mesh Method

As given data, we first allocate n nodal points in the domain of concern Ω:

$$P_i(x_i, y_i, r_i) \ \forall i \in \{1, \ldots, n\} \qquad \text{Equation (2.1)}$$

where $[x_i, y_i]$ are the Cartesian coordinates, and r_i is the nodal density information which is used to generate appropriate nodes as illustrated in Fig.1.

Free Mesh Method first allows each node to nominate a set of candidate nodes, C_i, with radius r_i, which is given by

$$C_i = \{ P_j(x_j, y_j) \mid l_{ij} < r_i, j \neq i, \forall j \in \{1, \ldots, n\} \} \qquad \text{Equation (2.2)}$$

where

$$l_{ij} = \sqrt{(x_j - x_i)^2 + (y_j - y_i)^2} \qquad \text{Equation (2.3)}$$

We will rewrite the set in terms of candidate nodes for further convenience.

$$C_i = \{ P_1^{c(i)}, \ldots, P_{n_c}^{c(i)} \} \qquad \text{Equation (2.4)}$$

where n_c is the number of candidate nodes($n_c \leq n$).

To select satellite nodes S_i from the candidate nodes C_i, or to create local elements L_i, we define the set of local segments (see Fig.2) stored in their length from the shortest with respect to node P_i as follows:

$$D_i \equiv \{ d_{jk} = \overline{P_j^{c(i)} P_k^{c(i)}} \mid j < k, \forall j, k \in \{1, \ldots, n_c\} \} \qquad \text{Equation (2.5)}$$

Algorithmically, local elements are generated by the following processes:

Let two domains Ω_{ijk}^{in} and Ω_{ijk}^{out}, be those inside and outside triangle $P_i P_j P_k$ with respect to segment d_{jk} respectively, at each segment d_{jk}. If there exists any node P_x in domain Ω_{ijk}^{in} illustrated in Fig.3, i.e., $\exists P_x \in \Omega_{ijk}^{in}$, then terminate the operation with respect to d_{jk}. Otherwise, in the domain Ω_{ijk}^{out} illustrated in Fig.4,

(I) $\exists P_x \in \Omega_{ijk}^{out} \rightarrow$ the operation with respect to d_{jk} is terminated.

If nodes exist in Ω_{ijk}^{out}, the length of segment l_{ix}, that has the least length out of the segments $d_{ix\gamma}$, $\forall\gamma\in\{1,...,m\}$, is defined:

(II) $\nexists P_{x1},...,P_{xm}\in\Omega_{ijk}^{out}\rightarrow l_{ix}\equiv\min\left\{l_{ix\gamma}|\forall\gamma\in\{1,...,m\}\right\}$

In case (II), the creation of local elements again differs:

(i) $l_{jk}\leq l_{ix} \rightarrow P_{xi},\forall i\in\{1,....,m\}$ is eliminated from candidate nodes.

(ii) $l_{jk}>l_{ix}\rightarrow$ the operation with respect to d_{jk} is terminated.

This process is done from the shortest to the longest and, by using the above selection algorithm, the satellite nodes for current satellite nodes S_i are given by

$$S_i=\{P_1^{s(i)},....,P_{n_s}^{s(i)}\} \quad \text{Equation (2.6)}$$

where n_s is the number of satellite nodes. ($n_s\leq n_c$)

Before generating the local element, the satellite nodes must be rearranged clock- (or counter clock-) wise:

$$S_i'=\{P_1^{s(i)'},....,P_{n_s}^{s(i)'}\} \quad \text{Equation (2.7)}$$

Finally temporary elements are generated from the current central node and two candidate nodes which are in neighbor as illustrated in Fig.5 . The number of temporary elements is n_e, and $n_e=n_s$ except for the boundaries.

After this selection, there is no inconsistency of triangulations in the analyzing domain as illustrated in Fig.6, even if the distribution of nodes is perfectly at random. In other words, this selection algorithm is a kind of mesh generation technique, which can be performed locally and independently on each node.

The node-by-node local stiffness matrix can be evaluated by each temporary element.

The local stiffness matrix of each temporary element is given by,

$$k^{e(i)}=\begin{bmatrix} k_i^{e(i)} \\ k_j^{e(i)} \\ k_k^{e(i)} \end{bmatrix} \quad \text{Equation (2.8)}$$

where $k_i^{e(i)}$ is the row vector of $k^{e(i)}$, to which the current central node is related, and $k_j^{e(i)}$ and $k_k^{e(i)}$ are the row vector of the current satellite nodes.

Then only the array ,which is related to the current central node, is added to the node-by-node local stiffness matrix, which is given by,

$$k^{L(i)}=\sum_1^{n_e} k_i^{e(i)} \quad \text{Equation (2.9)}$$

The global stiffness matrix is thus given by,

$$K=\sum_{i=1}^{n} k^{L(i)} \quad \text{Equation (2.10)}$$

Here, the node-by-node local stiffness matrix is the perfect i-th (current central node) array of the global stiffness matrix.

As a result, FMM is composed of two major processes. One is the construction of the global stiffness matrix as explained above, and the other is the derivation of the solution from the global stiffness matrix.

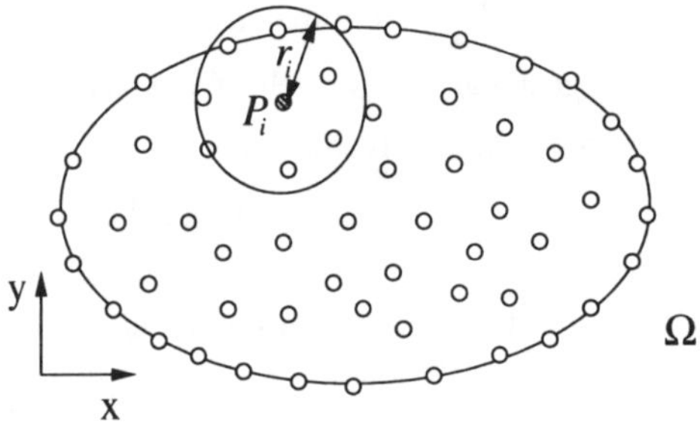

Figure 1. Given information before analysis

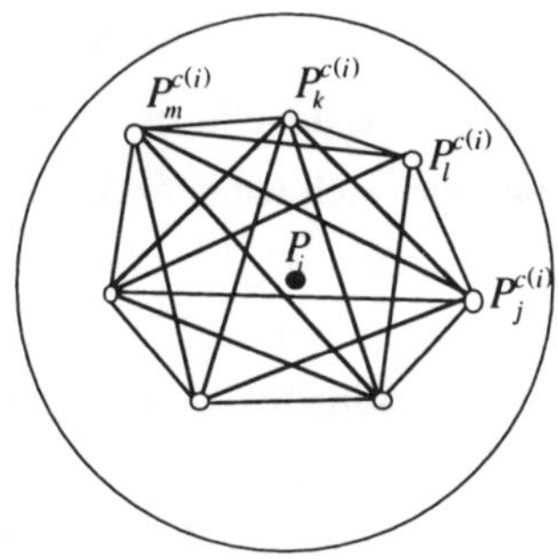

Figure 2. Segments connecting all the candidate nodes

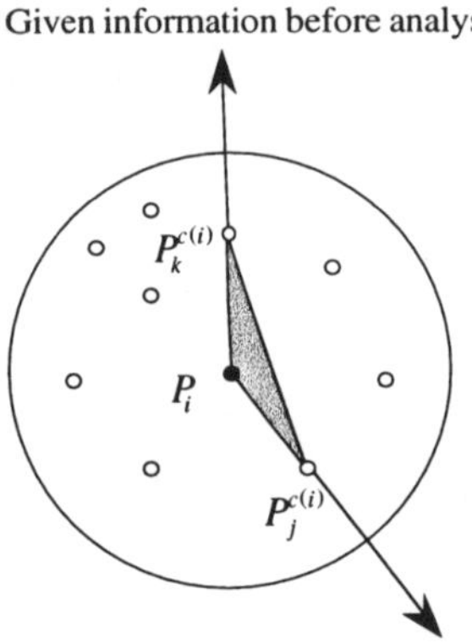

Figure 3. Inside region of triangle

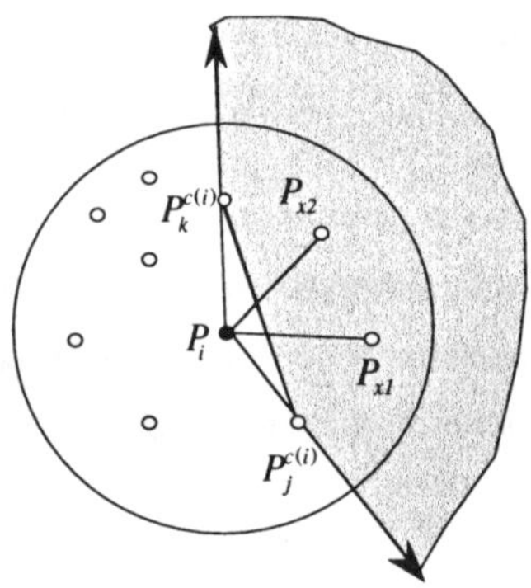

Figure 4. Outside region of triangle

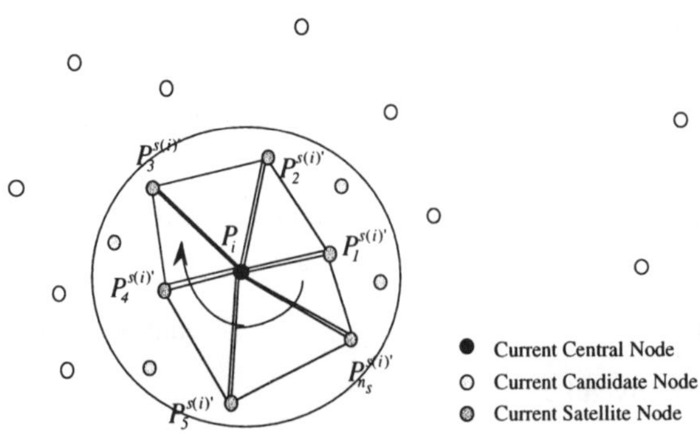

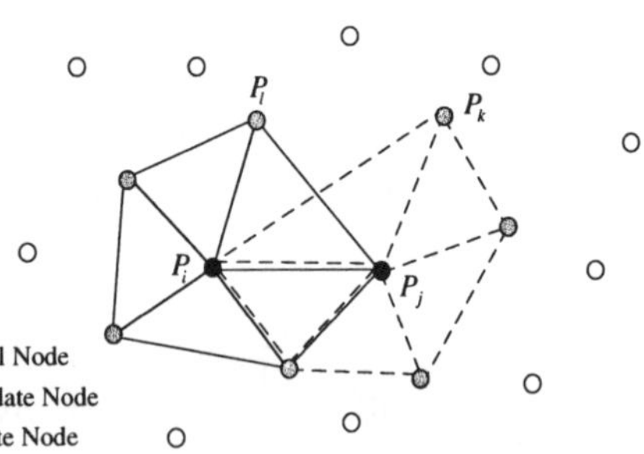

Figure 5. Local radial element arounl node P_i

Figure 6. Inconsistency in triangulation

3. Pre- and Post-Processing for Large Scale Analysis

3.1 PRE-PROCESSING

The domain to be analyzed is composed of a geometry model, analysis conditions and nodal data. As the nodal data consists of Cartesian coordinates and the nodal density information of each node as in equation (2.1), the pre-processing system distributes nodes in the analysis domain based on the nodal information after the user has designed a geometry model. For the appropriate distribution of nodes, fuzzy knowledge[9] is used. For example, let the original nodal patterns around holes or near crack tips be as shown in Fig.7. If a hole and a crack tip are so near that they may overlap and influence on each other, the resultant pattern takes into account both the patterns through the procedures explained in Fig.8. In detail, assuming that $w=Hole(x,y)$ and $u=Crack(x,y)$ are nodal density of two patterns near hole and around crack tip, respectively, the greater value of w and u is adopted as the final value of nodal density at the point $Overlap(x,y)$. Therefore the nodal density field over the whole analysis domain becomes the assembly of several nodal patterns.

After defining the nodal density field, the system must distribute nodes according to the field. For fast node generation the authors adopt the so-called bucket method which is a kind of hash sorting, and this bucket method will be introduced at the selection process of satellite nodes for each central node in FMM. A brief algorithm of the method is that the whole analysis domain is decomposed to small subdomains named "bucket", and nodes are generated bucket by bucket. When nodes, which are within a distance from a node, are searched, we need not check all nodes in the domain but only nodes in the appropriate bucket by using this method.

The procedure for node generation is as follows. First, a number of candidates for node, which is used in real analysis, are prepared. The distance between two neighboring candidates is set to be smaller than the distance representing nodal density defined around there. Each candidate for node is adopted as an input node to be generated when the following two criteria are satisfied: a) The candidate for nodes is inside the analysis domain; b) The distance of the candidate for node with the nearest node already generated is longer than the distance representing nodal density field. This length would be also used to search candidate nodes in FMM. Therefore r_i in equation (2.1) is a multiple of this radius. Finally we can get nodal information, the Cartesian coordinates and the nodal density information.

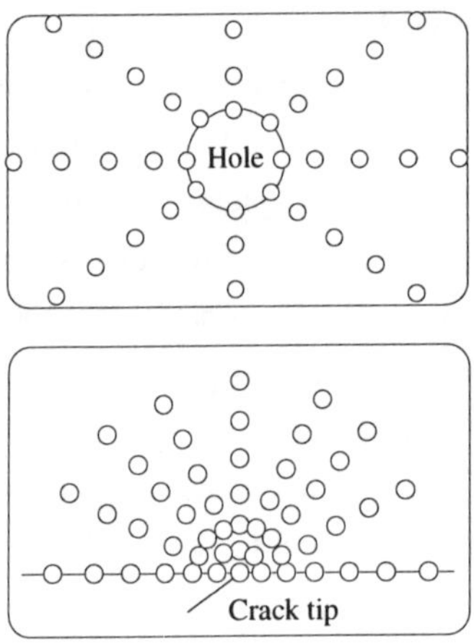

Figure 7. Local nodal patterns

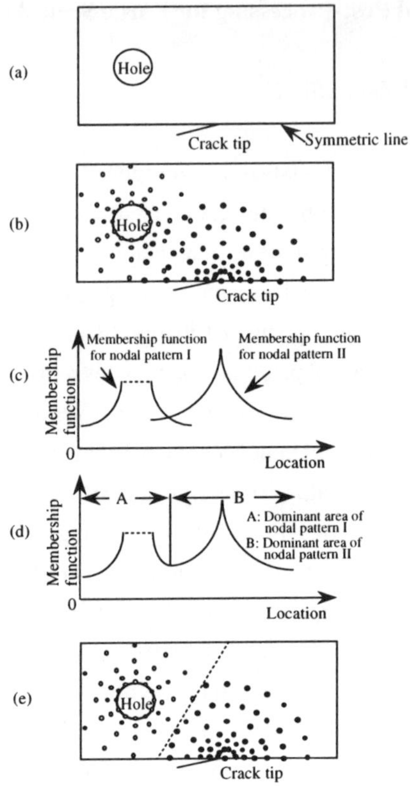

Figure 8. Superposition of nodal patterns based on fuzzy knowledge

3.2 POST-PROCESSING

In large scale analysis, post-processing is also a bottleneck because of its required memory size and visualization time. For instance, if a two-dimensional one-million d.o.f problem is solved by FEM, about two million triangles must be displayed. As the concerning module is not the triangle but the node in FMM, we only need to color nodes in the analysis domain if the number of nodes is considerably large. However, the CPU time and required memory size are still so large that displaying of the result requires approximately six minutes on Sun SparkStation 20 and that the output EPS file amounts to 50 MegaBytes for a two-dimensional one-million d.o.f steady state heat conduction problem. For avoiding these problems, we employ bucket visualization where the analysis result is displayed by coloring each bucket.

4. Parallelization of FMM

The parallelization of the computation is achieved by domain decomposition. In FMM, the domain of concern is more easily decomposed geometrically in terms of nodes by using recursive bisection than is done in terms of elements as illustrated in Fig.9. In addition, it is assumed that the processor used is MIMD (Multiple Instruction Multiple Data set) machine and each processor communicates by message passing style described by PVM (Parallel Virtual Machine) library.

When assembling the global stiffness matrix, no communication is necessary since at each node calculated are only its contributions to the global stiffness matrix using the coordinates data of neighbor nodes, which are stored *a priori*. Meanwhile, in the derivation of solution, a parallel iterative solver is adopted and the algorithm is the modified conjugate gradient algorithm[10] which was proposed by Aykanat.et.al for parallel computation.

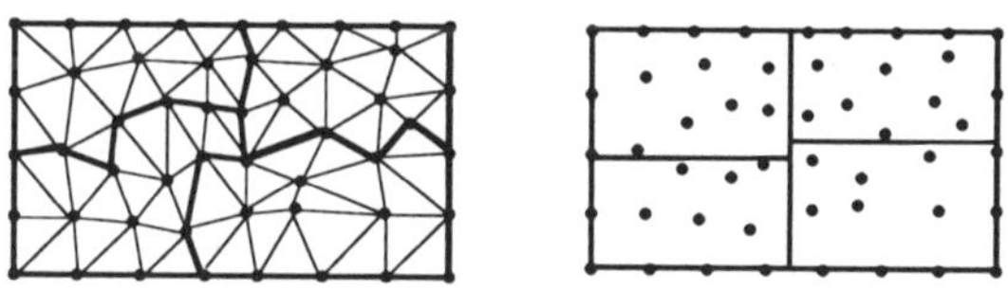

Figure 9. Domain partitioning in terms of element versus node

5. Numerical Example for Parallel Computing

About one million d.o.f heat conduction problem was solved on MIMD type parallel computer HITACHI SR2201. The domain analyzed and boundary conditions are shown in Fig.10. Distributed nodes were recursively decomposed to 4,16 and 64 subdomains.

Figure11 shows the result of temperature obtained from the parallel computation of 1,034,567 nodes model using 64 processors. Because the operating system of SR2201 is not so robust that computation time is not constant, we cannot discuss the parallel efficiency. However the computation time with 64 processors was approximately 20 minutes.

6. Conclusions

A parallel algorithm of FMM for MIMD type parallel computers and its pre- and post-processing have been proposed. A large scale two-dimensional heat conduction problem was solved on MIMD machine, HITACHI SR2201 efficiently, using the proposed algorithm. Conclusively the bottleneck for large scale computation of FEM, that is the mesh generation process, was avoided using parallel FMM and successfully applied to a large scale analysis with one million d.o.f.

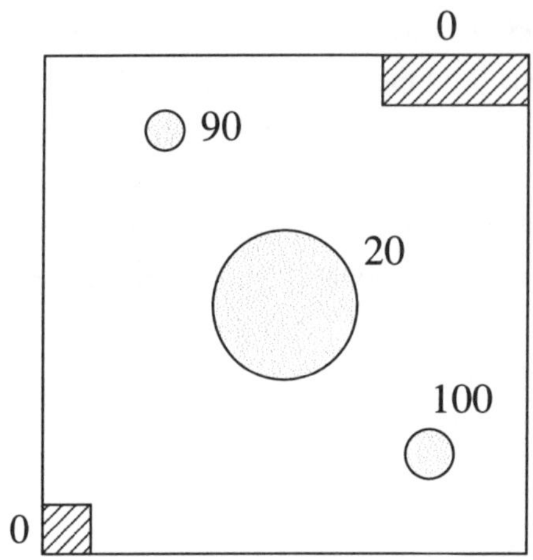

Figure 10. Analysis shape and boundary conditions

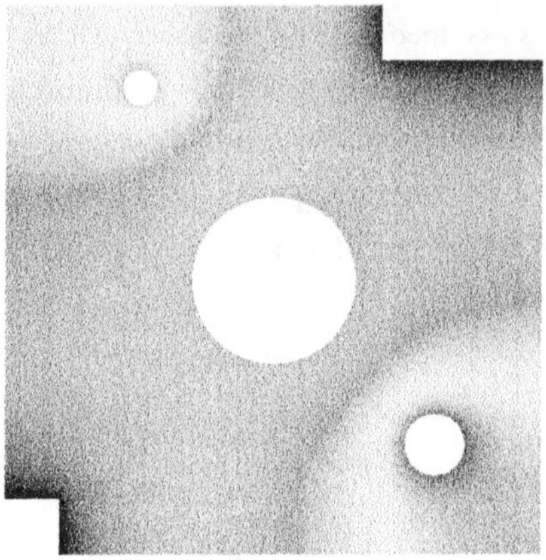

Figure 11. Analysis result on SR2201

REFERENCES

[1] E.Barragy , G.F.Carey (1988) A parallel element-by-element solution scheme, Int.J.Numer.Meth.Engrg., **26**, 2367-2382.

[2] C.Farhat, E.Wilson (1988) A parallel active column equation solver, Comp. and Struct., **28**, 289-304.

[3] G.Yagawa, N.Soneda, S.Yoshioka (1991) A large scale finite element analysis using domain decomposition method on a parallel computer, Comp. and Struct., **38**, 615-625.

[4] G.Yagawa, R.Shioya (1993) Parallel Finite Elements on a massively Parallel Computer with Domain Decomposition, Comp.Sys.Engrg., **2**, 427-449.

[5] B. Nayroles, G. Touzot, P. Villon (1992) Generating the Finite Element Method : Diffuse Approximation and Diffuse Elements, Comp.Mech., **10**, 307-318.

[6] T. Belytschko, Y. Y. Lu and L. Gu (1994) Element-Free Galerkin Methods, Int. J.Numer. Meth.Engrg., **37**, 339-256.

[7] W.K.Liu,S.Jun and Y.F.Zhang (1995) Reproducing Kernel Particle Methods, Int.J.Numer.Meth.Fluids., **20**, 1081-1106.

[8] G.Yagawa, T.Yamada (1996) Free Mesh Method: A kind of meshless finite element method,Int.J.Comp. Mech., **18**, 383-386.

[9] G.Yagawa, S.Yoshimura, N.Soneda, K.Nakao (1992) Automatic Two- and Three-dimensional Mesh Generation Based on Fuzzy Knowledge Processing, Int.J. Comp. Mech., **9**, 333-346.

[10] C.Aykanat, F.Ozguner, D.S.Scott (1990) Vectorization and parallelization of the conjugate gradient algorithm on hypercube-connected vector processors, Microprocessing and Microprogramming, **29**, 67-82.

MULTIGRID APPROACH TO ADAPTIVE ANALYSIS OF B.V. PROBLEMS BY THE MESHLESS GFDM

J. ORKISZ, P.LEZANSKI, P.PRZYBYLSKI
Cracow University of Technology
ul.Warszawska 24, 31-155 Kraków, Poland

ABSTRACT: Meshless methods are the subject of increasing interest nowadays. As such considered is here the adaptive finite difference method generalized for arbitrary irregular grids (GFDM). This is an integrated approach including original concepts of a'posteriori error analysis, solution smoothing mesh generation and modification as well as special emphasis posed on multigrid solution approach. Although the GFDM itself has been a well established method for years [15], its fully adaptive formulation has only recently been proposed and outlined [17,18,20]. The method is designed as a general and powerful tool of analysis of large and very large discrete boundary-value problems. A fully adaptive GFDM multigrid solution approach is outlined here. Presented are new concepts of prolongation, restriction and solution procedure.

1. Introduction

In the recent years more and more attention is paid to develop and apply discrete approach to analysis of boundary value problems based on a moving local approximation and unstructured grids. A variety of different names (see reviews [2,3]) is used for the so called meshless methods falling in such approach like: generalized finite difference (GFD) for arbitrary grids [7-9,12-20], kernel methods as smooth particle hydrodynamics and reproducing kernel particle, as diffuse element, element free Galerkin, h-p clouds meshless, partition of unity finite element, finite volume, finite point, particle-in-cell etc.. These names often describe similar or sometimes even identical approaches. The GDFM is the oldest [13,23] and, therefore, possibly the most developed of these methods. Following the earlier studies and the recent developments like error analysis, adaptivity and multigrid solution approach, the GFDM presents a general integrated solution tool displaying a variety of features. Some of them are briefly listed below.

(i) - Various formulations. The main idea of the FD approach is to replace e differential by difference operations. This concept may be applied to various formulations of boundary value problems. In the current state of development the GFDM may be used for solution of b.v. problems in such formulations as: local, global (given functional or variational principle), global in subdomains, mixed global and local (e.g. constrained optimization problem). Some of these formulations may be also presented as particular cases of the weighted residual ones. Combinations of the GFDM with other methods, e.g. FEM or BEM also may be considered. For sake of simplicity we assume here that in each of these formulations we deal with spaces providing functions sufficiently smooth in order to perform the necessary differentiations

H. A. Mang and F. G. Rammerstorfer (eds.),
IUTAM Symposium on Discretization Methods in Structural Mechanics, 173–180.

(ii) - Arbitrary irregular non structured grids may be used. Their generation is done by enforcing a'priori given local mesh density
(iii) - Moving local approximation is applied based on the truncated Taylor series and the weighted least squares fit. Moreover enhancement functions such as local asymptote singular solutions (e.g. jumps, corner solutions, etc.) may also be included.
(iv) - Consistency requirements are met, i.e., constant, linear and higher order functions when needed are exactly reproduced, due to interpolation nature of the local approximations used
(v) - Full adaptivity based on a local and global error control
(vi) - High computational efficiency addressed to solution of large b.v. problems

In this paper an overview of the adaptive GFDM multigrid solution approach is given. Presented are new concepts of error analysis and mesh modification as well as prolongation, restriction and solution approach.

Adaptivity is meant here in the "h" sense, i.e., due to appropriate mesh density modifications. However, a "p" type approach is also considered by using higher order GFD operators in a'posteriori analysis integrated into solution process.

2. A'posteriori Error Analysis

Considered are various criteria of the both local and global nature. Criteria worked out so far for the FEM [1,24], but expressed in terms of the GFDM might be used here. A general review of such a'posteriori error criteria is given in [1,7]. They are useful for various discrete methods. However, for the purpose of the adaptive GFDM special FD oriented criteria are proposed as well. They provide control of both magnitude of local residuum at selected points, and solution convergence at nodes common to the subsequent meshes used in the multigrid solution approach.

Residuum found by means of the local GFDM

Using the locally defined boundary value problem $\mathcal{L}\, u = f$ in Ω, $\mathcal{L}_b\, u = g$ on $\partial\Omega$, we may consider corresponding residuals $r \equiv \mathcal{L}u - f$, $\quad {}_b r \equiv \mathcal{L}_b u - g$ and require

$$\|r\| \le \eta_1 \|f\|, \qquad \|{}_b r\| \le \eta_2 \|g\|$$

where η_1, η_2 are threshold magnitudes assumed for admissible error level. The above norms may be taken either globally over the whole domain Ω, over subdomains Ω_i, $i=1,2,...,n$, or locally at any point P_i of this domain or its boundary.

Introducing local GFD representations of the operators $\mathcal{L}$ and $\mathcal{L}_b$ at any point P_i, based on star nodes $i+j$, $j=1,2,...,k$, we may approximately evaluate there the residuum r_i (or ${}_b r_i$)

$$r_i \approx \sum_{j=1}^{m_r} L_{i+j}^{(m_r)} u_{i+j} - f_i, \quad i = 1,2,...,n$$

at each considered point P_i by means of an appropriate GFD operator. Here m_r is a number of nodes in the star and $L_{i+j}^{(m_r)}$ are coefficients of the GFD operator used. The residuals r_i and ${}_b r_i$ may provide us with a useful error indication. However, the values of residuals evaluated in the way mentioned above depend on precision of both the GFD solution examined as well as a local GFD operator applied. This last dependence is inconvenient , but may be removed by taking into account higher order terms in the local approximation used.

We may expect that in 1D problems the largest residuals appear close to the mid points of neighboring nodes, while in 2D problems they may be found close to the centroids of triangles (quadrilaterals) generated on any arbitrary irregular grids. More detailed suggestions will be proposed when discussing mesh modification and the multigrid approach.

It is worth stressing that the above local residual error criteria does not need use of any reference solution, based on a'posteriori smoothing, as it is usually required by the other criteria. Here smoothing is built-in the generation procedure of the FD operators for $\mathcal{L}$ and $\mathcal{L}_b$ obtained by means of the moving weighted least squares approximation [10,15,22].

Convergence rates found at nodes common to subsequent meshes in the multigrid FD solution approach

A multigrid solution approach using more and more dense meshes is considered. We examine a solution convergence rate

$$\beta_i^k \equiv \frac{\left\| u_i^k - u_i^{k-1} \right\|}{\left\| u_i^k \right\|} \leq \eta_3$$

at those nodes P_i, $i = 1,2,\ldots$, that preserve the same locations in the subsequent meshes $\ldots,k\text{-}2,k\text{-}1,k,\ \ldots$; η_3 is an imposed error threshold value resulting from the required solution precision. Use of any other nodes is also possible, though less effective.

Using both error indicators r_i (and ${}_b r_i$) , and β_i^k defined above, we may generate series of more and more dense meshes, and establish an adaptive solution process. It is worth stressing that both ill-conditioned problems and very slowly convergent solutions may be detected and also controlled in this way.

The GFDM error indicators may be also applied to the adaptive solutions obtained by means of the global GFDM, or other discrete methods like the FEM as well.

3. A'posteriori Solution Smoothing

The moving weighted least squares approach [10,15,22] is used for smoothing purposes when a reference solution is required e.g. for error analysis. Either the direct GFD approximation procedure [8] may be applied here, or its specific version called the global-local method of the physically based approximation [5]. This last concept, worked out for smoothing of experimental data, is extended now [6] to numerical results. Advantage is taken from all information available for the problem in question, in order to provide the "very best" approximation of the given data.

4. Adaptive Mesh Generation and Modification

The adaptive approach considered here uses a mesh refinement concept. Basically any mesh generator, providing modifications based on a'posteriori solution errors, could be used for the adaptive GFDM solution approach. However more effective seems be using a special GFD oriented mesh generator. In order to control mesh modifications, it is convenient to define and use the notion of irregular mesh density. Advantage is taken from a simple idea of density of the regular squared mesh. Such density is in inverse proportion to the side of square assigned to a node [11,12,21]. In the case of irregular mesh we may use Voronoi tesselation in order to find Voronoi polygons assigned to each node. Converting each Voronoi polygon into an equivalent square of the same surface area Ω_i , we may determine the local mesh density ρ_i at a node i in the same way as if it was done for the regular mesh. Thus the relative local mesh density at a node i is defined like in [12]

$$\rho^{-1}{}_i = \rho^{-1}{}_{\max} + \frac{1}{2}\log_2\left(\frac{k\Omega_i}{\Omega_{\min}}\right)$$

Here, ρ_{max} is the largest density in the whole domain, corresponding to the Voronoi polygon with the smallest surface area Ω_{min}. For internal nodes $k=1$, for the boundary ones $k=2\pi/\alpha$, where α is the angle defining fraction of the polygon contained inside the domain ($\alpha=\pi$ for smooth boundary).

Mesh density $\rho(\mathbf{x})$ at an arbitrary point $\mathbf{x}$ of the domain Ω is obtained through interpolation of the nodal values, done on the optimal triangular Delaunay mesh (e.g. by using shape functions).

Mesh modifications may be done due to requirements imposed on mesh density upon results of a'posteriori error analysis.

Mesh refinement strategy

Proposed is here use of the Liszka's idea of "sieving" FD mesh generator [10], together with a new mesh modification strategy described below. This modification is based on a concept of generation of "correction filters" to the Liszka method sieve. They are found upon error indicators (residuals) defined above. The following are subsequent steps of the algorithm suggested:

(i) plan and generate the initial coarse mesh by the Liszka's method;
(ii) solve the boundary value problem in question, and find potential locations of new nodes using one step more dense sieve (*1/2* in 2D problems or *1* in 1D ones) than applied to the actual mesh;
(iii) examine the error indicators at potential locations of new nodes. Insert new nodes where admissible error norms (e.g. the average error value) are exceeded;
(iv) unless all error norms admissible for the final solution are satisfied, return to the second step of the algorithm.

In this way old nodes remain in their locations and new nodes are added. However, we may also wish to remove the old nodes sometimes, e.g. when they are totally surrounded by examined points with sufficiently low values of error indicators. Removed may be only those nodes, belonging to the actual mesh, that do not belong to one step less dense mesh (*-1/2* in 2D problems or *-1* in 1D ones), and are not prescribed as fixed ones.

It is worth noticing some features of the Liszka's type mesh generator

- mesh of any required mesh density p_i may be generated, by using criterion find the mesh of the minimal density satisfying constraint $\rho_i^{-1} - \rho_{\max}^{-1} \le p_i^{-1}$
- any given mesh may be converted into an equivalent Liszka type
- selection of nodes that belong to one step more or one step less dense meshes is immediate
- mesh may be used for both the GFDM and FEM purposes.

5. Multigrid Approach

Following the general idea of multigrid analysis [4], proposed, developed and applied here are new concepts of prolongation and restriction and solution algorithm. In the multigrid approach we deal simultaneously with many meshes varying from coarse to fine ones. They constitute a series. Usually, though not necessarily, each more fine mesh contains all nodes of the proceeding more coarse one.

The solution obtained for a coarser mesh is extended for a finer one by means of the so called „prolongation" procedure. Conversely residuum evaluated on a finer mesh is reduced to a coarser mesh by the „restriction" procedure. Concepts of the proposed prolongation and restriction are described below.

Prolongation

The prolongation procedure consists of three essential steps.

Step 1

Searched for points that are potential locations of the new nodes. These locations depend on the strategy adopted. In 2D domain the centroids of triangles taken from the optimal mesh triangulation may be considered. However, the best strategy, worked out so far, is to assume locations of new nodes obtained by means of the Liszka's type mesh generator after increase of the existing mesh density by ½ (in 2D) or *1* (in 1D).

Step 2

We examine the local residual r_i and $_br_i$, $i \in I_{\text{examined locations}}$ in all potential locations found in *step 1*, and insert new nodes wherever the conditions $\|r_i\| \le \eta_1 \|f_i\|$ and $\|_br_i\| \le \eta_2 \|g_i\|$ are violated. The residuum r_i (and $_br_i$) is approximately evaluated at each considered point by an appropriate GFD operator built only on nodes of the currently examined (old) mesh where a solution of the b.v. problem was found.

Step 3

Having assumed a new node P_i we built a new GFD operator ($L_{i+j}^{(m_p+1)}$) including this node and m_p old nodes. From the collocation requirement

$$\sum_{j=0}^{m_p} L_{i+j}^{(m_p+1)} u_{i+j} = f_i \qquad\qquad i \in I^{NEW}$$

we find the solution for the prolongation formula

$$u_i = \frac{1}{L_i^{(m_p+1)}}\left(f_i - \sum_{j=1}^{m_p} L_{i+j}^{(m_p+11)} u_{i+j}\right) \rightarrow \qquad\qquad u_i^{NEW} = \sum_j \alpha_{ij} u_j^{OLD} + b_i$$

It extends the solution u_{i+j} , $j = 1,\ldots,m_p$ found at the old nodes to a solution u_i required at the new node P_i. Here α_{ij} and b_i result from comparison of the formulae.

It is worth mentioning here, that prolongation formula of the same type may be also built, one by one, at each new node i , by using the local variations $\delta u_i^{NEW} \neq 0$, and $\delta u^{OLD} = 0$ in a considered variational formulation $\delta\Pi\left(u^{OLD}, u^{NEW}\right) = 0$. Such approach may be used not only in the GFD solution procedure but also in the FEM or BEM analysis as well.

Restriction

Using the most fine mesh we evaluate residuals

$$r = f - \mathcal{L}\, u$$

Having found these residuals we need to determine equivalent residuals for a subsequent coarser mesh. Proposed are the following basic assumptions

– Virtual work

$$\delta W = \int_\Omega r \delta u d\Omega$$

done by residuals („residual forces") r on virtual „displacements" δu is of the same magnitude for the both old and new meshes

– Virtual displacements of new nodes

$$\delta u_i^{NEW} = \sum_j \alpha_{ij} u_j^{OLD}$$

are found from the prolongation formula determined above.

Following these assumptions we may find a GFD evaluation of the virtual work done on *the new mesh*

$$\begin{aligned}\delta W &\approx \sum_l \Omega_l^{NEW} r_l^{NEW} \delta u_l^{NEW} + \sum_k \Omega_k^{NEW} r_k^{OLD} \delta u_k^{OLD} = \\ &= \sum_l \Omega_l^{NEW} r_l^{NEW} \sum_s \alpha_{ls} \delta u_s^{OLD} + \sum_k \Omega_k^{NEW} r_k^{OLD} \delta u_k^{OLD} = \\ &= \sum_k \left(\sum_l \Omega_l^{NEW} r_l^{NEW} \alpha_{lk} + \Omega_k^{NEW} r_k^{OLD} \right) \delta u_k^{OLD} =\end{aligned}$$

and on *the old mesh*

$$= \sum_k \tilde{\Omega}_k^{OLD} \tilde{r}_k^{OLD} \delta u_k^{OLD}$$

where $k \in I^{OLD}$, $l \in I^{NEW}$.

Virtual „displacements" δu_k^{OLD} are linearly independent from each other. Thus the residuals on the old mesh $\tilde{r}_k^{OLD}$ equivalent to the residuals r_l^{NEW}, r_k^{OLD} determined on the new mesh are

$$\tilde{r}_k^{OLD} = \left(\tilde{\Omega}_k^{OLD}\right)^{-1} \left(\sum_l \Omega_l^{NEW} r_l^{new} \alpha_{lk} + \Omega_k^{NEW} r_k^{OLD} \right)$$

In the above formulas

given are:

– residuals r^{NEW} and r^{OLD} for new and old nodes in the new mesh

– Voronoi polygons surface areas Ω^{NEW}, Ω^{OLD} for the new and old meshes

required are:

– residuals $\tilde{r}^{OLD}$ on the old mesh equivalent to those on the new one.

6. Multigrid adaptive solution procedure

In the multigrid approach considered here we carry the solution procedure, and at the same time we design subsequent meshes based on the results of a'posteriori error analysis. Therefore , sufficiently precise solution is needed not only on the finest mesh but also on intermediate ones. Thus the following steps of the general solution approach are proposed:

– design and generate an initial coarse mesh and find the solution of the boundary-value problem considered, exact for this mesh

– find one step more dense (and one step less dense as well) mesh (*± 1* for 1D and *±1/2* for 2D) , and determine in this way potential locations of new nodes (or nodes that might be removed)

– in these new locations find residuals based on a considered problem formulation and the moving weighted least squares approximation locally spanned over the old nodes only; adjust these residuals for higher order terms; decide upon magnitude of these residuals where to insert new nodes, and from where to remove the old ones

- find prolongation formulas from the old nodes to each new node; They are derived from the finite difference equations of the boundary value problem considered, written for each new node; each of these equations contains only one new node (central), all other nodes are the old ones
- use prolongation to extend the solution found for the initial mesh up to the currently finest one
- use the virtual work equivalence for old and new meshes in order to determine the restriction formula from the new to old nodes
- when a subsequent finest mesh is reached generate or modify the GFD stars and operators involving all new nodes. Find their adjustments for the higher order terms. Use these operators in order to find residuals. Return step by step to the preliminary coarse mesh using the restriction formulas. For this mesh find the exact solution for correction terms and use prolongation to extend them to the finest mesh. A smoothing step may be applied then followed by a next cycle of residual evaluation, restriction etc.
- control convergence of the solution using the exact results in the same nodes but for more and more fine meshes

7. Final Remarks

Proposed is here an adaptive multigrid approach to the meshless finite difference method generalized for arbitrary irregular grids. This is an integrated approach including original concepts of a'posteriori error analysis, solution smoothing, mesh generation and modification as well as multigrid solution procedure. The method is designed as a general and powerful tool of analysis of large and very large discrete boundary-value problems finally oriented on 3D ones.

A variety of 1D and 2D tests were done to examine convergence rate and solution effectiveness. The results obtained are very promising so far. The approach effectiveness increase is designed by implementation of concepts of distributed, and object-oriented programming [21]. Domain decomposition will be also considered. Two ways are defined, namely the basic mode and a fully distributed one. Both these modes can be used for concurrent work. The basic mode is preferred for smaller tasks while the fully distributed one is dedicated for solution of large problems. The approach is under current development.

8. References

1. Ainsworth M., Oden J.T., *A posteriori error estimation in finite element analysis*, Comput. Methods Appl. Mech. Engng. 142 (1997), 1-88.
2. Belytschko T., Krongauz Y., Organ D., Fleming M., Krysl P., *Meshless methods: An overview and recent developments*, Comput. Methods Appl. Mech. Engng. 139 (1996), 3-47.
3. Duarte C.A., A *Review of some Meshless methods to solve partial differential equations*, Technical Reports, 1995-06, TICAM, The Universities of Texas at Austin, May 1995.
4. Hackbusch W., *Multi-grid Methods and Applications*, Springer-Verlag, Berlin-Heidelberg-New York-Tokyo, 1992.
5. Karmowski W., Orkisz J., *A Physically Based Method of Enhancement of Experimental Data - Concepts, Formulation and Application to Identification of Residual Stresses*, Proc. IUTAM Symp. on Inverse Problems in Engng. Mech., Tokyo 1992; On Inverse Problems in Engineering Mechanics, Springer Verlag, 1993, 61-70.
6. Karmowski W., Orkisz J., *A'posteriori Error Estimation Based on Smoothing by the Global-Local Physically Based Approximation*, XIII Polish CMM, Poznań, 1997.

7. Krok J., Orkisz J., *A Unified Approach to the Adaptive FEM and FDM in Nonlinear Mechanics. Concepts and Tests.*, XIII Polish CMM, Poznań, May 1997.
8. Krok J., Orkisz J., *Application of the Generalized FD Approach to Stress Evaluation in the FE Solution*, Proc. Int. Conf. on Comp. Mech., Tokyo, 1986, 31-36.
9. Krok J., Orkisz J., *A Unified Approach to the FE and Generalized Variational FD in Nonlinear Mechanics, Concepts and Numerical Approach*, Int. Symp. on Discretization Methods in Structural Mechanics IUTAM/IACM, Vienna, Austria, 1989, Springer-Verlag, Berlin-Heidelberg, 1990, 353-362.
10. Lancaster P., Salkauskas K., *Surfaces Generated by Moving Least Squares Methods", Mathematics of Computation*, 155, 37(1981), 141-158.
11. Leżański P., Orkisz J., Przybylski P., *Mesh Generation for Adaptive Multigrid FDM and FEM Analysis*, XIII Polish CMM, Poznan, May 1997.
12. Liszka T., *An Automatic Generation of Irregular Grids in Two-dimensional Analysis* [in polish], Mechanika i Komputer, 4(1981), 181-186.
13. Liszka T.,Orkisz J., *Modified Finite Difference Methods at Arbitrary Irregular Meshes and its Application in Applied Mechanics*, Proc. of the 18th Polish Conf. On Mechanics of Solid, Wisla, Poland, 1976
14. Liszka T.,Orkisz J., *Finite Difference Methods of Arbitrary Irregular Meshes in Non-Linear Problems of Applied Mechanics*, 4th Int. Conference on Structural Mechanics in Reactor Technology, San Francisco, California, 1977.
15. Liszka T., Orkisz J., *The Finite Difference Method at Arbitrary Irregular Grids and its Applied Mechanics*, Comp.and Struct., 11(1980), 83-95.
16. Liszka T.,Orkisz J., *The Finite Difference Method for Arbitrary Irregular Meshes - a Variational Approach to Applied Mechanics Problems*, 2nd International Congress on Numerical Methods for Engineering, Paris, France, 1980, 277-288.
17. Orkisz J., *Adaptive Analysis of B.V. Problems by the Finite Difference Method at Arbitrary Irregular Meshes - Concept and Formulation*, 12th Symp.on the Unifaction of Analytical Comp. and Experimental Solution Methodologies, Worcester-Danvers, Mass., USA, 1995.
18. Orkisz J., *Finite Difference Method*, in *„Computer Methods in Solid Mechanics*', M.Kleiber ed., PWN, Warsaw, 1996.
19. Orkisz J. et al., *Numerical Analysis Methods*, in *„Fatigue Design Handbook*", R.C.Rice ed., Publ. SAE Inc., Warrendale, USA, 1989.
20. Orkisz J., *Adaptive Approach to the Finite Difference Method for Arbitrary Irregular Grids*, Interdisciplinary Symp. on Advances in Comp.Mech., Univ.of Texas Austin, 1997.
21. Przybylski P., *The Distributed Algorithm of a Unstructural Triangular Mesh Generator - Object-Oriented Approach*, XII Polish Conf.CMM, Warszawa- Zegrze, May 1995.
22. Shepard D., *A Two Dimensional Interpolation Function for Irregularly Spaced Points*, in ACM National Conference (1968), 517-524.
23. Wyatt M.J., Davies G., Snell C., *A New Difference Based Finite Element Method*, Instn. Engineers, 59, 2, 1975, 395-409.
24. Zienkiewicz O.C., Zhu J.Z., *A Simple Error Estimator and Adaptive Procedure for Practical Engineering Analysis*, Int.J.Num.Meth.Eng., 24(1987), 337-357.

HYBRID-TREFFTZ FINITE ELEMENT FORMULATIONS BASED ON THE FUNDAMENTAL SOLUTION

V. KOMPIŠ and J. BÚRY
Department of Mechanics, Faculty of Mechanical Engineering
University of Žilina, Vel'ký Diel
01026 Žilina, Slovakia

1. Introduction

For many years the efforts towards the development of more accurate, better performing and more robust finite elements (FE) attracted a large number of researchers. The common feature of the most known FE formulations is that some basic field equations, for example the equilibrium in isoparametric FE, are satisfied only in an average weighted residual (integral) sense over the element. In the Trefftz-type FE formulations all governing equations are satisfied a priori inside the element and the interelement continuity and the boundary conditions are then enforced in some weighted residual or pointwise sense. The formulation can be based on the approximation of the displacement [1, 2] or stress fields [3, 4]. The interelement continuity is enforced by using another frame function defined on the element boundaries only. This can be displacement or traction [1, 2] and we obtain the hybrid FE formulation, which has many attractive features comparing to the standard FE formulations. We can mention the integration over the element boundaries only in the stiffness matrix formulation, more general shape of the element, possibility to use coarse elements for local problems (local tractions, geometric peculiarities).

In this paper we present the hybrid-Trefftz displacement FE formulation in which displacements are approximated by the fundamental solution functions distributed along the element boundaries.

2. Interpolation Functions for Internal Displacements

In the hybrid-displacement finite element formulations the displacement field $\{u\}$ with components u_i should satisfy the governing equations inside the element. For linear isotropic problems the governing equations are the equilibrium equations expressed in

H. A. Mang and F. G. Rammerstorfer (eds.),
IUTAM Symposium on Discretization Methods in Structural Mechanics, 181–187.

displacements, the Lame-Navier equations.

$$(\lambda + \mu) u_{j,ji} + \mu u_{i,jj} = \bar{b}_i \tag{2.1}$$

where λ and μ are material constants (the Lame constants), $\bar{b}_i$ are body forces. $(i, j = 1, 2, 3)$ for 3D and $(i, j = 1, 2)$ for 2D plane strain problems. For plane stress problems

$$\lambda^* = \frac{2\lambda\mu}{\lambda + 2\mu} \tag{2.2}$$

is to insert instead of λ. The index after comma denotes the differentiation in corresponding co-ordinate direction and the Einstein's summation is used for repeated indices. The Cartesian co-ordinate system with co-ordinate components x_i is used. With a bar we denote functions, which will be supposed to be given. The displacements inside the elements can be written in the form

$$\{u\} = [B]\{q\} + \{\mathring{u}\} = \{u^h\} + \{\mathring{u}\} \tag{2.3}$$

where $[B]$ is a matrix of Trefftz functions (i.e. functions satisfying the homogeneous equations (2.1)). $\{q\}$ is a vector of unknown coefficients and $\{\mathring{u}\}$ is the particular solution of the problem (2.1) for the corresponding element field.

The Trefftz functions can be found in the polynomial forms (T-complete functions), which are simply expressed for 2D problems [1, 5, 6] and for 3D problems can be obtained using symbolic mathematics by expressing some terms of full polynomials by others from equilibrium conditions [7, 8], or they can be in the form of Kelvin's solution for a concentrated force (the Kupradze's functions given for 2D problems in [5, 6] and for 3D problems in [9-11]).

The polynomial approximation results in larger errors in the vicinity of the element boundaries especially in the stress and strain fields. The reason is in the higher order terms which increase with increasing co-ordinate. The stresses and displacements decay with the increasing distance from the acting load (Saint Venant principle). This is opposite to the polynomial behaviour. On the other side the Kupradze's functions have singularity in the point of acting force and so the points of application of the force are taken outside the elements based on numerical trials to avoid these singularities in approximation fields.

There is another possibility for choosing the Trefftz functions from the fundamental solutions, which are based on the Kelvin's solution. Such formulations are known from the boundary element formulations [11]. In this case the displacement field of the homogeneous solution inside the element is defined as

$$u_i(x) = \int_{\Gamma} U_{ij}(x, \tilde{x})\, \tilde{t}_j(\tilde{x})\, \mathrm{d}\Gamma(\tilde{x}) \tag{2.4}$$

With tilde we denote points or functions on the element boundaries, so $\tilde{x}$ is a point on the element boundary, $\tilde{t}_j$ is the distributed Kelvin traction (DKT) function and

$$U_{ij}(x, \tilde{x}) = -\frac{1}{16\pi(1-\nu)\mu r}\{(3 - 4\nu)\delta_{ij} + r_{,i}\, r_{,j}\} \tag{2.5a}$$

for 3D problems and

$$U_{ij}(x,\tilde{x}) = -\frac{1}{8\pi(1-\nu)\mu}\{(3-4\nu)\delta_{ij}\ln r - r_{,i}\,r_{,j}\} \tag{2.5b}$$

for 2D plane strain problems. For the plane stress problems we have to use

$$\nu^* = \nu/(1+\nu) \tag{2.6}$$

instead of ν. By $r(x,\tilde{x})$ we give the distance between the points x and $\tilde{x}$ and we denote

$$r_{,i} = \frac{\partial r}{\partial x_i(\tilde{x})} = \frac{r_i}{r} \quad \text{with} \quad r_i = x_i(\tilde{x}) - x_i(x) \tag{2.7}$$

δ_{ij} is the Kronecker delta.

The distributed tractions $\tilde{t}_j(\tilde{x})$ can be expressed by the shape functions of the local curve length or the local surface variables $(N^{(k)}(\xi))$ or $N^{(k)}(\xi,\eta)$ respectively) and corresponding nodal tractions p_j^k as

$$\tilde{t}_j(\tilde{x}(\xi,\eta)) = \sum_k N^{(k)}(\xi,\eta)\,p_j^k \tag{2.8a}$$

or in the matrix form

$$\{\tilde{t}\} = [N]\{p\} \tag{2.8b}$$

Then the coefficients of the vector $\{q\}$ in (2.3) correspond to the nodal tractions and the coefficients of the matrix $[B]$ are defined by the terms in the form

$$\int_\Gamma U_{ij}(x,\tilde{x}(\xi,\eta))\,N^{(k)}(\xi,\eta)\,J(\xi,\eta)\,\mathrm{d}\xi\,\mathrm{d}\eta \tag{2.9}$$

where J is the Jacobian of transformation from global to local co-ordinates.

Note that the tractions $\tilde{t}_j$ are fictitious quantities which introduce the loads of the full plane or space in order to obtain the same displacement and stress state as it is in the finite element and they are not equivalent to the real tractions for the element. They are discontinuous in the points where the element boundary is not smooth or contains discontinuity in the boundary tractions.

From displacements we can derive strains, stresses and also tractions along the element boundaries in the form

$$\{t\} = [S]\{q\} + \{\mathring{t}\} \tag{2.10}$$

where the matrix $[S]$ is derived from $[B]$ and the vector $\{\mathring{t}\}$ corresponds to the particular solution $\{\mathring{u}\}$ and is derived from it.

For DKT they are given by

$$t_i(x) = \int_\Gamma T_{ij}(x,\tilde{x})\,\tilde{t}_j(\tilde{x})\,\mathrm{d}\Gamma \tag{2.11}$$

where

$$T_{ij}(x,\tilde{x}) = -\frac{1}{8\pi(1-\nu)r^2}\{[(1-2\nu)\,\delta_{ij}+3r_{,i}\,r_{,j}]\,\frac{\partial r}{\partial n} - (1-2\nu)(r_{,i}\,n_j - r_{,j}\,n_i)\} \tag{2.12a}$$

for 3D problems and

$$T_{ij}(x,\tilde{x}) = -\frac{1}{4\pi(1-\nu)r^2}\{[(1-2\nu)\,\delta_{ij}+2r_{,i}\,r_{,j}]\,\frac{\partial r}{\partial n} - (1-2\nu)(r_{,i}\,n_j - r_{,j}\,n_i)\} \tag{2.12b}$$

for plane strain problems (again we have to insert (2.6) for plane stress problems). n is the outer normal to the element boundary and

$$\frac{\partial r}{\partial n} = r_{,i}\,n_i \tag{2.13}$$

Instead of the fictitious tractions we have also another choice to use distributed dislocations, or combination of both in order to obtain the regular integral presentation. This regular presentation introduces the solution of the half space / half plane. Readers interested in these formulations can find corresponding solutions in [12].

3. Hybrid-displacement FE Formulation

The displacement fields $\{u\}$ from (2.3) and corresponding tractions $\{t\}$ from (2.10) will be incompatible between the elements. To impose the compatibility we choose another independent displacement field $\{\tilde{u}\}$ defined on the interelement boundaries only and interpolated by means of shape functions $[N]$ and displacements $\{d\}$ in nodal points of the element as

$$\{\tilde{u}\} = [N]\{d\} \tag{3.1}$$

The compatibility between the internal variables $\{u\}$ and $\{t\}$ and corresponding boundary variables will be enforced by the variation [2]

$$-\int_{\Gamma_t}\delta\{u\}^{\mathrm{T}}(\{t\}-\{\bar{t}\})\,\mathrm{d}\Gamma+\int_{\Gamma_u}\delta\{t\}^{\mathrm{T}}(\{u\}-\{\bar{u}\})\,\mathrm{d}\Gamma+\int_{\Gamma_i}\delta\{t\}^{\mathrm{T}}(\{u\}-\{\tilde{u}\})\,\mathrm{d}\Gamma=0 \tag{3.2}$$

where Γ_t, Γ_u and Γ_i are parts of the element boundaries, where tractions $\bar{t}$ and displacements $\bar{u}$ are given and the interelement boundaries, respectively (mixed boundary conditions of Γ_t and Γ_u are also possible, e.g. for symmetry conditions). δ denotes the variation and the right superscript ^{T} the transposition. The variation

$$\sum_{\substack{\text{all}\\\text{elements}}}\int_{\Gamma_i}\delta\{\tilde{u}\}^{\mathrm{T}}\{t\}\,\mathrm{d}\Gamma=0 \tag{3.3}$$

enforces the equilibrium between the elements. Substituting (2.3) and (2.10) into (3.2) yields

$$[H]\{q\} = [G]\{d\} + \{p\} \tag{3.4}$$

where

$$
\begin{aligned}
[H] &= \int_{\Gamma_u \cup \Gamma_i} [S]^{\mathrm{T}}[B]\,\mathrm{d}\Gamma - \int_{\Gamma_t} [B]^{\mathrm{T}}[S]\,\mathrm{d}\Gamma \\
[G] &= \int_{\Gamma_i} [S]^{\mathrm{T}}[N]\,\mathrm{d}\Gamma \qquad (3.5) \\
\{p\} &= \int_{\Gamma_t} [B]^{\mathrm{T}}(\{\mathring{t}\} - \{\bar{t}\})\,\mathrm{d}\Gamma - \int_{\Gamma_u} [S]^{\mathrm{T}}(\{\mathring{u}\} - \{\bar{u}\})\,\mathrm{d}\Gamma - \int_{\Gamma_i} [S]^{\mathrm{T}}\{\mathring{u}\}\,\mathrm{d}\Gamma
\end{aligned}
$$

and from (3.3) we get

$$\sum_{\substack{\text{all}\\ \text{elements}}} [G]^{\mathrm{T}}\{q\} = \{\mathring{R}\} \qquad (3.6)$$

with

$$\{\mathring{R}\} = - \sum_{\substack{\text{all}\\ \text{elements}}} \int_{\Gamma_i} [N]^{\mathrm{T}}\{\mathring{t}\}\,\mathrm{d}\Gamma \qquad (3.7)$$

Substituting $\{q\}$ from (3.4) into (3.6) we obtain the final system of linear equations for the whole region

$$[K]\{D\} = \{R\} \qquad (3.8)$$

where the stiffness matrix and equivalent nodal loads are

$$[K] = \sum_{\substack{\text{all}\\ \text{elements}}} [G]^{\mathrm{T}}[H]^{-1}[G] \qquad (3.9)$$

$$\{R\} = \{\mathring{R}\} - \sum_{\substack{\text{all}\\ \text{elements}}} [G]^{\mathrm{T}}[H]^{-1}\{p\} \qquad (3.10)$$

and $\{D\}$ is the vector of nodal displacements of the whole region.

As the subsidiary displacements $\tilde{u}$ and thus also their nodal values are defined on the interelement boundaries, it reduces the problem. In such hybrid formulations also local load effects are better accounted as they can be included in the form of particular solutions which satisfy the boundary traction conditions [7].

We can find a very interesting hybrid FE formulation, when we suppose such distribution of tractions, which will give continuous displacements between the elements (i.e. $\tilde{u} = u$) and also the boundary conditions $u = \bar{u}$ on Γ_u and $t = \bar{t}$ on Γ_t. In this case the variational formulation (3.1) and (3.3) reduces to

$$\sum_{\substack{\text{all}\\ \text{elements}}} \int_{\Gamma_i} \delta\{u\}^{\mathrm{T}}\{t\}\,\mathrm{d}\Gamma = 0 \qquad (3.11)$$

In this case the displacement and traction fields will be connected by reciprocity theorem as it is by the direct BEM formulation [11]

$$\frac{1}{2}u_i(x) + \int_{\Gamma_e} T_{ij}(x,\tilde{x})\,u_j(\tilde{x})\,\mathrm{d}\Gamma(\tilde{x}) = \int_{\Gamma_e} U_{ij}(x,\tilde{x})\,t_j(\tilde{x})\,\mathrm{d}\Gamma(\tilde{x}) \qquad (3.12)$$

where $\Gamma_e = \Gamma_i \cup \Gamma_u \cup \Gamma_t$. Note that the first term (free term) in integral equation (3.12) is in this form only if the collocation point x lies on a smooth surface. If it is a corner point it has to be multiplied by a constant according to the angle of the corner [11].

The displacements will be given by shape functions $[N]$ and displacements in corresponding nodal points $\{d\}$ as

$$\{u\} = [N]\{d\} \tag{3.13}$$

As the tractions are not compatible between the elements and they are not known *a priori* in the integration points, their compatibility has to be enforced.

Using (3.13) the equation (3.12) can be formally written in the matrix form

$$[T]\{u\} = [U^i]\{t^i\} + [U^p]\{t^p\} \tag{3.14}$$

where the right superscripts p and i denote the parts of the matrix $[U]$ and the vector $\{t\}$ defined by the r.h.s. of (3.12) which correspond to the parts of the boundary with prescribed tractions and interelement boundaries (because of simplicity we do not suppose prescribed displacements except the rigid body components in this notation). The equation (3.14) can be solved for the interelement tractions on the element level

$$\{t^i\} = [U^i]^{-1}([T]\{u\} - [U^p]\{t^p\}) \tag{3.15}$$

From (3.11), (3.13) and (3.15) we obtain the resulting system of equations

$$\sum_{\substack{\text{all}\\\text{elements}}} [N]^{\mathrm{T}}[U^i]^{-1}[T][N]\{d\} = \sum_{\substack{\text{all}\\\text{elements}}} [N]^{\mathrm{T}}[U^i]^{-1}[U^p]\{t^p\} \tag{3.16}$$

We need not to use the same number of collocation points as we have defined the number of interelement integration points in the corresponding element. In that case (3.14) is solved in the least square sense and $[U^i]^{-1}$ will be replaced by $([U^i]^{\mathrm{T}}[U^i])^{-1}[U^i]^{\mathrm{T}}$.

It is well known that the singular boundary integral equation (SBIE) formulations require special care by the integration because of the singular terms, when the collocation point lies on the integration element and high gradients in the kernel functions (nearly singular terms), when it is placed in its vicinity. So, if we want to use the procedure described above, then the regularized BEM formulations [13] are to be used.

If the SBIE are used in the element formulation, then also the interelement tractions are to be expressed by shape functions $[M]$ and corresponding nodal point tractions $\{p\}$

$$\{t^i\} = [M]\{p\} \tag{3.17}$$

and $[U^i]^{-1}$ in (3.15) and (3.16) has to be replaced by $([M][U^i]^{\mathrm{T}}[U^i][M])^{-1}[U^i]^{\mathrm{T}}$.

The resulting system of equations (3.16) has similar structure as other FE systems. The elements formulated in this way can be connected to another type of finite elements if the same interpolating functions are used for the displacements on the corresponding sides of both neighbour elements.

Acknowledgement

The authors gratefully acknowledge the partial support of this research by the Slovak Grant Agency for Science (Grant No.1/3278/96).

References

1. Jirousek, J., Wróblewski, A.: T-elements: State of the art and future trends, *Archives of Comput. Meth. Eng.*, **3**, 323-434 (1996).
2. Jirousek, J., Zieliński, A.P.: Survey of Trefftz-type element formulations, *LSC Int. Report 95/06*, EPFL Lausanne, 1995.
3. Almeida, J.P.M, Freitas, J.A.T.: Alternative approach to the formulation of hybrid equilibrium finite elements, *Comp. & Struct.*, **40**, 1043-1047 (1991).
4. Maunder, E.A.W., Almeida, J.P.M., Ramsey, A.C.A.: A general formulation of equilibrium macro-elements with control of spurious kinematic modes, *Int. J. Num. Meth. Eng.*, **39**, 3175-3194 (1996).
5. Zieliński, A.P.: On trial functions applied in the generalized Trefftz method, *Advances in Eng. Software*, **24**, 147-155 (1995).
6. Szybiński, B., Zieliński, A.P.: Alternative T-complete systems of shape functions applied in analytical Trefftz finite elements, *Num. Meth. Part. Dif. eq.*, **11**, 375-388 (1995).
7. Kompiš, V., Kaukič, M., Žmindák, M.: Modelling of local effects by hybrid-displacement FE, *J. Comput. Appl. Math.*, **63**, 265-269 (1995).
8. Kompiš, V., Fraštia, L'.: Polynomial representation of hybrid finite elements, to be published in *Comp. Assist. Mech. Eng. Sci.*, 1997.
9. Kupradze, V.D. (ed.): *Three-dimensional Problems of the Mathematical Theory of Elasticity and Thermoelasticity*, North-Holland Amsterdam, 1979.
10. Leipholz, H.: *Einfuhrung in die Elastizitätstheorie*, in German, G.Braun Karlsruhe, 1968.
11. Balaš, J., Sládek, J., Sládek, V.: *Stress Analysis by Boundary Element Methods*, Elsevier, Amsterdam, 1989.
12. Rieder, G., Heise, U., Pahnke, U., Antes, H., Glahn, H., Kompiš, V.: *Computation of Elastic Stresses in Membranes and Plates with Arbitrary Curved Boundaries*, (in German), Westdeutscher Verlag, Leverkusen, 1976.
13. Tanaka, M., Sládek, V., Sládek, J.: Regularization techniques applied to boundary element method, *Appl. Mech. Rev.*,**47**, 457-499 (1994).

HYBRID NEURAL-NETWORK/COMPUTATIONAL PROGRAMS TO THE ANALYSIS OF ELASTIC-PLASTIC STRUCTURES

Z.WASZCZYSZYN, E.PABISEK and G.MUCHA
Cracow University of Technology
Institute of Computer Methods in Civil Engineering
ul. Warszawska 24, 31-155 Kraków, Poland

1. Introduction

In recent 7-8 years the Artificial Neural Networks (ANNs) have been widely introduced to structural engineering and especially to mechanics of structures and materials [1]. ANNs simulate biological neural networks very primitively. This concerns especially computer implementation of ANNs, called for short neurocomputing.

From among many features of neurocomputing, which distinguishe it from standard computer data processing, two features are worth to be mentioning. ANNs can be used to map input into output data without known relation between them. Neurocomputing can be also applied efficiently to the analysis of inverse problems.

The most frequently used are forward BackPropagation Neural networks (BPNs), cf. e.g. [2]. The BPN networks are trained and tested by means of specially selected patterns which are completed with pairs composed of known input/ouput data. After training BPNs can operate as very quick simulators.

BPNs can be efficiently used not only as separate simulators but they can be incorporated into computational programs as neural procedures. This idea leads to hybrid neurocomputing/computational systems associated with simple or more advanced problems of structural mechanics [3,4].

In the presented paper we focused our attention on neural simulators for constitutive equations. The BPNs were earlier used to formulate the stress-strain relation [5] or moment-curvature relation [6] on the base of experimental evidence. The inversion of uniaxial Ramberg-Osgood relation was performed in [7].

The aim of our paper is to apply a neural procedure in the hybrid program for the bending analysis of elastic-plastic beams by the finite difference method. The next problem concerns the consistent analysis of plane elastic-plastic stress state by the finite element method.

2. Analysis of bending of elastic-plastic beams

2.1. Basic equations

Linear geometrical and equilibrium equations are assumed:

H. A. Mang and F. G. Rammerstorfer (eds.),
IUTAM Symposium on Discretization Methods in Structural Mechanics, 189–198.

$$v' = \varphi(x), \quad \varphi' = -\kappa(x), \quad T' = -p(x), \quad M' = T(x), \tag{1}$$

where $(\ )' = d(\ )/dx$ and other variables are shown in Fig1:

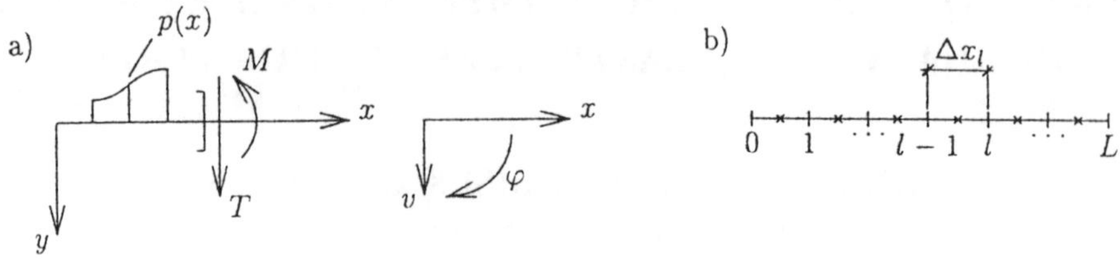

Figure 1: a) Load, cross-sectional forces and generalized displacements, b) Finite difference nodes along the beam axis

Eqs (1) are completed by physical relationship:

$$\kappa = \begin{cases} M/(EI) & - \ \textit{elastic deformation}, \\ F(M) & - \ \textit{elasto-plastic deformation}. \end{cases} \tag{2}$$

In a general case the elastic-plastic relationship $\kappa = F(M)$ cannot be expressed in an explicit form. This relation relies not only on material model $\sigma(\varepsilon)$ but also on the shape of beam cross-section.

In order to verify our results by existing analytical solutions the following assumptions are adopted: i) beam is of rectangular cross-section, ii) material is of symetric properties for tension/compression with perfect Bauschinger's effects (Fig.2a), iii) Bernoulli-Euler hypothesis of plane, perpendicular cross-section is valid (Fig.2b).

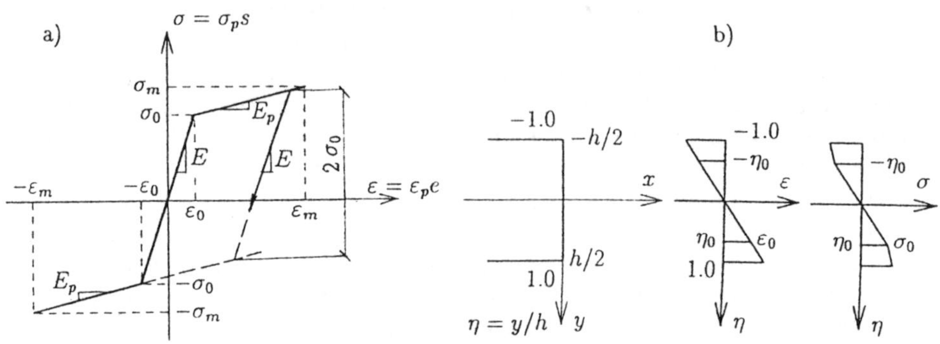

Figure 2: a) Relationship $\sigma(\varepsilon)$, b) Strain and stress distribution along the beam thickness

The bending moment and distribution of strains are defined by the classic formulae:

$$M = \iint_A \sigma y \, dA = \frac{bh^2}{4} \sigma_p \int_{-1}^{1} s\eta \, d\eta, \quad \varepsilon = \kappa y \ . \tag{3}$$

Introduction of dimenssionless variables

$$m=\frac{4M}{\sigma_p bh^2},\quad k=\frac{\kappa h}{2\varepsilon_p},\quad \alpha=\frac{E\varepsilon_p}{\sigma_o},\quad \chi=\frac{\alpha_p}{\alpha}=\frac{E_p}{E}, \tag{4}$$

where: ε_p, σ_p — reference strain and stress respectively, leads to the following formula:

$$m=\begin{cases}\dfrac{2}{3}\alpha\,, & for\ \eta_o \geq 1,\\[2ex] \dfrac{2}{3}\alpha\left[\chi k+(1-\chi)e_o\left(\dfrac{3}{2}-\dfrac{e_o^2}{2k^2}\right)\right], & for\ \eta_o < 1.\end{cases} \tag{5}$$

Formula (1) is written only for $m \geq 0$ but it can be easily extended on case $m < 0$.

The incremental formula, used in [8], can be derived on the basis of evident relation

$$\Delta m = m(k+\Delta k;\chi)-m(k;\chi)=f(\Delta k,k;\chi). \tag{6}$$

The finite difference scheme with a midpoint (Fig.1b) corresponds to the following formulae:

$$y_{l-1/2}\approx\frac{1}{2}(y_l+y_{l-1}),\quad y'_{l-1/2}\approx(y_l-y_{l-1})/\Delta x_l\,, \tag{7}$$

where: $y_l = v_l, \varphi_l, T_l, M_l$. After the boundary conditions are joined the set of $4*(L+1)$ algebraic equations can be easily formulated for the computation of y_l at nodes $l=0,1,\dots,L$ — cf.[9].

2.2. Neural simulations

Formula (5) is employed to compute patterns which were used for the training of the BPN network as shown schematically in Fig.3.

Figure 3: **Application of formulae (5) to computation of patterns for the BPN network**

After a number of numerical experiments the BPN network of architecture 1-10-1 was accepted for the fixed value of strain hardening parameter χ. In order to cover the range $k \in [0.67, 20.0]$ for $\varepsilon_p=\varepsilon_o$, $\sigma_p=\sigma_o$ four BPNs were assumed with the total number of training patterns L = 288. The computer program SNNS [10] was used to train the networks using the Rprop learning method (adaptive type method for evaluation the network learning parameter [2]). The training was limited to 2000 epochs (number of all patterns presentation). The testing on about 500 patterns proved that approximation error is below 1.0%.

The hybrid BPN/finite-difference program was used to the computation of two examples. The first example corresponds to the bending of a cantilever under a concentrated force applied to the free end. In case of the elastic perfect plastic material, i.e. for $\chi = 0$, the solution by hybrid program was compared with the analytical solution [11] and with the solution by finite difference with an iterative procedure to inverse the relation $m(k)$. The equilibrium paths $v(P)$ are shown in Fig.4.

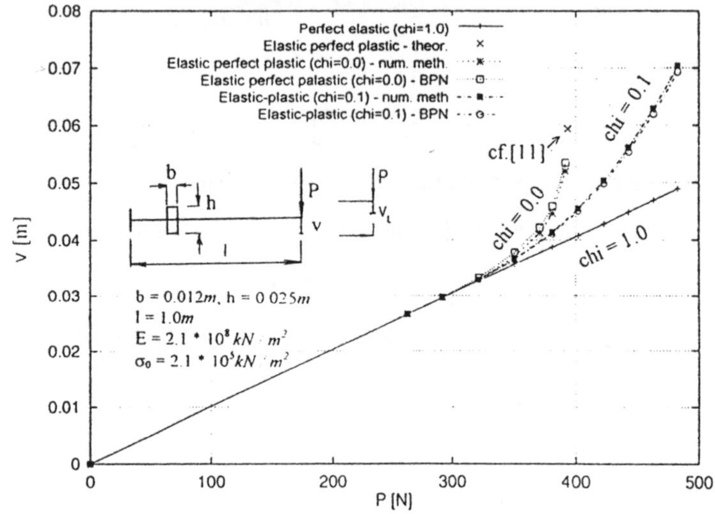

Figure 4. Equilibrium paths $v(P)$ for a cantilever under concentrated load

The second example, taken from [12], deals with the same cantilever as in Fig.4 but under a low cyclic loads. The descripancies between purely numerical and hybrid equlibrium paths shown in Fig.5 indicate that the accuracy of neural subroutines should be slightly improved.

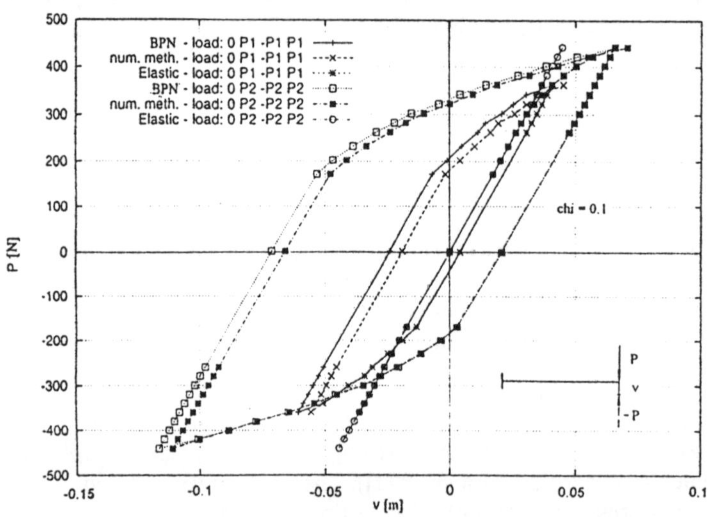

Figure 5: Equlibrium paths $P(v)$ for a cantilever under cyclic load

3. Analysis of plane strain state

3.1. Basic equations

Huber-Mises-Hencky yield condition is assumed in the following form [13]:

$$F \equiv \frac{1}{2}\sigma^T \boldsymbol{P}\sigma - \frac{1}{3}\sigma_e^2(\varepsilon_p) = 0 \ , \tag{8}$$

where:

$$\sigma = \{\sigma_x, \sigma_y, \tau_{xy}\}, \qquad \boldsymbol{P} = \begin{bmatrix} 2/3 & -1/3 & 0 \\ -1/3 & 2/3 & 0 \\ 0 & 0 & 2 \end{bmatrix}, \qquad \sigma_e = \sigma_o + H\varepsilon_p \ , \tag{9}$$

ε_p — Odqvist's parameter. Classical assumptions of the theory of plasticity are valid: i) splitting of strain increment $\Delta\varepsilon = \{\Delta\varepsilon_x, \Delta\varepsilon_y, \Delta\gamma_{xy}\}$ into elastic and elastic-plastic parts $\Delta\varepsilon^e$, $\Delta\varepsilon^p$, ii) associated flow rule:

$$\Delta\varepsilon = \Delta\varepsilon^e + \Delta\varepsilon^p \ , \qquad \Delta\varepsilon^p = \Delta\lambda \frac{\partial F}{\partial \sigma} \ . \tag{10}$$

The problem is how to compute the current stress σ^D and consistent modular matrix $\boldsymbol{E}_D^{ep}$ for known stress σ^A and strain increment $\Delta\varepsilon$. In Fig.6 a scheme of the consistent approach is shown. It depends on the elastic prediction of stress σ^* and orthogonal projection on the current yield surface:

$$F^D(\sigma^D, \Delta\lambda_D) = 0 \ . \tag{11}$$

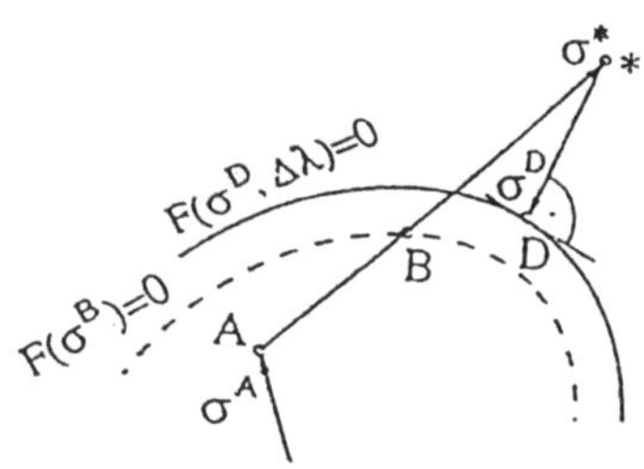

Figure 6: Elastic prediction and orthogonal projection on yield surface

The main problem is the solution of the highly nonlinear equation (11) with respect to increment of yielding parameter $\Delta\lambda_D$ (stress σ^D is computed simultaneously) since it enables us to calculate subsequent variables:

$$\Delta\lambda \;\rightarrow\; \boldsymbol{I}^D = (\boldsymbol{I} + \Delta\lambda_D \boldsymbol{E}\boldsymbol{P})^{-1}, \;\; \sigma^D = \boldsymbol{I}^D\sigma^* \ ,$$

$$\Delta\varepsilon_p^D = \Delta\lambda_D\left(\frac{2}{3}(\sigma^D)^T \boldsymbol{P}\sigma^D\right)^{1/2}, \quad \sigma_e^D = \sigma_e^B + H\,\Delta\varepsilon_p^D \ , \quad \boldsymbol{E}^D = \boldsymbol{I}^D\boldsymbol{E} \ , \tag{12}$$

$$\boldsymbol{s} = \boldsymbol{P}\sigma^D \ , \;\; \boldsymbol{a} = \boldsymbol{E}^D\boldsymbol{s} \ , \;\; A = \frac{4}{9}(\sigma_e^D)^2 H / \left(1 - \frac{2}{3}H\,\Delta\lambda_D\right), \;\; \boldsymbol{E}_D^{ep} = \boldsymbol{E}^D - \boldsymbol{a}\boldsymbol{a}^T / (A + \boldsymbol{s}^T\boldsymbol{a})$$

3.2. Neural simulation

Training and testing patterns are computed according to the scheme in Fig.6.

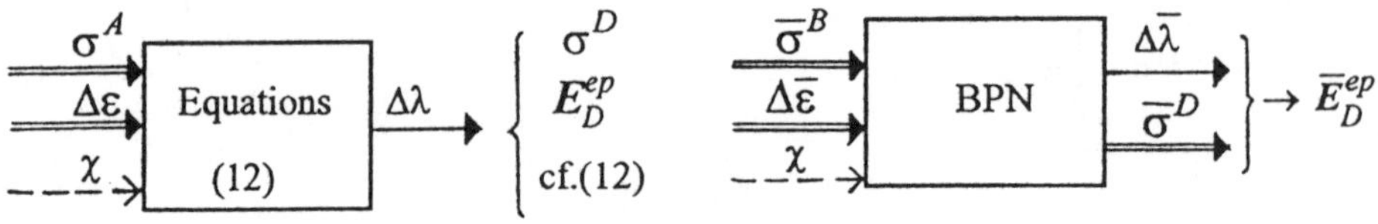

Figure 7: Application of equations (12) to formulation of patterns for BPN

In the neural simulator shown in Fig.7 the following dimenssionless variables are used:

$$\bar{\sigma} = \sigma / \sigma_e^B , \quad \Delta\bar{\varepsilon} = E\,\Delta\varepsilon / \sigma_e^B , \quad \Delta\bar{\lambda} = E\,\Delta\lambda , \quad \bar{E} = E / E , \; \chi = H / E , \tag{13}$$

where: σ_e^B - effective stress corresponding to the yield surface F^B in Fig.6.

Introduction of dimensionless variables makes it possible to be independent of current stress state and formulate so-called *objective patterns.* In the neural network (Fig.7) a surplus of output data is intentionally computed adding stress vector $\bar{\sigma}^D$ to $\Delta\bar{\lambda}$. Due to such a formulation one-to-one correspondence of input/output data is achieved.

In what follows two networks are formulated:

I) network BPN-I with patterns taken from a numerical example,

II) network BPN-II trained on objective patterns associated with the complete yield surface and different strain increments $\Delta\bar{\varepsilon}$. The patterns are randomly selected from among admissible points $\bar{\sigma}$ and increments $\Delta\bar{\varepsilon}$ (they should correspond to $\Delta\bar{\lambda} > 0$).

3.2.1. Numerical example and network BPN-I.

An example well known in literature [13] is shown in Fig.8. A tensioned plate with a circular hole and material with $\sigma - \varepsilon$ characteristics shown in Fig.8a was analysed by means of the finite element code ANKA [9]. Because of symetry a quarter of plate with FE mesh shown in Fig.8b was computed.

Equilibrium paths $\lambda(u_A)$ were computed under displacement control corresponding to the displacement increment $\Delta\tau \equiv \Delta u_A$. Different steps $\Delta u_A = 0.01, 0.02, 0.08\,\mathrm{mm}$ were kept constant during the computational process.

Table 1 shows numbers of training patterns L and testing patterns T for the network BPN-I. The patterns were selected from values which occur during the deformation of the plate, computed by means of the ANKA program for different steps $\Delta\tau$

TABLE 1. Networks and numbers of patterns

Networks	Architectures	L	T
BPN-I	6-20-4	1158	405
BPN-II	6-40-20-4	3777	3148

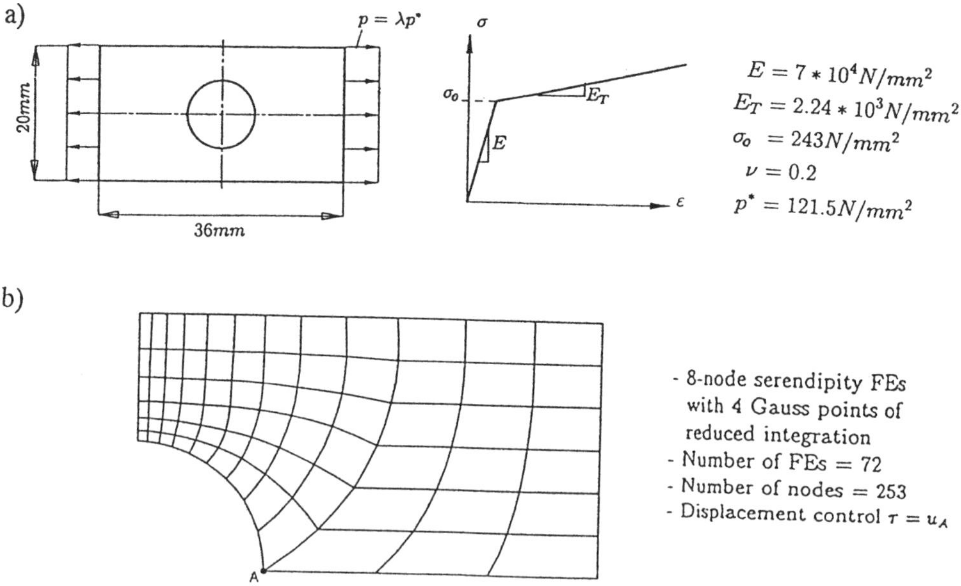

Figure 8: a) Data for plate and material, b) FE mesh for a quarter of plate

The network BPN-I was trained during 800 epochs. Then the network computer simulator, written in C, was incorporated into the ANKA program. In such a way the hybrid program ANKA-HI was implemented. The application of hybrid program gave very accurate results. Practically for all steps Δu_A the equilibrium paths $\lambda - u_A$ computed by ANKA-HI are in good agreement with the paths computed by means of the computational program ANKA (shown with continuous lines in Figures 9 and 10).

3.2.2. Objective patterns and network BPN-II

A disadvantage of the network BPN-I is that it can be useful to the analysis of only this example from which patterns were taken for training the network. In order to be independent of particular problems the points $\overline{\sigma}_j^B$ on the whole yield surface were selected together with associated, plastically active strain increments $(\Delta\overline{\varepsilon})_j^i$. In the following example 185 points were adopted, i.e. $j = 1,...,185$, and 12 strain increments, i.e. $i = 1,...12$, were associated with each point j. 251864 patterns were generated and from them the training and testing patterns of numbers $L = 3777$ and $T = 3148$ were randomly selected, cf. Table 1.

Similarly as in case of BPN-I the network BPN-II of structure 6-40-20-4 was trained using SNNS program [13] and Rprop learning method. Two networks BPN-II were trained for fixed values of the strain hardening parameters $\chi = 0.0,\ 0.032$. The average mean-square-error

$$MSE(P) = \frac{1}{P}\sum_{p=1}^{P}\sum_{i=1}^{4}\left(t_i^{(p)} - o_i^{(p)}\right)^2 \qquad (14)$$

was computed, where: $P = L, T$ — number of patterns, $t_i^{(p)}, o_i^{(p)}$ — target and computed output values for i-th exit and p-th pattern respectively.

In Table 2 the training and testing errors *MSE(L)* and *MSE(T)* are shown respectively after a different number of epochs. After training the PBN-II network was incorporated into the hybrid program ANKA-HII.

TABLE 2: Errors of training and testing for the network BPN-II at $\chi = 0.032$

Number of epochs	*MSE(L)* x 10^4	*MSE(T)* x10^4
2800	3.7	3.9
14800	1.4	1.6
40000	1.2	1.4

In Figures 9 and 10 there are shown equilibrium paths corresponding to the application of programs ANKA (continuous lines) and ANKA-HII (dashed lines). If the network BPN-II after 2800 training epochs is used than for smaller steps $\Delta u_A = 0.01$, 0.02 the equilibrium paths which result from the hybrid program ANKA-HII incline of the path computed by means of purelly computational FE program ANKA. The accuracy is significantly improved for all control steps $\Delta u_A = 0.01$, 0.02, 0.08 mm if the neural network is trained better (14800 epochs) .

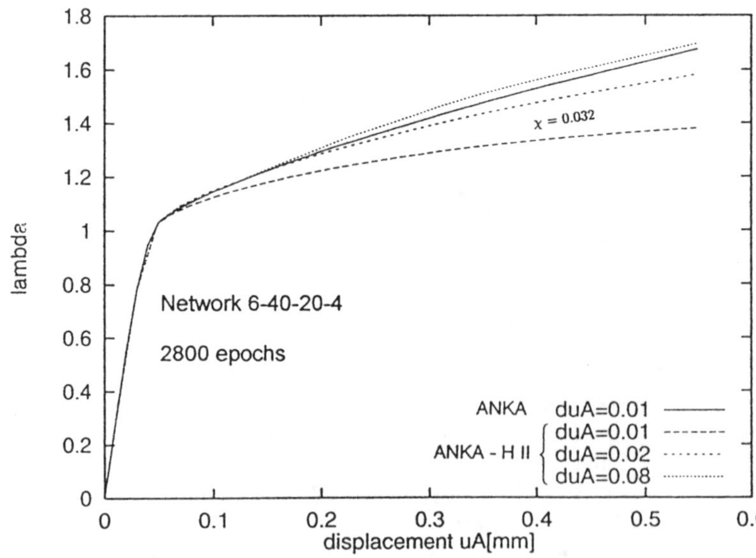

Figure 9. Equilibrium paths for the plate in Fig.8 computed by FE programs ANKA and NN/FE program ANKA-HII with the network BPN-II after 2800 epochs training

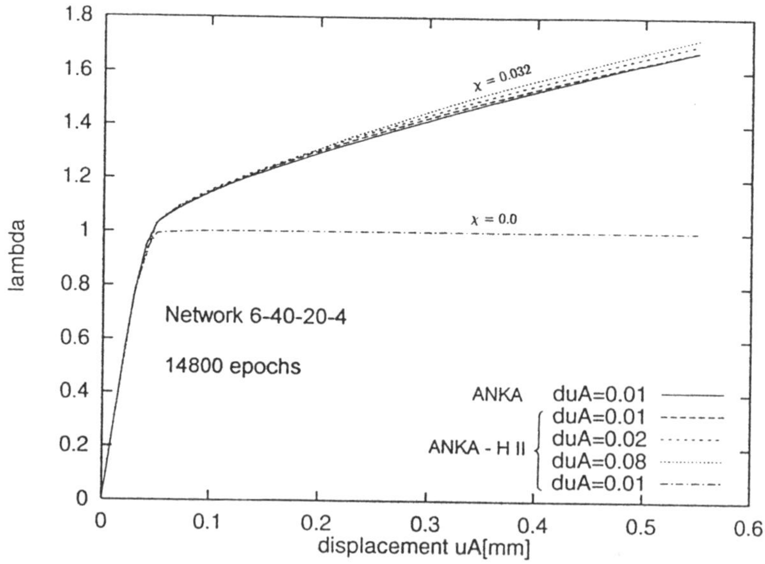

Figure 10: Equilibrium paths for the network BPN-II after 14800 epochs training

4. Conclusion and final remarks

Two pilot hybrid program implemented for simple problems of structural mechanics enable us to formulate the following conclusions:

- BPN networks can be used as quite accurate simulators for constitutive equations of elasto-plastic material;
- in special cases it is possible to compute objective patterns to train the networks independly of considered problems;
- validation of hybrid programs on structural level confirms their efficiency. Computational time of data processing by hybrid programs is slighly smaller than in cases of analysis of the same numerical examples using purely computational programs.

Neural procedures seem to be an efficient tool to analyse more complicated problems than those considered in the paper. More advanced hybrid programs concerning the analysis of plane frames and bending plates made of elasto-plastic materials are now developing in the Institute of Computer Methods in Civil Engineering of the Cracow University of Technology.

Financial support by the Polish Committee for Scientific Research, Grant No 7 T07E 056 11 on Application of Artificial Neural Network in Structural Engineering, is gratefully acknowledge.

5. References

1. Waszczyszyn Z.: Some recent and current problems of neurocomputing in civil and structural engineering, *Advances in Computational Structures Technology*, Topping B.H.V. (Ed.), Civil-Comp Press, Edinburgh, 1996, 43-58.

2. Rojas R.: *Neural Networks — A Systematic Introduction*, Springer-Verlag, Berlin-Heidelberg, 1996.

3. Hajela P. and Berke L.: Neurobiological computational models in structural analysis and design, *Comp. & Stru.*, **41** (1991), 657-667.

4. Szewczyk Z.P. and Noor A.K.: A hybrid neurocomputing/numerical strategy for nonlinear analysis, *Comp.& Stru.*, **58**, (1996), 661-677.

5. Ghaboussi J., Garret J.H.Jr and Wu X.: Knowledge-based modeling of material behavior with neural networks, *J.Eng.Mechanics*, **117**, (1991), 132-153.

6. Jadid M.N. and Fairbairn D.F: Neural-network applications in predicting moment-curvature parameters from experimental data, *Engng Applic. Artif. Intell.*, **9**, (1996), 309-319.

7. Yamamoto K.: Modeling and hysteretic behaviour with neural network and its application to nonlinear dynamic response analysis, *Application of Artificial Intelligence in Engineering*, Proc.7th Intern. Conf., 1992, 475-486.

8. Mucha G. and Waszczyszyn Z.: Application of artificial neural networks to bending analysis of elasto-plastic beams (in Polish), *Proc. XLII Polish Conference of Civil Engineering*, Kraków-Krynica 1996, Vol.2, 193-200.

9. Waszczyszyn Z, Cichoń Cz. and Radwańska M.: *Stability of Structures by Finite Element Methods*, Elsevier, 1994.

10. *SNNS (Stuttgart Neural Network Simulator)*, User's Manual, Version 4.1, Institut for Parallel and Distributed High Performance Systems, Rep. No 6/95, publ. University of Stuttgart, 1995.

11. Krzyś W. and Życzkowski M.: *Elasticity and Plasticity* (in Polish), PWN, Warsaw, 1962.

12. Mucha G.: Problem formulation used in application of BPNs on example of elastoplastic beam bending, *III Conference on Neural Networks and Their Applications*, Kule k/Częstochowy, October 13-17, 1997, (in press).

13. Ramm E. and Matzenmiller A.: Computational aspects of elasto-plasticity in shell analysis, *Computational Plasticity — Models, Software and Applications*, Owen D.R.J., Hinton E., Onate E. (Eds.), Pineridge Press, Swansea, 1987, P.I, 711-734.

CRACK DETECTION BY ELASTOSTATIC MEASUREMENTS

A Neural Network Approach based on BEM Modelling

G.E. STAVROULAKIS AND H. ANTES
Institute of Applied Mechanics,
Carolo Wilhelmina Technical University,
D–38023 Braunschweig, Germany

Abstract

A neural network approach is proposed for the treatment of inverse problems in linear and nonlinear elastostatics. The specific application concerns the identification of a crack in an elastic body with possible unilateral contact effects. The method is tested with numerical data produced by the multiregion boundary element method.

1. Introduction

The numerical solution of inverse problems in mechanics and, in particular, for nonlinear response is a relatively complicated problem (Natke 1993). In principle and for the variant of an output error identification procedure, one tries to minimize an appropriately defined error criterion between the measured response of the structure and the computed one, the latter being parametrized by a set of unknown parameters which define the required characteristics of the structure. For crack identification problems the number, the position and the shape of the crack may be used as parameters. Unfortunately different sensitivity of the response with respect to various parameters, possible nondifferentiabilities arising from highly nonlinear structural response (as, e.g., in unilateral contact problems) and the computational effort to solve a direct problem (accompanied, if possible, with a sensitivity analysis) makes the numerical solution of this optimization problem a difficult task. An indirect way is proposed here by using a restricted number of direct solutions to generate a set of learning data and, subsequently, by using neural network techniques for the solution of the inverse problem.

H. A. Mang and F. G. Rammerstorfer (eds.),
IUTAM Symposium on Discretization Methods in Structural Mechanics, 199–206.

For the direct problem of an elastic body containing a given crack under static loading a boundary element formulation is used. It leads to a system of linear equations for a classical crack model and to a nonsymmetric linear complementarity problem (L.C.P.) for a unilaterally constrained crack. The boundary element method is best suited to treat this kind of problems with only boundary nonlinearities (see, e.g., Antes and Panagiotopoulos (1992), Panagiotopoulos (1985)). Here, the direct LCP formulation of the unilateral contact problem with Boundary Elements is outlined and used in a code based on a implementation given by Brebbia and Dominguez (1989).

The inverse problem for nonlinear structural response, as posed previously, leads to ill–conditioned generally nonsmooth and nonconvex optimization problems. The direct numerical solution of this problem is a difficult task. For the identification of flaws and classical (bilateral) cracks, boundary element method techniques and classical minimization algorithms have been used in Tanaka and Masuda (1986), Nishimura and Kobayashi (1991), Mitra and Das (1992) and Mellings and Aliabadi (1994) among others. A stochastic approach based on Kalman filtering is used in Tosaka et al. (1995). Neural networks have been proposed in Tanaka et al. (1991) for dynamic, bilateral crack identification and for damage identification with eigenmodal data in Wu et al. (1992), Yoshimura et al. (1996). Finally the ultrasonic, nondestructive evaluation simulated by elastodynamic boundary elements has been treated by neural networks in Oishi et al. (1995).

In the problem studied here, the structure (state) is governed by a variational inequality. Thus, the identification problem, which can be written as an optimal design problem with a state governed by a variational inequality, is in general a nondifferentiable and possibly nonconvex problem (cf., e.g., Outrata and Zowe (1995), Stavroulakis (1995), Luo et al. (1996)). For the neural network treatment a set of direct problems have to be solved. The inverse output-parameter relation is reconstructed by a back propagation neural network, which is used in turn for the solution of the inverse problem, i.e., the prediction of the crack parameters for a given (measured) data. Since a closed crack is difficult (if impossible) to be identified and the considered unilateral crack model allows for this case, the heuristic of mixing several load cases is applied with some success (Stavroulakis and Antes 1997). Moreover, the preprocessing transformation proposed in Yoshimura et al. 1996 has also been used to enhance the performance of the neural network scheme.

2. Bilateral and Unilateral BEM Crack Analysis

From the matrix formulation of the boundary element method in elastostatics (cf. Brebbia and Dominguez (1989)):

$$\mathbf{Hu} = \mathbf{Gt}, \tag{1}$$

with, say, m boundary elements and n boundary nodes, the classical approach for the solution of bilaterally constrained structures uses the known boundary displacements or tractions to re–arrange system (1), and, finally, solve a nonsymmetric system of equations.

When assuming that on a certain part of the boundary unilateral contact relations hold, both the displacements and the tractions of that boundary must remain in the formulation. Hence, let $\mathbf{u}_c$, Q_c be the boundary nodal displacements and the boundary nodal tractions, respectively, at the unilateral (contact) boundary. After partitioning and treating the other boundary values by the above outlined classical way, one arrives from (1) at (f are the free or classical variables, c are the variables involved in the contact boundary):

$$\begin{bmatrix} \mathbf{H}_{ff} & \mathbf{H}_{fc} \\ \mathbf{H}_{cf} & \mathbf{H}_{cc} \end{bmatrix} \begin{bmatrix} \mathbf{x} \\ \mathbf{u}_{cN} \end{bmatrix} = \begin{bmatrix} \mathbf{f}_f \\ \mathbf{f}_c \end{bmatrix} + \begin{bmatrix} \mathbf{G}_{fc} \\ \mathbf{G}_{cc} \end{bmatrix} \mathbf{Q}_{cN}. \tag{2}$$

In accordance with the assumption of frictionless unilateral contact cracks, only the normal ($\mathbf{u}_{cN}$, $\mathbf{Q}_{cN}$) contact boundary contributions are of interest. Tangential displacements are unrestricted (thus they belong to vector $\mathbf{x}$) with zero tangential traction. The corresponding local coordinate transformation is not discussed here for brevity.

Equation (2) together with the unilateral contact relations:

$$\mathbf{u}_{cN} \leq \mathbf{0},\ \ \mathbf{Q}_{cN} \leq \mathbf{0},\ \ \mathbf{u}_{cN}^T \mathbf{Q}_{cN} = 0,\ \ \mathbf{Q}_{cT} = \mathbf{0}, \tag{3}$$

constitutes a Linear Complementarity Problem (L.C.P.).

2.1. MULTIPLE REGION BEM WITH INEQUALITY INTERFACES

For a two reqion structure, $\mu = \ A, B$, as the one arises in the classical treatment of crack problems by BEM, and on the assumption that the boundary/interface displacements and tractions are grouped in $\mathbf{u}_{Ac}$, $\mathbf{u}_{Bc}$, $\mathbf{f}_{Ac}$ and $\mathbf{f}_{Bc}$ (the remaining variables are treated in the classical way within vectors $\mathbf{X}_{Af}$ and $\mathbf{X}_{Bf}$), we have:

$$\begin{bmatrix} \mathbf{H}_{\mu ff} & \mathbf{H}_{\mu fc} \\ \mathbf{H}_{\mu cf} & \mathbf{H}_{\mu cc} \end{bmatrix} \begin{bmatrix} \mathbf{X}_{\mu f} \\ \mathbf{u}_{\mu c} \end{bmatrix} = \begin{bmatrix} \mathbf{f}_{\mu f} \\ \mathbf{f}_{\mu c} \end{bmatrix} + \begin{bmatrix} \mathbf{G}_{\mu fc} \\ \mathbf{G}_{\mu ff} \end{bmatrix} \mathbf{Q}_{\mu c},\ \mu = A, B. \tag{4}$$

Let us assumme that the quantities which participate in the contact interface relations, i.e., $\mathbf{u}_{Ac}, \mathbf{u}_{Bc}, \mathbf{Q}_{Ac}, \mathbf{Q}_{Bc}$,are splitted in two subvectors, the unilateral interface one denoted by a superscript (c) and the classical interface subvector denoted by a superscript (b). Then the set of interface relations is composed of the unilateral conditions (normal to the boundary contributions as in (3)):

$$\begin{aligned} \mathbf{u}_{Ac}^{(c)} &= \mathbf{u}_{Bc}^{(c)} + \mathbf{y},\ \mathbf{Q}_{Ac}^{(c)} = -\mathbf{Q}_{Bc}^{(c)} = \mathbf{Q}_c, \\ \mathbf{y} &\geq \mathbf{0},\ \mathbf{Q}_c \geq \mathbf{0},\ \mathbf{y}^T\mathbf{Q}_c = 0, \end{aligned} \tag{5}$$

and the bilateral ones

$$\mathbf{u}_{Ac}^{(b)} = \mathbf{u}_{Bc}^{(b)}, \mathbf{Q}_{Ac}^{(b)} = -\mathbf{Q}_{Bc}^{(b)}. \tag{6}$$

After matricial reformulation of relations (4), (5) and (6) one arives at the governing LCP (see, e.g., Panagiotopoulos 1985, Antes and Panagiotopoulos 1992 for more details).

3. Neural Network Solution of the Inverse Problem

Let a given structure be considered which contains an unknown crack, characterized by a set of parameters $\mathbf{z} = [z_1, \ldots, z_m]^T$. Here, the crack center coordinates, the length and the inclination of a rectilinear crack are considered. Let the response of the structural system for a given loading $\mathbf{p}^l$, $l = 1, \ldots, l_1$ and for a given crack configuration $\mathbf{z}$ be given by the vector $\tilde{\mathbf{r}}(\mathbf{z}, \mathbf{p}^l) = [\tilde{r}_1, \ldots, \tilde{r}_{m_1}]^T$, where m_1 is the total number of measurements. Boundary nodes' displacements are used here. The response of the examined structure with a known crack subjected to the same loading $\mathbf{p}^l$ is denoted by $\tilde{\mathbf{r}}_0(\mathbf{p}^l)$.

The inverse problem is formulated as an output minimization problem for a scalar performance error function:

$$\Phi(\mathbf{z}) = \sum_{l=1}^{l_1} \left(\|\tilde{\mathbf{r}}(\mathbf{z}, \mathbf{p}^l) - \tilde{\mathbf{r}}_0(\mathbf{p}^l)\| \right) \tag{7}$$

Here $\|\cdot\|$ is an appropriate (usually the L^2) norm in R^{m_1} (least square identification). The sum runs over all available loading cases.

For unilateral contact state problems function (7) is, in general, nonconvex and nondifferentiable (see, e.g., analogous considerations in Stavroulakis (1995), Luo et al. (1996)). In view of this complexity, for which classical algorithms have shown a poor numerical performance, the problem is attacked by a derivative free optimization procedure which is based on the neural network theory.

3.1. NEURAL NETWORK SOLUTION METHOD

Hierarchical, feed–forward neural networks trained by the back-propagation algorithm are used here to learn the relation

$$\tilde{\mathbf{r}}(\mathbf{z}, \mathbf{p}^l) \rightarrow \mathbf{z} \text{ (resp. } (\tilde{\mathbf{r}}(\mathbf{z}, \mathbf{p}^i),\ i = 1, 2, \ldots, l) \rightarrow \mathbf{z}), \tag{8}$$

for one (resp. several) loading vector(s) $\mathbf{p}^l$. Note that the inverse relation is directly considered in (8). These data have been produced here by the previously outlined BEM-LCP theory. For applications of the neural network method in damage identification through static data see, e.g., Wu et al. (1992) and through dynamic data see, e.g., Rhim and Lee (1995), Oishi et al. (1995).

The main idea behind the back propagation neural network model is that a specific norm between the network's output $\mathbf{z}_j'$, for inputs equal to $\tilde{\mathbf{r}}(\mathbf{z}_j, \mathbf{p}^l)$, respectively, and the known outputs $\mathbf{z}_j$, for $j = 1, \ldots, k$, is iteratively minimized during the learning phase. Thus, a neural network realizes a nonparametric structural identification method in the sense of Natke (1991).

After learning the nonlinear network, with fixed values, reproduces the relation $\tilde{\mathbf{r}} \rightarrow \mathbf{z}$, i.e., for a given set of measurements $\tilde{\mathbf{r}}$ the network gives a prediction for the variables which characterize the internal crack.

4. Numerical Examples

A plane stress plate with a crack is considered with quadratic three–node boundary elements. The material constants are the shear modulus $G = 100000.0$ and the Poisson's ratio 0.3, and the external dimensions of the plate are 100.00×100.00, all in compatible units. Each part of the plate is discretized by 84 boundary nodes. For the two-region modelling 29 of these nodes belong to the common interface between the two parts. From them 17 are classical, bilaterally connected interface nodes and 12 are unilateral ones on the two sides of the crack. The LCP-BEM algorithm is able to analyse open and partially or fully closed, frictionless cracks (see sample results in Fig. 1, where a given crack of length equal to 10 and boundary loadings with different orientations are assumed).

Let us consider a crack at a fixed position with unknown length. This unknown length can be identified by using measurements of the displacements at the upper (free) boundary of the plate. The performance of such an identification with a $58 - 10 - 1$ network is shown in Fig. 2.a. Here all displacements of the 29 upper boundary nodes are used as measurements (thus $2 \times 29 = 58$ is the total number of imputs). The uncknown length of the crack is the single output variable here.

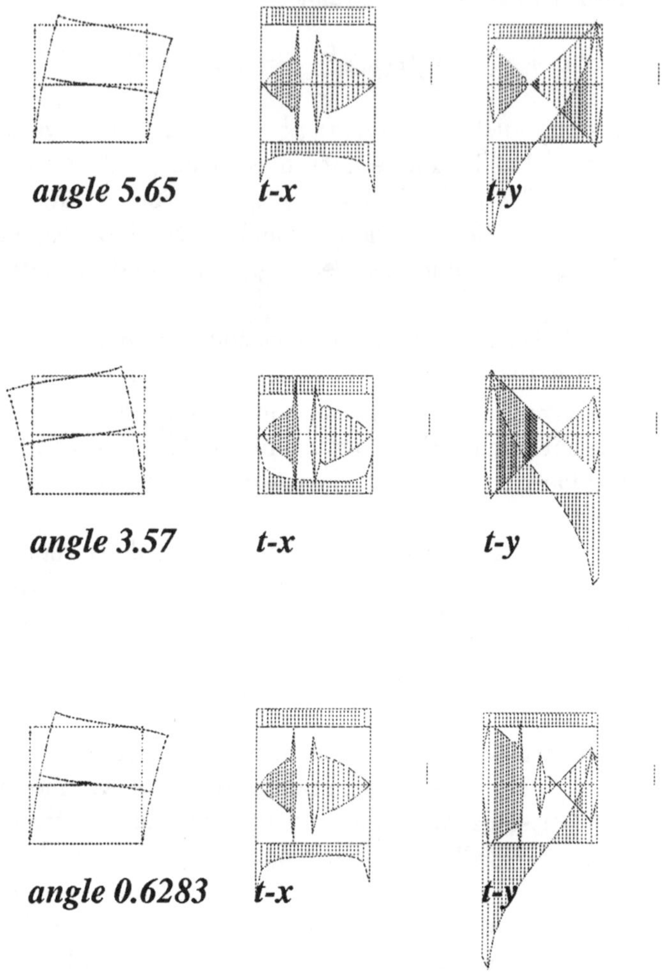

Figure 1. Deformed configuration, horizontal (t-x) and vertical (t-y) boundary tractions for various loadings and a given unilateral crack.

Analogously, the inclination of a crack with given length within the plate can be identified by a similar network (see Fig. 2.b).

For more complicated problems, concerning the identification of the position of a crack within the plate, see the sample results on Fig. 3 (see also Stavroulakis and Antes (1997)). Here a $58-50-50-2$ neural network

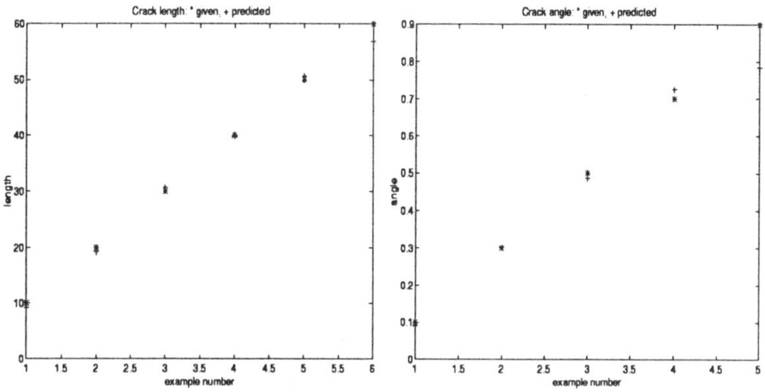

Figure 2. (a) Results for inclined cracks, (b) neural prediction of the crack inclination.

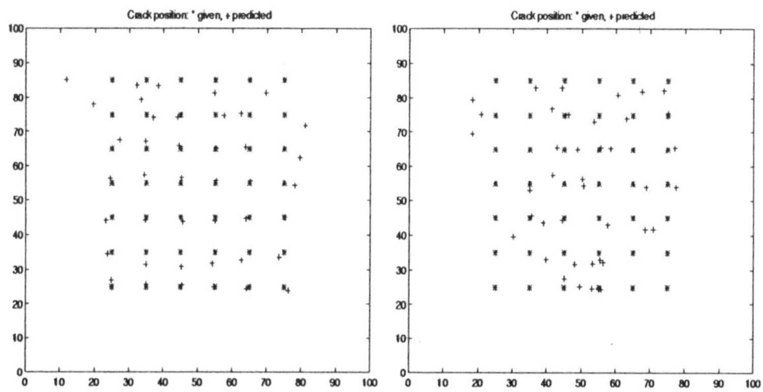

Figure 3. Neural identification of a crack position: (a) classical crack, (b) unilateral crack. All test cases for differenr crack positions are shown in one figure.

has been used. The imput variables are the same as in the previous example. The coordinates of the position of the center of a horizontal rectilinear crack are the two output variables.

For the first two problems learning took less than 1000 epochs (iterations) or about 1 minute within the MATLAB environment at an IBM RISC 6000 workstation. The last problem is considerably more demanding with typical requirements being 10000 epochs and $1 - 2$ hours.

Acknowledgements

The first author gratefully acknowledges the support of the European Union (Grant ERBFMBICT960987) and the hospitality of Prof. Dr. H. Antes.

References

Antes, H. and Panagiotopoulos, P.D. (1992) *An integral equation approach to the static and dynamic contact problems. Equality and inequality methods*. Birkhäuser Verlag, Boston, Basel, Stuttgart.

Brebbia, C.A. and Dominguez, J. (1989) *Boundary Elements. An introductory course*. Computational Mechanics Publications and McGraw–Hill Book Co.

Luo, Z.–Q., Pang, J.–S. and Ralph, D. (1996) *Mathematical Programs with Equilibrium Constraints*. Cambridge University Press, 1996.

Mellings S.C. and Aliabadi M.H. (1994) Three dimensional flaw identification using sensitivity analysis. In: *Boundary Element Method XVI*, Brebbia, C.A. (ed), pp. 149–156, Computational Mechanics Publications, Southampton, Boston.

Mitra, A.K. and Das, S. (1992) Solution of inverse problems by using the boundary element method. In: *Boundary Element Technology VII*, Brebbia,C.A., Ingber, M.S. (eds), pp. 721–731, Computational Mechanics Publ and Elsevier Appl. Science.

Natke H.G. (1991) *Einführung in Theorie und Praxis der Zeitreihen- und Modalanalyse*. 3rd Ed., Vieweg Verlag, Braunschweig, Wiesbaden.

Natke H.G. (1993) The state of the art of system identification. Emphasizing nonlinear systems and applications. In: *Structural Dynamics EURODYN'93*, Moan et al. (eds), Vol. 2, pp. 805–810, Balkema, Rotterdam.

Nishimura N. and Kobayashi S. (1991) A boundary integral equation method for an inverse problem related to crack detection, *Intern. Journal for Numerical Methods in Engineering*, **32**, pp. 1371–1387.

Oishi, A., Yamada, K. and Yoshimura, A., Yagawa G. (1995) Quantitative nondestructive evaluation with ultrasonic method using neural networks and computational mechanics, *Computational Mechanics*, **15**, pp. 521–533.

Outrata, J.V. and Zowe, J. (1995) A numerical approach to optimization problems with variational inequality constraints, *Mathematical Programming*, **68**, pp. 105–130.

Panagiotopoulos, P.D. (1985) *Inequality Problems in Mechanics and Applications, Convex and Nonconvex Enegy Functions*, Birkhäuser Verlag, Basel.

Rhim, J. and Lee S.W. (1995) A neural network approach for damage detection and identification of structures, *Computational Mechanics*, **16**, pp. 437–443.

Stavroulakis, G.E. (1995) Optimal prestress of cracked unilateral structures: finite element analysis of an optimal control problem for variational inequalities, *Computer Methods in Applied Mechanics and Engineering*, **123**, pp. 231–246.

Stavroulakis, G.E. and Antes, H. (1997) Nondestructive elastostatic identification of unilateral cracks through BEM and Neural Networks, *Computational Mechanics* (in press).

Tanaka, M. and Masuda, Y. (1986) An integral equation approach to inverse problems in structural mechanics. In: *Computational Mechanics 86*, Yagawa, G., Atluri, S.N. (eds), Vol. 2, pp. XI–15, XI–24, Springer.

Tanaka, M., Nakamura, M., Nakano, T., Ishikawa, H. (1991) Identification of defects by the elastodynamic boundary element method using noisy additional information. In: *Boundary Elements XIII*, Brebbia, C.A. Gipson, G.S. (eds), pp. 799–810, Computational Mechanics Publ. and Elsevier Appl. Science.

Tosaka, N., Utani, A. and Takahashi, H. (1995) Unknown defect identification in elastic field by boundary element method with filtering procedure, *Engineering Analysis with Boundary Elements*, **15**, pp. 207–215.

Wu, X., Ghaboussi, J. and Garett, Jr. J.H. (1992) Use of neural networks in detection of structural damage, *Computers and Structures*, **42**, pp. 649–659.

Yoshimura, S., Matsuda, A. and Yagawa, G. (1996) New regularization by transformation for neural network based inverse analyses and its application to structure identification, *Intern. Journal of Numerical Methods in Engineering*, **39**, pp. 3953–3968.

RECENT ADVANCES IN SENSITIVITY ANALYSIS FOR NONLINEAR STRUCTURAL MECHANICS PROBLEMS

AHMED K. NOOR AND JEANNE M. PETERS
Center for Advanced Computational Technology
University of Virginia
NASA Langley Research Center
Hampton, Virginia

1. Abstract

Recent developments in the sensitivity analysis for nonlinear structural mechanics and dynamics problems are reviewed. The activities are grouped into two general categories, namely, computational strategies and facilities for evaluating the sensitivity coefficients, and new applications. Brief description is given of the activities pertaining to the first category. More detailed discussion is presented for two recent applications: a) hierarchical sensitivity coefficients of the various response quantities of sandwich panels with composite face sheets with respect to the different laminate, layer and micromechanical characteristics of the face sheets and core; and b) sensitivity analysis of the large strain response of viscoplastic solids subjected to dynamic loading. Sample numerical results are presented and some of the future directions for research on sensitivity analysis of large-scale nonlinear structures are outlined.

2. Introduction

In recent years several attempts have been made to extend the domain of sensitivity analysis to nonlinear static and dynamic responses of structures. Some of these attempts are documented in a monograph [1], survey papers and conference proceedings [2-7]. The sensitivity information, in addition to being a necessary component in the automated optimum design of structures, is used to:

1) assess the effect of uncertainties, in the material and geometric parameters of the computational model, on the nonlinear response;

2) predict the changes in the nonlinear response due to changes in the parameters; and

3) generate an approximation for the nonlinear response of the structure (along with a rapid reanalysis technique). Other uses of sensitivity information will be outlined in subsequent sections.

The sensitivity analysis is performed by evaluating the derivatives of the response quantities with respect to the material and geometric coefficients of the structure. Henceforth, these derivatives will be referred to as sensitivity

H. A. Mang and F. G. Rammerstorfer (eds.),
IUTAM Symposium on Discretization Methods in Structural Mechanics, 207–216.

coefficients. Two general procedures are currently used for calculating the sensitivity coefficients of the nonlinear response of structures. The two approaches are the direct differentiation method, and the adjoint variable method. The first procedure is based on the implicit differentiation of the equations that describe the nonlinear response with respect to the desired parameters, and the solution of the resulting sensitivity equations. In the adjoint variable method an adjoint physical system is introduced whose solution permits rapid evaluation of the desired sensitivity coefficients. Both procedures can be applied to the governing discrete, semi-discrete or continuum equations of the structure (with a consequent change in the order of spatial discretization, temporal integration, and implicit differentiation).

Recent activities on sensitivity analysis for nonlinear structural response can be grouped into two general categories: a) computational strategies and facilities for evaluating sensitivity coefficients; and b) new applications, including composite structures [8]; flexible multibody systems [9]; path dependent problems for which the sensitivity coefficients depend also on the deformation history (e.g., viscoplastic response and frictional contact) [10-16]; and structural systems exhibiting probabilistic uncertainties. The presence of inequality constraints in contact problems implies non-differentiability. As a consequence, only directional sensitivities of the structural response with respect to problem parameters can be expected. The persistent contact conditions and the use of the penalty method alleviate the discontinuity problem. The present paper provides a brief description of some of the recent developments in computational strategies and facilities for sensitivity analysis for nonlinear structural response. The two applications are: a) hierarchical sensitivity analysis for sandwich panels with composite face sheets, and b) sensitivity analysis of porous viscoplastic solids subjected to dynamic loading, and are described in succeeding sections.

3. Brief Review of Work on Computational Strategies and Facilities

An automatic differentiation facility has been developed for evaluating the derivatives of functions defined by computer programs, exactly to within machine precision. The facility has the acronym ADIFOR (Automatic Differentiation of FORtran), and is described in [17]. The use of ADIFOR to evaluate the sensitivity coefficients from incremental/iterative forms of three-dimensional fluid flow problems is discussed in [18], and the additional facilities needed for ADIFOR to become competitive with hand-differentiated codes are listed in [19].

For large deflection static problems, an efficient reduced basis technique for evaluating the sensitivity of the nonlinear response was developed by the authors and their colleagues. The technique is based on approximating each of the response vector and its various order sensitivity coefficients, by a small number of basis (or global approximation) vectors. The Bubnov-Galerkin technique is then used to approximate each of the finite element equations governing the response and the sensitivity coefficients, by a small number of algebraic equations in the amplitudes of these vectors. Path derivatives (derivatives of the response vector with respect to path parameters, e.g., load parameters) are used as basis vectors for approximating the response. A combination of path derivatives and their derivatives with respect to the structural parameters is used for approximating the sensitivity coefficients. The technique was applied to geometrically nonlinear problems of composite panels and aircraft tires [20-21]. It was also extended to nonlinear vibration problems of composite panels [22].

Parallel computational strategies were developed for evaluating the sensitivity coefficients for static postbuckling and contact/impact response on distributed-memory computers. The strategies are applicable to any message-passing computational environment. The key elements of the strategy for geometrically nonlinear static problems are [23]: a) multiple-parameter reduced basis technique, b) a parallel sparse equation solver based on a nested dissection (or multilevel substructuring) node-ordering scheme, and c) a multilevel parallel procedure for evaluating the sensitivity coefficients. For contact/impact problems, an explicit central difference scheme is used for the temporal integration of the semi-discrete equations of motion, and the two key elements of the parallel strategy are [24]: a) an element-based domain decomposition technique, and b) a robust exchange algorithm for communicating information across subdomain interfaces.

Hybrid numerical/neurocomputing strategy has been developed for the evaluation of sensitivity coefficients of geometrically nonlinear structural response. In the hybrid strategy, multilayer feedforward neural networks are used to extend the range of validity of the sensitivity coefficients predicted by Padé approximants [25].

4. Hierarchical Sensitivity Coefficients of Sandwich Panels with Composite Face Sheets

The response characteristics of sandwich panels with composite face sheets are dependent on a hierarchy of interrelated parameters including panel parameters, effective face sheet layer and core properties, and micromechanical (core and face sheet fiber, matrix, interphase and interface) parameters. A study of the sensitivity of the response to variations in each of these parameters provides an insight into the importance of the parameter and helps in the development of materials to meet certain performance requirements. Moreover, in multiscale analysis, the hierarchical sensitivity coefficients can be used for identifying the extent of the regions for which a micromechanical analysis is needed to capture the phenomena occurring at the small length scales.

Three sets of sandwich parameters are considered herein; namely, panel, effective layer, and micromechanical parameters. The panel parameters include the extensional, bending-extensional, bending and transverse shear stiffnesses (components of the matrices $[A],[B],[D]$ and $[A_s]$); and the thermal forces and moments (components of the vectors $\{N_T\}$ and $\{M_T\}$). The effective layer parameters include the individual face sheet layer properties; elastic moduli E_L, E_T; shear moduli G_{LT}, G_{TT}; major Poisson's ratio ν_{LT}; coefficients of thermal expansion α_L, α_T; fiber orientation angle $\theta^{(\ell)}$; layer thickness $h^{(\ell)}$ where L and T refer to the longitudinal (fiber) and transverse directions, respectively. The layer parameters also include the effective core properties: elastic moduli E_{1c}, E_{2c}; shear moduli $G_{12c}, G_{13c}, G_{23c}$; Poisson's ratios $\nu_{12c}, \nu_{13c}, \nu_{23c}$; coefficient of thermal expansion α_c and core thickness $h^{(c)}$. The micromechanical parameters refer to the fiber, matrix and core material moduli $E_{1f}, E_{2f}, E_m, E_c, G_{12f}, G_m, G_c$; Poisson's ratios ν_{12f}, ν_{23f}; coefficients of thermal expansion $\alpha_{1f}, \alpha_{2f}, \alpha_m, \alpha_c$; the fiber volume fraction v_f of the face sheet layers; and the geometric parameters of the core $\ell_{1c}, \ell_{2c}, t_c$ and θ. The subscripts f, m and c denote the fiber, matrix and core property, respectively. The three sets of parameters will henceforth be referred to as $\lambda_i^{(p)}, \lambda_j^{(\ell)}, \lambda_k^{(m)}$ where superscripts p, ℓ and m refer to the panel, layer, and micromechanical parameters, respectively; and the indices i, j and k range from 1 to the number of parameters in each category.

The governing equations for the sensitivity coefficients are obtained by differentiating the nonlinear equations describing with respect to a typical parameter λ. The resulting equations are linear in the sensitivity coefficients.

The computational procedure consists of evaluating the sensitivity coefficients with respect to each of the panel parameters $\left\{\frac{\partial \mathbf{Z}}{\partial \lambda_i^{(p)}}\right\}$. The sensitivity coefficients with respect to the effective layer and micromechanical parameters are then obtained by forming the following linear combinations:

$$\left\{\frac{\partial \mathbf{Z}}{\partial \lambda_j^{(\ell)}}\right\} = \sum_i a_{ij} \left\{\frac{\partial \mathbf{Z}}{\partial \lambda_i^{(p)}}\right\} \tag{1}$$

and

$$\left\{\frac{\partial \mathbf{Z}}{\partial \lambda_k^{(m)}}\right\} = \sum_j b_{jk} \left\{\frac{\partial \mathbf{Z}}{\partial \lambda_j^{(\ell)}}\right\} = \sum_i c_{ik} \left\{\frac{\partial \mathbf{Z}}{\partial \lambda_i^{(p)}}\right\} \tag{2}$$

where

$$a_{ij} = \left\{\frac{\partial \lambda_i^{(p)}}{\partial \lambda_j^{(\ell)}}\right\} \tag{3}$$

$$b_{jk} = \left\{\frac{\partial \lambda_j^{(\ell)}}{\partial \lambda_k^{(m)}}\right\} \tag{4}$$

$$c_{ik} = \left\{\frac{\partial \lambda_i^{(p)}}{\partial \lambda_k^{(m)}}\right\} = \sum_j a_{ij} b_{jk} \tag{5}$$

The a_{ij} coefficients relate the panel stiffnesses to the effective properties of the individual layers and are obtained from the lamination theory. The b_{jk}

coefficients relate the effective layer properties to the constituent properties and are obtained from the micromechanical and core models; and the c_{ik} coefficients relate the panel stiffnesses to the micromechanical properties (see bottom part of Fig. 1). If the panel stiffnesses are uniform, and the constitutive relations of the panel, layer, and the constituents are linear, then the a_{ij}, b_{jk}, c_{ik} coefficients are constants and need to be generated only once for each panel, even when the response is nonlinear. For a curved sandwich panel with composite face sheets, the sensitivity coefficients with respect to panel stiffnesses, effective properties of the face sheets and core and micromechanical properties are shown in Fig. 2. The panel is subjected to a sequence of mechanical and thermal loadings: uniform pressure loading, monotonically increasing edge shortening, and then a temperature gradient through the thickness (see [26]).

5. Sensitivity Analysis of Porous Viscoplastic Solids Subjected to Dynamic Loading

The evaluation of the sensitivity coefficient of the nonlinear dynamic response entails the solution of an additional system of ordinary differential equations for the semi-discrete model. The sensitivity coefficients depend on the history of the response, the current values of the response quantities and the history of the sensitivity coefficients. Considerable simplification in the computational procedure results by exploiting the fact that response calculations are uncoupled from sensitivity calculations. If the response quantities are evaluated first, the sensitivity calculations reduce to solve a set of linear algebraic equations whose coefficients depend on the response quantities. For a rate problem such as that of porous viscoplastic solids with material strain and strain-rate hardening and thermal softening due to adiabatic heating, the increments of the sensitivity coefficients form a set of linear algebraic equations. This feature can be exploited in the sensitivity calculations of the constitutive model, wherein the increments of the matrix plastic strain are obtained in terms of response quantities and known sensitivity coefficients.

In the present study, the direct differentiation approach is used in conjunction with the automatic differentiation facility ADIFOR for evaluating the first- and second-order derivatives of the various response functions, with respect to the various geometric and material parameters. For complex material systems like the one considered herein, the sensitivity coefficients can be used in conjunction with experiments to refine the constitutive model and adjust the parameters used therein. For a notched specimen in a plane strain state subjected to an initial velocity, the first- and second-order sensitivity coefficients of the void volume fraction with respect to the material parameter q_2, at three different times, are shown in Fig. 3 (see [27]).

6. Future Directions for Research

Among the different aspects of sensitivity analysis which have high potential for research are the following:

• Extension to structural systems with uncertainties. If fuzzy sets are used to represent the uncertain parameters, then possibility distributions of the various sensitivity coefficients with respect to these parameters can be generated and bounds on these coefficients can be obtained.

• Extension of hierarchical sensitivity analysis of composite and sandwich structures to the subcomponent and the component levels (see top part of Fig. 1). A stiffened panel is an example of a subcomponent. The parameters at that level include the stiffener dimensions and spacing. A fuselage barrel section is an example of a component. The potential of using hierarchical sensitivity analysis to bridge the gap between structural design and material development can then be realized.

• Application to coupled field problems (e.g., coupled mechanical, thermal and electromagnetic fields for smart materials). The sensitivity coefficients can be used to identify the coupling coefficients which have the most impact on the response quantities of interest. Experiments can then be designed to determine these coefficients.

• Application to multilevel strategies for simulating damage propagation in structures. Sensitivity coefficients can help in the automatic updating of the model as the damage propagates.

7. Acknowledgments

This work was partially supported by Office of Naval Research Grant N00014-96-1-0640, Air Force Office of Scientific Research Grant F49620-96-1-0462, and NASA Headquarters Cooperative Agreement NCCW-0011.

8. References

1. Kleiber, M. et al.: *Parameter Sensitivity in Nonlinear Mechanics - Theory and Finite Element Computations*, Wiley, NY, 1997.
2. Haftka, R.T. and Adelman, H.M.: Recent developments in structural sensitivity analysis, *Structural Optimization* **1**(3) (1989), 137-151.
3. Haber, R.B., Tortorelli, D.A. and Vidal, C.A.: Design sensitivity analysis of nonlinear structures I: Large-deformation hyperelasticity and history-dependent material response, in M.P. Kamat (ed.), *Structural Optimization: Status and Promise*, Progress in Aeronautics and Astronautics Series, Vol. 150, American Institute of Aeronautics and Astronautics, Washington, DC, 1993, pp. 369-406.
4. Choi, K.K.: Design sensitivity of nonlinear structures - II, in M.P. Kamat (ed.), *Structural Optimization: Status and Promise*, Progress in Aeronautics and Astronautics Series, Vol. 150, American Institute of Aeronautics and Astronautics, Washington, DC, 1993, pp. 407-446.
5. Hinton, E. and Sienz, J.: Aspects of adaptive finite element analysis and structural optimization, in B.H.V. Topping and M. Papadrakakis (eds.), *Advances in Structural Optimization*, CIVIL-COMP, Ltd., Edinburgh, Scotland, 1994, pp. 1-25.
6. Kirsch, U.: Efficient sensitivity analysis for structural optimization, *Comp. Meth. Appl. Mech. Eng.*, 117(1-2) (1994), 143-156.
7. Tortorelli, D.A. and Michaleris, P.: Design sensitivity analysis: overview and review, *Inverse Problems in Eng.* **1** (1994), 71-105.

8. Noor, A.K.: Recent advances in the sensitivity analysis for the thermomechanical postbuckling of composite panels, *ASCE J. Eng. Mech.* **122**(4) (1996), 300-307.
9. Wasfy, T.M. and Noor, A.K.: Modeling and sensitivity analysis of multibody systems using new solid, shell and beam elements, *Comp. Meth. Appl. Mech. Eng.* **138** (1996), 187-211.
10. Tortorelli, D.A.: Sensitivity analysis for nonlinear constrained elastostatic systems, *Int. J. Num. Meth. Eng.* **33** (1992), 1643-1660.
11. Michaleris, P., Tortorelli, D.A. and Vidal, C.A.: Tangent operators and design sensitivity formulations for transient nonlinear coupled problems with applications to elastoplasticity, *Int. J. Num. Meth. Eng.* **37** (1994), 2471-2499.
12. Kulkarni, M. and Noor, A.K.: Sensitivity analysis of the nonlinear dynamic viscoplastic response of two-dimensional structures with respect to material parameters, *Int. J. Num. Meth. Eng.* **38** (1995), 183-198.
13. Kulkarni, M. and Noor, A.K.: Sensitivity analysis for the dynamic response of thermoviscoplastic shells of revolution, *Comp. Meth. Appl. Mech. Eng.* **129** (1996), 371-391.
14. Karaoglan, L. and Noor, A.K.: Dynamic sensitivity analysis of frictional contact/impact response of axisymmetric composite structures, *Comp. Meth. Appl. Mech. Eng.* **128** (1995), 169-190.
15. Kleiber, M. and Kowalczyk, P.: Sensitivity analysis in plane stress elasto-plasticity and elasto-viscoplasticity, *Comp. Meth. Appl. Mech. Eng.* **137** (1996), 395-409.
16. Kowalczyk, P. and Kleiber, M.: Parameter sensitivity for large deformation inelastic problems, *Comp. Assisted Mech. & Eng. Sci.* **4** (1997), 209-228.
17. Chinchalker, S.: The application of automatic differentiation to problems in engineering analysis, *Comp. Meth. Appl. Mech. Eng.* **118** (1994), 197-207.
18. Carle, A., Green, L.L., Bischof, C.H. and Newman, P.A.: Application of automatic differentiation in CFD, *Proc. 25th AIAA Fluid Dynamics Conf.*, June 20-23, 1994, Colorado Springs, CO (1994), AIAA Paper 94-2197.
19. Sherman, L.L., Taylor, A.C., III, Green, L.L., Newman, P.A., Hou, G.J.-W. and Korivi, V.M.: First- and second-order aerodynamic sensitivity derivatives via automatic differentiation with incremental iterative methods, *Fifth AIAA/UASF/NASA/ISSMO Sympo sium on Multidisciplinary Analysis and Optimization*, Sept. 7-9, 1994, Panama City Beach, FL (1994), AIAA Paper 94-4262.
20. Noor, A.K.: Recent advances and applications of reduction methods, *Applied Mechanics Reviews* **47**(5) (1994), 125-146.
21. Noor, A.K., Tanner, J.A. and Peters, J.M.: Reduced-basis technique for evaluating the sensitivity coefficients of the nonlinear tire response, *AIAA J.* **31**(2) (1993), 370-376.
22. Noor, A.K., Hadian, M.J. and Peters, J.M.: Reduced basis technique for evaluating the sensitivity of the nonlinear vibrational response of composite plates, *Comp. & Struct.* **52**(6) (1994), 1097-1105.
23. Watson, B.C. and Noor, A.K.: Sensitivity analysis large-deflection and postbuckling responses on distributed-memory computers, *Comp. Meth. Appl. Mech. Eng.* **129** (1996), 393-409.
24. Watson, B.C. and Noor, A.K.: Large-scale contact/impact simulation and sensitivity analysis on distributed-memory computers, *Comp. Meth. Appl. Mech. Eng.* **141** (1997), 373-388.
25. Szewczyk, Z.P. and Noor, A.K.: A hybrid numerical/neurocomputing strategy for sensitivity analysis of nonlinear structures, *Comp. & Struct.* **65**(6) (1997), 869-880.
26. Noor, A.K., et al.: Curved sandwich panels subjected to temperature gradient and mechanical loads, *ASCE J. Aerospace Div.* **10(4)** (1997), 143-161.

27. Noor, A.K., et al.: Sensitivity analysis for failure and damage in dynamically loaded tensile bars, *Comp. Meth. Appl. Mech. Eng.* **151(3-4)** (1998), 461-478.

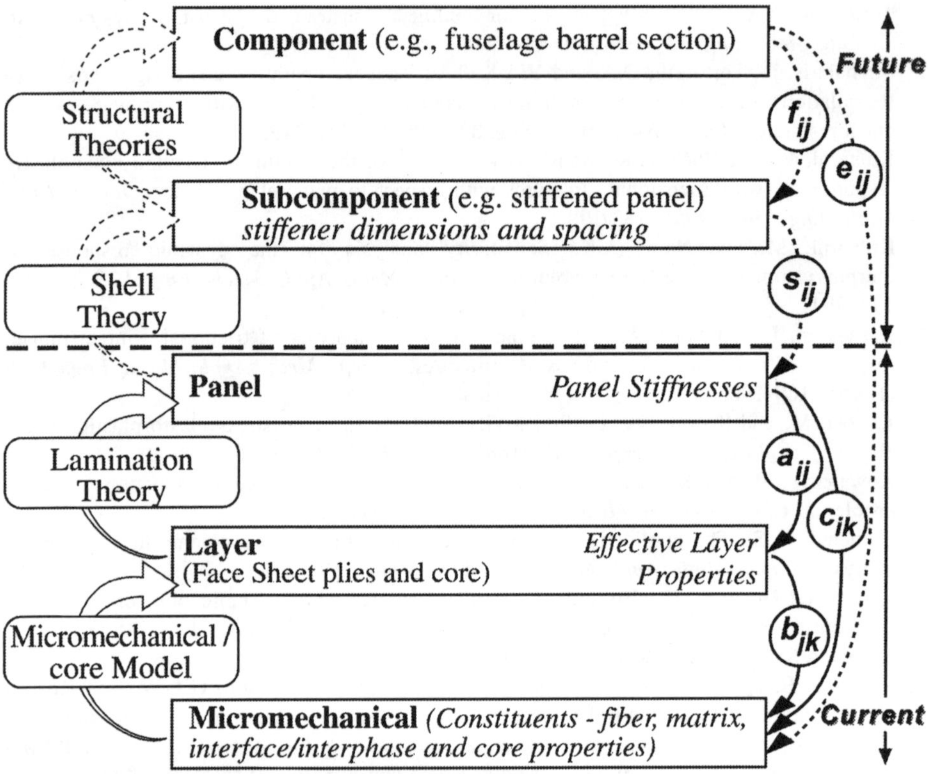

Figure 1 - Hierarchical sensitivity coefficients for sandwich structures with composite face sheets.

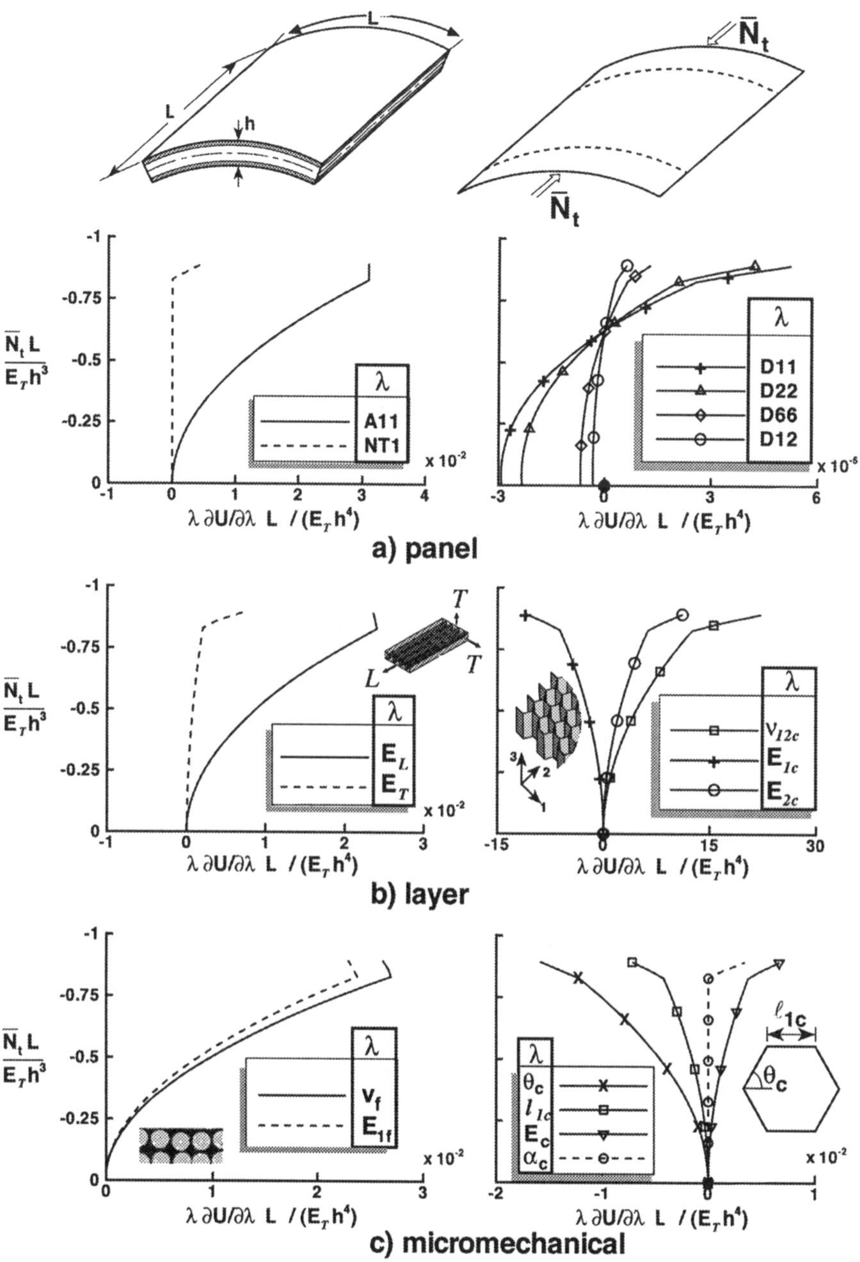

Figure 2 - Effect of loading on the normalized sensitivity coefficients of the total strain energy U with respect to panel, layer and micromechanical parameters. Cylindrical sandwich panels with honeycomb core and composite eight-layer face sheets, subjected to combined pressure loading, edge shortening and temperature gradient through the thickness (see [26]).

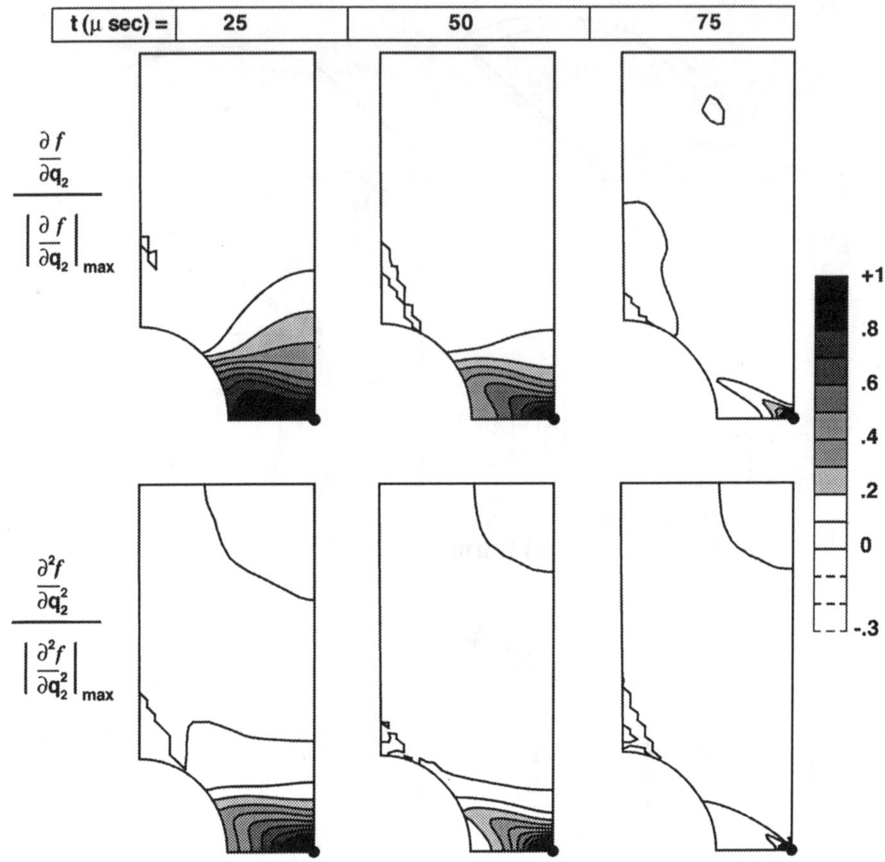

Figure 3 - Normalized contour plots for the first-order and second-order sensitivity coefficients of the void volume fraction f with respect to the material parameter q_2. $\Phi = \frac{\sigma_e^2}{\overline{\sigma}^2} + 2q_1 f \cosh\left(\frac{2q_2\sigma_h}{2\overline{\sigma}}\right) - 1 - q_1^2 f^2 = 0$**, where $\overline{\sigma}, \sigma_e$ and σ_h are the average strength of the matrix material, the macroscopic Mises stress and mean stress, respectively (see [27]).**

SOME ASPECTS OF FEM APPLICATION FOR SENSITIVITY ANALYSIS OF QUASI-BRITTLE FRACTURE CONDITIONS

N.V. BANICHUK AND V.V. SAURIN
The Institute for Problems in Mechanics, Russian Academy of Sciences
Pr. Vernadskogo 101, 117526, Moscow, Russia

Abstract. In this paper continuum expressions of the shape design sensitivity analysis are derived for elastic bodies with cracks. Derived sensitivity analysis relations are implemented in FEM algorithm. Important aspects of sensitivity analysis and FEM effective applications are discussed.

1. Introduction

A substantial literature has been devoted to the field of shape design sensitivity analysis (SDSA) and optimization problems with integral and local functionals (see for example Haug et al.,1986 and Banichuk,1992). Expressions for shape design sensitivity in terms of domain shape change have been derived in a continuous setting with the help of methods of classical calculus of variations and methods of optimization of distributed parameter systems. Shape design sensitivity information has been expressed as boundary integrals (boundary method of SDSA) or as domain integrals (domain method of SDSA) and evaluated using finite element analysis and boundary-element analysis. Fewer studies have been devoted to the important class of problems of construction and application of effective relations connecting the stress intensity factor and energy release rate with the variation of geometric parameters of the considered quasi-brittle bodies and structural elements with cracks. These relations and their implementation in FEM are of significant interest in fracture mechanics and optimal structural design. With their application it is possible to evaluate the sensitivity of a fracture criterion with respect to imperfect structural shape and non-ideal manufacture, to estimate structural parameters influence on crack

H. A. Mang and F. G. Rammerstorfer (eds.),
IUTAM Symposium on Discretization Methods in Structural Mechanics, 217–224.

propagation, to find shape modifications which increase the strength and help to solve the problems of optimal structural design. SDSA gives also important relevant information. An example of this statement is the subject of the paper (Fancello at al., 1993), where the growth of an initial crack was treated as a change in shape and it was shown how SDSA leads to a well known expression of Rice's path-independent integral and how this integral can be computed through domain integration, which gives much more accurate results when using FEM (see also Li at al.,1985).

This paper deals with the SDSA for quasi-brittle plane bodies and implementation of this analysis in numerical method. Special attention is devoted to basic relations of sensitivity analysis which are derived with the help of domain representation of the path-independent J-integral and the introduction of the adjoint system. A numerical technique for finding the sensitivity of the J-integral with respect to a wide class of boundary variations and specifically with respect to improved variations is worked out. Important aspects of SDSA realization related with FEM modelling, mesh adaptation and smoosing technique are considered. Numerical results of design sensitivity computations, performed for cracked plates loaded by in-plane forces, are presented.

2. Sensitivity of Quasi-Brittle Fracture Conditions with Respect to Shape Variations

Consider a deformed elastic body occupying a two-dimensional region Ω with boundary Γ . Surface tractions $\mathbf{T}$ are given on the boundary contour Γ_σ while zero displacements are assigned to the $\Gamma_u(\Gamma = \Gamma_\sigma + \Gamma_u)$. The body contains a crack modelled by a rectilinear notch Γ_f . The notch is traction free. The state of a stressed and deformed body is described by the following boundary value problem of the theory of elasticity

$$\begin{gathered} \nabla \cdot \boldsymbol{\sigma} = 0, \quad \boldsymbol{\sigma} = \mathbf{C} \cdot \cdot \boldsymbol{\epsilon}, \quad \boldsymbol{\epsilon} = \tfrac{1}{2}(\nabla \mathbf{u} + (\nabla \mathbf{u})^T) \\ (\boldsymbol{\sigma} \cdot \mathbf{n})_{\Gamma_\sigma} = \mathbf{T}, \quad (\mathbf{u})_{\Gamma_u} = 0, \quad (\boldsymbol{\sigma} \cdot \mathbf{n})_{\Gamma_f} = 0 \end{gathered} \tag{1}$$

where $\boldsymbol{\sigma}, \boldsymbol{\epsilon}$ and $\mathbf{u}$ - are respectively , the stress tensor, the strain tensor, the displacement vector, and by means of $\mathbf{C}$ and $\mathbf{n}$ in (1) we denote the elastic modulus tensor and the unit vector pointing in the direction of an outward normal to the boundary of the body Γ and to the boundary of the crack Γ_f, while $\mathbf{T}$ is a given vector function of the space coordinates. The basic relations (1) are derived under the assumptions that the laws of static are obeyed and the strains are small.

In the frame of the plane theory of elasticity the brittle fracture criteria can be written in the form $J \leq G_C$, where G_C is the given brittle strength constant and J is the path-independent integral. For our purposes it is important that the contour integral can be transformed (Li at al.,1985) to the domain integral expression (see Figure 1 -a))

$$J = \int_{\gamma} \left[W(\mathbf{n} \cdot \mathbf{l}) - (\boldsymbol{\sigma} \cdot \mathbf{n}) \cdot \frac{\partial \mathbf{u}}{\partial \mathbf{l}} \right] d\gamma = \int_{\Omega_J} \left[\frac{\partial \mathbf{u}}{\partial \mathbf{l}} \cdot \boldsymbol{\sigma} - W\mathbf{l} \right] \cdot \nabla \psi d\Omega_J \tag{2}$$

Here γ is any path beginning at the bottom crack face and ending on the top face as shown in Figure 1-a), $\mathbf{n}$ is the outward normal to γ, $d\gamma$ - the increment of arc length along γ, $\mathbf{l}$ - the unit vector tangential to the crack, $\Omega = \Omega_I + \Omega_J + \Omega_E$, ψ is a sufficiently smooth function in Ω_J that is unity on γ and vanishes on γ_a : $(\psi)_\gamma = 1, \quad (\psi)_{\gamma_a} = 0$, and Ω_J is a simply connected region enclosed by the contour $\gamma + \gamma_a + \Gamma_f^1 + \Gamma_f^2$. The energy release rate domain integral (2) is useful for modelling of quasi-brittle fracture conditions and give the possibility to diminish the loss of accuracy from numerical evaluation of the J integral. Domain integral expression (2) is also convenient for deriving design sensitivity analysis relations and will be used for this purpose in our investigations.

To evaluate the influence of shape variation on the value of the J-integral we perform the variation of the part Γ_v of the boundary Γ admitting the normal boundary variations $\delta t(P)$ for the boundary points $P \in \Gamma_v$ (see Figure 1-a). As a result we will obtain the increments $\delta\mathbf{u}$, $\delta\epsilon = 1/2(\nabla\delta\mathbf{u} + (\nabla\delta\mathbf{u})^T)$, $\delta\boldsymbol{\sigma}$ for displacements, strains and stresses. Let us derive the variation δJ with respect to the fields variations $\delta\mathbf{u}, \delta\boldsymbol{\sigma}$. For this purpose we vary the domain integral representation for the energy release rate and apply Gauss-Ostrogradskii divergence formula

$$\begin{aligned} \delta J = \int_{\Omega_J} \left(\frac{\partial \mathbf{u}}{\partial \mathbf{l}} \cdot \delta\boldsymbol{\sigma} \cdot \nabla\psi + \frac{\partial \delta\mathbf{u}}{\partial \mathbf{l}} \cdot \boldsymbol{\sigma} \cdot \nabla\psi - (\boldsymbol{\sigma} \cdot \cdot \delta\epsilon) \frac{\partial \psi}{\partial \mathbf{l}} \right) d\Omega_J = \\ \int_{\Omega_J} \left[\frac{\partial \mathbf{u}}{\partial \mathbf{l}} \cdot \delta\boldsymbol{\sigma} - \delta\mathbf{u} \cdot \frac{\partial \boldsymbol{\sigma}}{\partial \mathbf{l}} \right] \cdot \nabla\psi d\Omega_J + \\ \int_{\gamma+\gamma_a} \delta\mathbf{u} \cdot [\boldsymbol{\sigma} \cdot \nabla\psi(\mathbf{n} \cdot \mathbf{l}) - \boldsymbol{\sigma} \cdot \mathbf{n}(\mathbf{l} \cdot \nabla\psi)] d(\gamma + \gamma_a) + \\ \int_{\Gamma_f^1 + \Gamma_f^2} \delta\mathbf{u} \cdot [\boldsymbol{\sigma} \cdot \nabla\psi(\mathbf{n} \cdot \mathbf{l}) - \boldsymbol{\sigma} \cdot \mathbf{n}(\mathbf{l} \cdot \nabla\psi)] d(\Gamma_f^1 + \Gamma_f^2) \end{aligned} \tag{3}$$

To eliminate the derivatives of the displacements variations with respect to space coordinates let us transform the integral of the expressions in square brackets. Here $\mathbf{n}$ is the outward normal to the boundary $\gamma + \gamma_a + \Gamma_f^1 + \Gamma_f^2$. Let us evaluate the contour integral over $\gamma + \gamma_a$ in (3). For this purpose we note that the function ψ is constant on γ and γ_a. Hence the vector-gradient of ψ at points $P \in \gamma + \gamma_a$ is normal to the boundary of the domain Ω_J, i.e. $\nabla\psi = \kappa(P)\mathbf{n}, \quad P \in \gamma + \gamma_a$ where $\kappa(P)$ is scalar function, depending on the coordinates of the point P. This property stipulates vanishing of the contour integral over $\gamma + \gamma_a$ in (3). In fact, using this property we will have

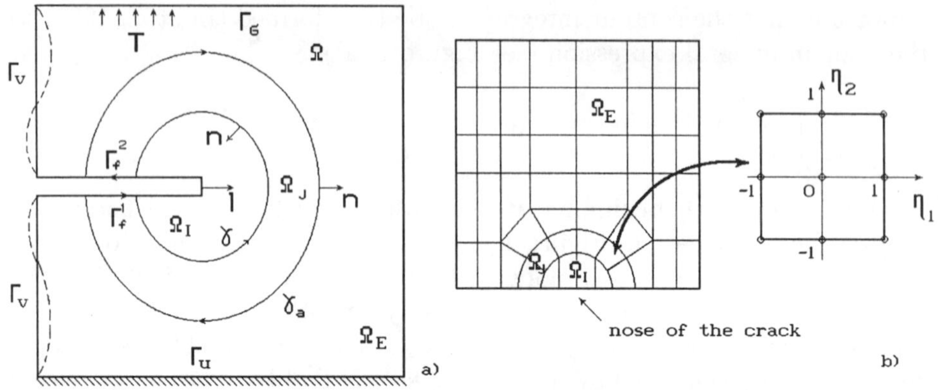

Figure 1. a) Cracked body with varied boundaries and domain decomposition for the evaluation of the J-integral; b) Finite-element mesh.

$\boldsymbol{\sigma} \cdot \nabla\psi(\mathbf{n} \cdot \mathbf{l}) = \boldsymbol{\sigma} \cdot \mathbf{n}(\mathbf{l} \cdot \nabla\psi) = \kappa\boldsymbol{\sigma} \cdot \mathbf{n}(\mathbf{l} \cdot \mathbf{n})$ and, consequently, the contour integral over $\gamma + \gamma_a$ in (3) is identically equal to zero. The contour integral over the part of the bottom crack face Γ_f^1 and the part of the top face Γ_f^2 in (3) (see Figure 1-a)) is equal to zero since both $\mathbf{n} \cdot \mathbf{l} = 0$ and $\boldsymbol{\sigma} \cdot \mathbf{n} = 0$ for the considered parts of the crack. Besides, let us define the function ψ such that $\nabla\psi(P) = 0$ for $P \in \Omega_I + \Omega_E$ and $\nabla\psi(P) = 0$ for $P \in \gamma + \gamma_a$. For such an extension of the definition of ψ the resulting expression for δJ can be written in the form

$$\delta J = \int\limits_{\Omega} \left(\frac{\partial \mathbf{u}}{\partial \mathbf{l}} \cdot \delta\boldsymbol{\sigma} - \delta\mathbf{u} \cdot \frac{\partial \boldsymbol{\sigma}}{\partial \mathbf{l}} \right) \cdot \nabla\psi d\Omega \tag{4}$$

The derived expression (4) for the variation of the domain J-integral depends on variations of the fields of stresses and displacements, that are determined by the normal boundary variations $\delta t(P)$ on the varied part Γ_v of the boundary ($P \in \Gamma_v$). Therefore, the determination of effective representations for δJ as a function of δt is reduced to finding the variations of displacements and stresses stipulated by the shape variations. In many cases direct finding of these relations proves to be very complicated. Therefore, for the considered cases we apply another approach to shape design sensitivity analysis based on the introduction of adjoint variables, augmented Lagrange functional and adjoint boundary value problems. This approach gives us the possibility to exlude the variations $\delta\mathbf{u}$, $\delta\boldsymbol{\sigma}$ in the expression for the varied functional in the effective manner. Let us construct the augmented Lagrange functional, introducing the adjoint variables $\mathbf{v}$ (vector of adjoint displacements), $\boldsymbol{\chi}$ (second rank tensor of adjoint strains) and adding the considered functional (4) to the integral

$$0 = \int_\Omega [\mathbf{v} \cdot (\nabla \cdot \boldsymbol{\sigma}) + \boldsymbol{\chi} \cdot \cdot (\boldsymbol{\sigma} - \tfrac{1}{2}\mathbf{C} \cdot \cdot (\nabla \mathbf{u} + (\nabla \mathbf{u})^T))] d\Omega \tag{5}$$

Taking into account the basic relations of the theory of elasticity, the variation of the domain integral (5) as it is shown in (Banichuk,1992), can be represented as

$$\begin{aligned} 0 = \int_\Omega \{ [\boldsymbol{\chi} - \tfrac{1}{2}(\nabla \mathbf{v} + (\nabla \mathbf{v})^T)] \cdot \cdot \delta\boldsymbol{\sigma} + [\nabla \cdot (\mathbf{C} \cdot \cdot \boldsymbol{\chi})] \cdot \delta \mathbf{u} \} d\Omega + \\ \int_{\Gamma + \Gamma_f} [\mathbf{v} \cdot \delta\boldsymbol{\sigma} \cdot \mathbf{n} - \delta \mathbf{u} \cdot (\mathbf{C} \cdot \cdot \boldsymbol{\chi}) \cdot \mathbf{n}] d\Gamma + \\ \int_{\Gamma_v} [\tfrac{\partial}{\partial \mathbf{n}}(\mathbf{v} \cdot \mathbf{T}) - 2H \mathbf{v} \cdot \mathbf{T} - \nabla \cdot (\mathbf{v} \cdot \boldsymbol{\sigma})] \delta t d\Gamma \end{aligned} \tag{6}$$

Here we suppose that the varied boundary $\Gamma_v \subset \Gamma_\sigma$ and there is no contact between the crack surfaces. In deriving sensitivity analysis relations the expression on the right side of (6) is added to the expression (4) for δJ. The summarized expression should not be a function of the variations $\delta\boldsymbol{\sigma}$ and $\delta\mathbf{u}$. To obtain this it is assumed that the adjoint variables $\mathbf{v}$ and $\boldsymbol{\chi}$ satisfy the differential equations and boundary conditions

$$\nabla \cdot (\mathbf{C} \cdot \cdot \boldsymbol{\chi}) + \mathbf{q}^0 = 0, \quad \boldsymbol{\chi} - \frac{1}{2}(\nabla \mathbf{v} + (\nabla \mathbf{v})^T) + \boldsymbol{\chi}^0 = 0 \tag{7}$$

$$((\mathbf{C} \cdot \cdot \boldsymbol{\chi}) \cdot \mathbf{n})_{\Gamma_\sigma} = 0, \quad (\mathbf{v})_{\Gamma_u} = 0, \quad ((\mathbf{C} \cdot \cdot \boldsymbol{\chi}) \cdot \mathbf{n})_{\Gamma_f} = 0$$

where vector $\mathbf{q}^0$ and second rank tensor $\boldsymbol{\chi}^0$ are determined by the following expressions

$$\mathbf{q}^0 = -(\frac{\partial \boldsymbol{\sigma}}{\partial \mathbf{l}}) \cdot \nabla \psi, \quad \boldsymbol{\chi}^0 = (\frac{\partial \mathbf{u}}{\partial \mathbf{l}}) \otimes \nabla \psi \tag{8}$$

The variation of the considered functional (basic relation of design sensitivity analysis) can now be presented in the form

$$\delta J = \int_{\Gamma_v} \Lambda \delta t d\Gamma, \quad \Lambda = \frac{\partial}{\partial \mathbf{n}}(\mathbf{v} \cdot \mathbf{T}) - 2H \mathbf{v} \cdot \mathbf{T} - \nabla \cdot (\mathbf{v} \cdot \boldsymbol{\sigma}) \tag{9}$$

For convenience of performing analytical investigations and computations it is reasonable to transform the boundary value problem (7) to the conventional form. Using new variables and notations $\mathbf{e} = \boldsymbol{\chi} + \boldsymbol{\chi}^0, \quad \mathbf{C} \cdot \cdot \boldsymbol{\chi}^0 = \mathbf{s}^0, \quad \mathbf{q} = \mathbf{q}^0 - \nabla \cdot \mathbf{s}^0$ and performing simple manipulations we have

$$\nabla \cdot \mathbf{s} + \mathbf{q} = 0, \quad \mathbf{s} = \mathbf{C} \cdot \cdot \mathbf{e}, \quad \mathbf{e} = \frac{1}{2}(\nabla \mathbf{v} + (\nabla \mathbf{v})^T) \tag{10}$$

$$(\mathbf{s} \cdot \mathbf{n})_{\Gamma_\sigma} = 0, \quad (\mathbf{v})_{\Gamma_u} = 0, \quad ((\mathbf{s} - \beta \mathbf{s}^0) \cdot \mathbf{n})_{\Gamma_f} = 0$$

where $\beta = 1$ for the parts Γ_f^1, Γ_f^2 of the boundary Γ_f and $\beta = 0$ for $\Gamma_f - \Gamma_f^1 - \Gamma_f^2$. Taking into account that $\nabla \cdot (\mathbf{v} \cdot \boldsymbol{\sigma}) = \boldsymbol{\sigma} \cdot \cdot \mathbf{e}$ and $\mathbf{e} = \chi$ on Γ_v $(\chi^0(P) = 0, \quad \nabla\psi = 0$ for $P \in \Gamma_v)$, we have

$$\Lambda = \frac{\partial}{\partial \mathbf{n}}(\mathbf{v} \cdot \mathbf{T}) - 2H\mathbf{v} \cdot \mathbf{T} - \boldsymbol{\sigma} \cdot \cdot \mathbf{e} = \frac{\partial}{\partial \mathbf{n}}(\mathbf{v} \cdot \mathbf{T}) - 2H\mathbf{v} \cdot \mathbf{T} - \boldsymbol{\sigma} \cdot \cdot \chi \quad (11)$$

If the varied boundary Γ_v is traction free ($\mathbf{T}(P) = 0$ for $P \in \Gamma_v$) then the expressions for Λ are reduced to

$$\Lambda = -\boldsymbol{\sigma} \cdot \cdot \mathbf{e} = -\boldsymbol{\sigma} \cdot \cdot \chi \quad (12)$$

If we vary δt in agreement with

$$\delta t = -\tau\Lambda, \quad \tau > 0 \quad (13)$$

then the value of J is decreased ($\delta J = -\tau \int \Lambda^2 d\Gamma_v \leq 0$).

3. Shape Design Sensitivity and Finite Element Analysis

We discuss the implementation of SDSA within the context of the displacement finite element method and in connection with the biquadratic shape function approximation. Using the isoparametric version of FEM and the notations proposed in (Li at al.,1985) We represent the shape functions $N_k \quad (k = 1, 2, ..., 9)$ in the isopaprametric space (η_1, η_2)

$$\begin{gathered} N_{3j-i-3} = f_i(\eta_1) f_j(\eta_2), \quad f_1 = \tfrac{1}{2}\eta(\eta - 1), \quad f_2 = (\eta - 1)(\eta + 1), \\ f_3 = \tfrac{1}{2}\eta(\eta + 1), \quad i, j = 1, 2, 3 \quad , \quad -1 \leq \eta_l \leq 1, \quad l = 1, 2 \end{gathered} \quad (14)$$

The coordinates x_1, x_2 in the physical space and the displacemets u_1, u_2 can be represented as

$$x_i = \sum_{k=1}^{9} N_k X_{ik}, \quad u_i = \sum_{k=1}^{9} N_k U_{ik}, \quad i = 1, 2 \quad (15)$$

Here X_{ik}, U_{ik} are the nodal coordinates and the nodal displacements, respectively. In finite element analysis the domains Ω_I, Ω_J and $\Omega_E \quad (\Omega = \Omega_I + \Omega_J + \Omega_E)$ are partitioned into elements

$$\Omega_k = \sum_{e=1}^{M_k} \Omega_k^e, \quad k = I, J, K \quad (16)$$

where M_k is the number of finite elements in zone Ω_k.

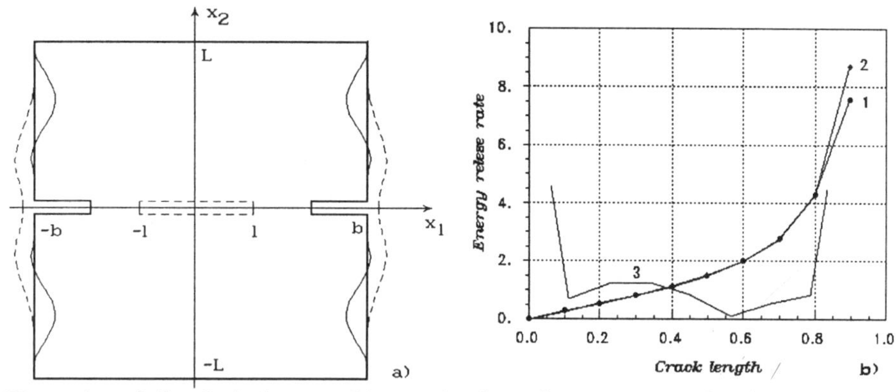

Figure 2. a) Cracked plates and improving boundary variations δt; b) Energy release rate depending on crack length and the relative error

The finite element mesh consists of the three distinct zones (see Figure 1-b)) corresponding the subdomains $\Omega_I, \Omega_J, \Omega_E$. The first is located in the circular subdomain Ω_I (tip zone) centered at the crack tip, the second is placed in the annular subdomains Ω_J (working zone) and the third is located in Ω_E (remote zone $\Omega_E = \Omega - \Omega_I - \Omega_J$). Taking into account the mesh structure and the conditions $\psi = 0$ for $\eta_1 = -1$ and $\psi = 1$ for $\eta_1 = 1$ we define the function ψ as $\psi = \frac{1}{4}(-\eta_1^3 + 3\eta_1 + 2)$. Function ψ is determined by this relation in Ω_J, and $\psi = 0$ in Ω_I, $\psi = 1$ in Ω_E.

Using FEM procedure we will have two algebraic systems for the global nodal displacements vector $\mathbf{U}$ and for the global nodal vector $\mathbf{V}$ of adjoint displacements

$$\mathbf{KU} = \mathbf{F}, \quad \mathbf{KV} = \mathbf{Q} \tag{17}$$

where $\mathbf{K}$ is the global stiffness matrix and $\mathbf{F}$, $\mathbf{Q}$ are the global vectors of nodal loads. Note that the adjoint problem for $\mathbf{V}$ differs from the direct analysis problem for $\mathbf{U}$ by means of the right hand side of the formulated equation.

Consider a rectangular plate with thickness h and in-plane sizes $2L, 2b$ ($-b \leq x_1 \leq b, \quad -L \leq x_2 \leq L$, see Figure 2-a)). The plate is loaded by the uniform static stretching stresses σ_0 applied to the parts $x_2 = L, \quad -b \leq x_1 \leq b$ and $x_2 = -L, \quad -b \leq x_1 \leq b$ of the boundary and has two symmetrical cracks (of the length $l > 0$) placed at $x_2 = 0, \quad -b \leq x_1 \leq l - b$ and $x_2 = 0, \quad b - l \leq x_1 \leq b$, where $l \leq b$. For computational purposes we take the parameters of the problem as h=1 mm, L=10 mm, b=10 mm, σ_0=1 N/mm^2, ν=0.3, E=7200000 N/mm^2. Here E is the Young's modulus, ν is the Poisson's ratio. Because of the symmetry it is sufficient to perform the computation for 1/4 part of the plate. The finite element mesh contained 635 elements and 2661 nodes. The values of the J-integral were computed for various values of crack - length - to width ratio l/b and compared with the asymptotic values given by the formula (Keer at al.,1973 and Koiter at al.,1973)

$$J = \frac{K_1^2}{E}, \quad K_1 = \sigma_\infty \sqrt{\pi l} \frac{1.12 - 0.61 l/b + 0.13 (l/b)^3}{\sqrt{1 - l/b}} \tag{18}$$

in the case $L = \infty$. The values $J \cdot 10^6$ obtained by the application of FEM (with L=30 mm) and formula (18) are shown by the curves 1,2 in Figure 2-b, respectively. The relative difference (in percents) between FEM and asymptotic results is shown by the curve 3 in Figure 2-b.

Sensitivity analysis was performed for the double notched plate and for the plate with one internal crack located at $x_2 = 0, \quad -l \leq x_1 \leq l$, taking the free sides $-L \leq x_2 \leq L, \quad x_1 \pm b$, and the varied boundaries Γ_v are shown in Figure 2-a by continuous (double notched plate) and dashed (plate with internal crack) curves in the case L=10 mm.

Conclusions. SDSA is an important branch of computational mechanics and can be effectively used in fracture mechanics and in analysis of structures optimized against brittle fracture. The work gives the approach to minimization of the energy release rate or stress intensity factor by means of variation of the elastic body's shape. The proposed approach can be considered as a basis for software development and gives us the effective method for solving important mechanical problems of stress intensity factor minimization by means of body shape modification.

Acknowledgement. The authors would like to express their appreciations to Prof. E.Stein and Dr. F.-J.Barthold for their discussion of the proposed approach to SDSA and obtained results and valuable comments.

Research is supported by Volkswagen-Stiftung in the frame of studies on "Adaptive Methods for Structural Analysis and Structural Optimization of Composites" under grant No. I/71113.

References

Haug, E.J., Choi, K.K. and Komkov, V. (1986) *Design Sensitivity Analysis of Structural Systems* . Academic Press, New York.

Banichuk, N.V., (1993) Shape Design Sensitivity Analysis for Optimization Problems with Local and Global Functionals, *Mech. Struct. and Mach.*, **Vol. 21 no. 3**, pp. 375–397.

Fancello, E. A., Taroco, E.O. and Feijoo, R.A. (1993) Shape Sensitivity Analysis in Fracture Mehanics, *in Structural Optimization 93 (J.Herskovits, ed.) Proceedings*, Vol. 2, pp. 239-248, Rio de Janeiro, Universidade Federal.

Li, F.Z., Shih, C.F. and Needleman, A. (1985) A comparison of methods for calculating energy release rates. *Engineering Fracture Mechanics*, **Vol. 21, no. 2**, pp. 405-421.

Keer, L.M. and Fridman, J.M. (1973) *Int. J. Eng. Sci.*, **no. 11**, pp.1265 -1275.

Koiter, W.T. and Benthem (1973): *in G.C.Sih (ed.), Mechanics of Fracture*, Noordhoff, Leyden, **Vol. 1**.

A SYMMETRIC COUPLED FE-BE METHOD FOR VIBRATING STRUCTURES INTERACTING WITH AN ACOUSTIC FLUID

Z. S. CHEN
Institute for Strength of Materials,
Vienna University of Technology, Austria

G. HOFSTETTER
Institute for Strength of Materials,
University of Innsbruck, Austria

AND

H. A. MANG
Institute for Strength of Materials,
Vienna University of Technology, Austria

Abstract. In order to take the interactions between the vibrations of an elastic structure and the sound field in the surrounding fluid into account, a standard finite element formulation for the dynamic behavior of elastic structures is coupled with a symmetric Galerkin formulation of the boundary element method for acoustic radiation and scattering. The coupled formulation is applied to the analysis of a microphone membrane.

1. Introduction

If the sound pressure in a fluid surrounding an elastic structure has an influence on the vibrations of the elastic structure, as may be the case, e. g., for thin membranes, then the interaction between the dynamic behavior of the elastic structure and the acoustic behavior of the surrounding fluid needs to be taken into account. The finite element method (FEM) is the most widely used numerical method for the simulation of the dynamic behavior of structures. For the numerical simulation of the radiation and scattering of sound, however, the boundary element method (BEM) is superior to the FEM in many cases. This holds especially for exterior problems, i. e., for the radiation of sound into an infinite domain, where the BEM offers the

H. A. Mang and F. G. Rammerstorfer (eds.),
IUTAM Symposium on Discretization Methods in Structural Mechanics, 225–232.

advantage of a considerably smaller discretization effort as compared with the FEM.

This is the reason why a FE-formulation for the dynamic behavior of elastic structures is coupled with a BE-formulation for the acoustic behavior of the surrounding fluid.

In order to preserve the symmetry of the coefficient matrix of the coupled problem, resulting from the FE-discretization of the structure and from the BE-discretization of the fluid, a symmetric Galerkin formulation of the BEM for acoustic radiation and scattering is employed in this paper. The latter was proposed in [1]. In this paper, the coupling procedure of the symmetric Galerkin formulation of the BEM for acoustic radiation and scattering with a standard FE-formulation for the vibrations of an elastic structure is described. The proposed formulation has been implemented into the commercial finite element program *NADwork* [2].

2. Elasto-Acoustic Coupling

2.1. FE-FORMULATION FOR THE ELASTIC STRUCTURE

Application of the finite element method (FEM) for the numerical simulation of harmonic structural vibrations yields the following algebraic system of equations

$$\left(\mathbf{K}^s + i\omega\mathbf{D}^s - \omega^2\mathbf{M}^s\right)\mathbf{u}^s = \mathbf{f}^s \,. \tag{1}$$

In (1) $\mathbf{K}^s$, $\mathbf{D}^s$ and $\mathbf{M}^s$ are the stiffness matrix, the damping matrix and the mass matrix of the elastic structure under consideration, $\mathbf{u}^s$ denotes the vector of generalized nodal displacement amplitudes, $\mathbf{f}^s$ is the vector of nodal force amplitudes, acting on the structure, and ω is the angular frequency. The superscript s denotes the elastic structure. The derivation of (1) can be found in textbooks on the FEM.

2.2. BE-FORMULATION FOR THE RADIATION OF SOUND

The BE-formulation is based on the Helmholtz differential equation for harmonic vibrations at a specified angular frequency ω

$$\nabla^2\phi + k^2\phi = 0 \,, \tag{2}$$

where ϕ denotes the potential of the sound field and k is the wave number.

In [1] an integral formulation of the Helmholtz differential equation for a structure, consisting of a three-dimensional closed part with the surface S_1 and a thin-walled open part with the middle surface S_2 was proposed. It is given as:

$$\int_{S_1} [G(\mathbf{y},\mathbf{x})A(\mathbf{x})A(\mathbf{y}) + \tau H(\mathbf{y},\mathbf{x})A(\mathbf{x}) + \tau H^T(\mathbf{y},\mathbf{x})A(\mathbf{y}) + E(\mathbf{y},\mathbf{x})]\phi(\mathbf{y})\, dS_y$$

$$+\int_{S_2}[G(\mathbf{y},\mathbf{x})A(\mathbf{x})\tilde{A}(\mathbf{y})+H(\mathbf{y},\mathbf{x})A(\mathbf{x})+\tau H^T(\mathbf{y},\mathbf{x})\tilde{A}(\mathbf{y})+\tau E(\mathbf{y},\mathbf{x})]\Delta\phi(\mathbf{y})\,dS_y$$
$$+\int_{S_2}[G(\mathbf{y},\mathbf{x})A(\mathbf{x})+\tau H^T(\mathbf{y},\mathbf{x})]\bar{A}(\mathbf{y})\bar{\phi}(\mathbf{y})\,dS_y$$
$$=\tau\alpha(\mathbf{x})v_0(\mathbf{x})-A(\mathbf{x})\phi_i(\mathbf{x})-\tau v_i(\mathbf{x})+\int_{S_1}[\tau G(\mathbf{y},\mathbf{x})A(\mathbf{x})+H^T(\mathbf{y},\mathbf{x})]v_0(\mathbf{y})\,dS_y\ ,$$
$$\mathbf{x}\in S_1\ ; \tag{3}$$

$$\frac{1}{2}\bar{A}(\mathbf{x})\Delta\phi(\mathbf{x})+\int_{S_2}[G(\mathbf{y},\mathbf{x})\tilde{A}(\mathbf{y})\tilde{A}(\mathbf{x})+H(\mathbf{y},\mathbf{x})\tilde{A}(\mathbf{x})+H^T(\mathbf{y},\mathbf{x})\tilde{A}(\mathbf{y})$$
$$+E(\mathbf{y},\mathbf{x})]\Delta\phi(\mathbf{y})dS_y+\int_{S_2}[G(\mathbf{y},\mathbf{x})\tilde{A}(\mathbf{x})+H^T(\mathbf{y},\mathbf{x})]\bar{A}(\mathbf{y})\bar{\phi}(\mathbf{y})\,dS_y$$
$$+\int_{S_1}[G(\mathbf{y},\mathbf{x})A(\mathbf{y})\tilde{A}(\mathbf{x})+\tau H(\mathbf{y},\mathbf{x})\tilde{A}(\mathbf{x})+H^T(\mathbf{y},\mathbf{x})A(\mathbf{y})+\tau E(\mathbf{y},\mathbf{x})]\phi(\mathbf{y})\,dS_y$$
$$=-\phi_i(\mathbf{x})\tilde{A}(\mathbf{x})+v_0(\mathbf{x})-v_i(\mathbf{x})+\tau\int_{S_1}[G(\mathbf{y},\mathbf{x})\tilde{A}(\mathbf{x})+H^T(\mathbf{y},\mathbf{x})]v_0(\mathbf{y})\,dS_y\ ,$$
$$\mathbf{x}\in S_2\ ; \tag{4}$$

and

$$-\frac{1}{2}\bar{A}(\mathbf{x})\bar{\phi}(\mathbf{x})+\int_{S_2}[G(\mathbf{y},\mathbf{x})\tilde{A}(\mathbf{y})+H(\mathbf{y},\mathbf{x})]\bar{A}(\mathbf{x})\Delta\phi(\mathbf{y})dS_y$$
$$+\int_{S_2}G(\mathbf{y},\mathbf{x})\bar{A}(\mathbf{y})\bar{A}(\mathbf{x})\bar{\phi}(\mathbf{y})\,dS_y+\int_{S_1}[G(\mathbf{y},\mathbf{x})A(\mathbf{y})+\tau H(\mathbf{y},\mathbf{x})]\bar{A}(\mathbf{x})\phi(\mathbf{y})\,dS_y$$
$$=-\phi_i(\mathbf{x})\bar{A}(\mathbf{x})+\tau\int_{S_1}G(\mathbf{y},\mathbf{x})\bar{A}(\mathbf{x})v_0(\mathbf{y})\,dS_y\ ,$$
$$\mathbf{x}\in S_2\ . \tag{5}$$

A detailed derivation of equations (3), (4) and (5) is contained in [1].

In (3) to (5) $\phi(\mathbf{y})$ denotes the potential of the sound field on S_1; ϕ_i and v_i are the potential and the particle velocity of an incident sound wave; v_0 denotes the velocity on the boundary of the vibrating structure in the direction of the normal $\mathbf{n}$ to the boundary; τ is a coefficient, which is equal to 1 for exterior acoustic problems and to -1 for interior acoustic problems; G(**y**,**x**) is the fundamental solution of the Helmholtz equation for an infinite three-dimensional domain; $H(\mathbf{y},\mathbf{x})$ and $H^T(\mathbf{y},\mathbf{x})$ are the derivatives of the fundamental solution $G(\mathbf{y},\mathbf{x})$ in the direction of $\mathbf{n}$ at $\mathbf{y}$ and at $\mathbf{x}$, respectively. The coefficient $\alpha(\mathbf{x})$ depends on the geometric properties at $\mathbf{x}$. $\Delta\phi(\mathbf{y})$ and $\bar{\phi}(\mathbf{y})$ are the difference and the sum of the potential of the sound field with respect to the right and to the left side of the middle surface S_2 of the thin-walled part of the structure. A results from acoustic admittance boundary conditions on S_1 and $\bar{A}$ and $\tilde{A}$ from acoustic admittance boundary conditions on the right and on the left side of S_2.

For the numerical solution of the integral equations (3), (4) and (5), the boundary of the three-dimensional closed part of the structure, S_1, and the middle surface of the thin-walled part of the structure, S_2, are discretized by means of isoparametric boundary elements. In order to obtain a symmetric formulation, the Galerkin method is applied, yielding the system of algebraic equations

$$\mathbf{K}^f \mathbf{q}^f = \mathbf{f}^f \ , \tag{6}$$

where $\mathbf{K}^f$ is a fully populated matrix with complex coefficients, $\mathbf{q}^f = \lfloor\ \phi \ \vdots\ \Delta\phi \ \vdots\ \bar{\phi}\ \rfloor^T$ denotes the vector of nodal values of the potential ϕ on S_1 and of the difference and the sum of the potential on the left and on the right side of S_2, and $\mathbf{f}^f$ is the vector resulting from incident sound waves and from vibrations of the elastic structure.

The formulation for acoustic radiation and scattering, presented in this section, is suited for uncoupled acoustic response, i. e., for situations where the sound field does not influence the structural vibrations. If, however, the structural vibrations are influenced by the surrounding acoustic fluid, then the FE-formulation for the structural vibrations has to be coupled with the BE-formulation for the acoustic response.

2.3. COUPLING PROCEDURE

Coupling of the FE-formulation for the vibrations of the elastic structure with the BE-formulation for the acoustic fluid requires continuity of the displacements and of the pressure for the coupling surface between the structure and the fluid. In particular, at the coupling surface the displacement component u_n^s of the elastic structure in the direction normal to this surface must be equal to the displacement component u_n^f of the fluid in the direction normal to the coupling surface. Moreover, the stress component p^s of the structure in the direction normal to the coupling surface must be equal to the acoustic pressure p^f in the fluid. Continuity of the displacement components normal to the coupling surface, u_n^s und u_n^f, yields continuity of the velocity components in the direction normal to the coupling surface, v_n^s und v_n^f.

In order to derive the coupling matrices, the vectors $\mathbf{f}^f$ in (6) and $\mathbf{f}^s$ in (1) are decomposed into two parts. Subdivision of $\mathbf{f}^f$ yields

$$\mathbf{f}^f = \mathbf{f}^{fi} + \mathbf{f}^{fs} \ , \tag{7}$$

with $\mathbf{f}^{fi}$ and $\mathbf{f}^{fs}$ denoting the vectors of nodal values, resulting from incident sound waves and from the vibrations of the structure, respectively. It follows from the Galerkin formulation of the equations (3), (4) and (5) that $\mathbf{f}^{fi}$ and $\mathbf{f}^{fs}$ can be split into three subvectors, yielding

$$\mathbf{f}^{fi} = \left\{ \begin{array}{c} \mathbf{f}_1^{fi} \\ \mathbf{f}_2^{fi} \\ \mathbf{f}_3^{fi} \end{array} \right\} = \left\{ \begin{array}{l} -\int_{S_1} \boldsymbol{\theta}_\phi(\mathbf{x})[A(\mathbf{x})\phi_i(\mathbf{x}) + \tau v_i(\mathbf{x})]dS_x \\ -\int_{S_2} \boldsymbol{\theta}_\phi(\mathbf{x})[\tilde{A}(\mathbf{x})\phi_i(\mathbf{x}) + v_i(\mathbf{x})]dS_x \\ -\int_{S_2} \boldsymbol{\theta}_\phi(\mathbf{x})\bar{A}(\mathbf{x})\phi_i(\mathbf{x})dS_x \end{array} \right\}, \tag{8}$$

where $\boldsymbol{\theta}_\phi$ denotes the vector of shape functions for the potential ϕ of the sound field and

$$\mathbf{f}^{fs} = \left\{ \begin{array}{c} \mathbf{f}_1^{fs} \\ \mathbf{f}_2^{fs} \\ \mathbf{f}_3^{fs} \end{array} \right\} \tag{9}$$

with

$$\begin{aligned} \mathbf{f}_1^{fs} &= \int_{S_1} \boldsymbol{\theta}_\phi(\mathbf{x}) \left\{ \tau\alpha(\mathbf{x})v_0(\mathbf{x}) + \int_{S_1} [\tau A(\mathbf{x})G(\mathbf{y},\mathbf{x}) + H^T(\mathbf{y},\mathbf{x})]v_0(\mathbf{y})dS_y \right\} dS_x \\ \mathbf{f}_2^{fs} &= \int_{S_2} \boldsymbol{\theta}_\phi(\mathbf{x}) \left\{ v_0(\mathbf{x}) + \tau \int_{S_1} [\tilde{A}(\mathbf{x})G(\mathbf{y},\mathbf{x}) + H^T(\mathbf{y},\mathbf{x})]v_0(\mathbf{y})dS_y \right\} dS_x \\ \mathbf{f}_3^{fs} &= \tau \int_{S_2} \boldsymbol{\theta}_\phi(\mathbf{x}) \int_{S_1} \bar{A}(\mathbf{x})G(\mathbf{y},\mathbf{x})v_0(\mathbf{y})dS_y dS_x \,. \end{aligned} \tag{10}$$

$\mathbf{f}_1^{fi}$ and $\mathbf{f}_1^{fs}$ refer to the boundary S_1 of the three-dimensional closed part of the structure, whereas $\mathbf{f}_2^{fi}$, $\mathbf{f}_2^{fs}$ and $\mathbf{f}_3^{fi}$, $\mathbf{f}_3^{fs}$ refer to the middle surface S_2 of the thin-walled part of the structure.

For harmonic vibrations, v_0 in (10) can be expressed in terms of the displacement amplitude u_n in the direction normal to the boundary or to the middle surface of the structure, yielding

$$\mathbf{f}^{fs} = -i\omega\mathbf{C}^{fs}\mathbf{u}^s = -i\omega \begin{bmatrix} \mathbf{C}_{11}^{fs} & \mathbf{0} \\ \mathbf{C}_{21}^{fs} & \mathbf{C}_{22}^{fs} \\ \mathbf{C}_{31}^{fs} & \mathbf{0} \end{bmatrix} \left\{ \begin{array}{c} \mathbf{u}_1^s \\ \cdots \\ \mathbf{u}_2^s \end{array} \right\}. \tag{11}$$

In (11) the vector of unknown nodal structural displacements has been split into the subvectors $\mathbf{u}_1^s$ and $\mathbf{u}_2^s$, referring to S_1 and S_2, respectively. The submatrices of the coupling matrix $\mathbf{C}^{fs}$ in (11) are obtained by means of (9) and (10).

By analogy to the subdivision of $\mathbf{f}^f$ according to (7), the right-hand side $\mathbf{f}^s$ in (1) is split into two parts

$$\mathbf{f}^s = \mathbf{f}^{ss} + \mathbf{f}^{sf} \tag{12}$$

with $\mathbf{f}^{sf}$ representing the nodal forces resulting from fluid loading of the structure, i. e., from the acoustic pressure, and with $\mathbf{f}^{ss}$ respresenting the nodal forces resulting from all other sources of structural loading without taking fluid-structure interaction into account.

For the three-dimensional closed part of the structure the sound pressure in the fluid contributes to $\mathbf{f}^{sf}$. For the thin-walled part of the structure a sound pressure of equal magnitude, acting on the right and on the left side of the middle surface S_2, does not produce any fluid-loading of the structure. Hence, only the difference of the sound pressure acting on these surfaces yields a contribution to $\mathbf{f}^{sf}$. Thus,

$$\mathbf{f}^{sf} = -\begin{bmatrix} \mathbf{C}_{11}^{sf} & \mathbf{0} & \mathbf{0} \\ \mathbf{0} & \mathbf{C}_{22}^{sf} & \mathbf{0} \end{bmatrix} \begin{Bmatrix} \mathbf{p} \\ \Delta\mathbf{p} \\ \bar{\mathbf{p}} \end{Bmatrix}, \tag{13}$$

where $\mathbf{p}$ and $\Delta\mathbf{p}$ denote the vectors of nodal values of p and Δp, respectively, and $\bar{\mathbf{p}}$ is defined by analogy to $\bar{\boldsymbol{\phi}}$. Using the relationship between the potential ϕ of the sound field and the sound pressure p, $\mathbf{p}$ and $\Delta\mathbf{p}$ can be expressed in terms of $\boldsymbol{\phi}$ and $\Delta\boldsymbol{\phi}$ yielding

$$\mathbf{f}^{sf} = -i\omega\rho \begin{bmatrix} \mathbf{C}_{11}^{sf} & \mathbf{0} & \mathbf{0} \\ \mathbf{0} & \mathbf{C}_{22}^{sf} & \mathbf{0} \end{bmatrix} \begin{Bmatrix} \boldsymbol{\phi} \\ \Delta\boldsymbol{\phi} \\ \bar{\boldsymbol{\phi}} \end{Bmatrix} = -i\omega\rho\mathbf{C}^{sf}\mathbf{q}^f . \tag{14}$$

Substitution of (11) into (7) and of the result into (6) and substitution of (14) into (12) and of the result into (1) yields the following coupled system of algebraic equations with the vectors $\mathbf{u}^s$ and $\mathbf{q}^f$, denoting the unknown nodal displacement amplitudes and the unknown nodal values of the potential of the sound field, respectively:

$$\begin{bmatrix} \mathbf{K}^s + i\omega\mathbf{D}^s - \omega^2\mathbf{M}^s & i\omega\rho\mathbf{C}^{sf} \\ i\omega\mathbf{C}^{fs} & \mathbf{K}^f \end{bmatrix} \begin{Bmatrix} \mathbf{u}^s \\ \mathbf{q}^f \end{Bmatrix} = \begin{Bmatrix} \mathbf{f}^{ss} \\ \mathbf{f}^{fi} \end{Bmatrix}. \tag{15}$$

In general, the coupled system of equations will not be symmetric. However, symmetry can be easily achieved, if fluid-structure interaction is only considered for thin-walled structures. In this case $\mathbf{C}^{sf}$ degenerates to

$$\mathbf{C}^{sf} = [\mathbf{C}_{22}^{sf} \vdots \mathbf{0}] = [(\mathbf{C}_{22}^{fs})^T \vdots \mathbf{0}] = (\mathbf{C}^{fs})^T. \tag{16}$$

The resulting coupled system of algebraic equations for thin-walled structures is solved either directly or by means of the method of modal superposition.

3. Numerical Study

The coupled formulation has been tested by a comparison of experimental results for microphone membranes, provided by the Austrian company AKG, with the numerically predicted response.

Fig. 1 shows the discretization of one quarter of the part of a microphone, containing the vibrating membrane. The numerical model for one quarter of the membrane consists of 420 coupled elasto-acoustic shell elements with 8 nodes each. The membrane was excited to vibrations at frequencies between 100 Hz and 10000 Hz. The sound pressure in the surrounding air was measured at several selected points.

Fig. 2 allows a comparison of the measured with the computed sound pressure for three points with the vertical distance H of 3 mm from the center of the membrane and with radial distances R from the center of the membrane of 0, 10 and 20 mm. Up to 4000 Hz the correspondence between the numerically predicted and the measured values is really good. At higher frequencies the correspondence between the numerical and the experimental results is less satisfactory. A possible reason for this discrepancy is the increasing influence of damping at higher frequencies, which was not taken into account in the analysis.

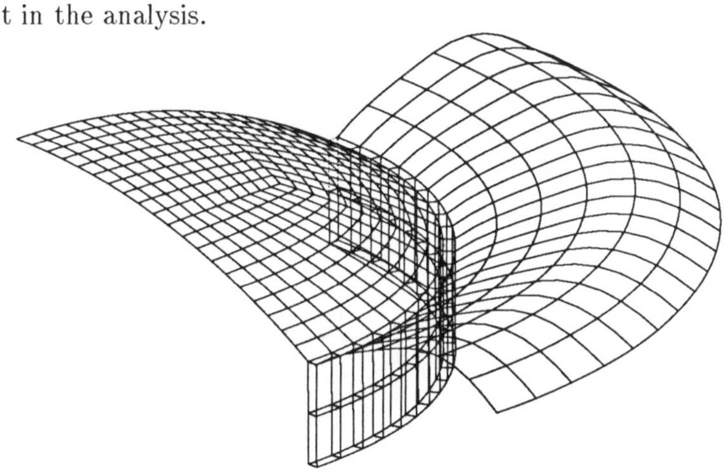

Figure 1. : Numerical model for one quarter of the membrane and the coil

Acknowledgements

Financial support for this research project, granted by the Austrian Company *NAD*, Numerical Analysis and Design GmbH., Landskrongasse 5, A-1010 Vienna, and the Austrian Forschungsförderungsfonds für die gewerbliche Wirtschaft is gratefully acknowledged. We are indebted to the Austrian company AKG for providing the experimental results for microphone membranes.

References

1. Chen, Z. S., Hofstetter, G., Mang, H. A. (1997), A symmetric Galerkin formulation of the Boundary Element Method for acoustic radiation and scattering, *Computational Acoustics*, **Vol. 5**, in press.

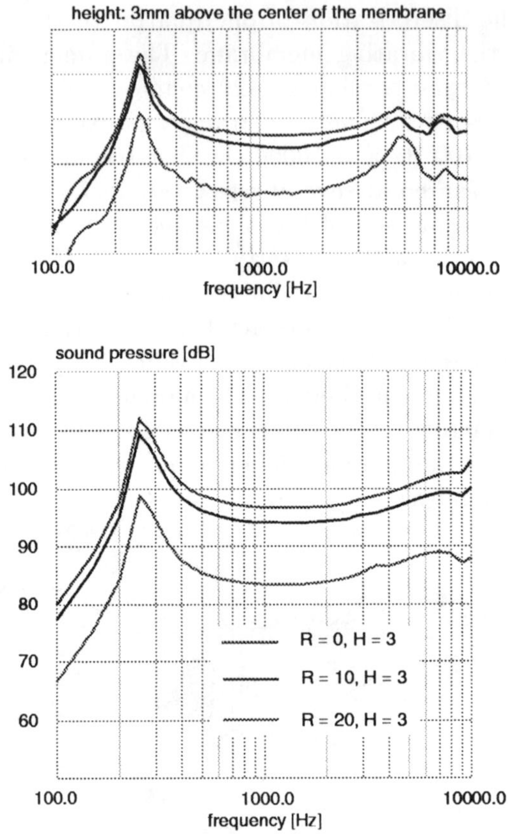

Figure 2. : Measured and computed values of the sound pressure

2. *NADwork* Release Notes, *NAD* Numerical Analysis and Design Gmbh, Landskrongasse 5, A-1010 Vienna, Austria.
3. Everstine, G. C., Henderson, F. M. (1990), Coupled finite element/boundary element approach for fluid structure interaction, *J. Acoust. Soc. Am.*, **Vol. 87(5)**, pp. 1938-1947.
4. Benthien, G. W., Schenck, H. A. (1991), Structural-Acoustic Coupling, in *BEM in Acoustics*, eds. R. D. Ciskowski and C. A. Brebbia, Computational Mechanics Publications, Southampton Boston, pp. 109-129.
5. Rajakumar, C., Ali, A., Yunus, S. M. (1992), A new acoustic interface element for fluid-structure interaction problems', *Int. J. Num. Meth. Eng.*, **Vol. 33**, pp. 369-386.
6. Mariem, J. B., Hamdi, M. A. (1987), A new boundary finite element method for fluid-structure interaction problems, *Int. J. Num. Meth. Eng.*, **Vol. 24**, pp. 1251-1267.
7. Jeans, R. A., Mathews, I. C. (1990), Solution of fluid-structure interaction problems using a coupled finite element and variational boundary element technique, *J. Acoust. Soc. Am.*, **Vol. 88(5)**, pp. 2459-2466.
8. Coyette, J. P. (1992), Formulation of a coupled FEM/direct BEM model based on a variational principle, in *Sysnoise Theoretical Manual*, Revision 4.4, Numerical Integration Technologie, Leuven Belgium.

DYNAMIC INELASTIC SOIL-STRUCTURE INTERACTION USING A HYBRID BEM/FEM SCHEME

G. D. PAVLATOS & D. E. BESKOS
Department of Civil Engineering, University of Patras GR-26500, Greece.

Abstract. A hybrid boundary element / finite element method (BEM/FEM) is developed for dynamic inelastic soil-structure interaction under conditions of plane strain. The time domain FEM is used to model the structure, as well as the part of the soil surrounding the structure, which is expected to behave inelastically, while the time domain elastodynamic BEM is used to model the remaining part of the soil extending to infinity, which is expected to behave elastically during the motion. The two methods are coupled together through equilibrium and compatibility at their interface. A special step-by-step time integration algorithm is developed which provides an accurate time history of the response. A numerical example involving seismic waves and Drucker-Prager elastoplastic soil is used to illustrate the method.

1. Introduction

Dynamic soil - structure interaction analysis usually involves local material nonlinearities and requires time domain numerical methods. The Finite Element Method (FEM) is ideal for dynamic problems involving nonlinearities, while the Boundary Element Method (BEM) is ideal for linear elastodynamic problems involving infinite or semi infinite domains. This is because in linear elastodynamic analysis the BEM requires only a boundary dicretization and satisfies automatically the radiation conditions at infinity for infinite or semi-infinite domains, while the FEM in dynamic analysis requires a boundary as well as an interior discretization and the creation of artificial boundaries for infinite or semi-infinite domains. However, in the presence of material nonlinearities, even the BEM requires an interior discretization (complete or partial). For more details on the above, one can consult the review articles of Beskos (1987,1997) for linear elastic dynamic analysis and Beskos (1995) for inelastic dynamic analysis by the BEM.

Thus, it seems natural to attempt to couple these two methods in order to construct a BEM/FEM hybrid scheme for dynamic inelastic soil-structure interaction analysis, which combines all their advantages and reduces or completely eliminates their disadvantages. The present paper describes a BEM/FEM scheme in the time domain for dynamic inelastic soil-structure interaction analysis under conditions of plane strain. The

H. A. Mang and F. G. Rammerstorfer (eds.),
IUTAM Symposium on Discretization Methods in Structural Mechanics, 233–240.

time domain FEM is used to model the structure as well as the part of the soil surrounding the structure, which is expected to behave inelastically, while the time domain elastodynamic BEM is used to model the remaining part of the soil extending to infinity, which is expected to behave elastically during the motion. The two methods are coupled together through equilibrium and compatibility at their interface. A special step-by-step time integration algorithm is developed which provides an accurate time history of the response to dynamic disturbances with many cycles during their duration of action, such as seismic ones.

The present work can be thought of as an improved and extended version of a previous work by the same authors (Pavlatos and Beskos 1994) on the first ever developed BEM/FEM scheme for dynamic inelastic analysis. That work was restricted to simple finite inelastic structures under simple transient forces and its direct application to general dynamic soil - structure interaction problems was found to produce results of low accuracy, especially for seismic disturbances characterized by many sign reversals during their time interval of action. An attempt of limited scope to study a simple dynamic inelastic soil-structure interaction problem by the BEM/FEM scheme of Pavlatos and Beskos (1994) can be found in the work of Pavlatos et all (1994). This paper presents a highly accurate hybrid BEM/FEM scheme for general dynamic inelastic structural analysis including soil-structure interaction.

Hybrid BEM/FEM schemes for elastodynamic structural analysis including soil-structure interaction in either the frequency or time domain have been developed by many authors in the past and a complete listing of them can be found in the review papers of Beskos (1987,1997).

Use of the BEM for the semi-infinite soil medium greatly reduces the size of the final stiffness matrix of the system. However, because BEM influence matrices are non-symmetric, the system stiffness matrix becomes also non-symmetric. Thus, the efficiency of the proposed hybrid scheme increases for problems with large BEM to FEM volume ratios, i.e., for problems of limited inelasticity.

2. Time Domain Elastodynamics by BEM

onsider the motion of a linearly elastic body (finite, infinite or semi-infinite) under conditions of plane strain subjected to external dynamic forces and/or seismic waves. The discretized BEM equations of motion for this body have the form (Pavlatos and Beskos 1994)

$$[T]_1\{u\}_n = [U]_1\{t\}_n + \{P\}_{1,n-1} \tag{1}$$

where $[T]_1$ and $[U]_1$ are the influence matrices at time step 1 with entries integrals over the boundary elements of the Stokes fundamental traction and displacement tensors T_{ij} and U_{ij}, respectively, $\{u\}_n$ and $\{t\}_n$ are the vectors of the nodal values of boundary displacements and tractions, respectively, at time step n and the vector $\{P\}_{1,n-1}$ represents the effect of the previous time steps (from 1 to n-1) and contains the effect of seimic waves, if there are any.

Application of the boundary conditions of the problem

$$\begin{aligned} u_j &= \tilde{u}_j \quad on \quad S_u \\ t_j &= \tilde{t}_j \quad on \quad S_t \end{aligned} \tag{2}$$

to equation (1), where tildes denote prescribed values and $S_u \cup S_t = S$, the surface of the body, and rearrangement for separation of known from unknown quantities result in

$$[A]_1\{x\}_n = \{B\}_n \tag{3}$$

which has to be solved in a stepwise fashion in time for the unknown boundary nodal values {x} of displacements and tractions. In this work 3-noded quadratic isoparametric line boundary elements are employed in conjunction with the displacement varying linearly with time and the traction being constant inside every time step.

3. Time Domain Elastoplastodynamics by FEM

Consider the small deformation and strain motion of an elastoplastic finite body under conditions of plane strain subjected to external dynamic forces and/or seismic waves. The time domain discretized FEM equations of motion for this body have the form (Pavlatos and Beskos 1994)

$$[M]\{\ddot{d}\} + \{p\} = \{f\} \tag{4}$$

where [M] is the mass matrix, {d},{p}and {f} are the nodal displacement vector, internal force vector (function of displacement) and applied force vector (external and or inertial), respectively and overdots indicate time differentiation.

Equation (4) is solved by a step-by-step time domain integration scheme to be developed in conjunction with an iterative scheme (Newton Raphson type) employed at every time step. Following Touhei and Ochmachi's (1993) work for the linear BEM/FEM coupling case, quadratic time interpolation functions for the displacements and internal force vectors between time stations t_{n-2}, t_{n-1} ,and t_n are assumed. i.e.,

$$\begin{aligned} d\{\xi\} &= \frac{1}{2}\xi(\xi-1)d_{n-2} + (1-\xi^2)d_{n-1} + \frac{1}{2}\xi(\xi+1)d_n \\ p\{\xi\} &= \frac{1}{2}\xi(\xi-1)p_{n-2} + (1-\xi^2)p_{n-1} + \frac{1}{2}\xi(\xi+1)p_n \end{aligned} \tag{5}$$

where ξ is the dimensionless parameter

$$\xi = \frac{t}{\Delta t} - n + 1 \tag{6}$$

Thus by time differentiation of the displacement (5) one has

$$\ddot{d}(\xi) = \frac{d_{n-2} - 2d_{n-1} + d_n}{\Delta t^2} \tag{7}$$

For external forces or tractions zero-order interpolation functions between time stations

t_{n-2}, t_{n-1} ,and t_n are assumed. i.e.,

$$
\begin{aligned}
f(\xi) &= f_{n-1} \quad \text{for} \quad t_{n-2} \le t \le t_{n-1} \\
f(\xi) &= f_n \quad \text{for} \quad t_{n-1} \le t \le t_n
\end{aligned} \tag{8}
$$

Application of the method of weighted residuals to equation (4) in the time interval $t_{n-2} \le t \le t_n$, which on account of (6) corresponds to $-1 \le \xi \le 1$, implies that

$$
\int_{-1}^{+1} w(\xi)\left[[M]\{\ddot{d}\} + \{p\} - \{f\}\right] d\xi = 0 \tag{9}
$$

where $w(\xi)$ is a weighting function.

Using expressions (5),(7) and (8) in equation (9), assuming $w(\xi)$ to be an even function of ξ and dividing the resulting expression by $\int_{-1}^{+1} w(\xi)d\xi$ one receives

$$
\begin{aligned}
\frac{[M]}{\Delta t^2}\{d\}_n + \beta\{p\}_n = \frac{-[M]}{\Delta t^2}\{d\}_{n-2} + \frac{2[M]}{\Delta t^2}\{d\}_{n-1} - \beta\{p\}_{n-2} \\
-(1-2\beta)\{p\}_{n-1} + \{f^*(\xi)\}
\end{aligned} \tag{10}
$$

where

$$
\beta = \frac{\int_{-1}^{+1} 0.5 w(\xi)\xi(1+\xi)d\xi}{\int_{-1}^{+1} w(\xi)d\xi} \quad , \quad \{f^*(\xi)\} = \frac{\int_{-1}^{+1} w(\xi)\{f(\xi)\}d\xi}{\int_{-1}^{+1} w(\xi)d\xi} \tag{11}
$$

Use of the relations $\{p\}_n=\{p\}_{n-1}+[K_T]\{\Delta d\}$ and $\{\Delta d\}=\{d\}_n-\{d\}_{n-1}$ in (10) results in

$$
\begin{aligned}
(\frac{[M]}{\Delta t^2} + \beta[K]_T)\{\Delta d\} = \frac{-[M]}{\Delta t^2}\{d\}_{n-2} + \frac{[M]}{\Delta t^2}\{d\}_{n-1} - \beta\{p\}_{n-2} \\
-(1-\beta)\{p\}_{n-1} + \{f^*(\xi)\}
\end{aligned} \tag{12}
$$

where $[K_T]=[K_T(\{d\})]$ is the tangent stiffness matrix, which is expressed in terms of the elastoplastic stiffness matrix $[D_{ep}]$ and the strain displacement matrix [B]. Explicit expressions for the $[D_{ep}]$ and [B] for the Drucker-Prager elastoplacticity used here to model the soil can be found, e.g. in Owen and Hinton (1980). The term $\{f^*(\xi)\}$ in the above equation can be replaced by $0.50[\{f\}_n+\{f\}_{n-1}]$ with the assumption of $w(\xi)=|\xi|$. The above step-by step time integration algorithm (12) provides the solution at time station t_n in terms of the known solutions at the two previous time stations t_{n-1} and t_{n-2}. In this work the weighting function is selected to be $w(\xi)=|\xi|$ resulting through (11) to β=0.25. The previous BEM/FEM algorithm of Pavlatos and Beskos (1994) uses Newmark's algorithm to solve equation (4), which was found to produce good results for only the first few cycles of vibrational response when used in the framework of the BEM/FEM hybrid scheme. Space discretization is done by 8-noded quadratic quadrilateral finite elements.

4. The Hybrid BEM/FEM Scheme

Consider the plane body of Figure 1 divided into two parts, B with surface S_B modeled by the BEM and F with surface S_F modeled by the FEM. These two parts are connected to each other through their interface C.

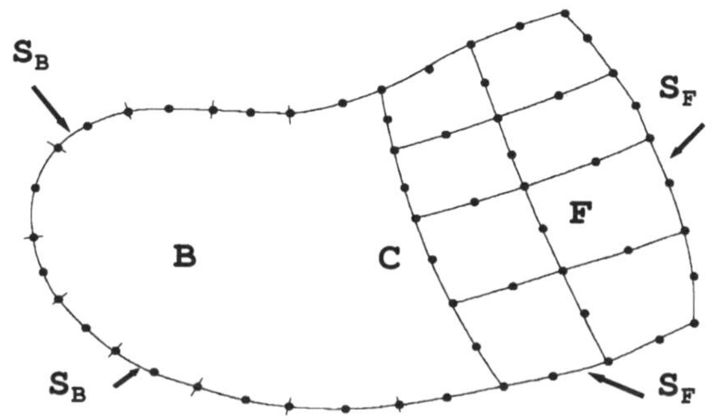

Fig. 1 Discritization of a plane body into boundary and finite elements

Equation (3) for the BEM part is written in partitioned form as

$$\begin{bmatrix} A_{11} & A_{12} & A_{13} & A_{14} \\ A_{21} & A_{22} & A_{23} & A_{24} \end{bmatrix} \begin{Bmatrix} \Delta u_n^B \\ \Delta u_n^C \\ t_n^B \\ t_n^C \end{Bmatrix} = \begin{Bmatrix} B_1 \\ B_2 \end{Bmatrix} \tag{13}$$

where superscripts B and C denote values at the boundary of part B and the interface C, respectively.

Equation (12) for the FEM part is written in partitioned form as

$$\begin{bmatrix} K_{11} & K_{12} & 0 \\ K_{21} & K_{22} & N^* \end{bmatrix} \begin{Bmatrix} \Delta d_n^F \\ \Delta d_n^C \\ \bar{t}_n^C \end{Bmatrix} = \begin{Bmatrix} W_1 \\ W_2 \end{Bmatrix} - \begin{Bmatrix} f_n^F \\ 0 \end{Bmatrix} \tag{14}$$

where [K] is the effective stiffness matrix, superscript F denotes values at the interior and boundary of part F, C values at the interface, vectors W_1 and W_2 consist of known solutions at time t_{n-1} and t_{n-2} and matrix $[N^*]$ transforms nodal tractions to nodal forces through the relation $\{f\}=[N^*]\{t\}$

Equilibrium of tractions and compatibility of displacements at the interface C dictates that

$$t_n^C + \bar{t}_n^C = 0 \quad , \quad \Delta u_n^C = \Delta d_n^C \tag{15}$$

Thus one can couple equations (13) and (14) with the aid of equations (15) and receive the combined BEM/FEM equation

$$\begin{bmatrix} K_{11} & K_{12} & 0 & 0 & 0 \\ K_{21} & K_{22} & 0 & 0 & -N^* \\ 0 & A_{12} & A_{11} & A_{13} & A_{14} \\ 0 & A_{22} & A_{21} & A_{23} & A_{24} \end{bmatrix} \begin{Bmatrix} \Delta u_n^F \\ \Delta u_n^C \\ \Delta u_n^B \\ t_n^B \\ t_n^C \end{Bmatrix} = \begin{Bmatrix} W_1 - f_n^F \\ W_2 \\ B_1 \\ B_2 \end{Bmatrix} \tag{16}$$

The computational procedure for the solution of equation (16) in a step-by-step fashion with iterations at every time step is similar to the one described in Pavlatos and Beskos (1994). However, the step-by-step integration algorithm used here is much more accurate than the one used in Pavlatos and Beskos (1994) and can very accurately handle seismic inputs with many hundreds of time steps. In addition, the present version treats problems of discontinuities of tractions at corner points on the BEM/FEM interface through the use of partially discontinuous boundary elements and problems involving subregions.

5. Numerical example

Consider a linear elastic structure AA'B'B bonded on the surface of the half-space soil medium under conditions of plane strain as shown in Figure 2. For reasons of consistency between the BEM/FEM and the FEM modeling in this comparison study, only the finite part CC'D'D of the soil is considered and the 1940 El-Centro earthquake (first 6 sec) with maximum acceleration 0.348g is applied at the bottom DD' (soil-bedrock interface) of the soil body as a vertical shear wave. The elastic properties of the structure are $E=1.438\ 10^7$ kN/m^2 and $v=0.17$, while its mass density $\rho=2400$ Kg/m^3. The soil is elastoplastic material of the Drucker-Prager type with $E=1.26\ 10^6$ kN/m^2, $v=0.40$, $\rho=1800$kg/m^3, cohesion $c=180$ kN/m^2 and friction angle $\phi=30^{\circ}$.

The system response is computed by the FEM of the present BEM/FEM hybrid scheme applied to the whole soil body, the FEM of the general purpose computer program ANSYS (1986) and the present BEM/FEM scheme. The FEM discretization is shown in Figure 2, while the BEM/FEM discretization in Figure 3, where only the structure and the portion of the soil expected to become inelastic during the motion are discretized by the FEM, while the remaining part of the soil is modeled by the BEM. When the BEM/FEM is used, the seismic input is applied at the bottom nodes in the form of displacements. Figures 4 depicts the horizontal component of displacement at point A' as computed by all three ways for elastic as well as elastoplastic soil behavior. In all cases the agreement between the three solutions is excellent.

6. Conclusions

A highly accurate BEM/FEM hybrid scheme in the time domain has been developed for dynamic/seismic inelastic soil-structure interaction problems under conditions of plane

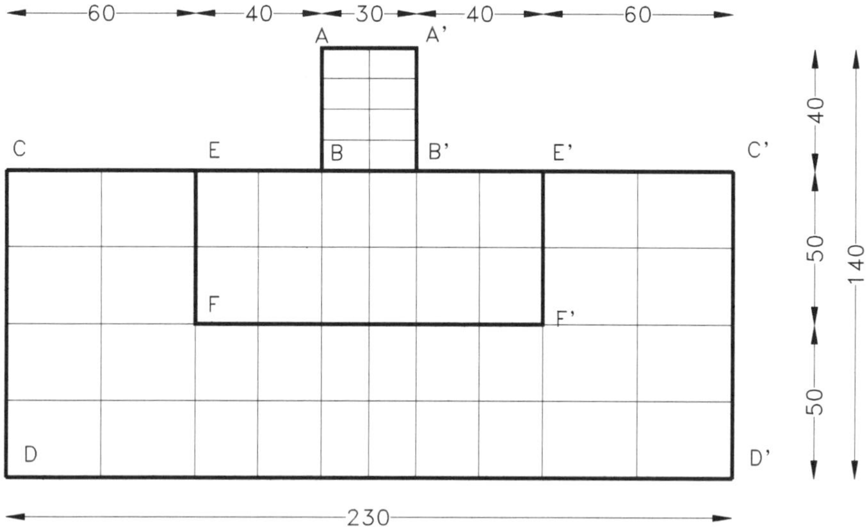

Fig. 2 Geometry and FEM discretization of the soil-structure system. The part of the soil EE'F'F is expected to have inelastic behaviour.

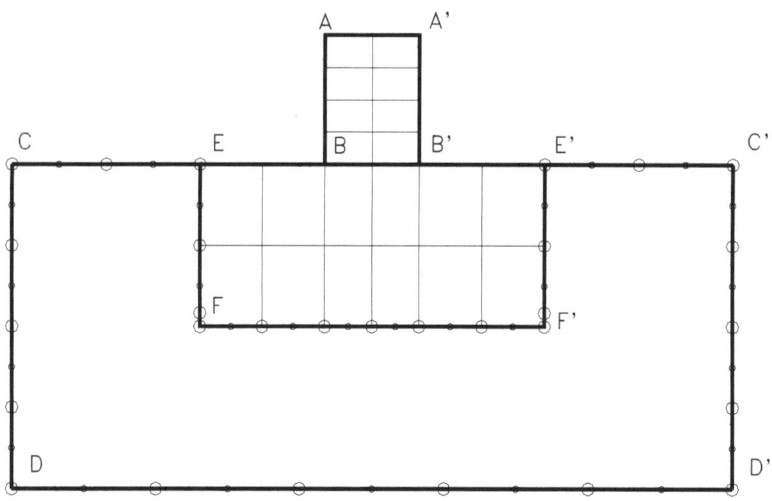

Fig. 3 BEM/FEM discretization of the soil-structure system.

strain. Its present version works only with problems involving soil deposits over bedrock and vertically propagating waves. Work is under way for extending the method to cases involving seismic waves of any kind and direction, more realistic inelastic soil models and unilateral contact conditions at the soil-structure interface.

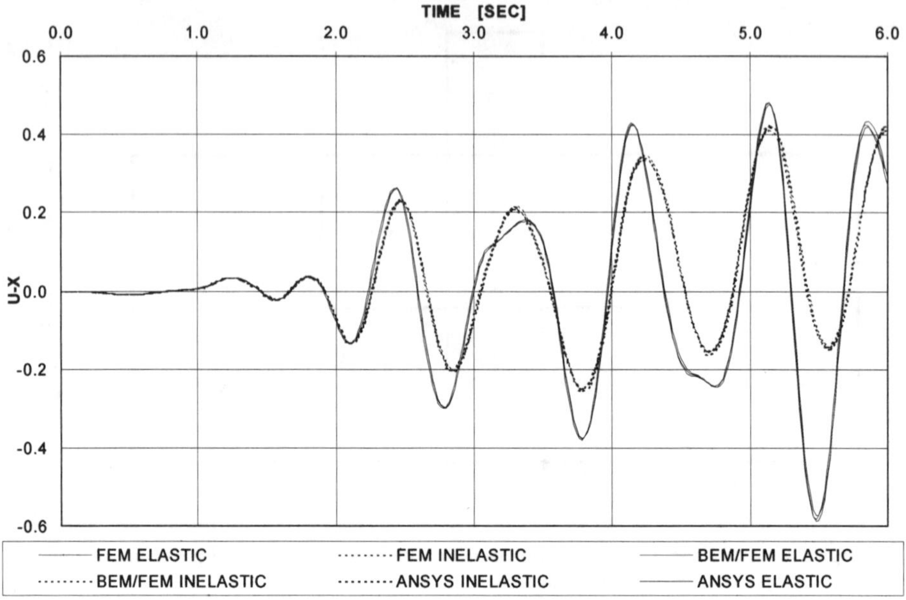

Fig. 4 Horizontal displacement history of node A' for elastic and inelastic cases.

7. Acknowledgements

The authors are grateful to the Greek Aseismic Planning and Protection Organization for supporting their research work on dynamic inelastic soil-structure interaction.

8. References

ANSYS (1986) Engineering Analysis, System User's Manual, Swanson Analysis System, Inc. ,Houston, Pennsylvania, USA.

Beskos, D.E. (1987) Boundary element methods in dynamic analysis, *Applied Mechanics Reviews* **40**,1-23.

Beskos, D.E. (1995) Dynamic inelastic structural analysis by boundary element methods, *Archives of Computational Methods in Engineering* **2**,1-33.

Beskos, D.E. (1997) Boundary element methods in dynamic analysis, Part II (1986-1987), *Applied Mechanics Reviews* **50**,149-197.

Owen, D.R.J. and Hinton, E. (1980), *Finite Elements in Plasticity: Theory and Practice*, Pineridge Press, Swansea, U.K.

Pavlatos, G.D. and Beskos, D.E. "Dynamic Elastoplastic Analysis by BEM/FEM", *Engineering Analysis with Boundary Elements* **14**, 51-63.

Pavlatos, G.D., Beskos D.E. and Karabalis, D.L. Dynamic soil-structure interaction: nonlinear material behavior, in M. Papadrakakis and B.H.V. Topping (eds.), *Advances in Simulation and Interaction*, Civil-Comp Press, Edinburgh, pp 23-29.

Touhei, T. and Ochmachi, T. (1993) A FE-BE method for dynamic analysis of dam-foundation-reservoir systems in the time domain, *Earthquake Engineering and Structural Dynamics* **22**, 195-209.

APPLICATION OF THE FINITE ELEMENT METHOD TO THE ANALYSIS OF AUTOMOBILE TIRES

H.-J. PAYER, G. MESCHKE AND H.A. MANG
Institute for Strength of Materials
Vienna University of Technology
Karlsplatz 13/E202, A-1040 Vienna, Austria

Abstract. The ingredients required for computationally efficient large-scale 3D FE contact analyses of automobile tires are presented. A hybrid finite element is employed for describing the nearly incompressible behaviour of rubber. The reinforcing cords are represented numerically by means of rebar elements. Strategies for reducing the amount of CPU time and disk space are reported including an iterative solver with preconditioning. Problems with the available commercial contact algorithm in the context of this 3D simulations and 2D analyses of tread blocks moving on snow suggested the implementation of a contact algorithm, at the moment for 2D problems, into the existing FE program. The derivation of the relations for the contact stiffness matrix and the force vector is described. Frictional effects are considered by means of a regularized Coulomb friction law. Details of the employed contact-search algorithm are given.

1. Motivation

It is useful to replace, at least in part, the empirical investigations performed for the design of tires by computer simulations based on modern continuum mechanics. This will result in a reduction of the costs and the time spent for the development of new tires. However, a realistic tire model requires consideration of large deformations and frictional contact, the nearly incompressible behaviour of rubber, the existence of reinforcing cord layers and the complicated geometry of the tread pattern. To render FE simulations of tires with a large number of degrees of freedom feasible, computationally efficient solution procedures must be employed.

2. 3D Simulations of Automobile Tires

2.1. 3D HYBRID FINITE ELEMENT FOR RUBBER

Rubber is a nearly incompressible material. This is taken into account by means of a split of the strain energy density of the Mooney material model into a deviatoric

H. A. Mang and F. G. Rammerstorfer (eds.),
IUTAM Symposium on Discretization Methods in Structural Mechanics, 241–248.

part W and a volumetric part U:

$$W(\bar{C}) = A_1(\bar{I}_1 - 3) + A_2(\bar{I}_2 - 3)\ , \qquad U(J) = K(J\, lnJ - J + 1)\ . \qquad (1)$$

$\bar{\boldsymbol{C}}$ is the deviatoric part of the right Cauchy-Green deformation tensor $\boldsymbol{C}$, $\bar{I}_1$ and $\bar{I}_2$ denote the first and the second invariant of $\bar{\boldsymbol{C}}$, A_1 and A_2 are material parameters, J is the determinant of the deformation gradient $\boldsymbol{F}$, K is the bulk modulus. The hybrid formulation is based on a 3-field Hu-Washizu variational principle [3], [7]. The displacements $\boldsymbol{u}$, the dilatation Θ and the hydrostatic pressure p are used as independent unknowns. Application of the stationarity condition to the respective functional

$$\Pi(\boldsymbol{u}, \Theta, p) = \Pi^{ext}(\boldsymbol{u}) + \int_V [U(\Theta) + W(\bar{\boldsymbol{C}}) + p(J - \Theta)] dV \qquad (2)$$

yields the Euler-Lagrange equations. Linearization and discretization of these equations give a system of linear algebraic equations. Employing a discontinuous approximation of Θ and p across the element boundaries allows to perform static condensation and results in a formulation in which the displacements are left as unknowns.

2.2. 3D FINITE ELEMENT FOR REINFORCING CORDS

Reinforcing fibres are represented numerically by means of rebar elements. The computational efficiency of these overlay-elements stems from the fact that several layers of fibres with arbitrary spatial orientation and different material behaviour can be represented within one single element without increasing the number of unknowns [1], [3].

Assuming hyperelasticity, the Second Piola-Kirchhoff stress S_R in the axial direction can be calculated from the strain energy density function W as follows:

$$S_R(\lambda_R, \lambda_L) = 2\boldsymbol{T}^T\, \partial_{\mathrm{C}} W(\lambda_R, \lambda_L)\, \boldsymbol{T}\ . \qquad (3)$$

The unit vector $\boldsymbol{T}$ defines the orientation of a single fibre in the reference configuration, λ_R and λ_L denote the axial and lateral stretch of the fibre. The final expression for the weak form of the equilibrium equations in the reference configuration is given by:

$$\delta_u \Pi = \int_V S_R \cdot \boldsymbol{T}^T \boldsymbol{F}^T GRAD\, \delta\boldsymbol{u}\, \boldsymbol{T}\ dV + \Pi^{ext}(\boldsymbol{u})\ . \qquad (4)$$

Linearization and discretization yield the respective contributions to the total system of linear algebraic equations.

2.3. ITERATIVE SOLUTION OF LINEAR SYSTEMS OF EQUATIONS

The coefficient matrices of the linear systems of equations are symmetric and positive definite. Hence, e.g., the method of conjugate gradients or the Lanzcos method can be employed for the solution. In order to accelerate the convergence of the iterative solvers, a preconditioning method is used. The linear system of equations $\boldsymbol{A}\,\boldsymbol{x} = \boldsymbol{b}$ is transformed to

$$\left(\bar{\boldsymbol{L}}^{-1}\,\boldsymbol{A}^{\star}\,\bar{\boldsymbol{L}}^{-T}\right)\left(\bar{\boldsymbol{L}}^{T}\,\boldsymbol{x}^{\star}\right) = \bar{\boldsymbol{L}}^{-1}\,\boldsymbol{b}^{\star}\,, \tag{5}$$

before applying the iterative solver. $\boldsymbol{x}$ and $\boldsymbol{b}$ are the vectors of increments of displacements and nodal forces. $\boldsymbol{A}^{\star}, \boldsymbol{b}^{\star}$ and $\boldsymbol{x}^{\star}$ are obtained from the original system of equations by scaling the diagonal elements of $\boldsymbol{A}$ to one. An incomplete Cholesky decomposition [5] gives the preconditioning matrix $\bar{\boldsymbol{L}}$, which is an approximation of the full Cholesky decomposition $\boldsymbol{L}$. In the course of an incomplete decomposition certain matrix elements are set equal to zero. This may cause a break-down of the decomposition process. Scaling the off-diagonal-elements of $\boldsymbol{A}^{\star}$ before performing the decomposition proved to be the best way to avoid such a break-down [6].

2.4. FURTHER STRATEGIES FOR ACCELERATING THE SOLUTION

A further reduction of CPU time and disk space can be achieved with a strategy called "rebar hiding". The rebar and the rubber elements are calculated by means of a user-defined subroutine. Normally, this subroutine returns the corresponding element matrices or vectors for one element, which is either a rebar or a rubber element. In the case of "rebar hiding" a call for this subroutine returns the contributions of the rubber element. If a rebar element is embedded in this element, also the rebar contributions are returned. As a consequence of this strategy no memory and disk space need to be reserved for the rebar elements. However, the rebar elements must be defined in an additional input file.

In the case of large-scale 3D analyses, the amount of main memory required for an in-core direct solution of the arising linear systems of equations in almost all cases exceeds the size of the main memory available on conventional computer systems and even on supercomputers. Therefore, an out-of-core solution procedure must be used, resulting in extensive I/O operations and in a deterioration of the over-all performance of the program. However, with the help of an iterative solver, even for large-scale analyses the solution step can usually be performed in-core.

Performing the iterative solution of the linear systems of equations in an external process makes possible to exploit this fact. The FE program only executes the steps of matrix assembly and stress recovery. After the assembly step the stiffness matrix is directly written to disk in a compressed form. Then, the external iterative solver is called. Apart from the fact that the solution step can be performed in-core, this strategy permits using different computer architectures. The steps of matrix assembly and stress recovery are characterized by very short vectors and, for the case of the out-of-core solution procedure, by extensive I/O operations. Thus, these steps should be performed on a scalar machine. The amount of main memory provided for the two steps can be set to a very small value. It must be just sufficient to execute the FE program. The iterative solver is well suited for the execution on a vector computer. Obviously, it is possible to perform the external solution step and the FE program on the same computer [3].

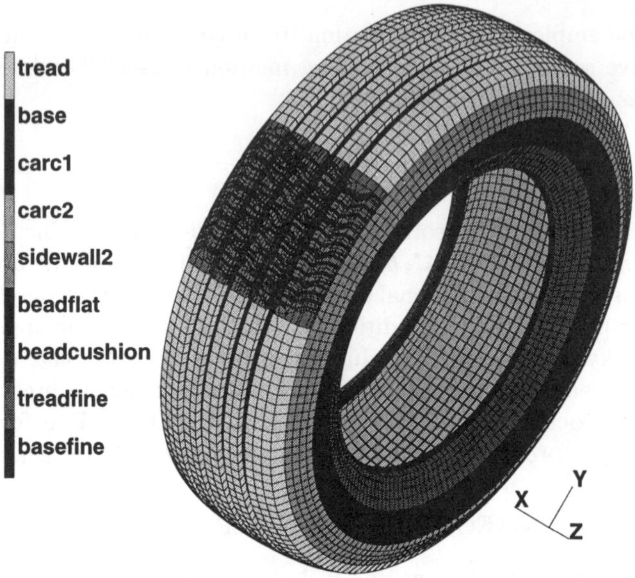

Figure 1. : Element sets of the fine tire mesh (142569 DOFs)

2.5. NUMERICAL EXAMPLES

Three large-scale simulations of an automobile tire were carried out with the FE program MARC to test the proposed strategies. Figure 1 shows the different element sets of the mesh for the tire with the largest number of unknowns (142569 DOFs), denoted as fine mesh. A coarser discretization (in the circumferential direction) was employed for the medium mesh (77085 DOFs). The coarse mesh (38760 DOFs) was obtained by means of symmetrization of one half of the medium mesh and by restriction of the analysis to this half. The mounting of the tire onto the rim was simulated by a horizontal displacement of the bead toe of 7.69 mm simul-

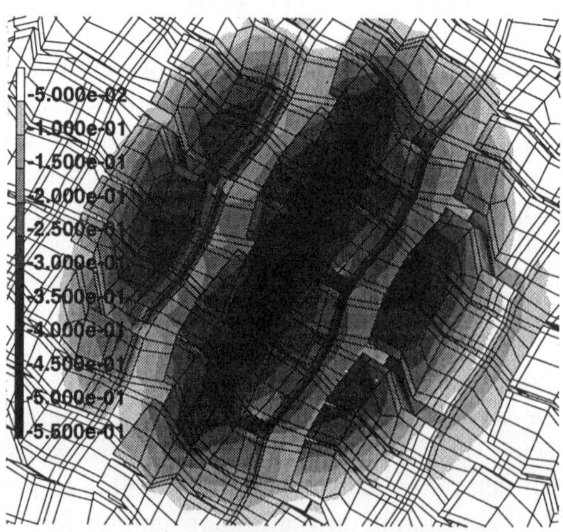

Figure 2. : Pressure distribution (in N/mm^2, contact displacement: 6.1 mm)

TABLE 1. : Disk space [MB] and CPU times [s] required for out-of-core analyses of the coarse mesh on an HP 9000/K200-2 (LMM ... method of Lagrange multipliers is employed for enforcement of incompressibility; hy ... hybrid formulation is used for rubber; reb-in (reb-ex) ... rebar elements are defined in MARC input file (in an additional input file); dir-so ... direct solver is employed; so-in (so-ex) ... internal (external) iterative solver is employed; t_a, t_c, t_s, t_r ... time required for assembly, compression, solution and recovery)

	LMM/reb-in/dir-so	hy/reb-in/dir-so	hy/reb-ex/so-in	hy/reb-ex/so-ex
Disk space	956	830	624	246
t_a/t_c	254	130	119/22	98/25
t_s	6441	4687	735	744
t_r	74	98	74	74

TABLE 2. : Main memory required for in-core computations [GB]

	coarse mesh	medium mesh	fine mesh
LMM/reb-in	1.041	3.238	-
hy/reb-in	0.597	1.859	-
hy/reb-ex	0.482	1.624	2.438

taneously to applying an internal pressure of 2.0 bar. Subsequently, a rigid plane was moved incrementally towards the tire axis taking frictional effects into account (friction coefficient: 0.5).

Figure 2 shows the pressure distribution for the fine mesh at a contact displacement of 6.1 mm. Employing the hybrid element formulation instead of a Lagrange multiplier method to enforce the incompressibility of rubber and using the "rebar hiding" strategy and the external iterative solver instead of the commercial direct solver resulted in a decrease of the disk space, required for an out-of-core analysis of the coarse tire model, from 956 MB to 246 MB (see table 1). The CPU times required for the different steps of a FE iteration step of this simulation are given in table 1. The amount of main memory which is necessary to perform an in-core solution of the three tire models is given in table 2. Making use of the external iterative solver instead of the commercial direct solver resulted in a reduction of the CPU time, required for the solution step, from 4687 seconds to 744 seconds, on an HP 9000/K200-2 for the case of hybrid elements (see table 1). For the fine mesh an average CPU time of 218 seconds was required for the matrix assembly, of 68 seconds for the compression of the stiffness matrix, of 2503 seconds for the external iterative solver and of 135 seconds for the stress recovery step on a DEC

TABLE 3. : CPU times [s] required on a DEC 8200 5/300

	t_a	t_c	t_s	t_r
coarse mesh/in-core/hy/reb-ex/so-in	41	4.4	459	32
medium mesh/in-core/hy/reb-ex/so-in	96	21	1357	73
fine mesh/out-of-core/hy/reb-ex/so-ex	218	68	2503	135

8200 5/300 (see table 3). These CPU times hold for the following situation: hybrid elements, "rebar hiding", external iterative solver.

3. Contact Algorithm for 2D Problems

3.1. EQUATIONS FOR CONSIDERATION OF CONTACT

Assuming that the contact tractions can be replaced by discrete contact forces t_N and t_T, the virtual contact work δA_c for one active contact pair is given by [8]:

$$\delta A_c = t_N \, \delta g_N + t_T \, \delta g_T \,. \tag{6}$$

Using linear shape functions for the interpolation of the geometry, the normal gap g_N and the relative tangential gap g_T can be calculated by means of:

$$g_N = (\boldsymbol{x}_s - \boldsymbol{x}_1) \cdot \boldsymbol{n} \,, \quad g_T = \bar{g}_T + (\alpha - \bar{\alpha}) \, \bar{l} \,, \tag{7}$$

where $\boldsymbol{n}(\boldsymbol{t})$ is the unit normal (tangential) vector, $\boldsymbol{x}_s(\boldsymbol{x}_1)$ indicates the current position of the slave node (first node point of the target segment), α is a dimensionless parameter coordinate. A bar over a letter symbol identifies a quantity which is already known. Variation of these two equations and application of the principle of virtual work yield the equation for the contact force vector:

$$\boldsymbol{r}_c = -\left[t_N \, \boldsymbol{N}_s + t_T \, \frac{\bar{l}}{l} \left(\boldsymbol{T}_s + \frac{g_N}{l} \, \boldsymbol{N}_0 \right) \right] . \tag{8}$$

$$\boldsymbol{N}_s^T = (\boldsymbol{n}, \; -(1-\alpha)\, \boldsymbol{n}, \; -\alpha \, \boldsymbol{n}) \,, \; \boldsymbol{N}_0^T = (\boldsymbol{0}, \; -\boldsymbol{n}, \; \boldsymbol{n}) \,, \tag{9}$$

$$\boldsymbol{T}_s^T = (\boldsymbol{t}, \; -(1-\alpha)\, \boldsymbol{t}, \; -\alpha \, \boldsymbol{t}) \,, \; \boldsymbol{T}_0^T = (\boldsymbol{0}, \; -\boldsymbol{t}, \; \boldsymbol{t}) \,. \tag{10}$$

Linearization of the virtual contact work gives the contact stiffness matrix which must be considered for one active contact pair in the total stiffness matrix:

$$\begin{aligned} \boldsymbol{k}_c \; = \; & \frac{\partial t_N}{\partial g_N} \boldsymbol{N}_s \boldsymbol{N}_s^T - \frac{t_N}{l} \left(\boldsymbol{T}_s \boldsymbol{N}_0^T + \boldsymbol{N}_0 \boldsymbol{T}_s^T + \frac{g_N}{l} \boldsymbol{N}_0 \boldsymbol{N}_0^T \right) + \\ & \frac{\partial t_T}{\partial g_T} \frac{\bar{l}^2}{l^2} \, \boldsymbol{A} + t_T \, \frac{\bar{l}}{l^2} \, \boldsymbol{B} + \frac{\partial t_T}{\partial g_N} \, \boldsymbol{C} \,, \end{aligned} \tag{11}$$

$$\bar{\boldsymbol{T}} = \boldsymbol{T}_s + \frac{g_N}{l} \boldsymbol{N}_0 \,, \boldsymbol{A} = \bar{\boldsymbol{T}} \bar{\boldsymbol{T}}^T \,, \boldsymbol{C} = \frac{\bar{l}}{l} \, \bar{\boldsymbol{T}} \boldsymbol{N}_s^T \,, \tag{12}$$

$$\boldsymbol{B} = \boldsymbol{N}_s \boldsymbol{N}_0^T + \boldsymbol{N}_0 \boldsymbol{N}_s^T - \boldsymbol{T}_s \boldsymbol{T}_0^T + \boldsymbol{T}_0 \boldsymbol{T}_s^T - 2 \, \frac{g_N}{l} \left(\boldsymbol{T}_0 \boldsymbol{N}_0^T + \boldsymbol{N}_0 \boldsymbol{T}_0^T \right) . \tag{13}$$

The regularized form of the Coulomb friction law is employed for the mathematical representation of friction [8]. For a state within the Coulomb friction cone, reversible tangential displacements can occur. In addition to the first two terms, the ones with the matrices $\boldsymbol{A}$ and $\boldsymbol{B}$ must be considered in the matrix $\boldsymbol{k}_c$. If the

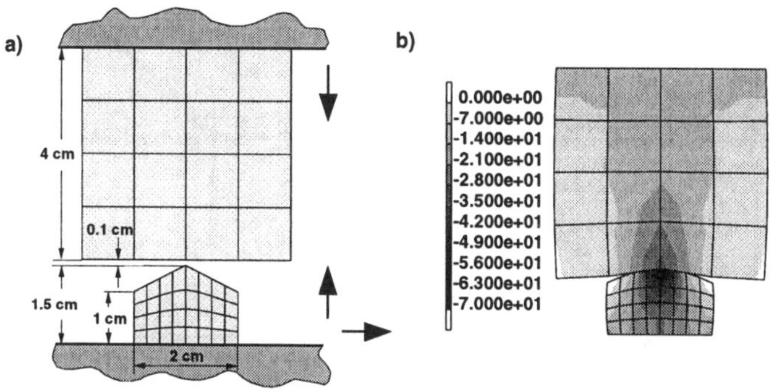

Figure 3. : Example #1, a) FE mesh, b) σ_{zz} [N/cm^2]

TABLE 4. : Largest absolute values of the residual components

	1	2	3	4	5	6	7	8
ABAQUS	0.12	0.74E-2	0.11E-2	0.15E-3	0.18E-4	0.23E-5	0.29E-06	0.36E-07
MARC	0.13E+1	0.24E-1	0.30E-3	0.18E-4	0.13E-5	0.11E-06	0.73E-08	0.63E-09
new module	0.95E-1	0.40E-1	0.14E-4	0.51E-11				

limit of the tangential force set by the slip function is exceeded, irreversible tangential displacements will occur. In this case the term with the unsymmetric matrix $\boldsymbol{C}$ must be considered in (11). The update of the tangential force is performed by means of the return map algorithm used in the theory of elastoplasticity.

3.2. SEARCH ALGORITHM

The contact search is divided into a global and a local search in order to reduce the amount of CPU time required. In the global search for each contact node point its nearest neighbouring node is determined. Only a region in the neighbourhood of the contact node point is taken into account in the search. In the local search the contact segments containing the node found in the global search are checked whether they are contacted by the contact node point [4].

3.3. NUMERICAL EXAMPLES

3.3.1. *Example #1*

To demonstrate the performance of the implemented contact algorithm the results of a contact simulation without consideration of frictional effects are given at first. The mesh employed is shown in figure 3 a). Linear-elastic material behaviour is assumed to hold for both bodies. The Young's modulus is set equal to 210 N/cm^2, the Poisson's ratio equal to 0.3. The two bodies are moved towards each other (12 $\times$ 0.025 cm), this movement is stopped and the wedge-shaped body is moved to the right side. Figure 3 b) shows the distribution of σ_{zz} at the end of the vertical movement. Table 4 shows the largest absolute values of the residual components for the respective increment for analyses with different FE programs and gives evidence for the superb convergence behaviour of the implemented contact algorithm.

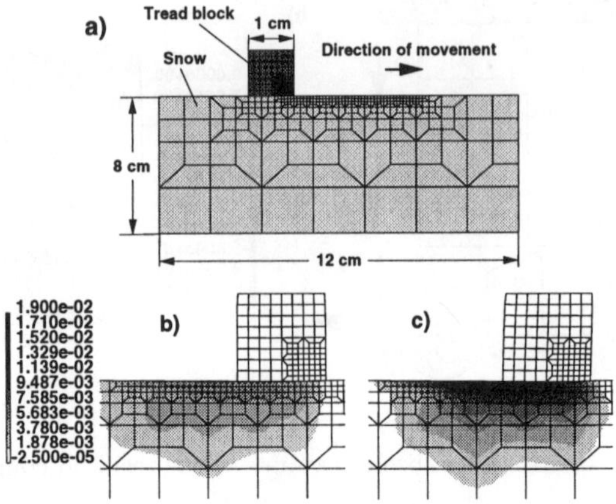

Figure 4. : Example #2, a) FE mesh, b) and c) equivalent deviatoric plastic strains of the snow (friction coefficient: 0.01 and 0.40)

3.3.2. *Example #2*

A tread block moving on a snow surface is investigated in this simulation. The FE mesh is shown in Figure 4 a). The rubber block was pressed against the snow surface (applied pressure: 20.0 N/cm^2) and moved horizontally (velocity: 1.282 cm/min). The incompressibility of rubber was enforced by means of Lagrange multipliers (Mooney parameters: A_1 = 91.5 N/cm^2 and A_2 = 21.5 N/cm^2). The material behaviour of snow was described by a viscoplastic material model [2]. Figure 4 b) and c) show the distribution of the deviatoric plastic strain for two different friction coefficients. Tensile strains indicate the failure of snow.

References

1. Helnwein, P., Liu, C.H., Meschke, G, Mang, H.A.: A New 3-D Finite Element Model for Cord-Reinforced Rubber Composites – Application to Analysis of Automobil Tires, *Finite Elements in Analysis and Design* **14** (1993), 1–16
2. Meschke, G.: A New Viscoplastic Model for Snow at Finite Strains, in D.R.J. Owen, E. Oñate and E. Hinton (eds.), *Computational Plasticity. Fundamentals and Applications*, Pineridge Press, Swansea, pp. 2295–2306, 1995
3. Meschke, G., Payer, H.-J., Mang, H.A.: 3D Simulations of Automobile Tires. Material Modelling, Mesh Generation and Solution Strategies, *Tire Science and Technology* (1997), in press
4. Oldenburg, M., Nilsson, L.: The Position Code Algorithm for Contact Searching, *International Journal for Numerical Methods in Engineering* **37** (1994), 359–386
5. Papadrakakis, M.: *Solving Large-Scale Problems in Mechanics*, Wiley & Sons, Chichester, 1993
6. Payer, H.-J., Mang, H.A.: Iterative Strategies for Solving Systems of Linear, Algebraic Equations Arising in 3D BE-FE Analyses of Tunnel Drivings, *Numerical Linear Algebra with Applications* **4** (1997), 239–268
7. Simo, J.C., Taylor, R.L., Pister, K.S.: Variational and Projection Methods of the Volume Constraint in Finite Deformation Plasticity, *Computational Methods in Applied Mechanics and Engineering* **51** (1985), 177–208
8. Wriggers, P.: Finite Element Algorithms for Contact Problems, *Archives of Computational Methods in Engineering* **2** (1995), 1–49

AUTOMATED MULTISCALE FRACTURE ANALYSIS

JACOB FISH, MARK S. SHEPHARD and MARK W. BEALL
Scientific Computation Research Center
Rensselaer Polytechnic Institute, Troy, NY 12180-3590 USA

Abstract. A procedure for automated crack propagation analysis in heterogeneous materials is presented. The system uses automated multiscale analysis which explicitly represents the microstructure in the vicinity of the crack front while using homogenized material properties elsewhere.

1. Introduction

To understand the failure of heterogeneous materials systems it is necessary to understand the interactions and failure of the individual constituents of the system. The local fields are dictated by the behavior of the structure. However, it is not feasible to analyze the entire structure explicitly representing the constituents throughout the domain. This problem is addressed by the application of adaptive multiple scale analysis methodologies which combine overall models (macromechanical) for the majority of the domain with local models (micromechanical) in critical regions. Section 2 outlines the techniques used to perform and adaptively control multiple scale analyses.

The simulation of discrete crack growth requires criteria to indicate when a crack will propagate, in which direction it will propagate, and how far it will propagate. Section 3 outlines the crack propagation criteria used in the present work.

Key to the multiple scale fracture simulation is the automatic generation of the models and meshes. Since the geometry evolves as the crack propagates, both the models and meshes must be updated. Sections 4 through 6 describe the procedures used to control models, generate meshes and reflect crack propagations. Section 7 demonstrates the application of the procedures described to crack growth in a unidirectional composite material. A comparison is made between the crack growth predicted with and without explicit consideration of the microstructure.

2. Adaptive Multigrid Method for Heterogeneous Structures

The multigrid technology with special inter-scale connection operators possesses superconvergent characteristics for the periodic heterogeneous medium (Fish and Belsky, 1995). The three-scale multigrid solver developed in (Fish and Belsky, 1995) automatically couples phenomena at the macro, meso, and micro levels as a part of the solution process.

The multigrid procedure shown in Figure 1 starts by performing several smoothing iterations on the micro-scale in the regions identified by homogenization error indicators (Fish, et al., 1994). Consequently, the higher frequency modes of error are damped out immediately. The remaining part of the solution error is smooth, and can be effectively eliminated on the coarse mesh. The mesh on the meso-scale (ply level) serves as a perfect mechanism for capturing the lower frequency response on the micro-scale (Fish and Belsky, 1995). Therefore, the residual on the micro-scale is restricted to the meso-scale, while the smooth part of the solution is captured on the meso-scale. The oscillatory part

H. A. Mang and F. G. Rammerstorfer (eds.),
IUTAM Symposium on Discretization Methods in Structural Mechanics, 249–256.

of the solution on the meso-scale is again damped out by smoothing. The lower frequency response on the meso-scale is resolved on the macro-mesh. The solution on the meso-scale is obtained by prolongating displacements from the macro-mesh back to the finite element mesh on the meso-scale and by adding the oscillatory part of the solution previously captured on the meso-scale. Likewise, the solution on the micro-scale is obtained by prolongating the smooth part of the solution from the meso-scale and by adding the oscillatory part that has been obtained by smoothing. This process is repeated until satisfactory accuracy is obtained.

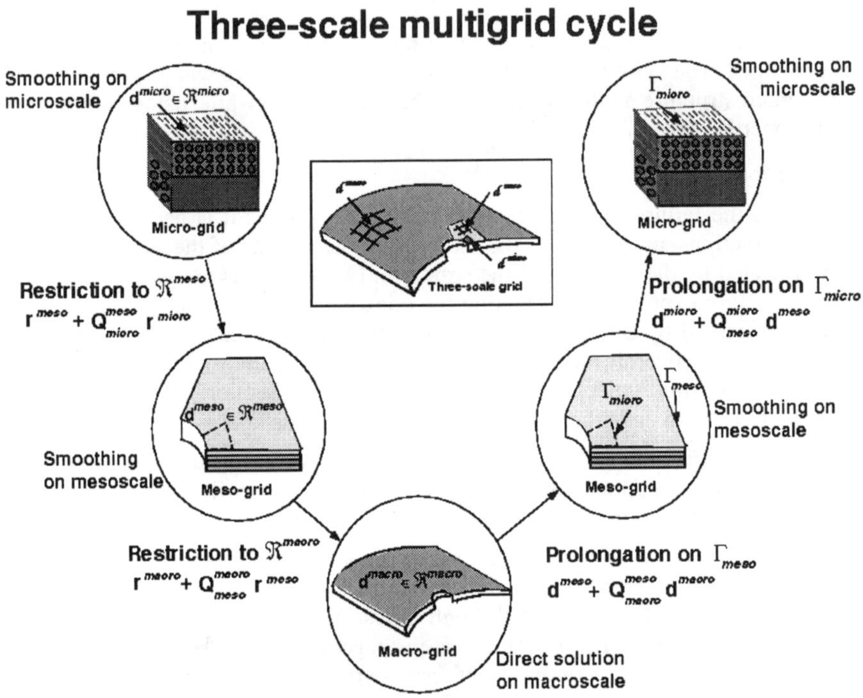

Figure 1. Three Scale Solution Process.

The derivation of the inter-scale transfer operators is based on the asymptotic solution expansion. The asymptotic forms of the prolongation and restriction operators were obtained by discretizing the corresponding asymptotic expansions. For unit cells of a finite size the regularization function was introduced in (Fish and Belsky, 1995) to obtain well-posed inter-scale transfer operator, termed as homogenization based operators.

The rate of convergence of the multigrid process has been studied in (Fish and Belsky, 1995). For problems in periodic 1-D heterogeneous media, the rate of convergence of the two-grid method with special inter-scale transfer operators is governed by:

$$\|e_{i+1}\| = \frac{q}{4-q}\|e_i\| \qquad q = \frac{\sqrt{D_1 D_2}}{0.5(D_1 + D_2)} \tag{1}$$

where $\|e_{i+1}\|$ is the norm of solution error in iteration i, D_i represent the stiffnesses of microconstituents. Note that if the material is homogeneous and the mesh is uniform, then $D_1 = D_2$ and we recover the classical two-grid estimate: $\|e_{i+1}\| = 1/3\ \|e_i\|$. Otherwise $q < 1$ resulting in: $\|e_{i+1}\|\ /\|e_i\| . < 1/3$.

Note that if the stiffness of a fiber is significantly higher than that of a matrix, i.e. $D_1 >> D_2$, the multigrid method converges in a single iteration. In multidimensions, convergence of the multigrid process for periodic heterogeneous media has been studied in (Fish and Belsky, 1995). This behavior of the multigrid method for heterogeneous media together with its *linear* dependence on the number of unknowns, makes it possible to solve large coupled global-local problems with the same amount of computational effort, or faster, than would be required to solve the corresponding uncoupled problem using direct solvers.

3. Crack Propagation Procedures

In the present study of crack propagation we adopt a global-local approach by which a 'local' description is used in the vicinity of the crack front while elsewhere the medium is treated as homogenized. Crack growth simulations are difficult to perform because of the need in continual geometry and computational mesh updates.

The crack propagation increment $\boldsymbol{x}_A - \boldsymbol{X}_A$ is defined in of terms its shape $\{\tilde{S}_A\}_{A=1}^{n}$ and amplitude α

$$\boldsymbol{x}_A = \boldsymbol{X}_A + \alpha\tilde{S}_A \tag{2}$$

where $\boldsymbol{X}_A, \boldsymbol{x}_A$ denote position of node A located at the crack front interface before and after incremental update respectively, and n is the number of nodes at the crack front interface. Prediction of crack trajectory requires determination of $\{\tilde{S}_A\}_{A=1}^{n}$ and α.

No generally accepted criterion exists for predicting trajectories of cracks in 3-D heterogeneous media. Among the popular phenomenological theories for predicting instantaneous angle of crack propagation in 2-D homogeneous media are maximum tangential stress criterion, maximum energy release criterion and minimum strain energy density criterion. The two major obstacles in applying these criteria to 3-D applications in heterogeneous media with arbitrary crack surfaces are: (i) crack growth depends not only on the geometry and boundary conditions but also on material characteristics of microconstituents, such as strength and fracture toughness, and (ii) asymptotic fields are generally not known for 3D heterogeneous media.

To overcome these difficulties we propose the following criteria:

(i) The preliminary crack propagation shape $\{\tilde{S}_A\}_{A=1}^{n}$ is defined on the basis of the following normalized criteria:

$$\|S_A\| = F_A/G_A \tag{3}$$

where $\|\ \|$ denotes Euclidean norm of a vector, F_A is either maximum energy release rate, maximum tangential stress, or minimal strain energy density. G_A is either fracture toughness, strength in tension or critical energy density, respectively.

(ii) The unit vector $\boldsymbol{e}_A = S_A/\|S_A\|$ defining the crack propagation direction at node A points in the direction of the maximum ratio F_A/G_A (among the elements connected to node A) projected onto the plane normal to the crack tip interface at node A.

(iii) Due to lack of crack tip asymptotic expansion, deformation and stress fields are either directly extracted from the finite element analysis or by postprocessing.

(iv) The crack propagation shape obtained from equation (3) could be oscillatory in nature and hence smoothing is necessary. The process of finding the smoothed crack propagation shape can be viewed as a constrained optimization problem, which states:

Find the smoothed crack propagation shape increment, $\|\tilde{S}_A\|$, such that

$$\int_L (\|S_A\| - \|\tilde{S}_A\|)^2 dL \rightarrow min \tag{4}$$

subjected to the smoothness constrained:

$$\int_L \left(\frac{\partial}{\partial L}\|\tilde{S}_A\|\right)^2 dL < \varepsilon^2 \tag{5}$$

where L is a coordinate along the crack tip, and ε is the smoothness parameter. Solution of the constrained optimization problem yields the following differential equation:

$$\left(I - \gamma\frac{\partial^2}{\partial L^2}\right)\|\tilde{S}_A\| = \|S_A\| \tag{6}$$

subjected to periodic boundary condition. Note that $\gamma \ll 1$ is defined by the user. Equation (6) can be solved using a finite element discretization of $\|\tilde{S}_A\|$.

Due to the history dependence of crack growth, the problem of evaluating the amplitude α can be stated in terms of a ordinary delay differential equation (Fish and Nath, 1993):

$$\frac{dx_A}{d\alpha} = \tilde{S}_A(\alpha) \tag{7}$$

The special nature of the ordinary delay differential equations has the effect of making the crack path smoother with the evolution of the right hand side vector in (7) is smooth. For problems in heterogeneous media G_A and thus $\tilde{S}_A$ are C^{-1} continuous functions in α and thus the optimal integrator for such a differential equation should be based on a type of integration scheme which does not rely on the smoothness of the crack path. The simplest form of such an integrator is based on a predictor-corrector scheme:

$$\begin{aligned} &Predictor \qquad && \bar{x}_A = X_A + \tilde{S}(X_A)\alpha \\ &Corrector && x_A = X_A + \frac{1}{2}(\tilde{S}(X_A) + \tilde{S}(\bar{x}_A))\alpha \end{aligned} \tag{8}$$

The normalized local truncation error β is given as:

$$\beta = \sqrt{\sum_{A=1}^{n} \frac{(x_A - \bar{x}_A)^2}{(x_A - X_A)^2}} = \sqrt{\sum_{A=1}^{n} \frac{(\tilde{S}(\bar{x}_A) - \tilde{S}(X_A))^2}{(\tilde{S}(\bar{x}_A) + \tilde{S}(X_A))^2}} \tag{9}$$

Crack propagation amplitude may increase or decrease to keep the magnitude of the normalized truncation error β below the user prescribed tolerance.

4. Problem Definition

The effective application of multiple scale analysis requires automation of the entire process. Without automation, the analysis is effectively impossible due to the time and effort required to update the models and meshes needed as the analysis progresses. The automation of a set of engineering analyses requires all analysis models, and their meshes, be derived from a single problem definition. The two key components of the problem definition are the domain description (the geometric model), and the analysis attributes defined in terms of it.

The geometric models constructed to support multiscale analyses vary from a representation of the middle surface of a composite structure to the micromechanical structure of the composite. The model construction and automatic mesh generation is supported by geometric modeling systems possessing representations of general combinations of volumes, surfaces and curves (Weiler, 1988). In addition to taking direct advantage of the massive development effort required to produce such geometric modelers, this approach can properly support all the geometry needs of automatic mesh generation (Shephard and Georges, 1992). Figure 2a shows the unit cell for a composite weave automatically constructed in the Parasolid solid modeler using basic manufacturing parameters defining the weave (Beall et al., 1996), (Shephard et al., 1995).

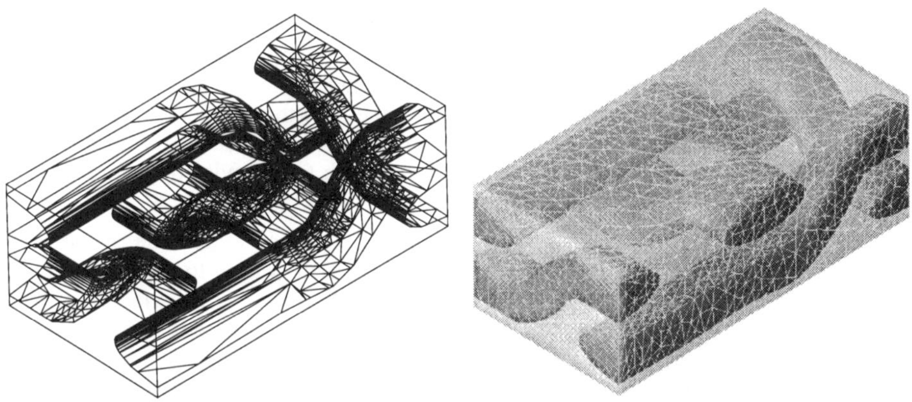

(a) Model of woven composite (b) Mesh of woven composite

Figure 2. Automatically constructed and meshed unit cell model.

Analysis attributes consist of the information past the geometric model needed to specify an analysis model and include material property, boundary condition, load, and initial condition information. A hierarchical organization structure allows the convenient collection of attributes.

5. Automatic Generation of Models and Meshes for Multiscale Analysis

The multiscale analysis used here has three levels. The top level is the component (macro) level, where all the material properties are considered to be homogenized. The bottom level is the microstructural level where all the details of the microstructure, and damage that is occurring to the microstructure, are represented. The middle level serves as a transition between these two levels and incorporates any macro level damage.

The problem definition consists of a solid model of the physical structure being analyzed (the macro model) with appropriate analysis attributes and idealization attributes that describe the layup, microstructure of the composite and initial crack geometry.

The domain of the micro model is the set of unit cells that encompass the critical areas of the model where it is necessary to resolve local fields. For the crack propagation problem this area is the unit cells that enclose the crack front. The steps to find this and construct the geometric domain for the crack propagation problem are given in Beall, et al. (1996). To support the analysis procedures, the mesh on the micro model boundary must be identical to that which is on the boundary of the unit cell used to calculate the homogenized material properties. The mesh on the interior of the model has no restrictions. This requirement is enforced by copying the mesh from the boundary of the unit cell to all the faces on the boundary of the model, then meshing the interior of the model using an automatic mesh generator (Shephard and Georges 1992).

A transition model connects the macro model, with its coarse mesh and homogenized material properties, to the micro model, with its fine mesh and explicit microstructure. The outer boundary of the transition model are the faces of the elements in the macro mesh to which the micro model is interior. The mesh for the transition model does not need to be compatible with that of the macro model, however the element faces on the boundary of the transition mesh must be on a single element face of the macro mesh.

6. Model and Mesh Updates for Crack Propagation

To simulate the crack growth, the analysis must update the representation of the crack to account for the growth predicted. This process is made efficient by performing local modifications to the models and meshes to reflect the growth of the crack. As the crack grows, it eventually approaches the boundary of the micro model, requiring the micro model domain and, thus, the transition model and mesh to be updated. The updates of the higher level models only happen occasionally.

As shown in Figure 3, the analysis provides the direction and magnitude of the crack growth at each vertex on the crack front. This information is used to define the "crack extension" which is the geometric representation of the new crack surface. To avoid placing unnecessary constraints on the growth of the crack the crack extension is defined independently of the mesh. It is defined as a set of triangular faces extending from the current crack front to new locations for the vertices on the crack front given by the analysis (Figure 3).

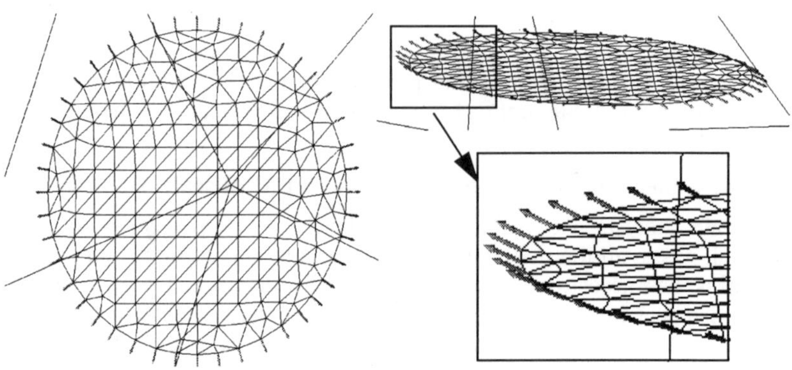

Figure 3. Analysis results showing predicted crack growth.

To construct the crack extension, the direction and magnitude of growth from each node on the crack front is added to the nodes current location. A triangular mesh is then constructed that extends from the old crack front to these new locations. The triangulation procedure must account for crack front expansion or contraction and for the interaction of the crack with any geometric model entities, such as intersecting with a fiber face.

The first step in growing the crack is to determine the interaction of the crack extension with the existing mesh. This is efficiently done using the mesh entity adjacency. A mesh modification procedure is applied to automatically modify the affected mesh to account for the crack propagation.

As the crack front moves though the micro model, the original selection of the domain for the micro model may no longer be acceptable due to either the crack approaching its boundary, or unit cells are no longer close to the crack front (making the micro model larger than it needs be). Unit cells are added and subtracted from the micro model as needed, which also forces updates to the transition model (Beall, et al., 1996).

7. Results

Two analyses are performed on a macro model loaded in off axis tension at an angle of 45 degrees to the crack surface. The material used is a unidirectional fiber reinforced composite. Both the fiber and matrix are taken as isotropic with the fiber properties: $E_f = 50000$, $\upsilon_f = 0.25$ and the matrix properties: $E_m = 10000$, $\upsilon_m = 0.25$. The first analysis was run using homogenized material properties, calculated from the unit cell of the composite, throughout the problem domain. The second analysis uses explicit microstructure in the vicinity of the crack front (Figure 4).

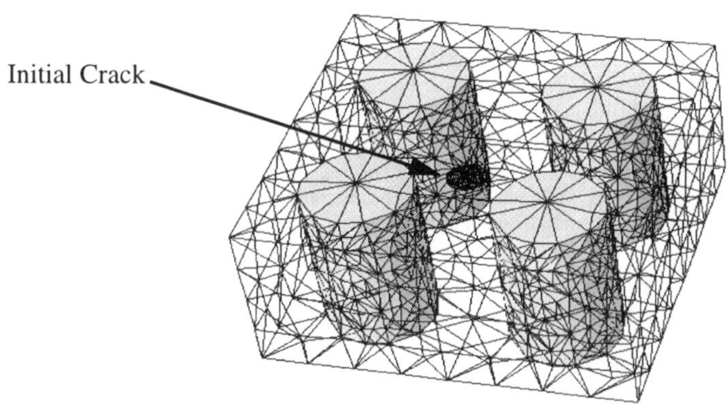

Figure 4. Micro scale model with crack location indicated.

Figure 5 shows the crack growth predicted from the two analyses. The crack in the homogeneous material is shown in the lighter grey and the crack for the heterogeneous material is shown in dark grey. The closest fiber is on the right side of the view.

The analysis with the homogeneous material exhibits the expected behavior in this situation. The crack is growing so that it becomes perpendicular to the direction of the loading. The final step shown in the results is right after the crack would have come in contact with the fibers.

In the inhomogeneous case the crack exhibits more complicated behavior. It can be clearly seen that the presence of the fibers is affecting the growth of the crack. As the crack is approaching the fiber on the right side it bends away from the fiber and becomes more parallel to it.

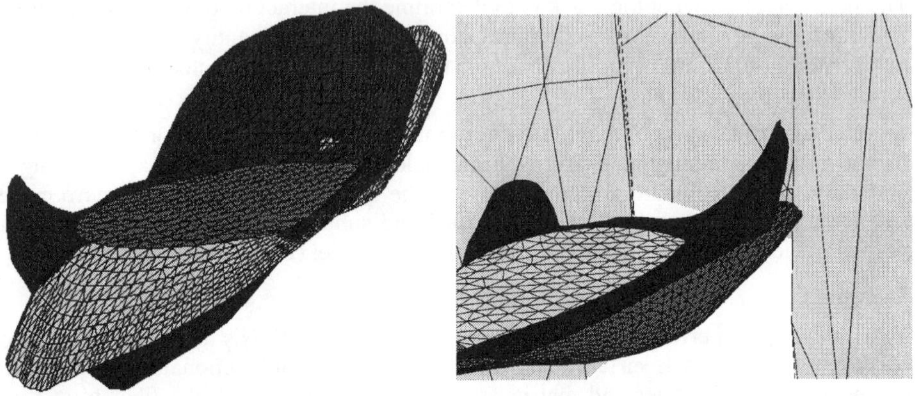

Figure 5. Difference in crack growth accounting for the microstructure. Light grey for homogeneous material, dark grey with the microstructure explicitly represented.

8. Acknowledgments

This research was supported by ARPA/ONR under grant N00014-92J-1779.

9. References

Beall, M.W., Belsky, V., Fish J. and Shephard, M.S. (1996) Automatic multiple scale fracture analysis, SCOREC Report 4-1994, Scientific Computation Research Center, RPI, Troy, NY, To appear in ASCE monograph on *High Performance Computing*.

Fish, J., Nayak, P. and Holmes, M.H. (1994) Microscale Reduction Error Indicators and Estimators for a Periodic Heterogeneous Medium, *Computational Mechanics*, **14**, 323-338.

Fish, J. and Belsky, V. (1995) Multigrid method for a periodic heterogeneous medium. Part I: Convergence studies in 1-D case; Part 2: multidimensions, *Comp. Meth. Appl. Mech. Engng.*, **126**, 1-16, 17-38.

Fish, J. and Nath, A. (1993) Adaptive and hierarchical modeling of fatigue crack propagation, *Int. J. For Num. Meth. in Engineering*, **36**, 2825-2836.

Shephard, M.S. (1988) The specification of physical attribute information for engineering analysis. *Engineering with Computers*, **4**, 145-155.

Shephard, M.S. and Georges, M.K. (1992) Reliability of automatic 3-D mesh generation. *Comp. Meth. Applied Mechanics and Engng*. **101**, 443-462.

Shephard, M.S., Beall, M.W., Garimella, R. and Wentorf, R. (1995) Automatic construction of 3-D models in multiscale analysis. *Computational Mechanics*, **17**(3), 196-207.

Weiler, K.J. (1988) The radial-edge structure: A topological representation for non-manifold geometric boundary representations. *Geometric Modeling for CAD Applications*, M.J. Wozny, H.W. McLaughlin and J. Encarnacao, eds., North Holland, 3-36.

MULTISCALE TREATMENT OF INHOMOGENEOUS MATERIALS BY FINITE ELEMENTS

H.J. BÖHM, C.M. CHIMANI, H.E. PETTERMANN, A.F. PLANKENSTEINER AND F.G. RAMMERSTORFER

Institute of Light Weight Structures and Aerospace Engineering
Vienna University of Technology
Gußhausstr. 27–29, A–1040 Vienna, Austria

Abstract. Some aspects of the Finite Element modeling of the thermomechanical behavior of structures and components made of inhomogeneous materials are discussed. Special attention is focused on micromechanically based material models, which may be used both at the mesoscale (e.g. to study clusters of reinforcing particles) and at the component level, i.e. the macroscale.

1. Introduction

For more than a decade considerable research and engineering interest has been evident in advanced materials such as metal matrix composites (MMCs), which are highly inhomogeneous at some microscale, contain at least one thermoelastoplastic constituent (or "phase"), and consequently show a nonlinear overall behavior. The requirements for their efficient and cost effective use have provided strong incentives to develop flexible numerical engineering approaches for analyzing the thermomechanical behavior of components and structures made of such inhomogeneous materials.

The present contribution describes such a formalism which is based on micromechanical considerations and was developed for use within the framework of the Finite Element (FE) method. It is shown that, in addition to its primary field of application at the structural level, the approach can also be used to advantage in other cases where the "smearing out" of inhomogeneities at some smaller length scale makes sense physically.

H. A. Mang and F. G. Rammerstorfer (eds.),
IUTAM Symposium on Discretization Methods in Structural Mechanics, 257–264.

2. A Micromechanically Based Material Model for FE Programs

2.1. REQUIREMENTS

From the point of view of carrying out FE analyses of the thermomechanical behavior of structures consisting of or containing inhomogeneous or composite materials, an obvious thing to do is to devise appropriate material models for use at the integration point level. Ideally, such models have to show a useful combination of accuracy, efficiency (i.e. low computational cost), and generality (i.e. an applicability over a wide range of inhomogeneous materials and loading conditions). In addition, it is desirable that they not only provide information on the overall behavior of an inhomogeneous material, but also on the microscale stress and strain fields in its constituents, thus allowing strength assessments at the phase level.

The material models for elastoplastic inhomogeneous materials reported in the literature tend to fall into two groups. On the one hand, semi-empirical material descriptions have been proposed that are extensions of "standard" material models modified to account for the specific behavior of materials such as MMCs, see e.g. (Xie and Adams, 1996). Such approaches tend to be computationally efficient, but restricted to a given class of inhomogeneous materials. They use inputs in terms of overall material parameters and do not provide predictions for the microscale states of the constituents. On the other hand, material descriptions may use data on the microscale topology and geometry of a given inhomogeneous material and on the thermomechanical behavior of its constituents. Depending on the complexity of the underlying micromechanical model, such approaches have shown a wide range of tradeoffs between computational cost, accuracy, and flexibility as exemplified by the works of Bahei–el–Din et al. (1981), Arenburg and Reddy (1991) as well as Lee and Ghosh (1996).

The method described in the following is based on a relatively simple micromechanical mean field model in order to allow 2/D analyses of composite structures to be carried out on engineering workstations.

2.2. MEAN FIELD APPROACHES

Mean field approaches in micromechanics are formulated in terms of the phase averages of the stresses and strains, $\varepsilon^{(p)}$ and $\sigma^{(p)}$, respectively, which for the thermoelastic case can be expressed in dependence on the applied constant overall strains and stresses, ε^a and σ^a, as well as the temperature difference ΔT with respect to some stress free reference temperature, as

$$\begin{aligned} \boldsymbol{\varepsilon}^{(p)} &= \frac{1}{V^{(p)}} \int_{\Omega^{(p)}} \boldsymbol{\varepsilon}(r)dV &= \bar{\mathbf{A}}^{(p)}\boldsymbol{\varepsilon}^a + \bar{\mathbf{a}}^{(p)}\Delta T \\ \boldsymbol{\sigma}^{(p)} &= \frac{1}{V^{(p)}} \int_{\Omega^{(p)}} \boldsymbol{\sigma}(r)dV &= \bar{\mathbf{B}}^{(p)}\boldsymbol{\sigma}^a + \bar{\mathbf{b}}^{(p)}\Delta T \quad . \end{aligned} \qquad (1)$$

Here (p) indicates the constituent, which for two-phase composite materials can be either the matrix (m) or the inclusions (i), $\Omega^{(p)}$ stands for the domain occupied by this constituent, and $V^{(p)}$ for the corresponding volume. $\bar{\mathbf{A}}^{(p)}$ and $\bar{\mathbf{B}}^{(p)}$ are known as the mechanical phase strain and stress concentration tensors for phase (p), respectively, with $\bar{\mathbf{a}}^{(p)}$ and $\bar{\mathbf{b}}^{(p)}$ being called the thermal strain and stress concentration tensors. The effective or overall elasticity tensor $\mathbf{E}^*$, compliance tensor $\mathbf{C}^*$, and thermal expansion tensor $\boldsymbol{\alpha}^*$ of the composite can be evaluated from the material properties of the constituents, from the inclusion inclusion volume fraction ξ, and from the concentration tensors by using relations such as

$$\begin{aligned}
\mathbf{E}^* &= \xi \mathbf{E}^{(i)} \bar{\mathbf{A}}^{(i)} + (1-\xi)\mathbf{E}^{(m)} \bar{\mathbf{A}}^{(m)} \\
\mathbf{C}^* &= \xi \mathbf{C}^{(i)} \bar{\mathbf{B}}^{(i)} + (1-\xi)\mathbf{C}^{(m)} \bar{\mathbf{B}}^{(m)} \\
\boldsymbol{\alpha}^* &= \xi[\mathbf{C}^{(i)}\bar{\mathbf{b}}^{(i)} + \boldsymbol{\alpha}^{(i)}] + (1-\xi)[\mathbf{C}^{(m)}\bar{\mathbf{b}}^{(m)} + \boldsymbol{\alpha}^{(m)}] \quad .
\end{aligned} \tag{2}$$

Because there are additional linkages between the concentration tensors, see e.g. (Benveniste and Dvorak, 1990), within mean field schemes the problem of finding the overall thermoelastic properties of a two-phase material reduces to obtaining one mechanical stress or strain concentration tensor.

A considerable number of mean field approaches have been reported in the literature, the most common being Mori–Tanaka methods (Mori and Tanaka, 1973) and two-phase self-consistent schemes (Hill, 1965), both of which are based on the work of Eshelby (1957) on dilute ellipsoidal inclusions. Mori–Tanaka approaches have the advantage of giving rise to explicit expressions for the concentration tensors and, because they describe microstructures consisting of aligned ellipsoidal inclusions in a matrix, they are well suited for use as material models for composite materials.

2.3. INCREMENTAL MORI–TANAKA METHOD

Mean field methods can be extended to the thermoelastoplastic range by formulating eqn. (1) in terms of strain, stress, and temperature rates, $\mathrm{d}\boldsymbol{\varepsilon}^{(p)}$, $\mathrm{d}\boldsymbol{\varepsilon}^{a}$, $\mathrm{d}\boldsymbol{\sigma}^{(p)}$, $\mathrm{d}\boldsymbol{\sigma}^{a}$, and $\mathrm{d}T$, respectively, and using instantaneous concentration tensors $\bar{\mathbf{A}}_{\mathrm{t}}^{(p)}$, $\bar{\mathbf{B}}_{\mathrm{t}}^{(p)}$, $\bar{\mathbf{a}}_{\mathrm{t}}^{(p)}$, and $\bar{\mathbf{b}}_{\mathrm{t}}^{(p)}$ to give

$$\begin{aligned}
\mathrm{d}\boldsymbol{\varepsilon}^{(p)} &= \bar{\mathbf{A}}_{\mathrm{t}}^{(p)} \mathrm{d}\boldsymbol{\varepsilon}^{a} + \bar{\mathbf{a}}_{\mathrm{t}}^{(p)} \mathrm{d}T \\
\mathrm{d}\boldsymbol{\sigma}^{(p)} &= \bar{\mathbf{B}}_{\mathrm{t}}^{(p)} \mathrm{d}\boldsymbol{\sigma}^{a} + \bar{\mathbf{b}}_{\mathrm{t}}^{(p)} \mathrm{d}T \quad .
\end{aligned} \tag{3}$$

For the case of thermoelastic inclusions in a thermoelastoplastic matrix the global instantaneous tensors then take the form

$$\begin{aligned}
\mathbf{E}_{\mathrm{t}}^* &= \mathbf{E}^{(i)} + (1-\xi)(\mathbf{E}_{\mathrm{t}}^{(m)} - \mathbf{E}^{(i)})\bar{\mathbf{A}}_{\mathrm{t}}^{(m)} \\
\boldsymbol{\alpha}_{\mathrm{t}}^* &= \mathbf{C}^{(i)}[\mathbf{e}^{(i)} + (1-\xi)(\bar{\mathbf{A}}_{\mathrm{t}}^{(m)})^{-1}(\mathbf{e}_{\mathrm{t}}^{(m)} - \mathbf{e}^{(i)})] \quad ,
\end{aligned} \tag{4}$$

where the specific thermal stress tensor is defined as $\mathbf{e}^{(p)} = -\mathbf{E}^{(p)}\boldsymbol{\alpha}^{(p)}$.

Using the formulation of Benveniste (1987), Mori–Tanaka expressions for the instantaneous concentration tensors can be obtained in the form

$$\begin{aligned}
\bar{\mathbf{A}}_{\rm t}^{(m)} &= \{(1-\xi)\mathbf{I} + \xi[\mathbf{I} + \mathbf{S}_{\rm t}\mathbf{C}_{\rm t}^{(m)}(\mathbf{E}^{(i)} - \mathbf{E}_{\rm t}^{(m)})]^{-1}\}^{-1} \\
\bar{\mathbf{a}}_{\rm t}^{(m)} &= (\mathbf{I} - \bar{\mathbf{A}}_{\rm t}^{(m)})(\mathbf{E}^{(i)} - \mathbf{E}_{\rm t}^{(m)})^{-1}(\mathbf{e}_{\rm t}^{(m)} - \mathbf{e}^{(i)}) \\
\bar{\mathbf{B}}_{\rm t}^{(m)} &= \{(1-\xi)\mathbf{I} + \xi[\mathbf{I} + \mathbf{E}_{\rm t}^{(m)}(\mathbf{I} - \mathbf{S}_{\rm t})(\mathbf{C}^{(i)} - \mathbf{C}_{\rm t}^{(m)})]^{-1}\}^{-1} \\
\bar{\mathbf{b}}_{\rm t}^{(m)} &= (\mathbf{I} - \bar{\mathbf{B}}_{\rm t}^{(m)})(\mathbf{C}^{(i)} - \mathbf{C}_{\rm t}^{(m)})^{-1}(\boldsymbol{\alpha}^{(m)} - \boldsymbol{\alpha}^{(i)}) \quad . \qquad (5)
\end{aligned}$$

Here $\mathbf{I}$ is the identity tensor and $\mathbf{S}_{\rm t}$ stands for the instantaneous Eshelby tensor, which depends on the instantaneous material properties of the matrix and on the aspect ratio of the inclusions. In the elastoplastic range it has to be evaluated numerically, e.g. by the procedure proposed by Gavazzi and Lagoudas (1990).

Compared to mean field approaches based on secant plasticity, incremental Mori–Tanaka (IMT) procedures have the advantage of not being restricted to radial loading at the microlevel, which is vital for their use as material models in FE programs. They show a tendency, however, towards overpredicting the overall hardening behavior of the inhomogeneous material, compare (Suquet, 1997). IMT-type procedures in the literature include those of Lagoudas et al. (1991) as well as Karayaka and Sehitoglu (1993).

The IMT algorithm given by eqns.(3)–(5) was implemented as a user supplied material module (UMAT) for the FE code ABAQUS (HKS, 1996), which requires that at each integration point and in each iteration the overall stress response and the elastoplastic tangent operator be evaluated for some prescribed strain increment. Because in the case of the IMT model the elastoplastic material behavior is directly defined only for the matrix and not for the overall response, a combination of a radial return mapping algorithm at the level of the elastoplastic matrix and an Euler backward iteration scheme was selected for this task. Appropriate provisions have to be made for resolving the thermal strains in the case of thermoelastoplastic behavior, because in inhomogeneous materials such as MMCs the thermal expansion response depends on the load history and — as is evident from eqn.(4) — on the instantaneous moduli of the matrix. For details of the algorithm, which is capable of handling temperature dependent material parameters, see (Pettermann, 1997).

The IMT UMAT has met the stated requirements and efficiency goals, but — as expected — is considerably slower than the thermoelastoplastic material models for homogeneous materials built into ABAQUS, the most time consuming operation being the evaluation of the instantaneous Eshelby tensor.

3. Applications

3.1. SELECTIVELY REINFORCED RING

As a first application of the IMT material model, results are presented for a selectively reinforced axisymmetric component consisting of an outer region of monolithic magnesium alloy surrounding an inner ring made of a carbon fiber/magnesium MMC with circumferentially arranged aligned fibers, see fig. 1a. The outer region was modeled using a standard thermoelastoplastic material model with linear hardening, whereas the IMT UMAT was employed for the insert. Temperature dependent material parameters for the AZ91 magnesium alloy, which is used both in the monolithic zone and for the matrix of the MMC, were taken from Aune and Westengen (1995), and the data for the T-300 carbon fibers follow Weeton et al. (1987).

Due to the markedly different thermal expansion behavior of the two zones, uniform temperature changes lead to high stresses in this component. Figure 1b shows the von Mises effective stresses in the monolithic outer region and the von Mises microstresses in the matrix of the MMC insert obtained by modeling the cooling down from 200°C to −50°C. The yield stress at the end temperature, 175MPa, is predicted to be exceeded in both monolithic magnesium and MMC insert.

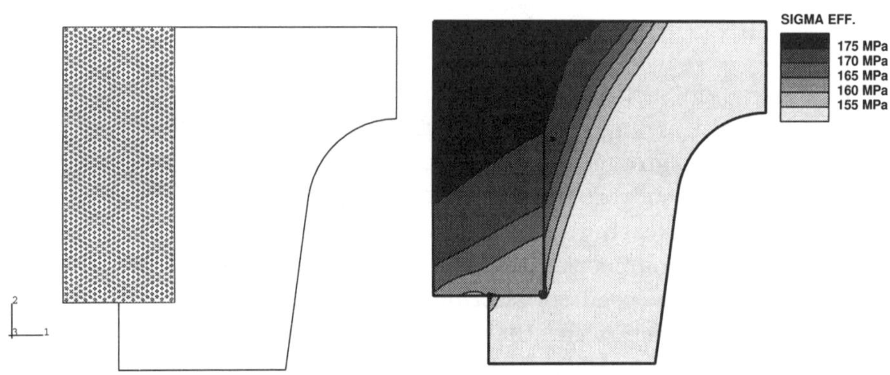

Figure 1. a) Cross section of the selectively reinforced component (the MMC insert is marked by dots) and b) predicted von Mises microstresses in the matrix of the MMC insert and von Mises stresses in the monolithic material.

A close inspection of fig. 1b shows zones of elevated stresses at the corner and the ends of the interface between monolithic metal and composite ring. Stress and strain concentrations (or even singularities) are typically predicted at such positions, because the IMT model treats the composite as a homogeneous material with effective properties and neglects the effects of the microstructure close to the interface. Here some care is required in interpreting the results obtained with homogenized material models such as the IMT, see the discussion in Chimani et al. (1997) for the elastic case.

3.2. CLUSTER-STRUCTURED HIGH-SPEED STEELS

The second application of the IMT UMAT demonstrates its use within a multiscale hierarchical modeling concept for studying the material response of cluster-structured High Speed Steels (HSSs), which show a highly heterogeneous arrangement of thermoelastic carbidic particles within a thermoelastoplastic martensitic–austenitic steel matrix as depicted in fig. 2.

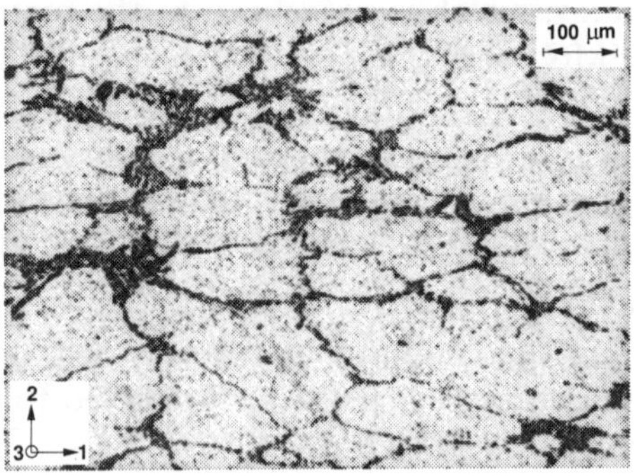

Figure 2. Micrograph of an HSS S6-5-2-5 showing agglomerations of carbidic particles in the steel matrix.

Metallographic sections were processed by digitizing and image processing to obtain arrangements of carbide-rich clusters embedded in an inclusion-poor background. These geometries were used to generate mesoscale unit cells, which were meshed with the preprocessor PATRAN (MSC, 1996). Because both carbide-rich and inclusion-poor zones can be viewed as particle reinforced composites, the IMT UMAT using appropriate inclusion volume fractions was used to describe their respective material behavior. The material parameters for the steel matrix and the carbidic particles follow the data given in Plankensteiner et al. (1997).

The behavior of the HSS under thermal and mechanical loading was studied. The use of the IMT material model allows the results of the computations to be evaluated both in the form of the overall response (i.e. performing macroscopical material characterization) and in terms of the position dependent microscale stresses and strains. Figure 3a shows the distribution of the phase averaged accumulated effective plastic microstrains in the matrix within the inclusion-poor regions predicted for cooling down from 600°C and subsequent uniaxial tensile loading to 3.0GPa in the 2-direction. Furthermore, the susceptibility of a given configuration to microscale damage by brittle cleavage, ductile failure, and interfacial decohesion can be

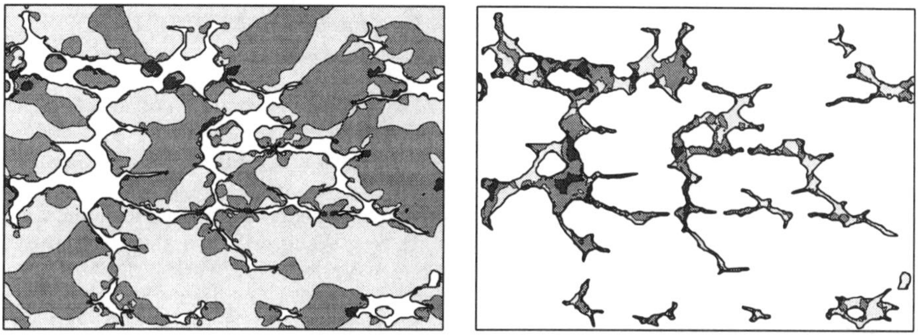

Figure 3. Distributions of a) the phase averaged accumulated effective plastic microstrains in the matrix within the inclusion-poor regions (fringe delimiters 0.25% (light) to 0.5% (dark)) and b) the phase averaged maximum principal microstresses in the particles within the carbidic clusters (fringe delimiters 3000MPa (light) to 3200MPa (dark)) predicted for an HSS S6-5-2-5 subjected to cooling down to room temperature from 600°C and subsequent uniaxial tensile loading to 3.0GPa in 2-direction.

stresses, ductile damage indicators (Fischer et al., 1995), and the maximum interfacial tractions (see e.g. Hill (1983) for their evaluation within mean field models), respectively. The predicted phase averaged maximum principal microstresses in the particles within the carbidic clusters are displayed in fig. 3b, the relatively smooth distribution being due to the presence of residual stresses from the cooling step.

4. Conclusions

An incremental Mori–Tanaka formalism for inhomogeneous materials consisting of thermoelastic inclusions in a thermoelastoplastic matrix was implemented as a material model for an FE code. Modifying the algorithm to handle thermoelastoplastic inclusions in a thermoelastic matrix is relatively straightforward, whereas extensions to materials with two or more nonlinear constituents would require larger modifications.

Acknowledgements

Parts of the present work are related to the Christian Doppler Laboratory "Micromechanics of Materials" and to the BRITE EURAM Project BE'95–1183. The financial support of the Austrian Ministry for Science and Transport under Grant Number GZ.49.935/3–II/4/94 is gratefully acknowledged.

References

Arenburg, R.T. and Reddy, J.N. (1991) Analysis of Metal-Matrix Composite Structures — I. Micromechanics Constitutive Theory, *Comput. Struct.* **40**, 1357–1368.

Aune, T.K. and Westengen, H. (1995) Magnesium Die Casting Properties, *Automotive Engineering*, August 1995, 87–91.

Bahei–el–Din, Y.A., Dvorak, G.J. and Utku, S. (1981) Finite Element Analysis of Elastic-Plastic Fibrous Composite Structures, *Comput. Struct.* **13**, 321–330.

Benveniste, Y. (1987) A New Approach to the Application of Mori–Tanaka's Theory in Composite Materials, *Mech. Mater.* **6**, 147–157.

Benveniste, Y. and Dvorak, G.J. (1990) On a Correspondence Between Mechanical and Thermal Effects in Two-Phase Composites, in G.J. Weng, M. Taya and H. Abé (eds.) in *Micromechanics and Inhomogeneity*, Springer–Verlag, New York, pp. 65–82.

Chimani, C.M., Böhm, H.J. and Rammerstorfer, F.G. (1997) On Stress Singularities at Free Edges of Bimaterial Junctions — A Micromechanical Study, *Scr. mater.* **36**, 943–947.

Eshelby, J.D. (1957) The Determination of the Elastic Field of an Ellipsoidal Inclusion and Related Problems, *Proc. Roy. Soc. London* **A241**, 376–396.

Fischer, F.D., Kolednik, O., Shan, G.X. and Rammerstorfer, F.G. (1995) A Note on Calibration of Ductile Failure Damage Indicators, *Int. J. Fract.* **73**, 345–357.

Gavazzi, A.C. and Lagoudas, D.C. (1990) On the Numerical Evaluation of Eshelby's Tensor and its Application to Elastoplastic Fibrous Composites, *Comput. Mech.* **7**, 12–19.

HKS (1996) *ABAQUS Standard V. 5.6.* Hibbitt, Karlsson & Sorensen Inc., Pawtucket, RI.

Hill, R. (1965) A Self Consistent Mechanics of Composite Materials, *J. Mech. Phys. Sol.* **13**, 213–222.

Hill, R. (1983) Interfacial Operators in the Mechanics of Composite Media, *J. Mech. Phys. Sol.* **31**, 347–357.

Karayaka, M. and Sehitoglu, H. (1993) Thermomechanical Deformation Modeling of Al2xxx-T4/SiC_p Composites, *Acta metall. mater.* **41**, 175–189.

Lagoudas, D.C., Gavazzi, A.C. and Nigam, H. (1991) Elastoplastic Behavior of Metal Matrix Composites Based on Incremental Plasticity and the Mori–Tanaka Averaging Scheme, *Comput. Mech.* **8**, 193–203.

Lee, K. and Ghosh, S. (1996) Small Deformation Multi-Scale Analysis of Heterogeneous Materials with the Voronoi Cell Finite Element Model and Homogenization Theory, *Comput. Mater. Sci.* **7**, 131–146.

Mori, T. and Tanaka, K. (1973) Average Stress in the Matrix and Average Elastic Energy of Materials with Misfitting Inclusions, *Acta metall.* **21**, 571–574.

MSC (1996) *MSC/PATRAN V. 6.2.* MacNeal–Schwendler Corp., Los Angeles, CA.

Pettermann, H.E. (1997) *Derivation and Finite Element Implementation of Constitutive Material Laws for Multiphase Composites Based on Mori–Tanaka Approaches.* Doctoral Thesis, Vienna University of Technology.

Plankensteiner, A.F., Böhm, H.J., Rammerstorfer, F.G., Buryachenko, V.A. and Hackl, G. (1997) Modeling of Layer-Structured High Speed Steel, *Acta mater.* **45**, 1875–1887.

Suquet, P. (1997) Effective Properties of Nonlinear Composites, in P. Suquet (ed.) *Continuum Micromechanics*, Springer–Verlag, Vienna, pp. 197–264.

Weeton, J.W., Peters, D.M. and Thomas, K.L. (1987) *Engineer's Guide to Composite Materials*. ASM, Metals Park, OH.

Xie, M. and Adams, D.F. (1996) A Nonlinear Finite Element Analysis for Composite Materials, *Fin. Elem. Anal. Design* **22**, 211–223.

GLOBAL NON-LINEAR BEHAVIOUR OF PERIODIC COMPOSITE MATERIALS

B.A. SCHREFLER, U. GALVANETTO, C. PELLEGRINO
Department of Constructions and Transportations, University of Padua
Via Marzolo, 9 - 35131 Padova, Italy
F. OHMENHÄUSER
Institute for Statics and Dynamics of Aerospace Structures (ISD)
Pfaffenwaldring, 27-University of Stuttgart, Germany

Abstract

This paper aims to define a homogenised constitutive relation for the global behaviour of periodic composite structures in the case of non-linear material components. Special emphasis is put on the description of the generality of the algorithm which, in principle, can be applied to any kind of nonlinear material behaviour affecting the representative volume element. The method is currently restricted to plane situations with monotonic proportional loading.

1. Introduction

Composite materials find a large variety of applications in modern structural and mechanical engineering. Buildings, cars, ships, aircrafts almost any modern human construction is, at least partially, made of some composite material. In many situations these materials present a *periodic* structure: a composite material is called (spatially) *periodic* if it is possible to decompose it in a large number of equal elementary components or *cells of periodicity*. The characteristic size of the single cell of periodicity is assumed much smaller than the geometrical dimensions of the structure which is therefore composed by a large number of cells.

If we consider a periodic body Ω and a generic mechanical or geometric property a (for example the constitutive tensor) we can write:

$$\text{if } \boldsymbol{x} \in \Omega \text{ and } (\boldsymbol{x} + \boldsymbol{y}) \in \Omega \Rightarrow a(\boldsymbol{x} + \boldsymbol{y}) = a(\boldsymbol{x}) \tag{1}$$

In Eq. (1) $\boldsymbol{y}$ is the (*geometric*) *period* of the structure, i.e. the smallest vector for which equation (1) is true. It is apparent that a detailed finite element discretization of such kind of structures would require an enormous computational cost. To reduce this computational effort some homogenisation techniques were introduced, with acceptable results, to solve linear problems [1], [2].

In the study of the global properties of a composite medium two different length scales can

H. A. Mang and F. G. Rammerstorfer (eds.),
IUTAM Symposium on Discretization Methods in Structural Mechanics, 265–272.

be considered [3]: the *macroscopic* scale, relative to the whole structure, in which the dimensions of the heterogeneities are very small, and the *microscopic* scale, relative to the single cell of periodicity, which is the scale of the heterogeneities.

Hence, displacement, strain and stress distributions, in heterogeneous media, can be treated as the sum of two contributions: an averaged term, regarding the global structure, and a periodic fluctuating contribution (with zero mean value) regarding the local heterogeneity. Linear problems can be solved by means of the application of the principle of superposition of the effects through an asymptotic expansion of the displacement, strain and stress fields in the two scale variables. In the numerical applications only the first two or three terms of the expansion are usually taken into account [4].
This approach, based on the principle of superposition, is not possible in the case of non-linear problems. This paper deals with the definition of a homogenised constitutive relation for the global *non-linear behaviour of periodic composite bodies.* The proposed method allows the description of a large class of different constitutive behaviours, two of which are taken into account:

- elasto-plastic components with perfect bonding
- elasto-plastic components with the possibility of relative slip between different components.

The method is currently restricted to plane situations with monotonic proportional loading but we believe that it can be extended to more generic situations.
It should be noted that the present approach clearly differs from those already presented in literature [5], [6], [7], [8].

2. Basic Assumptions

The homogenised (or macroscopic, or effective, or global) constitutive law is determined by investigating the mechanical behaviour of the *representative volume element (RVE)* which, in the case of periodic composite structures, is the cell of periodicity. Such a RVE plays, in the mechanics of composite materials, the same role as the classical elementary volume of continuum mechanics, therefore a homogenised approach cannot be successful if the size of the cell is not small compared to that of the structure. In a macroscopic point X it is usual to distinguish macroscopic (or global) quantities, which are characteristic of the homogeneous equivalent medium and microscopic (or local) quantities which exist in the real heterogeneous material. In the following *microscopic* quantities are indicated with small letters while capital letters are used to indicate *macroscopic* quantities. In particular:
σ_{ij} = microscopic (or local) stress tensor
e_{ij} = microscopic (or local) strain tensor
Σ_{ij} = macroscopic (or global) stress tensor
E_{ij} = macroscopic (or global) strain tensor
It results from classical arguments on oscillating functions that the macroscopic stress and strain tensors must be the averages of the microscopic corresponding quantities:

$$\Sigma_{ij} = \frac{1}{V}\int_V \sigma_{ij}\, dV \qquad (2.a) \qquad\qquad E_{ij} = \frac{1}{V}\int_V e_{ij}\, dV \qquad (2.b)$$

where V is the volume of the representative volume element.
A homogenisation technique allows to define a constitutive relation between the global stresses and the global strains by using Eqs. (2), the constitutive relations of the single components, the interface properties between the components and the geometrical shape of the RVE. Such a constitutive relation can be assumed as the *constitutive law of the homogenised medium*.

3. Numerical Constitutive Law

In this section we outline the steps which allow the definition of a 'numerical' constitutive law for the global behaviour of the composite [9], [10]. The method is not dependent on the constitutive laws of the single components and therefore seems to be applicable to a large variety of different cases. We also observe that the method can be applied to composite materials statistically homogeneous and not necessarily periodic. The only requirement is that the RVE be a prismatic volume. The numerical construction of the global constitutive law is presented in the case of small strains and plane stress for $E_{12} = 0$, but it is clearly applicable to the full planar case.
Following the unit-cell approach [11] the cell of periodicity is part of an infinite periodic medium among cells which deform in the same manner.
We call E_{11} and E_{22} the global strains imposed to the cell. Each loading path is characterised by a fixed ratio E_{11}/E_{22}. Starting from a cell in a natural state, i.e. in which no strain and no stress are present, a large elastic step is firstly applied to reach the global linear elastic frontier, then a large number of smaller kinematic increments which generate considerable plastic deformations in the cell are applied. At the end of each step the homogenised stresses Σ_{11} and Σ_{22} are computed, by means of Eq. (2), and plotted for a fixed ratio E_{11}/E_{22} and for each step of the load history. Therefore we have one point, in the stress space, for each load step. These points are called *interpolation points*: here the behaviour of the homogenised material (and precisely the value of the homogenised strains E_{11} and E_{22} and of the corresponding stresses Σ_{11} and Σ_{22}) is known.
Repeating the procedure for several different values of the ratio E_{11}/E_{22} we know the behaviour of the homogenised material in a discrete number of points in the stress space and for a discrete variety of load situations.
At this point we introduce a simplifying hypothesis: we assume that all the interpolation points, on different loading paths, characterised by the same number of steps, are on the same plastic surface, i.e. they are labelled by the same value of an internal variable k. In this manner by connecting points relative to the corresponding steps of different loading paths it is possible to construct a series of plastic surfaces generated by the numerical experiments [9], [10] (see Figure 1).
Remark:
In the case of *full plane stress* ($E_{12} \neq 0$) the imposed displacements would be linear instead of constant along the edges of the cell and the paths of homogenised stresses would be three-dimensional curves in the stress space.
Observing the loading paths we note that outside the linear elastic region the material conserves the capability of supporting increasing stresses, therefore, in general, the homogenised material exhibits some kind of hardening behaviour (even if the single

components are perfectly plastic). The behaviour of the composite is linear up to the most interior curve, after reaching it a flow rule and a hardening law have to be defined in some way. Each point in the plane Σ_{11}-Σ_{22} in the elasto-plastic range, can be indicated by the values of two variables: a value of k and a value of the ratio E_{11}/E_{22}. We use such values to extract a flow rule and a hardening law for the effective material. Since the stress in the material components is limited so also the global stress has to be contained in a region of the space Σ_{11}-Σ_{22}. For the homogenised elasto-plastic material this *consistency condition* is given by:

$$\Sigma(\boldsymbol{x}) \in \mathrm{P}^{\mathrm{eff}}(\boldsymbol{x}) \qquad \text{where} \qquad \mathrm{P}^{\mathrm{eff}}(\boldsymbol{x}) = \{\Sigma \mid f(\Sigma) \leq \Sigma_0(\boldsymbol{x})\} \tag{3}$$

is the effective plastic domain:
$\Sigma_0(\boldsymbol{x})$ is the global yield stress which is a function of the position $\boldsymbol{x}$ and of the stress state in the stress plane. Once the function $f(\Sigma)$ has been defined, the yield stress $\Sigma_0(\boldsymbol{x})$ is known at the interpolation points and its value, in a generic point of the plane, is obtained by simple linear interpolation among the four points at the corners of the region (*patch*) where the current stress state lies (see Figure 2).
The *flow rule* is written as:

$$\boldsymbol{R}{:=}\ -\Sigma^{tr}+\Sigma+\dot{k}\,\boldsymbol{Dm}=\boldsymbol{0}, \quad \dot{k} \geq 0 \tag{4}$$

where:

$\boldsymbol{D}$ = elastic constitutive tensor obtained with the elastic homogenisation [1]
$\dot{k}$ = increment of the plastic flow
$\boldsymbol{m}$ = 'flow direction'
Σ^{tr}, Σ = global stresses

It is clearly possible to evaluate the 'flow direction' at each interpolation point, in fact (see Figure 2) starting from any interpolation point at the level $(n\text{-}1)$ it is possible to multiply the fixed strain increment $\Delta\boldsymbol{E}$ by the elastic effective tensor, in this way the trial stress Σ^{tr} can be computed. Since the material in some part of the unit cell has undergone some plastic deformation the strain $\Delta\boldsymbol{E}$ generates the new stress Σ^{n} which can be utilised to compute the flow direction at the interpolation point at the level $(n\text{-}1)$:

$$\dot{k}\,\boldsymbol{Dm} = \Sigma^{tr} - \Sigma^{n} \Rightarrow \boldsymbol{m} = \frac{1}{\dot{k}}\boldsymbol{D}^{-1}\,[\Sigma^{tr} - \Sigma^{n}] = \frac{1}{\dot{k}}\,[\Delta\boldsymbol{E} - \boldsymbol{D}^{-1}\,\Sigma^{n}] \tag{5}$$

It is worth to observe that since the value of $\dot{k}$ has been arbitrarily chosen to be 0.1, the quantity $\boldsymbol{m}$ does not give only an information about the flow direction but also about the entity of the plastic flow.

Remark:
For $\boldsymbol{m}$ the interpolated value is taken and not the derivative of the consistency condition, so that associative and non-associative plasticity can be taken into account.

Once consistency condition and flow rule have been fully defined, the elasto-plastic problem can be solved using the Newton-Raphson algorithm and the Euler backward method.

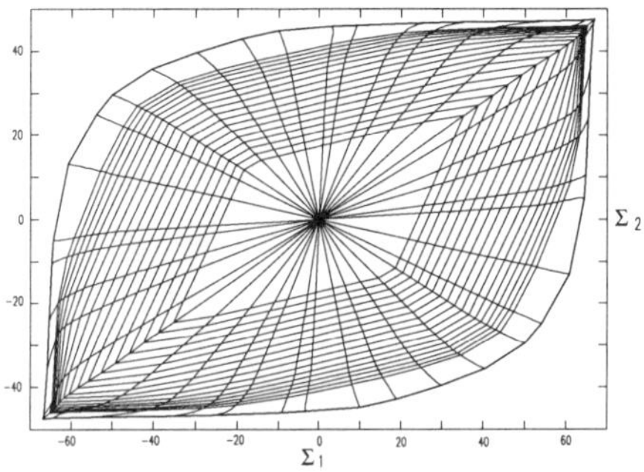

Figure 1. Grid of interpolation points and plastic surfaces

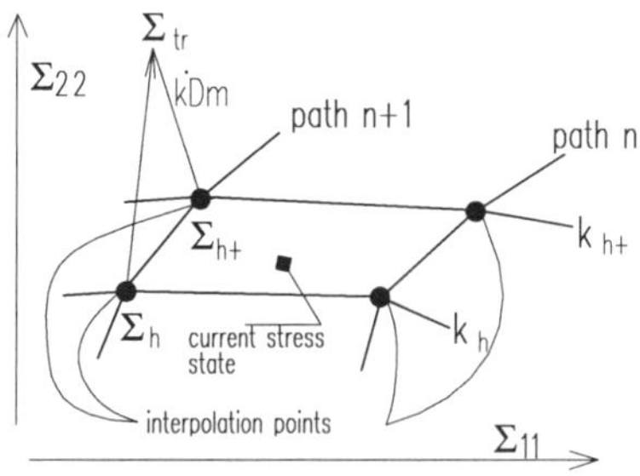

Figure 2. Interpolation of the flow direction.

4. Numerical Application

Figure 3 shows schematically the cross section of a group of strands in cable in conduit superconductors. Cable in conduit multistage high current superconductors can contain a great quantity of individual strands in a multistage bundle where each strand is supported by repeated cross-overs with other strands. Due to the complex geometry it is a very heavy task to carry out a three-dimensional discretisation of the whole conductor considering the single wires, hence homogenisation will be used to study the global behaviour of the bundle of strands. The plane stress case, which is taken into consideration here, cannot represent the fully 3D behaviour of the structure but it posseses some key features of it as the possibility of relative tangential motions.

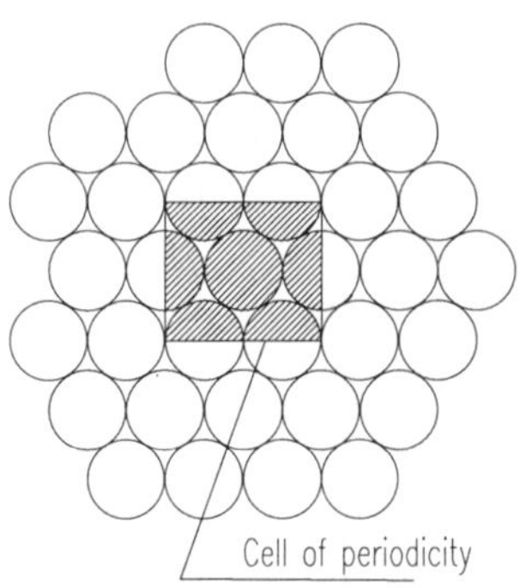

Figure 3. Periodic structure composed by strands

The superconducting strands are composed of bronze, copper and Nb_3Sn so that the mechanical characteristics of the superconductor are:

E=128000 MPa, ν=0.34, σ_Y=80 MPa

It is also necessary to model the behaviour of the strands at their interface.
Following [12], in the present work the condition for stick-slip is considered to be dependent on the level of shear stress at the interface. If the shear stress is larger than a specified value, slip will occur. The present model may represent sliding controlled by a large residual compression. In these cases *slip is governed by the plastic straining of the interface*.

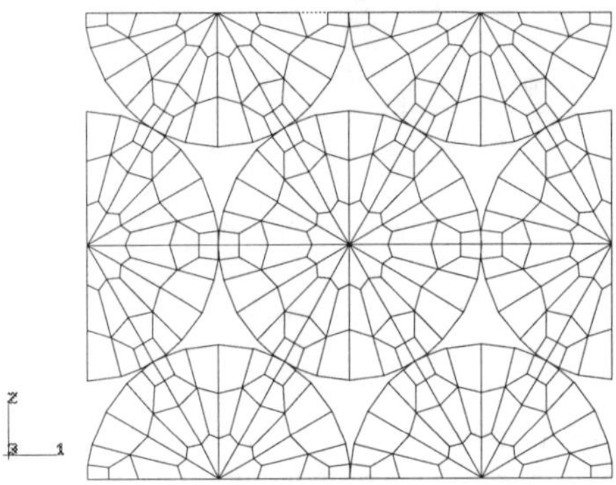

Figure 4. Discretisation of the cell of periodicity

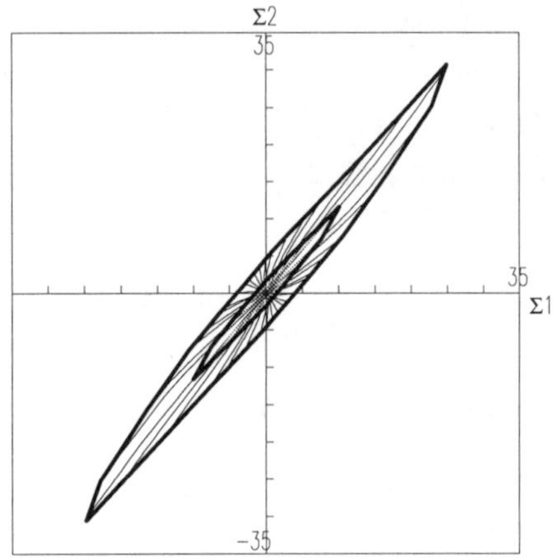

Figure 5. Yield surfaces and loading paths

The interface zone is considered as a thin bonded layer of perfectly plastic material between two phases of the composite. *No debonding* is permitted i.e. the layer of interface cannot be subjected to plastic strains which increase the thickness of the layer.
The perfectly plastic interfacial layer is characterised by a tensile yield stress σ_i. The yield stress in shear is then $\tau_i = \sigma_i / \sqrt{3}$. When the value of the shear stress in the interfacial layer is larger than τ_i "large" shear strain will occur. This plastic strain simulates the slip phenomenon between the constituents.
An acceptable simulation of the sliding strongly depends on the choice of the material constants and the thickness of the interfacial layer.
The cell of periodicity, indicated in Figure 3, is discretised with 4-node plane stress elements (Figure 4); following the procedure described in Section 3 we construct the "plastic" surfaces which, in this case, indicate different levels of sliding (i.e. of yielding of the interfacial layer). In Figure 5 we show the first plastic surface, the 25th plastic surface (after 25 non-linear kinematic steps) and the different loading paths.
To verify the attendibility of the procedure a group of superconducting strands subjected to uniform compression was anlysed. The vertical displacements of the bottom edge and the horizontal displacement of the right and left edge are restrained to zero. A constant distribution of vertical displacements is applied to the top edge.
The problem is solved using a rough discretization (48 nodes and 35 elements) with the numerical constitutive law (described in Section 3) which takes into account the non-linearities due to the elasto-plastic behaviour of the components and the stick-slip between

the strands.

The vertical reactions at the bottom edge are compared to those of an equivalent model with a finite element discretization which describes the real material distribution and the real mechanical characteristics of the single components. This discretization consists of 17963 nodes and 12565 elements. In Figure 6a, b, c we show the boundary conditions and the discretisation of the heterogeneous and of the homogenised model.

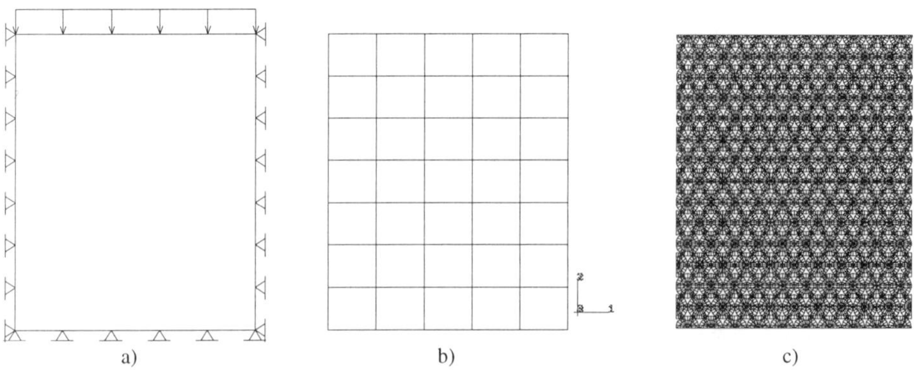

Figure 6 a,b,c. Boundary conditions (a), homogeneous model (b) and heterogeneous model (c)

In Figure 7 the slip in a single cell of the real deformed structure is shown.

The comparison between the vertical reactions in the two models is finally shown in Figure 8 (the reactions of the heterogeneous model are integrated along the restrained boundary in order to have quantities comparable with those obtained with the homogeneous model).

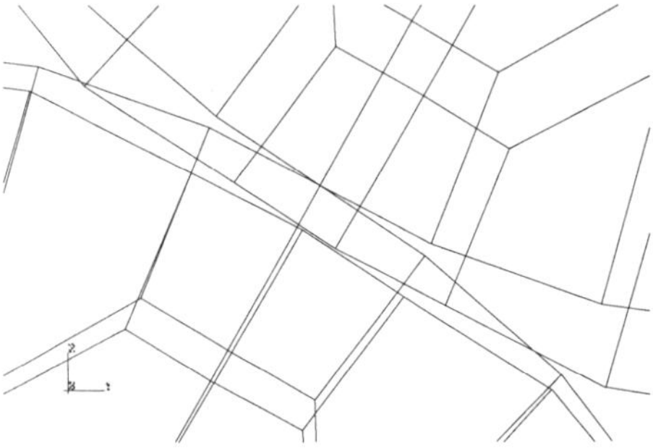

Figure 7. Slip between the strands

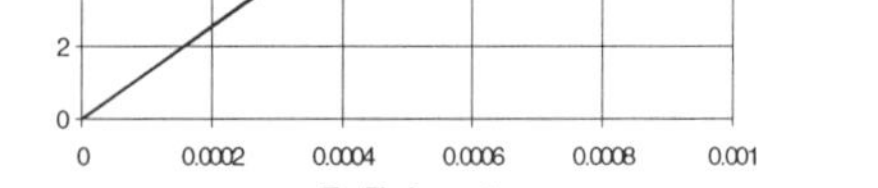

Figure 8. Comparison between the reactions

5. Conclusions

A homogenised constitutive relation for periodic composite media with non-linear material components has been defined. In particular, non-linearity due to elasto-plastic components

and to the possibility of slip between components are taken into account.
The method is valid for small strains and for plane problems with monotonic proportional loading.
Such a procedure allows a considerable reduction of the computational effort needed for finite element analyses of real composite structures.
The advantages are the following:
- the method closely follows the mechanical behaviour of the structure
- it is applicable both to associative and non-associative plasticity
- it allows to take into account sliding between different materials in a simple manner.

A drawback is the large number of numerical experiments required to obtain a sufficiently dense grid of interpolation points.
We believe that one of the main limits of the method presented in this paper is the hypothesis of monotonic proportional loading which greatly restricts the possibility of application of the proposed algorithm. Such a drawback can be overcome by means of the introduction of 'local' numerical experiments as proposed in [8].

6. References

1. Lefik, M. and Schrefler, B.A. (1994) Application of the homogenisation method to the analysis of superconducting coils, *Fusion Engng. Des.* **24**, 231-255.
2. Sanchez-Palencia, E. (1980) *Non-homogeneous media and vibration theory,* Springer Verlag, Berlin.
3. Suquet, P.M. (1985) Elements of homogenisation for inelastic solid mechanics, in Lecture Notes in Physics 272, *Homogenisation techniques for composite media*, Springer Verlag, Berlin.
4. Schrefler, B.A., and Lefik, M., Galvanetto, U., Correctors in a beam model for unidirectional composites, to be published on *Mech. Comp. Mat. Struct.*
5. Suquet, P.M. Effective properties of non-linear composites, *Lecture Notes of the CISM course on Continuum Micromechanics*, Udine, Italy, 2-6 September 1996.
6. Dvorak, G.J., Bahei-El-Din, Y.A. and Wafa, A. (1994) Implementation of the transformation field analysis for inelastic composite materials, *Computational Mechanics* **14**, 201-228.
7. Lourenco, P.B. (1995) *Computational strategies for masonry structures*, PhD Thesis Delft University of Technology.
8. Miehe, C. and Schoeder, J. (1997), Aspects of computational homogenisation analysis of polycristalline materials, in D.R.J. Owen, E. Onate and E. Hinton (eds.) *Computational Plasticity: Fundamentals and Applications*, CIMNE, Barcelona, 1004-1011.
9. Galvanetto, U., Ohmenhäuser, F. and Schrefler, B.A., A homogenised constitutive law for periodic composite materials with elasto-plastic components, to appear in *Composite Structures.*
10. Pellegrino, C., Galvanetto, U., and Schrefler, B.A., Computational techniques for periodic composite materials, in D.R.J. Owen, E. Onate and E. Hinton (eds.) *Computational Plasticity: Fundamentals and Applications*, CIMNE, Barcelona, 1229-1236.
11. Tvergaard, V. (1990) Analysis of tensile properties for whisker-reinforced metal-matrix composites *Acta metall. Mater.* **38**, 185-194.
12. Zahl, D.B. and McMeeking, R.M. (1994) The effect of interfacial properties on the flow strength of discontinuous reinforced metal matrix composites, *Mech. Comp. Mat. Struct.* **1**, 31-52.

DISCRETIZATION CONCEPTS FOR SOLID - SOLID PHASE TRANSFORMING MATERIALS

G. REISNER AND F. D. FISCHER
Institute of Mechanics
Montanuniversität Leoben
A-8700 Leoben, Austria

1. Introduction

The exploitation of the martensitic transformation (solid - solid phase transition) as a tool to improve the balance between strength and ductility becomes increasingly important for several materials, e.g. zirconia reinforced ceramics, shape memory alloys and steels exhibiting **TR**ansformation **I**nduced **P**lasticity (TRIP-steels). For the understanding of the effect of the phase transformation on the mechanical behavior of the material the interaction of the local stress state with the strain incompatibility caused by the phase change has to be investigated. If the phase transformation occures in the plastic regime of the material behavior it seems difficult to give analytic, explicit constitutive relations for the flow- as well as for the transformation behavior of the material. The finite element method allows to monitor the development of the local stress and strain states and to evaluate the stress dependent parts of the driving force of the phase transformation. The implementation of a thermodynamic transformation criterion to the finite element method also makes possible to follow the transformation process and thus to predict the kinetics of the solid-solid phase change.

2. Transformation criterion

Since we will deal with martensitic transformation in steels only, some simplifications may be made considering the experimental evidence at hand:

H. A. Mang and F. G. Rammerstorfer (eds.),
IUTAM Symposium on Discretization Methods in Structural Mechanics, 273–280.

- The mechanical framework uses a small strain approximation.
- The phenomenological crystallographic theory of the martensitic transformation by Wechsler, Lieberman and Read (WLR) (Wechsler *et al.*, 1953) proved to be valid in a variety of steels leading to a transformation eigenstrain tensor $\underline{\underline{\epsilon}}^{\mathrm{T}}$.
- The WLR theory assumes a kinematically coherent interface S (habit plane) between the martensite and the austenite. For the special microstructure of the martensite in the materials considered in this work, the interface is parallel to the habit plane nearly everywhere (except at the corners) of the martensitic layers.
- For the nucleation and at least for the early stage of growth of a martensite region the surface energy plays an important role. It determines the shape of the transforming region. In the cases studied in this work, the surface energy causes the material to form plates (bands) of martensite.
- As discussed by Nishyama, the temperature T shows no jump across the interface, $[\![T]\!] = 0$ (Nishiyama, 1978).

Starting from the thermodynamic force acting on a moving interface, F, (Eshelby, 1990; Roitburd, 1978; Kondaurov and Nikitin, 1986; Heidug and Lehner, 1985; Abeyaratne and Knowles, 1990; Abeyaratne and Knowles, 1993) a transformation criterion may be derived by comparison of the total energy balance for a specimen befor and after the transformation of a microregion V_μ:

$$\int_{V_\mu} \underline{\underline{\sigma}}|_{t_s} : \underline{\underline{\epsilon}}^{\mathrm{T}}\, dV + \int_{V_\mu} \rho \Delta\varphi_{\mathrm{chem}}\, dV = F_{\mathrm{c}} + \Delta\Gamma + W_{\mathrm{pl}}^{\tau} + W_{\mathrm{el}}^{\tau}. \tag{1}$$

For details of the derivation of Eq. 1 see Fischer 1997 or Fischer and Reisner 1997. The sum of the external stress state $\underline{\underline{\Sigma}}$ and the internal stress state $\underline{\underline{\sigma}}^{\mathrm{int}}$ yields the local stress state $\underline{\underline{\sigma}}$ at the beginning of the transformation of V_μ. The additional internal stress state $\underline{\underline{\tau}}$ represents the stress fluctuation introduced by the martensitic transformation of V_μ. $\int_{V_\mu} \underline{\underline{\sigma}}|_{t_s} : \underline{\underline{\epsilon}}^{\mathrm{T}}\, dV$ denotes the mechanical driving force (MDF). The chemical driving force (CDF) is given by $\int_{V_\mu} \rho\Delta\varphi_{\mathrm{chem}}\, dV$. The dragging force of the martensitic transformation comprise the critical thermodynamic force F_{c} necessary to move an interface, the increase in interfacial energy $\Delta\Gamma$, the plastic work due to $\underline{\underline{\tau}}$ only, $W_{\mathrm{pl}}^{\tau} = \int_V \int_{t_s}^{t_f} \underline{\underline{\tau}} : \dot{\underline{\underline{\epsilon}}}^{\mathrm{pl}}\, d\tilde{t}\, dV$, and the elastic strain energy due to $\underline{\underline{\tau}}$, $W_{\mathrm{el}}^{\tau} = \frac{1}{2} \int_V \underline{\underline{\tau}} : \underline{\underline{\underline{\underline{L}}}}^{-1} : \underline{\underline{\tau}}\, dV$. W_{pl}^{τ} and W_{el}^{τ} are calculated by integrating over the volume of the whole specimen and are, therefore, nonlocal entities reflecting the contributions of the nontransforming material surrounding V_μ to the transformation barrier. To facilitate the comparison of the LTC for varying volume fractions of newly formed martensite, V_μ, one can recast Eq. 1 by using energy densities instead:

$$MDF' + CDF' = F_c' + \Delta\Gamma' + (W_{\mathrm{pl}}^{\mathcal{T}})' + (W_{\mathrm{el}}^{\mathcal{T}})', \tag{2}$$

where X' denotes the density of entity X, i.e. its specific value per unit volume of transforming material.

For low alloyed TRIP-steels with a mean carbon content of 1.3 wt.% of the retained austenite (RA), room temperature is approximately 80 K above M_s (Andrews, 1965). The CDF' depends mainly on the temperature and on the chemical composition (carbon content). It consists of a contribution due to the difference in the chemical free energies of the *fcc* and the *bcc* crystal structure of pure Fe (Kaufman and Cohen, 1958) and of a second part accounting for the role of the carbon content (Nishiyama, 1978). In the RA grains a large number of martensite bands are formed so that even within the smallest volume of consideration, V_0, which represents a small part of a RA grain, several bands do exist. We therefore assume, that all grains show the same good amount of self accommodation of the stress free transformation strain. Thus, $(W_{\mathrm{pl}}^{\mathcal{T}})'$ and $(W_{\mathrm{el}}^{\mathcal{T}})'$ are considered not to change in the course of the strain induced martensitic transformation (SIMT). Since only one grain size of the RA is assumed, $\Delta\Gamma'$ can be treated as constant, too. Thus, at M_s (i.e. $MDF' = 0$) one can write:

$$CDF'|_{M_\mathrm{s}} = F_c' + \Delta\Gamma' + (W_{\mathrm{pl}}^{\mathcal{T}})' + (W_{\mathrm{el}}^{\mathcal{T}})', \tag{3}$$

and, therefore, the LTC at room temperature for the low alloyed TRIP steel reads:

$$MDF'|_{300\,\mathrm{K}} = CDF'|_{M_\mathrm{s}} - CDF'|_{300\,\mathrm{K}}. \tag{4}$$

The difference $CDF'|_{M_\mathrm{s}} - CDF'|_{300\,\mathrm{K}}$ depends on the carbon content (which alters also the M_s) and on the temperature. The value of $CDF'|_{M_\mathrm{s}} - CDF'|_{300\,\mathrm{K}}$ entering the model is calculated from (Nishiyama, 1978; Kaufman and Cohen, 1958) as 27.4 MPa. The influence of the carbon content on the yield stress and on the work hardening is larger for martensite than for austenite (Tamura and Wayman, 1992). The carbon content of the RA is very important for the stability against SIMT. It determines the chemical driving force, the stress free transformation strain (via its influence on the lattice parameters) and the flow behavior of the RA and the martensite.

An isothermal and an adiabatic process are considered as the two limiting cases in the simulation of a tensile test. In the adiabatic case, the latent heat of the martensitic transformation (MT) would cause a temperature rise inside the RA grain by 200 K if the grain transforms completely and no heat conduction is considered. An increase in temperature by 100 K reduces the chemical driving force by 43.9 MPa. Hence, the adiabatic case gives a lower bound for the transformation kinetics. The isothermal case yields the upper bound for the kinetics. The influence of the temperature on the mechanical properties is neglected. In both cases no heat conduction is considered in this model.

3. Geometrical model

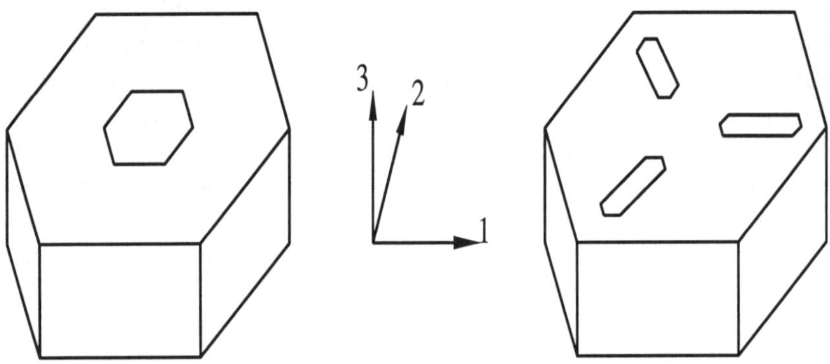

Figure 1. Sketch of the hexagonal unit cell with the inclusion-type (left) and the network type (right) of microstructure

Fig. 1 depicts the two unit cells used in the investigation of the load type sensitivity of SIMT. The unit cell with the single inclusion represents the inclusion-type of retained austenite (RA), whereas in the other unit cell three inclusions form a network like structure of RA grains. In both cases the inital volume fraction of RA is 0.1. On the surface of the hexagonal prism periodic boundary conditions are prescribed by enforcing that the displacements of all corresponding nodes on opposing faces of the prism differ by a constant displacement vector. The analysis with the finite element code ABAQUS (Hibbit *et al.*, 1995) is performed in a fully three-dimensional way to account for the 24 distinct martensitic variants. However, a plane geometrical model is used for the sake of simplicity. In direction 3 this representatve volume element (RVE) behaves like a parallel composite. All selected types of loading act in the transverse plane of the model (directions 1 and 2). Thus a slab of the material is modeled under generalized plane strain conditions. The same carbon content of 1.3 wt.% is assumed for all RA grains, whereas the lattice orientation varies for the single grains in the network type of microstructure.

4. Material model

Since all phases are elastically isotropic in this simulation, the lattice orientation alters the transformation strain with respect to the global coordinate system only. The tensile and the compressive loading acts in direction 2. The rolling process is simulated by a plane strain compression with compressive loading in direction 2 and no displacements of the surfaces in direction 3. The pure shear stress state (torsion) is accomplished by compressive loading in direction 1 and tensile loading in direction 2.

To calculate the amount of transformation in a certain load increment the elastic predictor-radial return method used in J_2-plasticity models is modified (Hibbit *et al.*, 1995; Daves, 1994). The strain increment associated with the load increment initially is interpreted as an elastic strain increment leading to the total elastic strain $\underline{\underline{\epsilon}}^{\mathrm{el}}_{\mathrm{trial}}$. The corresponding trial stress state can be either the new stress state at the end of the load increment in the case of an elastic increment or it will be relaxed by martensitic transformation and/or yielding. The martensitic transformation occurs approximately at one third of the speed of sound and it does not necessitate a dislocation movement (Verhoeven, 1975). Thus, the numerical algorithm considers first the martensitic transformation to reduce the trial stress state. If the equivalent stress is larger than the actual flow stress, a second radial return occurs by plastic deformation.

The *MDF* at the end of the martensitic transformation for $\underline{\underline{\Sigma}} + \Delta\underline{\underline{\Sigma}}$ depends on the increment of the volume fraction of martensite per unit volume of austenite, $\Delta\xi'$, formed in the actual load increment:

$$MDF' = (\underline{\underline{\epsilon}}^{\mathrm{el}}_{\mathrm{trial}} - \Delta\xi' \underline{\underline{\epsilon}}^{\mathrm{T}}) : \underline{\underline{\underline{\underline{L}}}} : \underline{\underline{\epsilon}}^{\mathrm{T}}, \tag{5}$$

where it is assumed that only the increment in the transformation strain alters the elastic trial strain state. $(\underline{\underline{\epsilon}}^{\mathrm{el}}_{\mathrm{trial}} - \Delta\xi' \underline{\underline{\epsilon}}^{\mathrm{T}}) : \underline{\underline{\underline{\underline{L}}}}$ denotes the local stress state at the end of the martensitic transformation, but prior to a possible stress relaxation by plastification. On the other hand, the LTC given by Eq. 1 has to be fulfilled for the MDF' at the end of the martensitic transformation at an external load of $\underline{\underline{\Sigma}} + \Delta\underline{\underline{\Sigma}}$. This leads to a linear expression for $\Delta\xi'$ ensuring that the TC in Eq. 1 is not violated at any time.

The transformation occurs only in the austenite having an actual relative volume fraction of $(1-\xi)$, where ξ is the relative amount of martensite in the partially transformed RA grain. Hence, the increment of the relative martensitic volume fraction, $\Delta\xi$, in the considered region V_0 can be written as:

$$\Delta\xi = (1-\xi)\frac{\underline{\underline{\epsilon}}^{\mathrm{el}}_{\mathrm{trial}} : \underline{\underline{\underline{\underline{L}}}} : \underline{\underline{\epsilon}}^{\mathrm{T}} - (CDF'|_{M_s} - CDF'|_{300\,\mathrm{K}})}{\underline{\underline{\epsilon}}^{\mathrm{T}} : \underline{\underline{\underline{\underline{L}}}} : \underline{\underline{\epsilon}}^{\mathrm{T}}}. \tag{6}$$

All phases are modeled with an isotropic, Ludwik-type of hardening behavior using J_2-plasticity (Ludwik, 1909):

$$\Sigma = \Sigma_0 + K \cdot \varphi^n \tag{7}$$

The mechanical properties assigned to the matrix phase are chosen to represent a mixture of ferrite and bainite and are similar to those of a ferritic/bainitic steel. The actual flow stress of the transforming particle is calculated from a linear rule of mixture.

The material input data are summarized in Table 4.

TABLE 1. Material input data for the numerical simulation of the SIMT in a low alloyed TRIP steel. Σ_0 denotes the yield strength. K and n are hardening parameters and ν is the Possion's ratio.

Phase	Σ_0 [MPa]	K [MPa]	n	ν	Young's Modulus [GPa]
Matrix	300	1000	0.2	0.3	210
Austenite	650	1200	0.4	0.3	210
Martensite	1100	1500	0.5	0.3	210

5. Results and discussion

There is a marked load case sensitivity on the transformation kinetics, which is more pronounced for the inclusion-type microstructure since in this case only one lattice orientation of the RA is present. The pure shear type of loading (torsion) yields the highest relative amount of martensite. For simple tensile and compressive loading the transformation kinetics does not differ significantly. However, plane strain compression (rolling) causes only little transformation. In the case of hydrostatic compression the model predicts no transformation at all. For the various stress states distinct martensite variants are predicted. For the network-type microstructure the SIMT under torsion and plane strain compression is calculated. The difference in these two loading cases is smaller than for the inclusion-type microstructure.

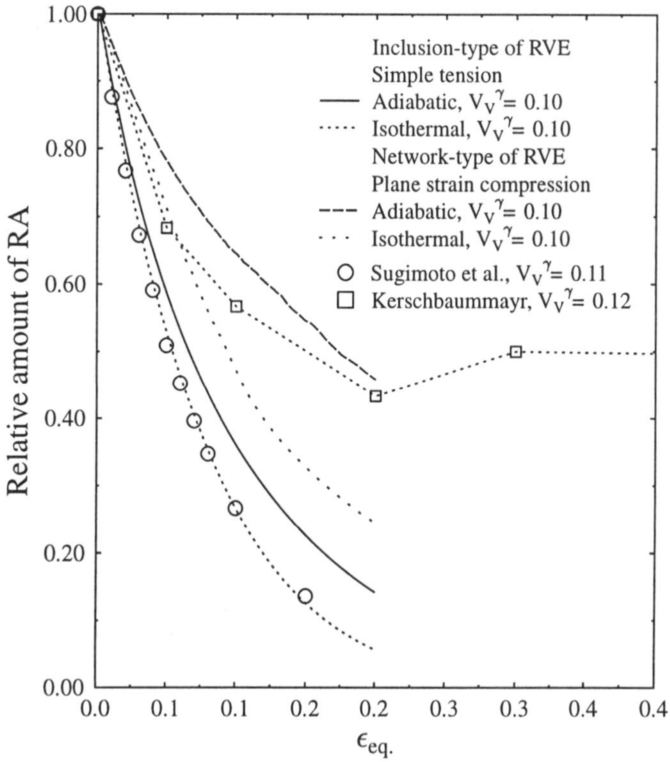

Figure 2. Comparison of experimental data with numerical simulation for isothermal and adiabatic transformation conditions.

The effect of the latent heat on the martensitic transformation is depicted in Fig. 2. The graph shows the relative volume fraction of RA for the isothermal and the adiabatic transformation conditions for tensile and plane strain compressive loading. Under isothermal conditions more martensite is formed than under adiabatic conditions at identical stress levels. The austenite content in a 15 CrNi 6 steel as a function of global strain (thickness reduction) is obtained by cold rolling experiments (Kerschbaummayr, 1996). Each data point corresponds to a rolling pass. The network-type of RVE is subjected to plane strain compression for the simulation of the rolling process. The measured volume content of austenite lies between the calculated curves of the RA volume fraction for the adiabatic and the isothermal conditions, respectively. The numerical results of the tensile test of the inclusion-type of microstructure are compared with the experimental values, taken from Sugimoto (Sugimoto *et al.*, 1992) for this type of loading. Sugimoto investigated a low alloyed TRIP steel alloyed with C, Si and Mn. The mean carbon content of the RA is 1.3 wt %, its initial volume fraction is 0.11. The flow behavior of the composite is close to the one of the matrix phase used in our calculations.

References

Abeyaratne, R. and Knowles, J. K. (1990) On the driving traction acting on a surface of strain discontinuity in a continuum, *J. Mech. Phys. Solids* **Vol. no. 38,** pp. 345-360.

Abeyaratne, R. and Knowles, J. K. (1993) A continuum model of the thermoelastic solid capable of undergoing phase transitions, *J. Mech. Phys. Solids* **Vol. no. 41,** (1993) 541-571.

Andrews, K. W. (1965) Empirical formulae for the calculation of some transformation temperatures, *JISI*, July 1965, pp. 721-727.

Daves, W. (1994) *Mikro- und Makromechanische Simulationen des Deformationsverhaltens von Stählen unter Berücksichtigung von Umwandlungs- und Diffusionsvorgängen.* VDI Verlag, Reihe 18, Nr. 141, Düsseldorf.

Eshelby, J. D. (1970) Energy relations on the energy-momentum tensor in continuum mechanics. In *Inelastic Behavior of Solids.* McGraw-Hill, New-York.

Fischer, F. D. (1997) Modelling and Simulation of TRIP in Elasto-plastic Materials. In *Mechanics of Solids with Phase Canges.* CISM courses and lectures no. 368., Springer Verlag, Wien, New-York (in print).

Fischer, F.D. and Reisner G. (1997) A Criterion for the Martensitic Transformation of a Microregion in an Elastic-Plastic Material, submitted to *Acta mater.*

Heidug, W., and Lehner, F. K. (1985) Thermodynamics of coherent phase transformations in nonhydrostatically stressed solids, *PAGEOPH* **Vol. no. 123,** pp. 91-98.

Hibbit, Karlsson and Sorensen (1995) *ABAQUS Version 5.5, Theory Manual,* Hibbit, Karlsson & Sorensen, Pawtucket, RI.

Ludwik, P. (1909) *Elemente der technischen Mechanik.* Julius Springer, Berlin.

Kaufman, L. and Cohen, M. (1958) The martensitic transformation in the iron-nickel system, *Journal of Metals*, October 1958, pp. 165-173.

Kerschbaummayr, P. (1996) *Wärmebehandlung und Umwandlungsverhalten von niedrig legierten TRIP-Stählen,* Master thesis, Montanuniversität Leoben.

Kondaurov, V. I., and Nikitin, L. V. (1986) Phase transition of the first kind in an elastic-viscoplastic medium, *Izv. AN SSSR, Mekhanika Tverdogo Tela* **Vol. no. 21,** pp. 130-139.

Nishiyama, Z. (1978) *Martensitic Transformations*, Academic Press, New-York.

Roitburd, A. L. (1978) Martensitic transformation as a typical phase transformation in solids, *Solid State Physics* Academic Press, New-York.

Sugimoto, K., Kobayashi, M. and Hashimoto, S. (1992) Ductility and strain-induced transformation in a high-strength transformation induced plasticity-aided dual-phase steel. *Met. Trans. A* **Vol. no. 23A,** (1992) 3085-3091.

Tamura, I. and Wayman, C. M. (1992) *Martensite*, ASM International, USA.

Verhoeven, J. D. (1975) *Fundamentals of Physical Metallurgy.* John Wiley and Sons, New-York.

Wechsler, M. S., Lieberman, D. S. and Read, T. A. (1953) On the theory on the formation of martensite, *Journal of Metals* **Vol. no. 197,** pp. 1503-1515.

STRATEGIES FOR THE FINITE ELEMENT ANALYSIS OF NONLINEAR COMPOSITE STRUCTURES

W. WAGNER AND F. GRUTTMANN
Institute for Structural Analysis
University of Karlsruhe (TH)
PO Box 6980
D–76128 Karlsruhe
Fed. Rep. of Germany

Abstract. Composite materials are often used in light weighted thin–walled structures. Accurate and efficient element formulations have to be used to describe the 2D– and especially the complicated 3D–stress state in thickness direction. In this paper we discuss different 2D- and 3D-models for composite structures.

1. Introduction

Composite materials are often used in light weighted thin–walled structures. Besides the global structural behaviour frequently local damage effects, e.g. delamination have to be considered. To study such problems the complicated 3D–stress state - especially in thickness direction - has to be investigated. Thus, in general a numerical calculation of the problem e.g. with finite elements is necessary. In this paper we want to discuss different finite element strategies to model thin composite structures. An overview on such strategies is given in Figure 1.

The first model uses classical shell elements based on the first order shear deformation theory with inextensible director vector which is sufficient for the description of the global behaviour of structures. Refined 6 or 7 parameter theories are applied if intersections occur or if thickness changes have to be taken into account. Secondly 3D-brick elements with an accurate numerical integration procedure can be used (linear elements with extensions or at least quadratic elements). Such formulations will be discussed later in detail. The third alternative formulation are brick–type shell elements

H. A. Mang and F. G. Rammerstorfer (eds.),
IUTAM Symposium on Discretization Methods in Structural Mechanics, 281–289.

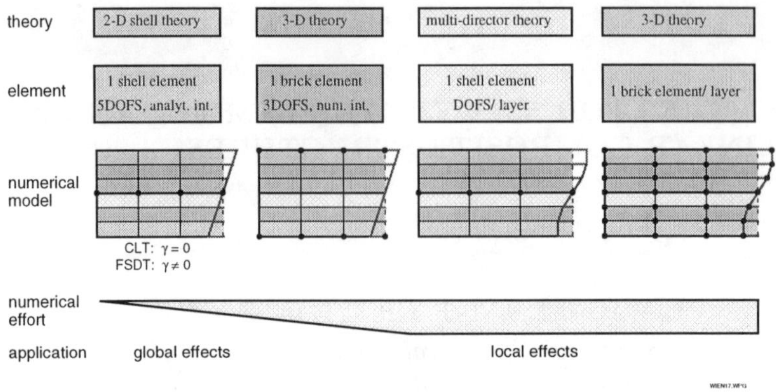

Figure 1. Finite element modeling of thin composite structures

where only displacements at bottom and top surface of the shell have been introduced, see e.g. Parisch [9].

Models which consider the local 3D–behaviour of thin composite structures are discussed in the following. The fourth method is the use of a shell-like 2D–modeling based on the introduction of a multi–director theory, see e.g. [1]–[3]. Here a director for each layer is introduced which allows the description of warping. If changes of thickness for each layer are possible, layerwise constant stresses in thickness direction can be described. 4-node elements are successfully used. Finally the fifth method is to apply a 3D–model, where for each layer at least one 3D–element is introduced. The basic element is a displacement based 8–node brick element which shows the well known poor behaviour for thin structures. A refined element behaviour can be achieved with tri–quadratic shape functions, but here on element level already 20 or 27 nodes are necessary. Other refinements are based on assumed strain methods, see e.g. [4], [5], and are introduced to improve especially the bending behaviour for thin beam and shell structures significantly.

In this paper, an overview on all this geometrically nonlinear formulations is given. Some examples demonstrate the range of applicability of the different models.

2. First Order Shear Deformation Theory

Elements based on first order shear deformation theories are quite standard. Theoretical and numerical details for such formulations can be found elsewhere, e.g. [7].

3. Multi–Director Formulation

The n **physical** layers of thickness h^j of the laminate can be described in the multi–director formulation with N **numerical** layers with thickness h^i. Thus each physical layer can be subdivided into several sublayers or vice versa several layers can be summarized to an equivalent numerical layer.
The initial geometry is described by an arbitrary reference surface and a normal vector as in standard shell theories. Thus the position vector $\mathbf{X}_0$ of the reference surface Ω is labeled with convective coordinates Θ^α. An orthonormal basis system $\mathbf{t}_k(\Theta^\alpha)$ is attached to this surface where $\mathbf{t}_3$ is the normal vector and Θ^3 the coordinate in thickness direction ($h_u \leq \Theta^3 \leq h_o$). The transformations between the different base systems is achieved using a proper orthogonal tensor $\mathbf{R}_0$

$$\mathbf{t}_k(\Theta^\alpha) = \mathbf{R}_0(\Theta^\alpha)\,\mathbf{e}_k\,. \tag{1}$$

Introducing a displacement field $\mathbf{u}(\Theta^i)$ the position vectors of the reference and the current configuration are given by

$$\begin{aligned} \mathbf{X}(\Theta^\alpha,\Theta^3) &= \mathbf{X}_0(\Theta^\alpha) + \Theta^3\,\mathbf{t}_3(\Theta^\alpha) \\ \mathbf{x}(\Theta^\alpha,\Theta^3) &= \mathbf{X}(\Theta^\alpha,\Theta^3) + \mathbf{u}(\Theta^\alpha,\Theta^3)\,. \end{aligned} \tag{2}$$

For the displacement field $\mathbf{u}(\Theta^i)$ in shell space we assume a multiplicative decomposition with independent functions for the shape in thickness direction and functions defined on the reference surface of the shell. The displacement vector of the numerical layer i is interpolated through the thickness by, see Ref. [1]

$$\begin{aligned} \mathbf{u}(\Theta^\alpha,\Theta^3) &= \sum_{l=1}^{m} \phi_l^i(\Theta^3)\,\bar{\mathbf{u}}_l^i(\Theta^\alpha) = \mathbf{\Phi}^i(\Theta^3)\,\bar{\mathbf{u}}^i(\Theta^\alpha) \\ \bar{\mathbf{u}}^i(\Theta^\alpha) &= [\bar{\mathbf{u}}_1^i, \bar{\mathbf{u}}_2^i, \quad \dots \quad, \bar{\mathbf{u}}_m^i]^T \quad (2 \leq m \leq 4)\,. \end{aligned} \tag{3}$$

The shape functions are arranged in a matrix

$$\mathbf{\Phi}^i = [\phi_1^i\mathbf{1}, \phi_2^i\mathbf{1}, \quad \dots \quad, \phi_m^i\mathbf{1}] \tag{4}$$

with hierarchical functions ϕ_l^i up to third order.
Thus eq. (3) allows warping of cross sections and changes of thickness.
The covariant base vectors of the reference configuration are introduced by $\mathbf{G}_i = \partial\mathbf{X}/\partial\Theta^i$ whereas the dual base vectors $\mathbf{G}^i$ are defined by $\mathbf{G}_i \cdot \mathbf{G}^j = \delta_i^{\,j}$. Accordingly, one obtains the convected base vectors $\mathbf{g}_i$ and $\mathbf{g}^j$ of the current configuration.

Top and bottom surface of the shell are loaded with loads $\hat{\mathbf{p}} = \hat{p}^k\, \mathbf{e}_k$, whereas body forces are neglected for simplicity. The principle of virtual work in a material description reads

$$\mathbf{g}(\mathbf{u}, \delta\mathbf{u}) = \int\limits_{(\Omega)} [\int\limits_{(\Theta^3)} \mathbf{S} \cdot \delta\mathbf{E}\, J d\Theta^3]\, d\Theta^2 d\Theta^1 - \int\limits_{(\Omega_\sigma)} \hat{\mathbf{p}} \cdot \delta\mathbf{u}\, d\Omega_\sigma = 0\,. \tag{5}$$

with $J = (\mathbf{X}_{,1} \times \mathbf{X}_{,2}) \cdot \mathbf{X}_{,3}$. Note that integration in (5) has to be performed considering the different material parameters and fiber angles through the thickness. The covariant components of the Green–Lagrangian strain tensor $\mathbf{E} = E_{kl}\, \mathbf{G}^k \otimes \mathbf{G}^l$ and the associated variation are obtained by partial differentiation of the position vectors with respect to the convective coordinates Θ^i

$$\begin{aligned} E_{kl} &= \tfrac{1}{2}(\mathbf{x}_{,k} \cdot \mathbf{x}_{,l} - \mathbf{X}_{,k} \cdot \mathbf{X}_{,l}) \\ \delta E_{kl} &= \tfrac{1}{2}(\delta\mathbf{x}_{,k} \cdot \mathbf{x}_{,l} + \delta\mathbf{x}_{,l} \cdot \mathbf{x}_{,k})\,. \end{aligned} \tag{6}$$

The work conjugate stresses S^{kl} are the contravariant components of the Second Piola–Kirchhoff stress tensor $\mathbf{S} = S^{kl}\mathbf{G}_k \otimes \mathbf{G}_l$. Each ply is considered as a homogeneous orthotropic medium, where the axes of orthotropy coincide with the material principal axes. Hence the stresses of the physical layer j are given by the material law

$$\mathbf{S} = \mathbf{C}\,\mathbf{E} \tag{7}$$

An associated 4–node isoparametric shell element can be described straight forward. Shear locking is avoided using a procedure similar to the so called 'Bathe–Dvorkin–Approach'. Details can be found in [2], [3].
With such elements the stress concentrations at free edges of plate and shell structures can be calculated. This is demonstrated with the following example of a cylindrical shell panel under single load. Discretized is one quarter of the shell by 25 × 25 elements with 6 numerical layers for a layup of [90°/0°/90°]. The geometrical and material data are

$$\begin{array}{lllllllll} R &=& 2540\,mm & \Theta &=& 0.1\,rad & \nu &=& 0.25 \\ b &=& 254\,mm & E_1 &=& 3300\,MPa & G_{12} &=& 660\,MPa\;. \\ h &=& 12.7\,mm & E_2 &=& 1100\,MPa & G_{23} &=& 450\,MPa \end{array}$$

In Figure 2 the system together with the stress distribution along line 1–2 for the shear stress S^{13} and the normal stresses S^{33} in the numerical layer 4 (upper part of physical layer 2) are shown.

4. Enhanced 8–Node Brick Element Formulation

Details of the following formulation of an enhanced 8–node brick element can be found in Ref. [4-6]. Here only the main ideas are pointed out. The

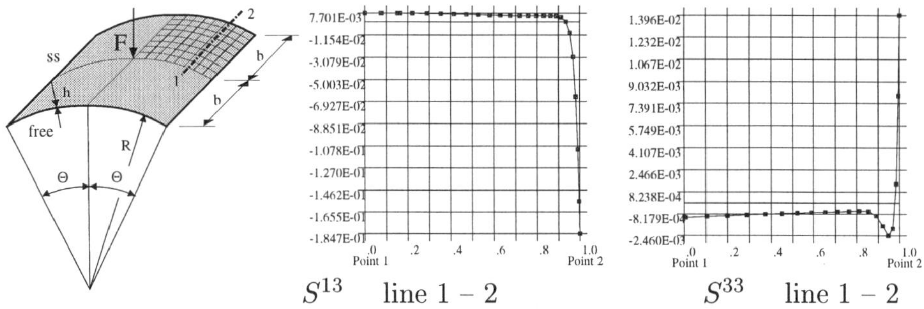

Figure 2. Cylindrical panel under single load

variational framework for the enhanced assumed strain method is the following three field variational functional (Lagrangian description)

$$\Pi = \hat{\Pi}(\mathbf{u}, \tilde{\mathbf{E}}, \mathbf{S}) = \Pi_{int}(\mathbf{u}, \tilde{\mathbf{E}}, \mathbf{S}) + \Pi_{ext}(\mathbf{u}) \ . \tag{8}$$

The three independent variables are the displacement field $\mathbf{u}$, the enhanced strain field $\tilde{\mathbf{E}}$ and the second Piola Kirchhoff stress tensor $\mathbf{S}$. The idea of the EAS–method is to enrich the compatible (nonlinear Green–Lagrangean) strains $\mathbf{E}^c$ with additional enhanced strains $\tilde{\mathbf{E}}$.

Based on the assumption that the material is homogeneous and hyperelastic the internal potential can be written as

$$\Pi_{int} = \int_{\mathcal{B}_0} [W_{0S}(\mathbf{E}^c + \tilde{\mathbf{E}}) - \mathbf{S} : \tilde{\mathbf{E}}] \quad dV \, , \tag{9}$$

where W_{0S} is the stored energy function of the initial reference configuration.

The first variation $\delta\Pi$ of the functional (8) is obtained via the directional derivative

$$\delta\Pi = D\Pi(\mathbf{u}, \tilde{\mathbf{E}}, \mathbf{S}) \cdot (\delta\mathbf{u}, \delta\tilde{\mathbf{E}}, \delta\mathbf{S}) \tag{10}$$

Furthermore the second variation $\Delta\delta\Pi$ can be derived, see Ref. [6].

The associated finite element formulation bases on standard isoparametric concepts. Here, coordinates, displacements and virtual compatible strains are introduced on element level (e) as

$$\mathbf{x}_e = \sum_{I=1}^{8} N_I \mathbf{x}_I, \quad \mathbf{u}_e = \sum_{I=1}^{8} N_I \mathbf{v}_I, \quad \delta\mathbf{E}_e^c = \sum_{I=1}^{8} \mathbf{B}_I \, \delta\mathbf{v}_I \ . \tag{11}$$

The enhanced strains are introduced using matrix notation, see e.g. Ref. [4]

$$\tilde{\mathbf{E}}_e = \frac{det \ \mathbf{J}_0}{det \ \mathbf{J}} \ \mathbf{T}_0^{-T} \tilde{\varepsilon}_e \tag{12}$$

with the Jacobians $\mathbf{J}$ and $\mathbf{J}_0 = \mathbf{J}|_{\xi=0,\,\eta=0,\,\zeta=0}$. For $\mathbf{T}_0 = \mathbf{T}_0(\mathbf{J}_0)$, see e.g. Ref. [5].

Subsequently, we define the enhanced strain field interpolation in isoparametric space on element level

$$\tilde{\varepsilon}_e = \mathbf{M}(\xi,\,\eta,\,\zeta)\,\boldsymbol{\beta}, \quad \boldsymbol{\beta} \in \mathcal{R}^{n_\varepsilon}\,. \tag{13}$$

Here $\boldsymbol{\beta}$ is the vector of internal strain parameters. The interpolation matrix $\mathbf{M}(\xi,\,\eta,\,\zeta)$ is derived using an orthogonality condition. Assuming $\delta\mathbf{S}_e = \delta\mathbf{S}_e^0$, thus element–wise constant interpolation, it holds

$$\int\limits_{B_e} \delta\mathbf{S}_e^0 \cdot \tilde{\mathbf{E}}_e\, dV = \int\limits_{B_e} \delta\mathbf{S}_e^0\, \frac{det\ \mathbf{J}_0}{det\ \mathbf{J}}\, \mathbf{T}_0^{-T}\, \tilde{\varepsilon}_e\, det\mathbf{J}\, d\xi\, d\eta\, d\zeta = 0\,. \tag{14}$$

Suppressing the constant values ($\delta\mathbf{S}_e^0$, $det\ \mathbf{J}_0$, $\mathbf{T}_0^{-T}$) the condition reduces to

$$\int\limits_{B_e} \mathbf{M}(\xi,\,\eta,\,\zeta)\,\boldsymbol{\beta}\, d\xi\, d\eta\, d\zeta = 0\,. \tag{15}$$

The interpolation matrix $\mathbf{M}$ has to satisfy the patch test (see e.g. Ref. [4].) We use the matrix $\mathbf{M}$ defined in Ref. [5] (there for the infinitesimal theory). $\mathbf{M}$ expands the whole strain field up to a complete trilinear field. Thus we end up for the enhanced strains

$$\tilde{\mathbf{E}}_e = \mathbf{G}_E(\xi,\,\eta,\,\zeta)\,\boldsymbol{\beta} = \frac{det\ \mathbf{J}_0}{det\ \mathbf{J}}\, \mathbf{T}_0^{-T}\, \mathbf{M}(\xi,\,\eta,\,\zeta)\,\boldsymbol{\beta}. \tag{16}$$

Furthermore we introduce the 2nd Piola Kirchhoff stress tensor and the rank four elasticity tensor from the stored energy function W_{0S} by

$$\hat{\mathbf{S}} = \frac{\partial W_{0S}}{\partial \mathbf{E}}, \qquad \mathbf{C} = \frac{\partial^2 W_{0S}}{\partial \mathbf{E} \partial \mathbf{E}}\,. \tag{17}$$

The definitions for $\hat{\mathbf{S}}$ and $\mathbf{C}$ are the same as for the multi–director theory. Based on the above mentioned orthogonality condition for the element stresses and the enhanced interpolations it is possible to eliminate the stress field $\mathbf{S}$ from the variational equations.

Finally in the associated finite element formulation the following vectors and matrices occur on element level:

$$\mathbf{f}_e^{int} = \int\limits_{\mathcal{B}_e} \mathbf{B}^T\, \hat{\mathbf{S}}\, dV \qquad \mathbf{f}_e^{ext} = \int\limits_{\mathcal{B}_e} \mathbf{N}^T\, \rho_0 \mathbf{b}\, dV + \int\limits_{\partial\mathcal{B}_e} \mathbf{N}^T\, \bar{\mathbf{t}}\, dA \tag{18}$$

$$
\begin{aligned}
\mathbf{K}_{Te} &= \int\limits_{\mathcal{B}_e} (\mathbf{B}^T\, \mathbf{C}\, \mathbf{B} + \mathbf{K}_G)\, dV & \mathbf{A}_e &= \int\limits_{\mathcal{B}_e} \mathbf{G}_E^T\, \mathbf{C}\, \mathbf{B}\, dV \\
\mathbf{L}_e &= \int\limits_{\mathcal{B}_e} \mathbf{G}_E^T\, \mathbf{C}\, \mathbf{G}_E\, dV & \mathbf{g}_e &= \int\limits_{\mathcal{B}_e} \mathbf{G}_E^T\, \hat{\mathbf{S}}\, dV
\end{aligned}
\tag{19}
$$

The initial stress matrix $\int\limits_{\mathcal{B}_e} \mathbf{K}_G\, dV$ can be derived from $\int\limits_{\mathcal{B}_0} \hat{\mathbf{S}} \cdot \delta\Delta\mathbf{E}^c\, dV$, see eq. (10). Finally we obtain the discrete system of equations:

$$
\begin{pmatrix} \mathbf{K}_{Te} & \mathbf{A}_e^T \\ \mathbf{A}_e & \mathbf{L}_e \end{pmatrix} \begin{pmatrix} \Delta\mathbf{v} \\ \Delta\boldsymbol{\beta} \end{pmatrix} = \begin{pmatrix} -\mathbf{f}_e^{int} + \mathbf{f}_e^{ext} \\ -\mathbf{g}_e \end{pmatrix}, \tag{20}
$$

where the internal degrees of freedom $\boldsymbol{\beta}$ can be eliminated on element level. Thus we end up with the generalized displacement problem.

Elements derived from the above given formulas can be introduced in two ways for composite structures. Firstly they can be used in a layerwise formulation, similar to the multi–director formulation. This is possible if at least one element is chosen for each layer. Secondly such elements can be presented in a layered formulation which means that we have one element in thickness direction. Hence integration is performed by a double isoparametric map. Details on this mapping procedure can be found in Ref. [10].

4.1. EXAMPLE FOR A LAYERWISE APPLICATION

The nonlinear behaviour of a clamped cylindrical shell panel of composite material under a uniform load for a cross ply $[0^o, 90^o]$ has been analyzed by Reddy ([8]) using a displacement finite element model based on the von Kármán equations. Geometrical and material data are:

$$
\begin{array}{lllllllll}
R & = & 2540 \;\; in & E_1 & = & 25 \cdot 10^6 \;\; psi & G_{12} & = & 0.5 \cdot 10^6 \;\; psi \\
a & = & 254 \;\; in & E_2 & = & 1 \cdot 10^6 \;\; psi & G_{13} & = & 0.5 \cdot 10^6 \;\; psi \\
h & = & 2.54 \;\; in & \nu & = & 0.25 & G_{23} & = & 0.2 \cdot 10^6 \;\; psi
\end{array}
$$

We run this example to compare the results of standard shell elements (derived for a finite rotation theory, see Ref. [7]) with the enhanced 3D–brick element, where we use two elements over the shell thickness, one for each ply). In addition results for 20–node elements are shown. The load deflection curves for the uniform load versus the center deflection are depicted in Figure 3 using 8*8 and 16*16 finite element meshes for one quarter of the shell.

Due to the boundary conditions (clamped along the edges) a certain number of elements is necessary to give accurate results, especially in the range where large changes in the center displacement at nearly constant external

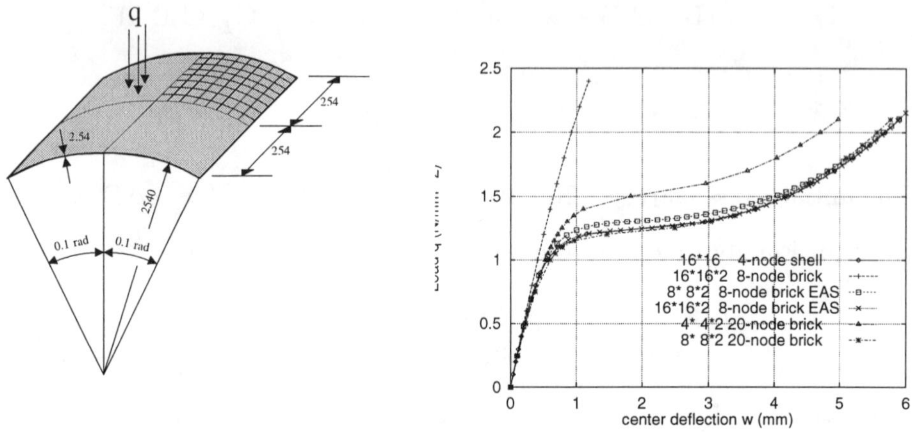

Figure 3. Load deflection curves for a shell panel under uniform load

loads occur. The results between shell elements and 3D–elements are in excellent agreement over the whole range of deformation.

4.2. EXAMPLE FOR A LAYERED APPLICATION

The following hyperboloidal shell with a cross ply $[0^o, 90^o]$ section has been calculated by Başar et. al. [11]. Own investigations with a simple finite rotation shell model, [7], confirm the given results. Here we show that a 3D–formulation leads to the same results. System and material data are given in Figure 4. The formula for the radius has to be extended due to the fact that we have to describe the inner and outer surface of the shell.

5. Conclusions

It has been shown that thin composite shell structures can be modeled using different type of finite elements due to the required accuracy.

6. References

1. Robbins, D. H. and Reddy, J. N. (1993) Modeling of thick composites using a layer–wise laminate theory, *Int. J. Num. Meth. Eng.* **36**, 655–677.

2. Gruttmann F. and Wagner, W. (1994) On the Numerical Analysis of Local Effects in Composite Structures, *Composite Structures* **29**, 1–12.

3. Gruttmann, F. and Wagner, W. (1996) Coupling of 2D– and 3D–Composite Shell Elements in Linear and Nonlinear Applications, *Comp. Meth. Appl. Mech. Eng.* **129**, 271-287.

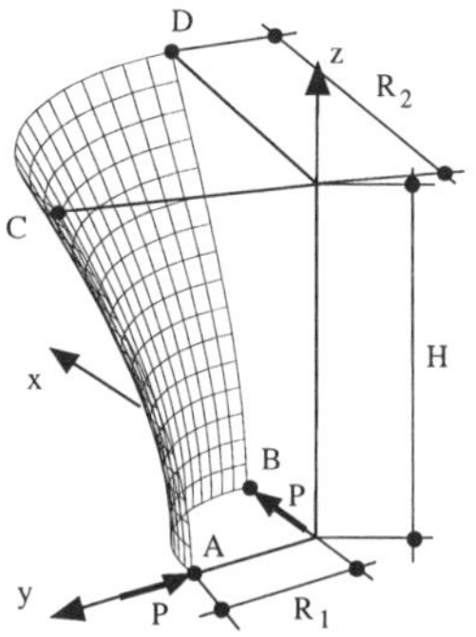

$$
\begin{aligned}
R_1 &= 7.5\ m & E_1 &= 40 \cdot 10^6\ N/mm^2 \\
R_2 &= 15\ m & E_2 &= 1.0 \cdot 10^6\ N/mm^2 \\
H &= 20\ m & \nu &= 0.25 \\
h &= 0.04\ m & G_{12} &= 0.6 \cdot 10^6\ N/mm^2 \\
h_i &= 0.013\ m & G_{23} &= 0.6 \cdot 10^6\ N/mm^2 \\
P &= 5\ kN & & \\
R(z) &= R_1 \frac{\sqrt{3}}{20} \sqrt{\left(\frac{400}{3} + z^2\right)} \ \boxed{\pm \frac{h}{2\,cos\alpha}} & & \\
\alpha &= arctan\left(R_1 \frac{\sqrt{3}}{20} \frac{z}{\sqrt{\left(\frac{400}{3}+z^2\right)}}\right) & &
\end{aligned}
$$

Figure 4. Hyperboloidal shell

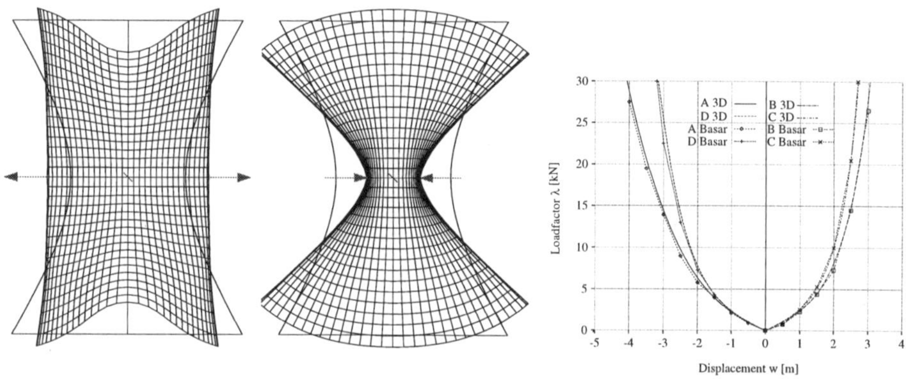

Figure 5. Hyperboloid at $\lambda = 30$ and load–deflection curves for A,B,C,D

4. Simo, J.C. and Rifai, D. (1990) A Class Of Mixed Assumed Strain Methods And The Method Of Incompatible Modes, *Int. J. Num. Meth. Eng.* **29**, 1595–1638.

5. Andelfinger, U. and Ramm, E. (1993) EAS-Elements for Two-Dimensional, Three-Dimensional, Plate and Shell Structures and their Equivalence to HR-Elements, *Int. J. Num. Meth. Eng.* **36**, 1311–1337.

6. Klinkel, S. and Wagner, W. (1998) A Geometrical Nonlinear Brick Element based on the EAS–Method, accepted for publication in *Int. J. Num. Meth. Eng.*

7. Wagner, W. and Gruttmann, F. (1994) A Simple Finite Rotation Formulation for Composite Shell Elements, *Engineering Computations* **11**, 145–176.

8. Reddy, J.N. and Chandrashekhara, K. (1985) Nonlinear Analysis of Laminated Shells including Transverse Shear Strains, *AIAA Journal* **23**, 440–441.

9. Parisch, H. (1995) A Continuum–Based Shell Theory for non–linear Applications, *Int. J. Num. Meth. Eng.* **38**, 1855–1883.

10 Klinkel, S. (1998) An effective geometrical nonlinear 3D–element with anisotropic composite material, to be published in *ZAMM*.

11 Başar, Y. , Ding, Y. , Schultz, R. (1993) Refined Shear–Deformation Models for Composite Laminates with Finite Rotations, *Int. J. Sol. Struct.* **30**, 2611–2638.

COMPOSITE BEAM DYNAMICS UNDER CONDITIONS OF INELASTIC INTERFACE SLIP

C. ADAM*, H. IRSCHIK**, F. ZIEGLER*
** Department of Civil Engineering, Technical University of Vienna*
Wiedner Hauptstr. 8-10/E201, A-1040 Vienna, Austria
*** Institute of Mechanics and Machine Design,*
Johannes-Kepler-University Linz, A-4040 Linz-Auhof, Austria

Abstract. The dynamic flexural behavior of symmetrically designed three-layer beams is analyzed under severe conditions of inelastic slip developments in the physical interfaces. Layerwise continuous and linear in-plane displacement fields are implemented, and as such, model both the global and the local inelastic response of laminate beams. By proper definition of an effective cross-sectional rotation the complex problem reduces to the simpler case of a homogenized shear-deformable beam, with effective stiffness and boundary conditions. Inelastic defects in the physical interfaces are equivalent to eigenstrains in an otherwise identical but elastic background structure of the homogenized beam with proper effective virgin stiffness. Mathematically, a multiple field approach in the elastic background results. Since the incremental response of the background is considered to be linear within a given time step, solution methods of the linear theory of flexural vibrations become applicable.

1. Introduction

Numerous laminate theories have been developed to describe approximately the kinematics and the stress states of composite structures. If the layers are connected continuously by means of strong adhesives, the mechanical assumption of rigid contact between the layers is reasonable. For that case, basically two classes of theories can be distinguished: the equivalent-single-layer theories and the layerwise laminate theories, [1]. The latter category is derived by admitting a separate displacement field within the individual layers of the composite, see e.g. [2], [3]. Alternatively, the extension of homogeneous beam and plate theories is based on one displacement expansion throughout the thickness of the laminate that results in equivalent-single-layer theories, e.g. [4], [5]. Such theories cannot accurately model laminates made of dissimilar material layers.

In some frequently used structures, such as in layered wood systems connected with nails or layers that are glued with weak adhesives, rigid bonds between the layers

H. A. Mang and F. G. Rammerstorfer (eds.),
IUTAM Symposium on Discretization Methods in Structural Mechanics, 291–298.

cannot be achieved. Due to relative deformation of the connection an interlayer slip occurs, that significantly can affect both strength and deformation of the layered structure. Assuming that the Bernoulli-Euler hypothesis holds for each layer separately, and a linear constitutive equation between the horizontal slip and the interlaminar shear force applies, a sixth-order initial-boundary value problem results. Linear analysis of layered beams with partial or flexible connection is well established in Refs. [6], [7], [8]. An extension to steel and concrete composite beams with a nonlinear shear force-slip relationship is undertaken in [9]. Alternatively, if the interlayer joints are modeled as additional layers, a proper layerwise laminate theory can be used to analyze such composites.

In the present paper an inelastic layerwise laminate theory, developed in [10], [11], is adopted to discuss the flexural response of symmetrically designed composite beam structures where viscoplastic slip may develop in the physical interfaces. According to the layerwise laminate theories, the governing equations are derived by the application of the Timoshenko theory of shear-deformable beams to each individual layer. The continuity of the transverse shear stress across the interfaces is considered by definition of the interlaminar shear stress, and hence, by means of the generalized Hooke's law. An effective cross-sectional rotation is introduced subsequently, which reduces the complex problem to the simpler case of an equivalent homogeneous beam with effective stiffness and with a corresponding set of boundary conditions. Furthermore, the dynamic response of the homogenized beam due to imposed inelastic strains in the physical interfaces is analyzed. Within a multiple field approach these additional fields of strain can be interpreted conveniently as eigenstrains acting in the background beam, [12]. The nonlinear problem turns incrementally in two linear problems, where the first one is simply the response of the associated linear elastic background structure to the given external loads. The second part accounts incrementally for the effects of the physical nonlinearities in the structure. Since the response in both cases is linear within a given time step, solution methods of the linear theory of flexural vibrations are applied.

2. Governing Equations

Symmetrically designed three layer composite beams of span l in principal bending about the y-axis are considered. Each layer has a constant rectangular cross-sectional area A_i and modulus of elasticity E_i. The adhesive physical interfaces are modeled as additional layers. The investigation is limited to materials, which exhibit the same behavior in tension and compression. The governing equations are derived by applying the assumptions of Timoshenko's theory of shear deformable beams to each individual layer. Consequently, the displacement field in the i-th layer is assumed to be of the form, [13],

$$u_i = u_i^{(0)} + z\,\psi_i\,, \quad w_i = w\,, \qquad i = 1, \dots, 5\,, \tag{1}$$

where u_i represents the horizontal displacement at vertical distance z from the central axis, $u_i^{(0)}$ is the portion of u_i at $z=0$, ψ_i denotes the cross-sectional rotation of the i-th layer and w is the lateral deflection common to all layer axes. The components $u_i^{(0)}$ $(i=1,2,4,5)$ can be expressed in terms of the cross-sectional rotations in order to satisfy the interface displacement continuity relations and the symmetry requirements, [3],

$$u_1^{(0)}=-u_5^{(0)}=u_2^{(0)}+(h_2+\tfrac{1}{2}h_3)(\psi_1-\psi_2)\ ,\ u_2^{(0)}=-u_4^{(0)}=\tfrac{1}{2}h_3(\psi_2-\psi_3)\ ,\ u_3^{(0)}=0\ ,$$
$$\psi_1=\psi_5\ ,\ \psi_2=\psi_4\ . \tag{2}$$

In Eqs. (1) and (2) the numbers 1 and 5 refer to the upper and lower face, respectively, 3 corresponds to quantities of the core, and 2 and 4 refer to the physical interfaces. h_2 and h_3 denote the thickness of the physical interfaces and of the core, respectively.

Within a geometrically linearized theory, linear strain-displacement relations apply, [10], [14]. The normal stress component σ_{zz} is neglected. The remaining stress components $\sigma_j\equiv\sigma_{xxj}$, $\tau_j\equiv\tau_{xzj}\equiv\tau_{zxj}$, of the interface layers are related to the strain and shear angle by the generalized Hooke's law, in general the rate form applies,

$$\dot{\sigma}_j=E_j(\dot{\varepsilon}_j-\dot{\bar{\varepsilon}}_j)\ ,\quad \dot{\tau}_j=G_j(\dot{\gamma}_j-\dot{\bar{\gamma}}_j)\ ,\qquad j=2,4\ , \tag{3}$$

where G_j is the shear modulus and the imposed fields $\bar{\varepsilon}_j$ and $\bar{\gamma}_j$ denote the inelastic strain and inelastic shear angle, respectively, which are assumed to be constant across the small thickness of the adhesive layers.

A simplified boundary value problem results by enforcing the shear stress continuity according to the generalized Hooke's law, [3], [10],

$$G_j(\psi_j+w_{,x}-\bar{\gamma}_j)=G_k(\psi_k+w_{,x})\ ,\qquad j=2,4\ ;\ k=1,3,5\ . \tag{4}$$

The equations of motion are derived following the lines of [10], [11]. Thereby, both, the longitudinal as well as the rotatory inertia are neglected, thus, limiting the analysis to the lower frequency band of structural dynamics. By definition of an effective cross-sectional rotation ψ_e, see [3], [10], the motion of the higher order problem of a composite beam with inelastic interface slip can be described in full analogy to the lower order engineering theory of a homogeneous shear-deformable background beam with effective flexural stiffness B_e, effective shear stiffness S_e and mass per unit length μ forced by the given loads q and m and by the two fields: the effective inelastic curvature $\bar{\kappa}_e$ and the effective inelastic shear angle $\bar{\gamma}_e$,

$$S_e(\psi_e+w_{,x})-B_e\,\psi_{e,xx}=m-B_e\,\bar{\kappa}_{e,x}+S_e\,\bar{\gamma}_e\ ,\ \mu\ddot{w}-S_e(\psi_{e,x}+w_{,xx})=q-S_e\,\bar{\gamma}_{e,x}\ . \tag{5}$$

Classical homogeneous boundary conditions are specified in analogy to the homogeneous shear-deformable beam, [10]. In (5), the effective properties and the effective eigenstrains are formally derived,

$$\mu=\sum_{i=1}^{5}\rho_i A_i\ ,\ B_e=\sum_{i=1}^{5}E_i C_i\ ,\ S_e=\frac{B_e}{\alpha}\delta\ ,\ \bar{\kappa}_e=\frac{2E_2C_2\bar{\varepsilon}_2}{B_e(h_2+h_3)}\ ,\ \bar{\gamma}_e=\frac{\beta}{B_e}\bar{\gamma}_2\ , \tag{6}$$

$$\delta=\kappa^2 G_3\sum_{i=1}^{5}A_i\ ,\ D_i=A_i z_{si}\ ,\ \beta=E_2(D_2h_3+2C_2)-2E_1D_1h_2\ ,$$

$$\alpha=G_3\sum_{i=1}^{5}\frac{E_i}{G_i}C_i+E_1D_1\left[h_3\left(\frac{G_3}{G_2}-1\right)+(2h_2+h_3)\left(\frac{G_3}{G_1}-\frac{G_3}{G_2}\right)\right]+E_2D_2h_3\left(\frac{G_3}{G_2}-1\right)\ . \tag{7}$$

C_i is the moment of inertia of the i-th layer with respect to the y-axis, and z_{si} denotes the vertical coordinate of the centeroid of the i-th layer. The factor κ^2 is a shear coefficient. The proper choice of its value is discussed in [13]. ψ_e is connected to the layer deformation by means of the total shear force and the bending moment,

$$Q=S_e(\psi_e+w_{,x}-\bar{\gamma}_e)\ ,\ M=B_e(\psi_{e,x}-\bar{\kappa}_e)\ , \tag{8}$$

relations that lead to the following expression for ψ_e, [3], [10],

$$\psi_e=\frac{1}{B_e}[\alpha\psi_3+(\alpha-B_e)w_{,x}]+\bar{\gamma}_e\ . \tag{9}$$

The coupled set of differential equations (5) is solved considering the actual boundary conditions for the deflection w and the effective cross-sectional rotation ψ_e in an incremental procedure. Subsequently, the cross-sectional rotations of the core and of the faces are to be determined. Decomposition of equation (9) yields the cross-sectional rotation of the core,

$$\psi_3=\frac{1}{\alpha}[B_e(\psi_e-\bar{\gamma}_e)-(\alpha-B_e)w_{,x}]\ . \tag{10}$$

The cross-sectional rotations of the faces and of the interlayer joints are calculated from Eq. (4),

$$\psi_1=\psi_5=\frac{G_3}{G_1}(\psi_3+w_{,x})-w_{,x}\ ,\ \psi_2=\psi_4=\frac{G_3}{G_2}(\psi_3+w_{,x})-w_{,x}+\bar{\gamma}_2\ . \tag{11}$$

3. Constitutive Relations

The rate dependent plastic strain distribution in the physical interfaces is determined by a viscoplastic law, which is similar to that of Perzyna, [15]. For plane stress (σ, τ) it reads,

$$\dot{\varepsilon}^p = \frac{2\,\Gamma}{k} \langle \Phi(F) \rangle \, \sigma \ , \quad \dot{\gamma}^p = \frac{12\,\Gamma}{k} \langle \Phi(F) \rangle \, \tau \ , \quad \Phi(F) = F^m \ , \tag{12}$$

with the yield surface

$$F = (\sigma^2 + 3\tau^2) / k^2 - 1 \ . \tag{13}$$

In Eq. (12), $\langle . \rangle$ denotes Macauley's bracket, Γ and m are constant viscosity parameters and k is the constant radius of the static yield surface. The equivalent eigenstrain distribution in the interlayer joints is obtained from the equivalence $\bar{\varepsilon}_2 \equiv \varepsilon^p$, $\bar{\gamma}_2 \equiv \gamma^p$ in the absence of ductile damage. The latter may be considered in a smeared manner using the concept of effective stress, [16], see also [17] and [18].

4. Discretization and Numerical Solution

In the present analysis the solution of the coupled set of equations of motion (5) is derived by superposition of two linear elastic contributions, $w = w^e + w^*$, $\psi_e = \psi_e^e + \psi_e^*$, where w^e, ψ_e^e are the deflection and effective cross-sectional rotation, respectively, of the elastic homogenized background beam due to the given loads q and m, while w^*, ψ_e^* are the deflection and effective cross-sectional rotation, respectively, produced in the homogeneous background by the imposed effective curvature $\bar{\kappa}_e$ and by the imposed effective shear angle $\bar{\gamma}_e$. The response w^e, ψ_e^e thus is evaluated in advance by the well-known procedure of linear elastodynamics. In simple cases that linear response of the linear elastic background beam is given in analytic form for the whole observation time. Since the distribution of imposed fields of eigenstrains in the background beam is not known in advance and depends on the current state of overall stress and strain, w^*, ψ_e^* have to determined incrementally by stepping the time and updating the strength of the effective eigenstrains iteratively in each time step.

In general, the solution of the equations of motion (5) exists in a modal form. This representation of the solution, and especially of its derivatives, is not desirable because the quasistatic response may contain singularities and discontinuities due to sudden changes in the load history, which are poorly modeled through a finite modal series approximation. Hence, a separate treatment of the quasistatic and of the complementary dynamic response gives higher numerical accuracy for the same number of summations. Further, the analysis is based on an integral equation formulation. The complementary dynamic increments of the deflection due to effective eigenstrains are obtained from the following integral equation, [10],

$$\Delta \overset{*}{w}_D(x) = \int_l \int_0^{\Delta t} [\, \tilde{M}_D^q(\xi,x,\Delta t-\bar{\tau})\, \Delta \ddot{\bar{\kappa}}_e(\xi,\bar{\tau}) + \tilde{Q}_D^q(\xi,x,\Delta t-\bar{\tau})\, \Delta \ddot{\bar{\gamma}}_e(\xi,\bar{\tau}) \,]\, d\tau\, d\xi + \Delta \overset{0\,*}{w_D}(x) \,. \quad (14)$$

In (14) the complementary dynamic Green's functions $\tilde{M}_D^q(\xi,x,\Delta t)$ and $\tilde{Q}_D^q(\xi,x,\Delta t)$ are the bending moment and the shear force in the cross-section at ξ and at time Δt that are produced by an instantaneous transverse unit force in x applied at time $\Delta t = 0$, respectively. The complementary dynamic Green's functions of undamped vibrations of such a homogeneous shear beam with effective properties are given by infinite modal series, see e.g. [10]. The function $\Delta \overset{0\,*}{w_D}$ represents the portion of the deflection referring to the initial conditions to be prescribed at the beginning of the time interval $\Delta t = 0$, at time instant t_a. Integral equation (14) is solved assuming a linear variation of the imposed effective eigenstrains $\Delta \bar{\kappa}_e$, $\Delta \bar{\gamma}_e$ within the time increment $\Delta t = t_{a+1} - t_a$.

In a first step, i.e. in the first time interval, the linear elastic increments Δw^e, $\Delta \psi_e^e$ due to the given loads are determined. The second step consists of the iterative computation of the increments of the effective eigenstrains $\Delta \bar{\kappa}_e$, $\Delta \bar{\gamma}_e$. The length of the physical interfaces equal to the length l is divided in elements and the discrete values of the strain increments $\Delta \varepsilon_2$, $\Delta \gamma_2$ are determined. The constitutive equations (12) render a first estimate of the eigenstrain increments $\Delta \bar{\varepsilon}_2^{(1)}$, $\Delta \bar{\gamma}_2^{(1)}$. Further, effective quantities of the eigenstrain distribution $\Delta \bar{\kappa}_e$, $\Delta \bar{\gamma}_e$ are determined elementwise, Eq. (6). Substituting these amounts into the corresponding solution equations, static part and Eq. (14), yields a first estimate $\Delta w^{*(1)}$, $\Delta \psi_e^{*(1)}$, and consequently the total deformation $\Delta w^{(1)} = \Delta w^e + \Delta w^{*(1)}$, $\Delta \psi_e^{(1)} = \Delta \psi_e^e + \Delta \psi_e^{*(1)}$. These results are the initial values for the second iterative step. The procedure is repeated and continued in time. The algorithm was found stable in many applications.

5. Illustrative Example

The proposed multiple field procedure is applied to a simply supported beam with rectangular cross-section composed of three layers and two adhesive interlayers. The eigenfrequencies and the normalized mode shapes of such a simply supported shear deformable beam with effective stiffness can be found e.g. in [10]. The formation of interface slip is modeled by considering the ductility of the adhesive layers in the form a material of the ideally elastic - perfectly viscoplastic type, Eq. (12), while, for simplicity's sake, the core and the faces are assumed to remain unlimited elastic. The numerical results are compared with those predicted when the adhesive interlayers behave elastic, likewise to the core material, in the applied range of loading. In all subsequent calculations the mechanical properties of the composite are characterized by the following parameters: longitudinal wave speed $\sqrt{E_j/\rho_j} = 5040\, m/s$, Poisson's ratio $\nu_j = 0.3$, ductility parameter $E_k/k_k = 1000$, viscosity parameters $\Gamma_k T_1 = 4.96 \cdot 10^{-4}$ and $m_k = 0.4$, and the ratios $E_j/E_3 = 23.0$, $\nu_j/\nu_3 = 0.75$, $\rho_j/\rho_3 = 1.94$, $E_3 = E_k$, $\nu_3 = \nu_k$, $\rho_3 = \rho_k$ $(j = 1,5; k = 2,4)$, and with $T_1 = 2\pi/\omega_1$ denoting the linear fundamental period. Prescribing Young's modulus of the faces by $E_1 = 68.9 \cdot 10^9\, N/m^2$ renders the

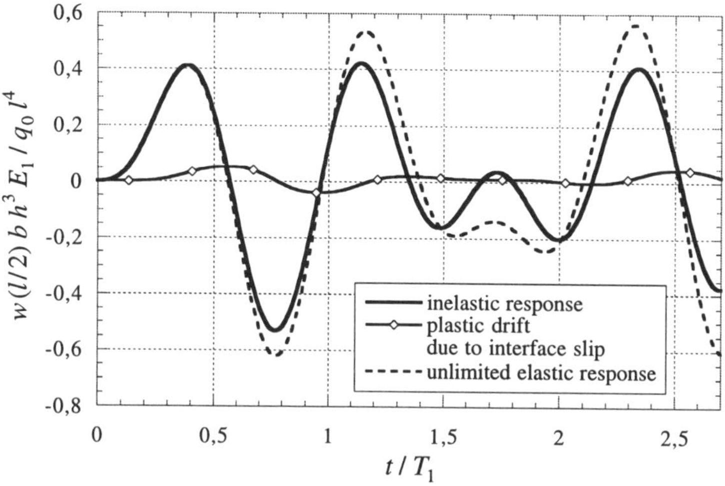

Figure 1. Lateral deflection and plastic drift at mid-span

foregoing parameters such that they refer to a laminate beam, whose faces consist of aluminum and the core of a thermoplastic material. The material of the interface layers has the same elastic stiffness as the core, however, with a much lower yield stress. The dimension of the beam is determined by the ratios $h_1 = h_5 = h/3$, $h_3 = 19h/60$, $h_2 = h_4 = h/120$, $b_i = b$ $(i = 1, .., 5)$, $h/b = 1.5$, and $l/h = 4$. The shear coefficient is chosen to be $\kappa^2 = 0.83$ so that the linear fundamental frequency of the equivalent shear deformable beam and that derived by the plane stress theory takes on the same value. At time instant $t = 0$ the composite is subjected to a distributed load according to a sine-half wave over the span, which varies harmonically in time, $q(x,t) = q_0 \sin(\pi x/l) \times$

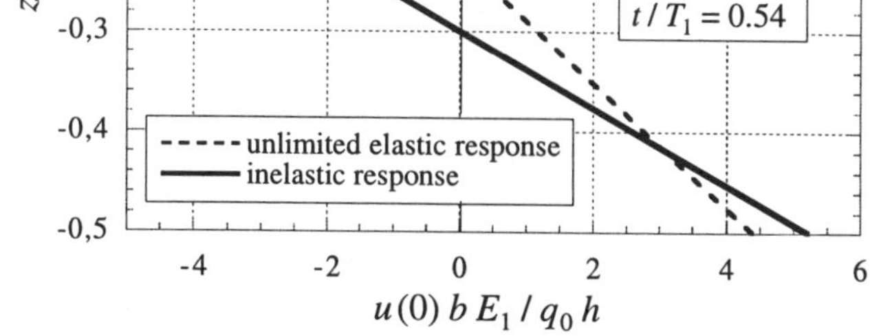

Figure 2. Horizontal displacement at $x = 0$ for specified time instant

$\times \sin \nu t$, $t \geq 0$. The ratio of excitation frequency versus linear fundamental eigenfrequency is chosen to be $\nu / \omega_1 = 1.58$, the amplitude of the applied force is determined by the non-dimensional ratio $\bar{q}_0 = q_0 \, l^2 / (E_1 \, b \, h^2) = 52.2 \cdot 10^{-4}$.

Figure 1 shows the normalized time evolution of the plastic drift that is due to the interface slip and of the inelastic mid-span deflection. For comparisons sake the unlimited elastic response is also included. Phase shift and damping by energy dissipation due to the inelastic slip can be observed. Also the distribution of the horizontal displacement at the boundary $x = 0$ (Figure 2) illustrates the jump-effect of the inelastic slip.

References

1. Reddy, J.N.: An evaluation of equivalent-single-layer and layerwise theories of composite laminates, *Composite Structures* **25** (1993), 21 - 35.
2. Swift, G.W., and Heller, R.A.: Layered beam analysis, *J. Engng. Mech. Div. ASCE* **100** (1974), 267 - 282.
3. Heuer, R.: Static and dynamic analysis of transversely isotropic, moderately thick sandwich beams by analogy, *Acta Mechanica* **91** (1992), 1 - 9.
4. Whitney, J.M., and Pagano, N.J.: Shear deformation in heterogeneous anisotropic plates, *J. Appl. Mech.* **37** (1970), 1031 - 1036.
5. Reddy, J.N.: A simple higher-order theory for laminated composite plates, *J. Appl. Mech.* **51** (1984), 745 - 752.
6. Goodman, J.R., and Popov, E.P.: Layered beam systems with interlayer slip, *J. Struct. Div., ASCE* **94** (1968), 2535 - 2547.
7. Girhammar, U.A., and Pan, D.: Dynamic analysis of composite members with interlayer slip, *Int. J. Solids Structures* **30** (1993), 797 - 823.
8. Adam, C., Heuer, R., and Jeschko, A.: Flexural vibrations of elastic composite beams with interlayer slip, *Acta Mechanica*, in press.
9. Mistakidis, E.S., Thomopoulos, K., Avdelas, A., and Panagiotopoulos, P.D.: On the nonmonotone slip effect in the shear connectors of composite beams, *Int. J. Eng. Analysis and Design* **1** (1994), 395 - 409.
10. Adam, C., and Ziegler, F.: Forced flexural vibrations of elastic-plastic composite beams with thick layers, *Composites Part B* **28B** (1997), 201 - 213.
11. Adam, C., and Ziegler, F.: Dynamic response of elastic-viscoplastic sandwich beams with asymmetrically arranged thick layers, in Y.A. Bahei-El-Din and G.J. Dvorak (eds.), Proc. IUTAM Symp. on *Transformation Problems in Composite and Active Materials*, Cairo, Egypt 1997, in press.
12. Irschik, H., and Ziegler, F.: Dynamic processes in structural thermo-viscoplasticity, *Appl. Mech. Rev. AMR* **48** (1995), 301 - 316.
13. Yu, Y.-Y.: *Vibrations of Elastic Plates*, Springer, New York, 1995.
14. Ziegler, F.: *Mechanics of Solids and Fluids*, 2nd ed., Springer, New York, 1995.
15. Perzyna, P.: The constitutive equations for rate sensitive plastic materials, *Quart. Appl. Math.* **20** (1963), 321 - 332.
16. Lemaitre, J.: *A Course on Damage Mechanics*, Springer, Berlin, 1992.
17. Fotiu, P.A., Irschik, H., and Ziegler, F.: Micromechanical foundations of dynamic plasticity with applications to damaging structures, in O. Brüller , V. Mannl and J. Najar (eds.), Advances in Continuum Mechanics, pp. 338 - 349, Springer , Berlin 1991.
18. Adam, C., and Ziegler, F.: Moderately large forced oblique vibrations of elastic-viscoplastic deteriorating slightly curved beams, *Arch. Appl. Mech.* **67** (1997), 375 - 392.

FINITE ELEMENT ANALYSIS OF DELAMINATION GROWTH

S. RINDERKNECHT AND B. KRÖPLIN
Institute for Statics and Dynamics of Aerospace Structures, University of Stuttgart, Pfaffenwaldring 27, D-70550 Stuttgart

Abstract

Different approaches for the analysis of delamination growth are discussed on the basis of a two-dimensional macroscale finite element model. Linear-elastic fracture mechanics and damage mechanics are compared which may be used in conjunction with two fundamentally different control methods in order to simulate delamination growth. Although the model does not allow an accurate representation of the local kinematics in the vicinity of the delamination front, it is well-suited for the prediction of qualitative delamination growth phenomena. Furthermore, it is discussed that even currently available three-dimensional macroscale models do not yet allow a general prediction of the quantitative delamination growth behaviour, due to the difficulty of incorporating complex micromechanical mechanisms.

1. Introduction

Delamination in composite laminates, which is characterized by the debonding of two adjacent laminae, represents a very frequently observed damage mode in commercial aircraft service [1]. Both stationary and growing delaminations can cause a dramatic reduction of structural stiffness and may even lead to the total failure of the component.

Although delamination growth is dominated by micromechanical failure mechanisms [2], real-life components may only be modeled on the macroscopic level due to the high computational effort. For the same reason, even three-dimensional macroscale models are so far only applied when delamination onset is investigated. Here, different approaches for the analysis of delamination growth are thus discussed using the two-dimensional finite

H. A. Mang and F. G. Rammerstorfer (eds.),
IUTAM Symposium on Discretization Methods in Structural Mechanics, 299–306.

element model [3] which is based on the Reissner-Mindlin plate theory. Although a consideration of the three-dimensional stress state in the vicinity of the delamination front is not possible in this macromechanical approach, qualitative delamination growth phenomena may be described rather well.

Owing to the fact that delamination growth preferentially occurs in brittle composites [4], (linear-elastic) fracture mechanics as well as damage mechanics are used to model the material failure.

2. Load-Deflection Analysis for Stationary Delaminations

So far, load-deflection analyses considering plane stationary delaminations are either performed with three-dimensional finite element models (e.g. [5, 6]) which require the usage of several elements over the laminate thickness, or on the other hand, two-dimensional models are applied (e.g. [7, 8, 9, 10]) which partly show significant restrictions with respect to delamination geometry (only elliptical delaminations), delamination buckling (only local buckling) and contact conditions between the delaminated laminate parts.

With the two-dimensional delamination process element (Fig. 1) used here, the entire structure may be modeled with only one element over the laminate thickness because the damage state can be specified in the material law of the process layer

$$\boldsymbol{\sigma} = (1 - s) \cdot \boldsymbol{C\varepsilon} \qquad \text{with} \quad 0 \leq s \leq 1 , \tag{1}$$

where $\boldsymbol{\sigma}$ denotes the stress vector, $\boldsymbol{\varepsilon}$ the strain vector and $\boldsymbol{C}$ the material stiffness matrix of the undamaged material. The damage parameter is then set to $s = 0$ in the undamaged area and to $s = 1$ in the delaminated area.

Furthermore, the above-mentioned restrictions do not have to be made for this model and contact may be considered between the baselaminate and the sublaminate. Sublaminate and baselaminate are modeled as Reissner-Mindlin plates while the process layer may represent either a continuum with linear shape functions in the direction of thickness t or an interface with $t \to 0$. Thus, the damage zone can be considered as three- or two-dimensional. The principal kinematics of this element are depicted in Fig. 2. For an undamaged plate, sublaminate (SL), baselaminate (BL) and process layer (PL) show independent rotations while the process layer additionally allows for a thickness change. In the case of a delaminated state, the stiffness of the process layer vanishes, resulting in a totally independent behaviour of baselaminate and sublaminate.

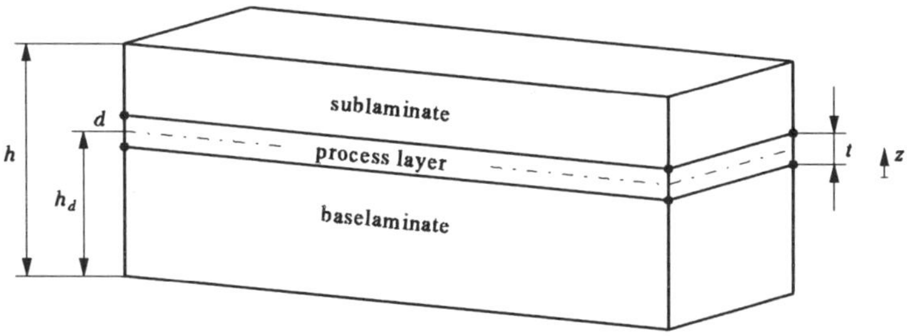

Figure 1. Delamination process element

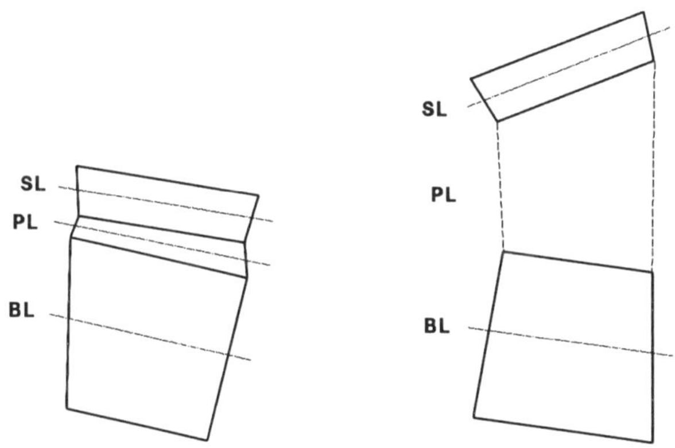

Figure 2. Element kinematics for undamaged and delaminated state

3. Analysis of Delamination Onset

The analysis of delamination onset represents a post-processing procedure of the load-deflection analysis for stationary delaminations.

In the framework of fracture mechanics, the modified crack closure method [11] is well-suited for the calculation of the energy release rate G:

$$-G = \frac{d\Pi}{dS} = \frac{\text{change of total energy (global)}}{\text{increase of crack surface } (dS > 0)} \tag{2}$$

Additionally, this method allows for the well-known mode separation ($G = G_I + G_{II} + G_{III}$) of technical fracture mechanics. This mode separation represents an engineering approach rather than a strict physical derivation because the original physical fracture mechanics theory [12] only covers the correlation of atomic bond forces with the critical total energy release rate G_c.

In conjunction with macroscale models, it is important to remark that the increasing crack surface is always assumed to be smooth, i.e. the surface roughness which can be observed fractographically is neglected. Owing to micromechanical mechanisms, the surface roughness may be severely influenced depending on material properties, stacking sequence and loading conditions. Hence, the actual crack surface used in the physical fracture mechanics theory may be significantly underestimated, leading to an overestimation of the energy release rates. This may explain the smaller critical energy release rate for the normal mode (I) in comparison with the shear modes (II, III) because experimental observations [2] indeed show significantly rougher surfaces for shear loading.

However, with the finite element used here, the mode separation leads to unreliable results because the local kinematics in the immediate vicinity of the delamination front cannot be handled accurately. Nevertheless, the total energy release rate G only depends on the global energy released by the entire structure and is thus not greatly affected by local errors along the front.

In the framework of damage mechanics, the micro cracks are related to the damage parameter s. The energy release rate Y depends on the local strains $\boldsymbol{\varepsilon}$ and is derived on basis of the thermodynamical potential:

$$-Y = \frac{\partial \psi_i(s, \boldsymbol{\varepsilon})}{\partial s} = \frac{\text{change of internal energy density (local)}}{\text{increase of damage parameter } (\partial s > 0)} \tag{3}$$

This damage mechanics energy release rate may also be used for the delamination onset analysis and even permits a mathematical exact correlation with the fracture mechanics energy release rate [13]:

$$G = \lim_{t \to 0} (t \cdot Y) \qquad (t = \text{width of assumed crack band}) \tag{4}$$

Both approaches may be applied using the presented finite element model. However, to avoid singularities, $t > 0$ has to be assumed and thus numerical calculations only allow an approximate correlation ($G \approx t \cdot Y$ for small t). In conjunction with the finite element method, a decreasing t requires an increasing mesh refinement in the vicinity of the delamination front. Thus, the direct calculation of the fracture mechanics energy release rate using the modified crack closure method is more efficient.

4. Simulation of Delamination Growth

The simulation of a growing delamination, i.e. the inclusion of the physical nonlinearity of material failure in the load-deflection analysis may be achieved in two fundamentally different ways.

In the front control method, the position of the delamination front is moved depending on the local loading state in order to find the primarily unknown new location of the delamination front. Thus, the delamination front represents a one-dimensional control domain (B1, see Fig. 3)

$$\begin{array}{ll} \text{Initial configuration of B1:} & f(x,y)=0 \quad \text{(1D-curve)} \\ & \Downarrow \ \text{change of loading state in B1} \\ \text{Configuration change of B1:} & \Delta f(x,y) \quad \text{(delamination growth)}\ . \end{array} \tag{5}$$

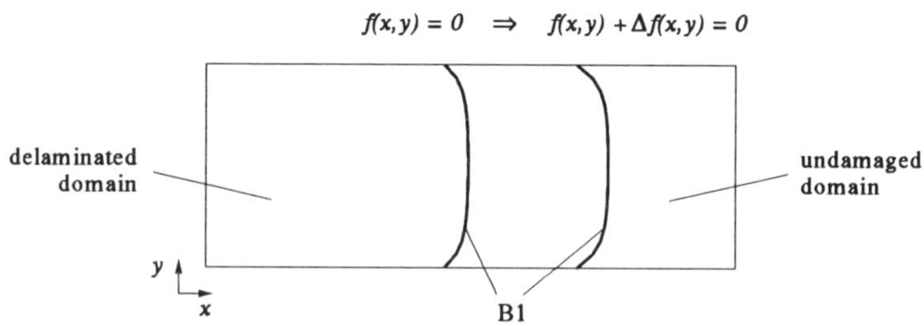

Figure 3. Front control method

This front control method may be realized by applying a moving mesh technique [14] which allows the application of both fracture mechanics and damage mechanics growth criteria. For damage mechanics approaches, the necessary mesh refinement zone is moved with the delamination front.

In the material control method, the idea is to start from the initial damage state and to change the damage parameters in the material points according to the local strain state in order to determine the unknown new distribution of the damage parameter. Thus, the whole plate surface represents a two-dimensional control domain (B2, see Fig. 4):

$$\begin{array}{ll} \text{Initial damage state in B2:} & s(x,y) \quad \text{(2D-distribution)} \\ & \Downarrow \ \text{change of strain state in B2} \\ \text{Change of damage state in B2:} & \Delta s(x,y) \quad \text{(delamination growth)}\ . \end{array} \tag{6}$$

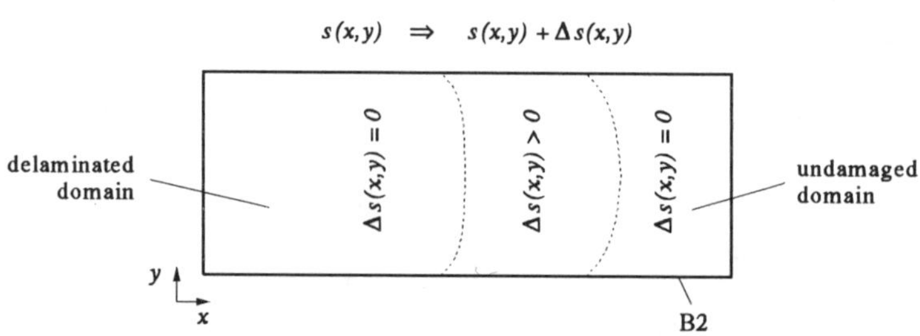

Figure 4. Material control method

This method corresponds with the classical physical nonlinear formulation. Owing to the fact that fracture mechanics energy release rates may only be calculated at the delamination front and not in general material points, the direct application of fracture mechanics is not possible. Applying a damage mechanics approach with sudden-drop material behaviour as well as an infinitely thin crack band ($t \rightarrow 0$), however, represents the limit case of fracture mechanics.

A damage mechanics approach based on [15] has already been applied to the delamination process element. However, it has been found that very fine meshes must be used in the vicinity of the delamination front when classical Gauss integration is applied. As investigations in [16] show, this drawback may be overcome by using the Simpson integration with a large number of integration points.

5. Numerical Results

With the static investigation of an end notched flexure (ENF)-specimen, it may be shown that the presented model is well-suited for describing qualitative delamination growth phenomena. Similar to [17], this specimen is assumed to be of C12K/R6376 graphite/epoxy material with a lay-up of $[\pm 30/(\mp 30)_2/(\pm 30)_2/\mp 30/delam./\mp 30/(\pm 30)_2/(\mp 30)_2/\pm 30]$, a ply thickness of 0.127 mm or a total laminate thickness of 3.048 mm respectively, a width of 25 mm, a total length of 101.6 mm, and an initial delamination length of 28.6 mm. As indicated in Fig. 5, the ENF-specimen is modeled under three point bending conditions with simply supported end edges and a uniformly distributed center load.

In Fig. 5, it can also be seen that the delamination growth simulation is carried out by using the moving mesh technique. Both the initial growth position 0 as well as the advancing growth positions 1 to 3 correspond with meshes of identical topology. On the projected surfaces underneath, the light gray shaded areas represent the initial delamination, while the dark gray shaded areas represent the grown delamination. Thus, it can be seen that the application of the Griffith growth criterion ($G_c = 1.4\,\text{N/mm}$) leads to a curved delamination front, where increasing growth is observed from the center towards the edges. Furthermore, Fig. 6 shows stable delamination growth up to growth position 2 and a sudden change to unstable growth leading to total failure unless the load is reduced. This transition point is caused by the beginning of delamination growth at the front center i.e. delamination growth occurring along the entire front. Finally, these simulated phenomena of front curvature and growth stability characteristics are in good qualitative agreement with experimental results [17].

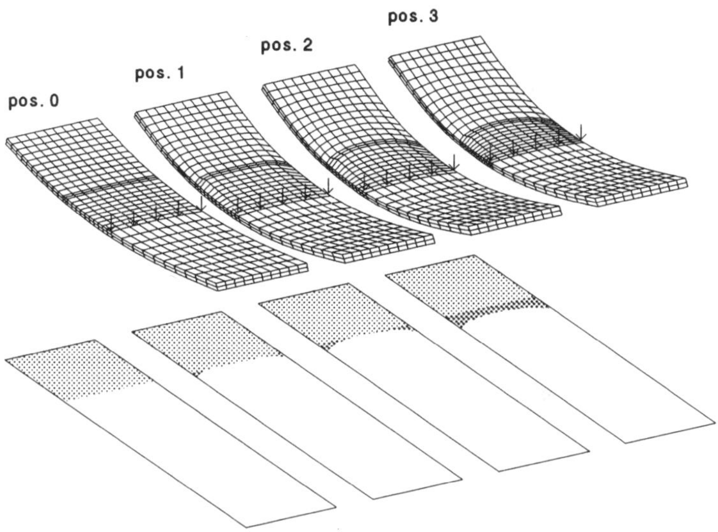

Figure 5. Magnified structural deformation and mesh movement for ENF-specimen

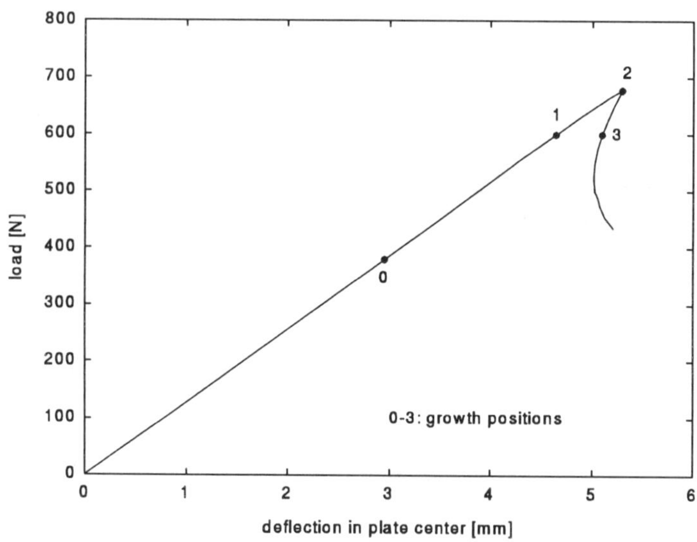

Figure 6. Load-deflection behaviour of ENF-specimen

6. Conclusions

Different approaches for the analysis of delaminated structures are discussed on basis of a two-dimensional macroscale model. For both delamination onset and growth analysis, fracture and damage mechanics may be applied. In conjunction with the finite element method, however, fracture mechanics is much more efficient as long as standard integration schemes are used. Although the presented model is restricted to growth criteria of the Griffith type, qualitative delamination growth phenomena may be described rather well. Owing to the relevance of micromechanical mechanisms, it is still an open question whether three-dimensional macroscale models are really suited to additionally allow a reliable quantitative delamination growth prediction.

References

1. Miller, A.G., Lovell, D.T., (1994), The Evolution of an Aerospace Material: Influence of Design, Manufacturing and In-Service Performance, *Composite Structures*, **27**, 193-206.
2. S.N. Chatterjee, W.A. Dick, R.B. Pipes (1986) Mixed-Mode Delamination Fracture in Laminated Composites, *Composites Science and Technology*, **25**, 49-67.
3. Rinderknecht, S., Kröplin, B. (1995) A Finite Element Model for the Delamination in Composite Plates, *Mechanics of Composite Materials and Structures*, **2**, 19-47.
4. W.L. Bradley, R.N. Cohen (1985) Matrix Deformation and Fracture in Graphite-Reinforced Epoxies, *Delamination and Debonding of Materials, ASTM STP 876*, 389-410.
5. J.D. Whitcomb (1992) Analysis of a Laminate with a Postbuckled Embedded Delamination, Including Contact Effects, *Journal of Composite Materials*, **26**, 1523-1535.
6. R. Krüger, S. Rinderknecht, C. Hänsel, M. König (1996) Computational Structural Analysis and Testing: An Approach to Understand Delamination Growth, Interlaminar Fracture of Composites, *Key Engineering Materials*, **120-121**, 181-202.
7. H. Chai, C.D. Babcock (1985) Two-Dimensional Modelling of Compressive Failure in Delaminated Plates, *Journal of Composite Materials*, **19**, 67-98.
8. B. Storåkers, B. Andersson (1988) Nonlinear Plate Theory Applied to Delamination in Composites, *Journal of Mechanics and Physics of Solids*, **36**, 689-719.
9. B. Cochelin, M. Potier-Ferry (1991) A Numerical Model for Buckling and Growth of Delaminations in Composite Laminates, *Computational Methods in Applied Mechanics and Engineering*, **89**, 361-380.
10. S.O. Peck, G.S. Springer (1991) The Behavior of Delaminations in Composite Plates – Analytical and Experimental Results, *Journal of Composite Materials*, **25**, 907-929.
11. E.F. Rybicki, M.F. Kanninen (1977) A Finite Element Calculation of Stress Intensity Factors by Modified Crack Closure Integral, *Engineering Fracture Mechanics*, **9**, 931-938.
12. H.G. Hahn (1976) *Bruchmechanik*, B.G. Teubner Stuttgart.
13. S. Rinderknecht, B. Kröplin (1994) Calculation of Delamination Growth with Fracture and Damage Mechanics, *Recent Developments in Finite Element Analysis*, T.J.R. Hughes, E. Onate, O.C. Zienkiewicz (Editors), 181-190.
14. S. Rinderknecht, B. Kröplin (1994) Delamination Growth Simulation with a Moving Mesh Technique, *Advances in Non-Linear Finite Element Methods*, B.H.V. Topping and M. Papadrakakis (Editors), 187-197.
15. O. Allix, P. Ladevèze (1992) Interlaminar Interface Modelling for the Prediction of Delamination, *Composite Structures*, **22**, 235-242.
16. N. Point, E. Sacco (1996) A Delamination Model for Laminated Composites, *Int. Journal of Solids and Structures*, **33**, 483-509.
17. B.D. Davidson, C.S. Altonen, J.J. Polaha (1994) *Effect of Stacking Sequence on Delamination Toughness and Delamination Growth Behaviour in Composite End-Notched Flexure Tests*, Technical Report No. MAME-94-402, Department of Mechanical, Aerospace and Manufacturing Engineering, Syracuse University.

NUMERICAL SOLUTION OF PROBLEMS OF CYCLIC PLASTICITY

I. BABUŠKA AND Y. LI
Texas Institute for Computational and Applied Mathematics
The University of Texas at Austin, Austin, TX 78712, USA

1. Introduction

Computational problem of cyclic plasticity which leads to reliable results for use in engineering faces formidable difficulties. These difficulties relate to the following aspects

a) The aim of computational analysis.

b) The selection of the constitutive model and assessment of its reliability for the given goals.

c) Mathematical formulation of the problem and analysis of its properties, such as existence and uniqueness of the solution, its dependence on the selected model, regularity of the solution, etc.

d) Numerical solution of the mathematical problem, its convergence, a priori and a posteriori error estimation of the data of interest, adaptive approaches etc.

We will discuss these aspects in a simple one dimensional setting. We will analyze two problems

A) The problem of the uniform unit rod $(0 < x < 1)$ when the displacements $U(0,t) = 0$ and $U(1,t) = g(t)$ are given.

B) The problem of the uniform unit rod clamped on both ends, and loaded with the body force $f(x,t)$ for $0 < x < 1$, $0 \leq t \leq T$.

We will address only the rate independent problem when the inertia can be neglected, and will also assume that the displacement and strain are small, so that the linear relation between the displacement and strain is used and the necking phenomenon is excluded.

H. A. Mang and F. G. Rammerstorfer (eds.),
IUTAM Symposium on Discretization Methods in Structural Mechanics, 307–314.

2. The aim of computation problem

Usually we are interested in various failure effects, which needs to formulate the failure criteria, which are not reliably available. Hence, in this presentation we will assume that only the stress or (and) strain is (are) of interest.

In the problem A, only the stress which is constant along the rod is of interest.

3. The selection of the constitutive model

This part is the most crucial. There are many models suggested in the literature. We will restrict ourself to the models which satisfy the Drucker's postulate, although in many instances it is not satisfied.

Further we will restrict ourself mostly to the determinate model, although the fully stochastic models likely are better for practical purposes. There are many models of this class suggested in the literature, we will restrict ourself only to the models which are based on the existence of the internal parameters $\boldsymbol{\alpha} \in \mathbf{R}^{\mathrm{m}}$, which characterize the memory of the material. These models are derived usually from the micro mechanics or engineering heuristics. In one dimensional setting, the elastic region of the material can be described by a yield surface of the set

$$\mathcal{S} = \Big\{ (\sigma, \boldsymbol{\alpha}) \in \mathbf{R}^1 \times \mathbf{R}^{\mathrm{m}} \quad | \quad \mathrm{F}(\sigma, \boldsymbol{\alpha}) = \mathrm{Z}_0 \}$$

where the function $F(\sigma, \boldsymbol{\alpha})$ satisfies the following three assumptions:

A1) $F(\sigma, \boldsymbol{\alpha})$ is a convex and piecewise analytic function of the variables σ and $\boldsymbol{\alpha}$.

A2) $F(0, \mathbf{0}) = 0$.

A3) There exist constants $\Gamma, \gamma > 0$ such that $\gamma \leq \left|\frac{\partial F}{\partial \sigma}\right|, \left|\frac{\partial F}{\partial \boldsymbol{\alpha}}\right| \leq \Gamma$ uniformly on the set $\mathcal{S}$.

Then using Drucker's postulate, the constitutive and evolution equations for the stress and internal parameters are given by the following two sets of ODEs

$$\begin{cases} \dot{\sigma} = E\,\dot{\epsilon} \\ \dot{\boldsymbol{\alpha}} = \mathbf{0} \end{cases} \quad \text{if} \quad \begin{array}{l} F(\sigma, \boldsymbol{\alpha}) < Z_0 \quad \text{or} \\ F(\sigma, \boldsymbol{\alpha}) = Z_0 \quad \text{and} \quad \frac{\partial F}{\partial \sigma}\,\dot{\sigma} \leq 0 \end{array} \tag{1}$$

for the elastic deformation, or

$$\begin{cases} \dot{\sigma} = \dfrac{E\,\frac{\partial F}{\partial \boldsymbol{\alpha}} \cdot \frac{\partial F}{\partial \boldsymbol{\alpha}}}{\frac{\partial F}{\partial \boldsymbol{\alpha}} \cdot \frac{\partial F}{\partial \boldsymbol{\alpha}} + E\,(\frac{\partial F}{\partial \sigma})^2}\,\dot{\epsilon} \\ \\ \dot{\boldsymbol{\alpha}} = -\dfrac{E\,\frac{\partial F}{\partial \sigma}\,\dot{\epsilon}}{\frac{\partial F}{\partial \boldsymbol{\alpha}} \cdot \frac{\partial F}{\partial \boldsymbol{\alpha}} + E\,(\frac{\partial F}{\partial \sigma})^2}\,\dfrac{\partial F}{\partial \boldsymbol{\alpha}} \end{cases} \quad \text{if} \quad \begin{array}{l} F(\sigma, \boldsymbol{\alpha}) = Z_0 \quad \text{and} \\ \\ \frac{\partial F}{\partial \sigma}\,\dot{\sigma} > 0 \end{array} \tag{2}$$

for the plastic deformation.

This class of models guarantees the existence and uniqueness of the solution when it is used (see [1] or [10] for the details of the proof). Many classical models such as kinematic, isotropic, or combination of both etc. could be written in this form (cf. e.g. [2]). Some other engineering models do not fall directly into this frame, but they could be modified so that the modified versions do fall into this frame. Let us mention, for example, one of the Chaboche Law (see [2] and [6] for details). The model is then characterized by its mathematical form and its numerical parameters. These parameters are of two types, the initial value parameters, and the law parameters. For example, the classical kinematic law has two law parameters (modulus of elasticity and plasticity) and one initial value parameter (initial yield stress).

In the literature, many very complex laws were suggested, but the major problem is to obtain the numerical values of the parameters. The parameter values are usually derived from experimental data. In [2] we made large experimental study of the problem A for the aluminum alloy when random samples were taken from the sheet (1/4 inch) with the goal to see the adequacies of various models. As quantitative characterization of the accuracy of the law, we introduce the error indicator

$$\theta\left(T,\epsilon(t)\right)=\frac{\|\,\sigma_1(t)-\sigma_2(t)\,\|_{L^\infty(0,T)}}{\frac{1}{2}\,\|\,\sigma_1(t)+\sigma_2(t)\,\|_{L^\infty(0,T)}} \tag{3}$$

where $\sigma_1(t)$ was the measured stress and $\sigma_2(t)$ the computed stress corresponding to a same strain input $\epsilon(t)$. (The $\|\cdot\|_{L^\infty}$ norm was replaced by maximum over fine uniform mesh in time t). For $\epsilon(t)$, we used cyclic load with different mean and the same amplitude (sinusoidal or random) with 1,000 reversals. The optimal parameters were computed as the ones which minimize $\theta\left(T,\epsilon(t)\right)$ (thus they depend on $\epsilon(t)$). Finally we can take the average of those optimal parameters over various $\epsilon(t)$, or use the analogue of (3) for a group of data sets.

Another accuracy measure was based on the criteria of constitutive relation error suggested in [7]. The results of our numerical analysis indicate that the criteria (3) serves better than the constitutive relation error, and we recommend the use of criteria (3).

As illustration, in Table 1 we give the mean, standard deviation, maximum and minimum of θ for the class of $\epsilon(t)$ when optimal parameters for every particular one was used for different laws, and also when parameter values are derived from some metal handbooks. The original Chaboche law can not be cast into the class of constitutive models we mentioned above. Nevertheless, we were able to find an approximation of the Chaboche law, which also belongs to this class of models, and was denoted as B-L law (for more details about this model, we refer to [2]).

TABLE 1. Statistic data for the accuracy of various constitutive models

	Optimal parameters				Handbook parameters			
	Mean	StdDv	Max	Min	Mean	StdDv	Max	Min
Kinematic	24.3%	3.05%	29.5%	17.8%	33.2%	2.66%	39.9%	29.5%
Isotropic	17.8%	2.19%	26.6%	13.8%	36.6%	3.05%	41.5%	32.4%
Mixture	17.5%	2.44%	22.1%	13.2%	26.7%	2.29%	32.6%	22.4%
Chaboche	12.3%	1.69%	15.2%	9.35%	21.4%	2.69%	29.0%	17.4%
B-L	11.2%	1.53%	15.0%	8.68%	–	–	–	–

We see that the error indicator is large. For more details we refer to [2]. These results are for one dimensional case only. Obviously for higher dimensional cases the error indicator will be larger.

Remark 1: There are many constitutive laws in the literature which are based on the internal parameters (see [12] [14]) or other types as hysteresis (see [8] [9]).

4. Mathematical formulation and its properties

In one dimension, as presented here, mathematical formulation is straight forward which combines equilibrium equation with the constitutive law. The problem then is an evolution problem involving time derivatives and has in general a character of ODE in Banach space.

In [1] and [10], it has been proven that under the assumption about the constitutive law formulation presented in section 2, the solution exists, and is unique (in one and higher dimensional setting). In [8] and [9] other class of constitutive laws were analyzed, and the existence and the uniqueness of the solution for quasi-static and dynamic problems have been proven. In [5], with one dimensional setting, the regularity of the solution for problem B with general kinematic law was analyzed. Unfortunately, in general very little is known about the regularity of the solution in two and three dimensions.

In [1] some statement about the continuous dependence on the data, such as the loading force and the initial values of the internal parameters, are given. Nevertheless, very little is known about the relation between the error indicator of the law as formulated above and the accuracy (e.g. of the stress or strain of the problem B, or even more complicated one). Problems of this type are of great importance, especially in light of large error indicator which is unavailable in practice. In [2], we did some statistical analysis of the experimental data which indicates that the fully stochastic formulation and analysis should be used. However, there are no theoretical results available.

In one dimensional setting, we analyzed the relation between the error indicator for the law and the accuracy of the computed data of the problem B with the boundary conditions of the form $U(0,t) = 0$ and $U(1,t) = g(t)$ (or $\sigma(1,t) = h(t)$), and the loading force:

$$f_1(x\,,t) = A\,x\,\sin(t) \quad \& \quad f_2(x\,,t) = B\,x\,\sin(t) + C\,(1 - \cos(e\,t))\,(x - 3\,x^2).$$

We came to the following conclusions:
1) Different criteria and parameter identification schemes will usually lead to different optimal parameters. Group fitting criteria and the constitutive relation error are not suitable for the purpose of reliability analysis.
2) The accuracy of the computed data is closely related to the type of the imposed boundary conditions of the problem. All the laws perform extremely poorly for problems with prescribed stress boundary conditions. Better accuracy can be achieved when stress is treated as the independent variable in the parameter identification procedure, which indicates that the criteria for parameter determination depends also upon the goal of the reliability analysis.
3) Some quantities, such as strain, are comparatively more unreliable than others, and error of 80% or more in strain is very typical.
4) All constitutive laws we have discussed here significantly increase the inherent uncertainty characterized by the reproducibility.
5) Parameter values derived from handbook are very unreliable, and the results computed from these parameters is inaccurate by 40% or more.
6) For problem B, the discretization error of the numerical solution (e.g. FEM solution) is usually much smaller than the error caused by the inaccuracy of the constitutive law. To get better results, some more accurate laws should be used.

5. Numerical solution of the problem

As mentioned above the plasticity formulation is of the evolution type where the solution is an incremental one. In [10], [11] and literature quoted there some finite element methods were proposed and the convergence proven. In contrast to the usual approach convergence has been shown for higher order elements too (in one and higher dimensions). In [10] we proved the convergence of the p-version (resp. h-p version) under some restriction assumption in the one dimensional setting.

Nevertheless, because of the lack of the regularity of the solution, the use of the higher order elements on the uniform (non-adaptive) mesh does not pay, since the rate of convergence does not improve to offset the higher cost. For more details, we refer to [3]. This indicates that an adaptive procedure is essential.

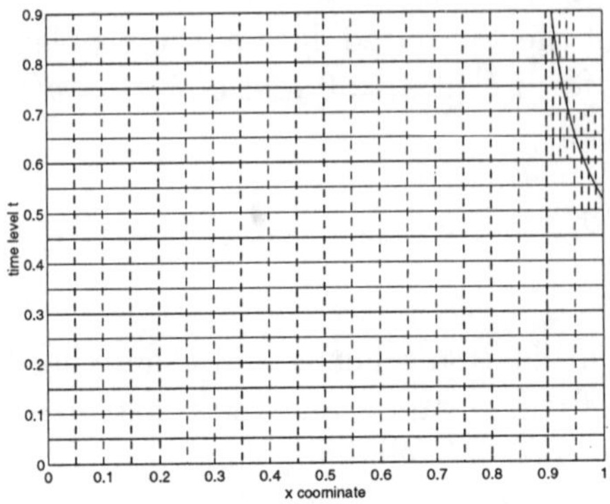

FIG 1. Adaptive mesh refined in the neighborhood of the front.

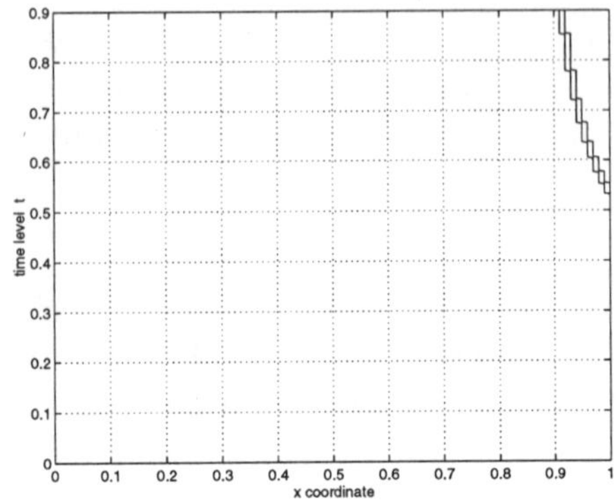

FIG 2. Position of the front computed from the numerical solution.

Because the similarity of the evolution type of the plasticity problem with ODE, we developed an adaptive procedure which controls the error in the increments (in energy norm). The increments are smooth (under the assumption that $f_t(x,t)$ is smooth as the function of x) except in the neighborhood of the "front" where the modulus of elasticity is changing (yield). This is the same for the internal parameters (for which the same FE mesh is used). The effect of the proper selection of the adaptive scheme is shown in one dimensional setting for the problem B with $f(x,t) = 6\,x\,\sin(t)$ and the simple kinematic law. In this case the analytic form of the solution can be found over the time interval $0 \leq t \leq \pi/3.5$. In Fig 1 we show the picture of the front and the mesh which is refined in the neighborhood of the front. In Table 2, we show the relative error in stress and strain (Err_Sig and Err_Eps respectively) measured in L^2 norm for the uniform mesh and elements of degree $1 \leq Dp \leq 4$, and the properly refined mesh at the time level $t = 0.5834$ (with time step size 0.009). The column labeled N.O.E. indicates the maximum total number of elements used in the computation. In the procedure of mesh refinement, every element of the original size, say h, in the neighborhood of the front was refined into four smaller elements of the size $h/4$, which can be thought as a simple model of adaptive mesh refinement. In the computation, uniform polynomial degree was used over all the elements. Better convergence rate can be achieved if some h-p version adaptive schemes were used. Table 3 shows the results for uniform mesh and uniform polynomial degree, properly refined mesh with uniform polynomial degree, and properly refined mesh with nonuniform polynomial degree. For the last case, polynomials of degree 3 and 2 were used over the smooth region and the neighborhood of the front respectively.

TABLE 2. Accuracy for the uniform mesh and properly refined mesh.

	Uniform Mesh			Properly refined mesh		
	N.O.E.	Err_Sig	Err_Eps	N.O.E.	Err_Sig	Err_Eps
Dp=1	26	4.3055%	11.626%	26	4.8254%	5.5562%
Dp=2	26	1.1075%	2.3019%	26	0.0608%	0.4085%
Dp=3	26	0.0505%	1.1075%	26	0.0043%	0.1977%
Dp=4	26	0.0158%	0.5066%	26	0.0015%	0.0925%
Dp=1	46	2.4314%	6.1915%	46	2.6007%	2.9423%
Dp=2	46	0.0165%	0.9583%	46	0.0156%	0.1945%
Dp=3	46	0.0162%	0.4770%	46	0.0011%	0.0670%
Dp=4	46	0.0051%	0.2196%	46	0.0008%	0.0527%

TABLE 3. Accuracy for the uniform mesh and properly refined mesh with uniformly distributed p values, and refined mesh with non-uniform p values.

	P_{smooth}	P_{front}	N.O.E.	D.O.F.	Err_Sig	Err_Eps
Uniform h & p	3	3	46	414	0.01615%	0.4770%
Refined h & unif. p	3	3	46	414	0.00035%	0.0299%
Both non-uniform	3	2	58	402	0.00018%	0.0140%

In Fig 2 we show the position of the front computed from the use of element of degree $p = 1$ and uniform mesh of size h. We note that the position of the front cannot be established by this uniform mesh with the accuracy higher than $2h$. This strip of width $2h$ is shown in the Fig 2. In [4] we designed a general adaptive algorithm in the one dimension setting which can be generalized to higher dimensions with the error indicator for H^1 and energy norm. As was mentioned above the plasticity problem is an evolution one and for fixed (spatial) mesh and certain constitutive laws (e.g. kinematic law), the solution can be solved by some ODE solver with constraints, such as Ddassl. This was done in [13].

6. Conclusions

The cyclic plasticity problem has large intrinsic uncertainty in the constitutive law as well in the aims of computations. The major problem is not only to have the form of the law but also the numerical parameter values. Complicated laws without numerical values of of parameters have very limited value for practical computations.

Some basic results about the mathematical properties are available, but many more are needed especially the influence of the uncertainties in the constitutive laws on the computed results. Further there are no sufficient results related to the stochastic formulation as well as probabilistic description of the constitutive laws.

Numerical treatment of the plasticity problem has to be adaptive, and higher order elements are preferable in smooth regions. If no adaptive mesh is used, then elements of degree 2 are preferable by our experience. The adaptive procedure is based on the accuracy measure which has to be related to the aim of computation. Here many questions are open. One approach is to control the accuracy of the increments as it is done in the numerical solution of ODE's.

References

1. Eric Bonnetier, *Ph.D. Dissertation*, Institute for Physical Science & Technology and Department of Mathematics. The University of Maryland at College Park. 1988.
2. I. Babuška, K. L. Jerina, Y. Li & P. Smith, *Quantitative Assessment of the Accuracy of Constitutive Laws for Plasticity with an Emphasis on Cyclic Deformation*, Proceedings of the Symposium on Parameter Estimation for Modern Constitutive Equations at the 1993 ASME Winter Annual Meeting, ASME, New York, 1993.
3. I. Babuška & Y. Li, *On High Order Finite Element Method for Plasticity Problems*, TICAM Report 96–22, The University of Texas at Austin, May 1996.
4. I. Babuška & Y. Li, *On h-p Adaptive Finite Element Method for Plasticity Problems*, in preparation.
5. I. Babuška & P. Shi, *Regularity of solutions to a one dimensional plasticity model*, TICAM Report 96–16, The University of Texas at Austin, April 1996.
6. J.L. Chaboche, *Time independent Constitutive Theories for Cyclic Plasticity*, International Journal of Plasticity, Vol. 2, No. 2, pp. 149-188, 1986.
7. L. Gallimard, P. Ladevèze & J. Pelle, *Error in constitutive relation and time error indicator for nonlinear computation*, Numerical Methods in Engineering'96, John Wiley & Sons Ltd, 1996.
8. W. Han & B. D. Reddy, *Computational plasticity: the variational basis and numerical analysis*, Computational Mechanics Advances **2**,No. 2, 1995.
9. P. Krejći, *Hysteresis, Convexity and Dissipation in Hyperbolic Equations*, Springer, 1996.
10. Y. Li & I. Babuška, *A convergence analysis of a P–version FEM for 1–D elasto–plasticity problems*, SIAM Journal on Numerical Analysis, Vol. 33, No.2, pp. 809-842, April 1996.
11. Y. Li & I. Babuska, *A convergence analysis of an h–version high order FEM for 2–D elasto–plasticity problems*, will appear in SIAM Journal on Numerical Analysis July 1997.
12. J. Lubliner, *Plasticity Theory*, Macmillan Publishing Company, New York, 1990.
13. P. Shi & I. Babuška, *Analysis and computation of a cyclic plasticity model by aid of Ddassl*, will appear in Computational Mechanics.
14. M. Zyczkowski, *Combined Loading in the Theory of Plasticity*, PWN-Polish Scientific Publisher, Warszawa, 1981.

PERFORMANCE OF ENHANCED LOW-ORDER ELEMENTS FOR PLASTIC CONTINUA

RENÉ DE BORST, AREND E. GROEN, OTTO M. HEERES

Koiter Institute Delft / Faculty of Civil Engineering
Delft University of Technology
P.O. Box 5048, 2600 GA Delft, The Netherlands

1. Introduction

Because of reasons of economy and easy mesh generation and post-processing, a huge effort has been invested in developing high-performance finite elements with interpolation functions of low order. Unfortunately, there are a number of cases where low-order, displacement-based elements perform poorly. A well-known example is the finite element modelling of incompressible material behaviour, where elements with low-order interpolations of displacements tend to 'lock' completely, thus yielding an overstiff solution (Nagtegaal *et al.* 1974). Many solutions have been put forward since, including special arrangements of elements (Nagtegaal *et al.* 1974), reduced or selective integration (Zienkiewicz and Taylor 1989), which is closely related to the $\bar{\mathbf{B}}$ approach (Hughes 1980), the use of higher-order displacement models (Sloan and Randolph 1982, de Borst and Vermeer 1984), mixed approaches with an independent interpolation of displacements and pressures (Sussman and Bathe 1987, Van den Bogert *et al.* 1991), and most recently, the Enhanced Assumed Strain concept (Simo and Rifai 1990).

While the phenomenon of 'element locking' in isochoric deformations has received much attention in the past, it has not been recognised at a large scale that dilatant or contractant plastic flow imposes essentially the same kinematic constraint upon the elements. As will be argued below in a more rigorous manner, plastic dilatancy or contractancy essentially imposes a pointwise relation between the shear deformation and the volumetric deformation, thus causing locking behaviour. The implications for low-order quadrilateral and triangular plane-strain and axisymmetric elements will be assessed as well as some solution strategies.

2. Basic Notions in Soil Plasticity

One of the most frequently used yield criteria for soils and rocks is the classical Coulomb criterion. Expressed in terms of principal stresses, the Mohr-Coulomb yield function reads:

H. A. Mang and F. G. Rammerstorfer (eds.),
IUTAM Symposium on Discretization Methods in Structural Mechanics, 315–322.

$$f = \tfrac{1}{2}(\sigma_3 - \sigma_1) + \tfrac{1}{2}(\sigma_3 + \sigma_1)\sin\phi - c\cos\phi \tag{1}$$

with ϕ the angle of internal friction, c the cohesion of the material and $\sigma_1 \leq \sigma_2 \leq \sigma_3$. A non-associative flow rule is obtained by adopting

$$\dot{\boldsymbol{\varepsilon}}^{\mathrm{p}} = \dot{\lambda}\,\frac{\partial g}{\partial \boldsymbol{\sigma}} \tag{2}$$

with g a plastic potential function that resembles the yield function f:

$$g = \tfrac{1}{2}(\sigma_3 - \sigma_1) + \tfrac{1}{2}(\sigma_3 + \sigma_1)\sin\psi - \text{const.} \tag{3}$$

ψ is an additional material constant, which is commonly named the angle of dilatancy. The angle of dilatancy controls the amount of plastic volume change. Defining the volumetric plastic strain rate as

$$\dot{\varepsilon}_{\mathrm{v}}^{\mathrm{p}} = \dot{\varepsilon}_1^{\mathrm{p}} + \dot{\varepsilon}_2^{\mathrm{p}} + \dot{\varepsilon}_3^{\mathrm{p}} \tag{4}$$

and introducing the rate of plastic shear deformation

$$\dot{\gamma}^{\mathrm{p}} = \dot{\varepsilon}_3^{\mathrm{p}} - \dot{\varepsilon}_1^{\mathrm{p}} \tag{5}$$

one can use eqs (2) and (3) to derive that

$$\dot{\varepsilon}_{\mathrm{v}}^{\mathrm{p}} = \dot{\gamma}^{\mathrm{p}}\sin\psi \tag{6}$$

which shows that the angle of dilatancy ψ sets the ratio between the rate of plastic shear deformation and the rate of plastic volume change. For $\psi > 0$ an irreversible increase of volume occurs, while for $\psi < 0$ a decrease is predicted (plastic contraction). $\psi = 0$ is the special case of plastically volume-preserving (isochoric) flow.

3. Low-Order Quadrilateral Elements

In ideal plasticity, the behaviour at a limit state requires that the stress field remains stationary. In consideration of the injective relation that exists between the stress rate $\dot{\boldsymbol{\sigma}}$ and the elastic strain rate $\dot{\boldsymbol{\varepsilon}}^{\mathrm{e}}$, the elastic strain rates must vanish at this point and relation (6) changes into

$$\dot{\varepsilon}_{\mathrm{v}} = \dot{\gamma}\sin\psi \tag{7}$$

which effectively imposes a kinematic constraint on the possible velocity field. It is emphasised that this constraint condition applies irrespective of the value of ψ, and that $\psi = 0$ (volume-preserving plastic flow) is just a special case.

Now, we choose the principal axes of the strain rate tensor to coincide with the local ξ,η-coordinate system of an element. This choice is permissible, since under planar deformations $\dot{\varepsilon}_{\mathrm{v}}$ and $\dot{\gamma}$ are both invariant. For the Mohr-Coulomb yield function resembling plastic potential (3), eq. (7) then specialises as

$$(1 - \sin\psi)\dot{\varepsilon}_{\xi} + (1 + \sin\psi)\dot{\varepsilon}_{\eta} + \dot{\varepsilon}_{\zeta} = 0 \tag{8}$$

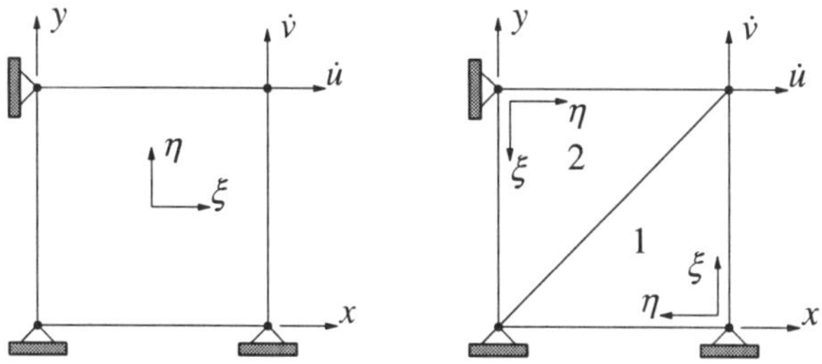

Figure 1. Element subject to non-uniform shear. Left: Quadrilateral. Right: Triangles.

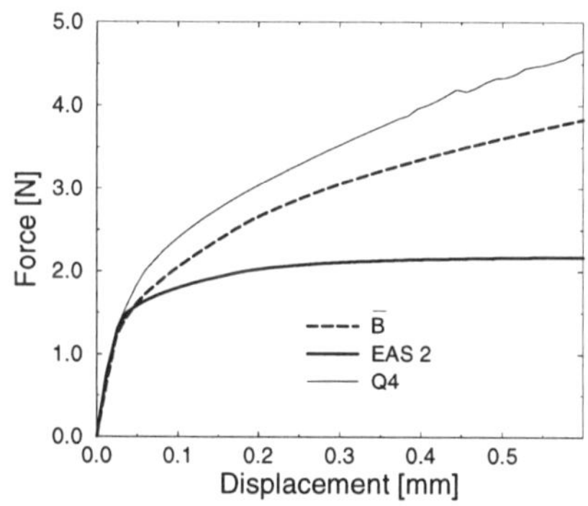

Figure 2. Force-displacement curves for single element with dilatant plastic flow.

We now consider the four-noded plane strain element of Figure 1. This element configuration can be thought of as representative for an arbitrary domain of finite elements with boundary conditions (Hughes 1987). The velocities within the element are interpolated in a standard isoparametric fashion, so that

$$\dot{u}(\xi,\eta) = \tfrac{1}{4}(1+\xi)(1+\eta)\dot{u} \tag{9a}$$

$$\dot{v}(\xi,\eta) = \tfrac{1}{4}(1+\xi)(1+\eta)\dot{v} \tag{9b}$$

and the normal strain rates within the element are obtained by standard differentiation as:

$$\begin{bmatrix} \dot{\varepsilon}_\xi \\ \dot{\varepsilon}_\eta \\ \dot{\varepsilon}_\zeta \end{bmatrix} = \frac{1}{4}\begin{bmatrix} 1+\eta & 0 \\ 0 & 1+\xi \\ 0 & 0 \end{bmatrix}\begin{bmatrix} \dot{u} \\ \dot{v} \end{bmatrix} \tag{10}$$

Upon substitution of expressions (10) for the strain rates into the kinematic constraint (8), the following restriction upon the velocity field ensues:

$$[(1-\sin\psi)\dot{u} + (1+\sin\psi)\dot{v}] + (1-\sin\psi)\dot{u}\eta + (1+\sin\psi)\dot{v}\xi = 0 \tag{11}$$

The term between square brackets sets the ratio between the horizontal velocity $\dot{u}$ and the vertical velocity $\dot{v}$ of the right upper node of the element. Its vanishing is a direct reflection of the kinematic constraint imposed on the possible velocity field by the constitutive relation. Since this term must be zero, vanishing of the entire identity (11) can only be achieved for arbitrary pairs (ξ,η) if $\dot{u}=\dot{v}=0$. This implies that the element is not able to deform, which is known as volumetric locking. This observation holds for all values of ψ, including the isochoric case ($\psi = 0$).

Next, we consider the four-noded element with selective integration on the dilatational strain rate. This approach can be cast within the so-called $\bar{\mathbf{B}}$-concept (Hughes 1980) and the normal strain rates are redefined as

$$\begin{bmatrix} \dot{\varepsilon}_\xi \\ \dot{\varepsilon}_\eta \\ \dot{\varepsilon}_\zeta \end{bmatrix} = \frac{1}{4} \begin{bmatrix} 1+\frac{2}{3}\eta & -\frac{1}{3}\xi \\ -\frac{1}{3}\eta & 1+\frac{2}{3}\xi \\ -\frac{1}{3}\eta & -\frac{1}{3}\xi \end{bmatrix} \begin{bmatrix} \dot{u} \\ \dot{v} \end{bmatrix} \tag{12}$$

Substitution of this strain rate field in the kinematic constraint (8) leads to

$$[(1-\sin\psi)\dot{u} + (1+\sin\psi)\dot{v}] - \sin\psi(\dot{u}\eta + \dot{v}\xi) = 0 \tag{13}$$

Obviously, this condition can only be satisfied for arbitrary pairs (ξ, η) when $\psi = 0$, the special case of plastically volume preserving flow. For arbitrary values of ψ, that is for dilatant plastic flow ($\psi > 0$), or contractant plasticity ($\psi < 0$), $\dot{u}$ and $\dot{v}$ must vanish identically, which means that the $\bar{\mathbf{B}}$ element locks for the general case of $\psi \neq 0$, and is only effective for the case of isochoric plastic flow. This is demonstrated for the elementary test of Figure 2, where a single element is subjected to non-uniform shear. An angle of internal friction $\phi = 30^\circ$ and a dilatancy angle of $\psi = 20^\circ$ have been adopted. Obviously, the standard four-noded element locks completely, but also the $\bar{\mathbf{B}}$ element does not result in a yield plateau, Figure 2. The latter result confirms the conclusion drawn from eq. (13).

We now augment the strain rates by defining additional strain rate fields (EAS elements, Simo and Rifai 1990). The simplest possible enrichment for the normal strain rate field is to set

$$\begin{bmatrix} \dot{\varepsilon}_\xi \\ \dot{\varepsilon}_\eta \\ \dot{\varepsilon}_\zeta \end{bmatrix} = \frac{1}{4} \begin{bmatrix} 1+\eta & 0 \\ 0 & 1+\xi \\ 0 & 0 \end{bmatrix} \begin{bmatrix} \dot{u} \\ \dot{v} \end{bmatrix} + \begin{bmatrix} \xi & 0 \\ 0 & \eta \\ 0 & 0 \end{bmatrix} \begin{bmatrix} \dot{\alpha}_1 \\ \dot{\alpha}_2 \end{bmatrix} \tag{14}$$

with $\dot{\alpha}_1, \dot{\alpha}_2$ additional, incompatible strain rate variables. Substitution of the strain rate field (14) in the kinematic constraint yields

$$[(1-\sin\psi)\dot{u} + (1+\sin\psi)\dot{v}] + (1-\sin\psi)\eta(\dot{u} + 4\dot{\alpha}_2) + (1+\sin\psi)\xi(\dot{v} + 4\dot{\alpha}_1) = 0 \tag{15}$$

which is satisfied for arbitrary pairs (ξ, η) if

$$(1-\sin\psi)\dot{u} + (1+\sin\psi)\dot{v} = 0 \quad , \quad \dot{u} + 4\dot{\alpha}_2 = 0 \quad , \quad \dot{v} + 4\dot{\alpha}_1 = 0 \tag{16}$$

Accordingly, an element is obtained that is free of volumetric locking effects for *all* values of ψ. This is confirmed by a non-uniform shear test on the single element configuration of Figure 1. Figure 2 shows that this element, which will henceforth be called EAS 2 since two extra variables are used to enrich the strain rate field, indeed captures a proper limit load.

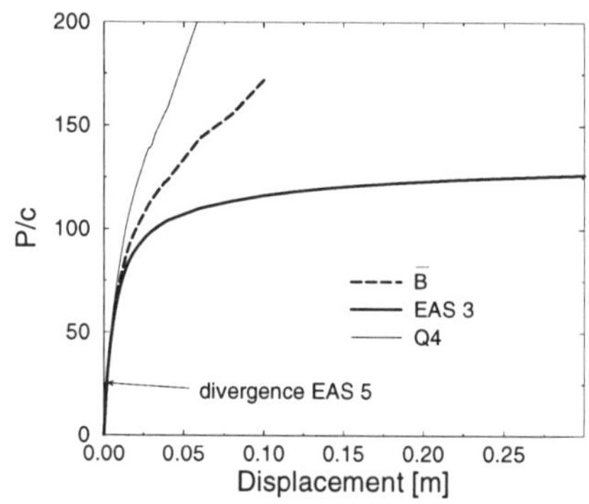

Figure 3. Load-settlement curve for axisymmetric punch problem on dilatant soil.

Figure 3 shows the response obtained for a rigid disc which is being pushed into a halfspace. The boundary conditions are such that symmetry conditions apply at the left boundary while the lower and right boundaries are fully restrained. Loading is applied by direct displacement control of the disc. Non-associative Drucker-Prager ideal plasticity is used with a Young's modulus $E = 50$ MPa, a Poisson's ratio $\nu = 0.25$, a cohesion $c = 2$ kPa, a friction angle $\phi = 30^{\circ}$ and a dilatancy angle $\psi = 10^{\circ}$ (dilatant flow). The standard four-noded element and - as predicted for dilatant plastic flow - the selectively integrated $\bar{\mathbf{B}}$ element show significant locking. The EAS 3 element (the axisymmetric version of the EAS 2 element) captures the limit load properly without giving convergence difficulties. On the other hand, the EAS 5 element, in which the shear strains are enhanced in addition to the normal strains, originally introduced by Simo and Rifai (1990), behaves in an unstable manner and diverges in an early stage of the loading process. The reason for this behaviour has been explained by de Borst and Groen (1995).

4. Low-Order Triangular Elements

We next turn our attention to simple, three-noded triangles. In their standard form, these elements are known to show a pronounced locking behaviour for isochoric conditions (e.g., Nagtegaal *et al.* 1974). It is demonstrated easily that, similar to the standard four-noded quadrilateral elements, they also lock for dilatant/contractant soils. We first consider element 1 of Figure 1. It directly follows that the strain rates $\dot{\varepsilon}_x$ in the global x , y coordinate system are given by

$$\begin{bmatrix} \dot{\varepsilon}_x \\ \dot{\varepsilon}_y \end{bmatrix} = \begin{bmatrix} 0 & 0 \\ 0 & 1 \end{bmatrix} \begin{bmatrix} \dot{u} \\ \dot{v} \end{bmatrix} \tag{17}$$

In an identical fashion we obtain by simple inspection for element 2:

$$\begin{bmatrix} \dot{\varepsilon}_x \\ \dot{\varepsilon}_y \end{bmatrix} = \begin{bmatrix} 1 & 0 \\ 0 & 0 \end{bmatrix} \begin{bmatrix} \dot{u} \\ \dot{v} \end{bmatrix} \tag{18}$$

We now substitute the strain rate fields for both elements in the kinematic constraint (8), but now formulated in the x, y-coordinate system:

$$(1-\sin\psi)\dot{\varepsilon}_x+(1+\sin\psi)\dot{\varepsilon}_y+\dot{\varepsilon}_z=0 \tag{19}$$

We obtain for element 1

$$(1+\sin\psi)\dot{v}=0 \tag{20}$$

and for element 2

$$(1-\sin\psi)\dot{u}=0 \tag{21}$$

respectively, so that the (trivial) conclusion can be drawn that the only solution is $(\dot{u}, \dot{v})=(0,0)$, which implies locking.

Turning now attention to possible enrichments of the three-noded triangle that eliminate the above locking behaviour, we first cast the EAS-concept as for instance in eq. (14) for the quadrilateral four-noded element, into a general format:

$$\dot{\tilde{\boldsymbol{\varepsilon}}} = \mathbf{B}\dot{\mathbf{u}} + \mathbf{M}\dot{\boldsymbol{\alpha}} \tag{22}$$

with $\dot{\tilde{\boldsymbol{\varepsilon}}}$ the enriched strain rates, $\mathbf{B}$ the standard strain-nodal displacement matrix and $\mathbf{M}$ a matrix that contains the interpolation polynomials of the incompatible strain rates that are assembled in the vector $\dot{\boldsymbol{\alpha}}$. Similar to the case of the quadrilateral element we consider a two-parameter enhancement of the normal strains $\dot{\varepsilon}_\xi$ and $\dot{\varepsilon}_\eta$. For the case of a linear polynomial we have in the local ξ, η coordinate system

$$\mathbf{M}_{\xi\eta}=\begin{bmatrix} a_1+b_1\xi+c_1\eta & 0 \\ 0 & 0 \\ 0 & a_2+b_2\xi+c_2\eta \end{bmatrix} \tag{23}$$

which allows for possible different interpolations in both directions. Of course, application of different interpolations would lead to a directional dependence. For both elements we obtain after transformation to the global x, y system:

$$\mathbf{M}=\begin{bmatrix} a_1+b_1x+c_1y & 0 \\ 0 & 0 \\ 0 & a_2+b_2x+c_2y \end{bmatrix} \tag{24}$$

and substitution of the enhanced strain rates $\dot{\tilde{\boldsymbol{\varepsilon}}}$ into the kinematic constraint (19) gives

$$(1+\sin\psi)\dot{v} + a_1(1-\sin\psi)\dot{\alpha}_{1,1}+a_2(1+\sin\psi)\dot{\alpha}_{2,1} = 0 \tag{25a}$$

$$b_1(1-\sin\psi)\dot{\alpha}_{1,1}+b_2(1+\sin\psi)\dot{\alpha}_{2,1} = 0 \tag{25b}$$

$$c_1(1-\sin\psi)\dot{\alpha}_{1,1}+c_2(1+\sin\psi)\dot{\alpha}_{2,1} = 0 \tag{25c}$$

and

$$(1-\sin\psi)\dot{u} + a_1(1-\sin\psi)\dot{\alpha}_{1,2} + a_2(1+\sin\psi)\dot{\alpha}_{2,2} = 0 \tag{25d}$$

$$b_1(1-\sin\psi)\dot{\alpha}_{1,2} + b_2(1+\sin\psi)\dot{\alpha}_{2,2} = 0 \tag{25e}$$

$$c_1(1-\sin\psi)\dot{\alpha}_{1,2} + c_2(1+\sin\psi)\dot{\alpha}_{2,2} = 0 \tag{25f}$$

for elements 1 and 2, respectively. We can combine eqs (25) to give

$$\dot{v} = (a_1 b_2/b_1 - a_2)\dot{\alpha}_{2,1} \tag{26a}$$

$$\dot{u} = (a_1 b_2/b_1 - a_2)\dot{\alpha}_{2,2} \tag{26b}$$

$$c_1 b_2 - c_2 b_1 = 0 \tag{26c}$$

An important issue regarding incompatible elements as discussed here is whether the patch test is satisfied (Taylor *et al.* 1986). For the EAS concept the patch test requires that

$$\int_V \delta(\mathbf{M}\dot{\boldsymbol{\alpha}})^{\mathrm{T}} \boldsymbol{\sigma} \mathrm{d}V = 0 \tag{27}$$

When the additional strain rates $\dot{\alpha}_i$ are transformed to the global coordinate system with a constant isoparametric mapping, this condition reduces to

$$\int_V \mathbf{M} \mathrm{d}V = \mathbf{0} \tag{28}$$

Application of this condition to linear interpolation polynomials $a_i + b_i\xi + c_i\eta$ the following constraint ensues

$$\tfrac{1}{2}a_i + \tfrac{1}{6}b_i + \tfrac{1}{6}c_i = 0 \tag{29}$$

Upon substitution of eq. (29), which results from the patch condition, into eqs (26a,b) and utilising eq. (26c), we arrive at the result that

$$\dot{u} = \dot{v} = 0 \tag{30}$$

and thus, locking occurs for all possible two-parameter enhancements with linear polynomials that satisfy the patch test.

5. Concluding Remarks

In this contribution it has been argued that plasticity models in which irreversible volume changes can occur, impose a constraint upon the velocity field, similar to isochoric deformations. Indeed, 'volumetric locking' occurs also now, and even occurs for some cases which are conventionally regarded as locking-free. A prominent example is the four-noded quadrilateral plane-strain element with selective integration.

A possible remedy is the EAS concept, proposed by Simo and Rifai (1990). However, caution should be exercised, since the enhancement of shear strains may lead to unstable element behaviour and to premature termination of the calculation (de Borst and Groen 1995). For this reason it is recommended to enhance only the normal strains for this purpose. A similar attempt to improve the three-noded plane-strain triangular element was not successful, as it appeared not possible to satisfy the patch test with relatively simple enhancements.

6. References

de Borst, R. and Vermeer, P.A. (1984) Possibilities and limitations of finite elements for limit analysis, *Geotechnique* **34** , 199-210.

de Borst, R. and Groen, A.E. (1995) Some observations on element performance in isochoric and dilatant plastic flow, *Int. J. Num. Meth. Eng.* **38** , 2887-2906.

Hughes, T.J.R. (1980) Generalization of selective integration procedures to anisotropic and nonlinear media, *Int. J. Num. Meth. Eng.,* **15** , 1413-1418.

Hughes, T.J.R. (1987) *The Finite Element Method,* Prentice-Hall, Englewood Cliffs, New Jersey.

Nagtegaal, J.C., Parks, D.M. and Rice, J.R. (1974) On numerically accurate finite element solutions in the fully plastic range, *Comp. Meth. Appl. Mech. Eng.* **4** , 153-177.

Sloan, S.W. and Randolph, M.F. (1982) Numerical prediction of collapse loads using finite element methods, *Int. J. Num. Anal. Meth. Geomech.* **6** , 47-76.

Simo, J.C. and Rifai, M.S. (1990) A class of mixed assumed strain methods and the method of incompatible modes, *Int. J. Num. Meth. Eng.* **29** , 1595-1638.

Sussman, T. and Bathe, K.-J. (1987) A finite element formulation for nonlinear incompressible elastic and inelastic analysis, *Comput. Struct.* **29** , 357-409.

Taylor, R.L., Simo, J.C., Zienkiewicz, O.C. and Chan, A.C. (1986) The patch test - a condition for assessing FEM convergence, *Int. J. Num. Meth. Eng.,* **22** , 39-62.

Van den Bogert, P.A.J., de Borst, R., Luiten, G.T. and Zeilmaker, J. (1991) Robust finite elements for 3D analysis of rubber-like materials, *Eng. Comput.* **8** , 3-17.

Zienkiewicz, O.C. and Taylor, R.L. (1989) *The Finite Element Method,* McGraw-Hill, New Jersey.

MESH ALIGNMENT STRATEGY FOR LOCALISATION PROBLEMS

N. BIĆANIĆ AND C. J. PEARCE
University of Glasgow
Department of Civil Engineering
GLASGOW, UK

AND

L. J. SLUYS
Delft University of Technology
Department of Civil Engineering
DELFT, THE NETHERLANDS

Abstract. Localised failure predictions for concrete and other quasi brittle materials have often been found to be influenced by the adopted discretisation. A simple mesh alignment strategy is presented, whereby a smooth internal mesh boundary is continuously traced to reflect the development of a localised failure mode. The construction and the orientation of the internal band is found by tracing a ridge of the scalar field corresponding to the intensity of incremental total strain.

1. Introduction

Various studies indicate that the localised failure predictions of concrete can be influenced by the adopted finite element mesh or finite element type chosen for the analysis, especially when lower order elements are employed. Just as an 'unbiased' finite element mesh may not be capable of resolving the localised failure mode, or it may have the effect of smearing the localisation over a number of elements so resulting in an apparently more diffuse failure, it is also possible that such an arbitrary mesh may artificially influence the numerically predicted failure mode by providing a kinematically more preferable solution.

Mesh densification and alignment are frequently used for improving the ability of a finite element mesh to realise the correct failure mode. The

H. A. Mang and F. G. Rammerstorfer (eds.),
IUTAM Symposium on Discretization Methods in Structural Mechanics, 323–330.

increased resolution provided by mesh densification only does not necessarily reduce the diffuse nature of a localisation zone, whereas the use of correct mesh alignment enables a proper localisation within a finite band of elements, but is restricted to simple uniform states of stress where the localisation needs to be triggered by an imperfection. Such proper alignment of the finite element mesh is generally not possible in advance and therefore it is necessary to improve the mesh successively as the solution progresses. Adaptive meshing strategies, whereby each new mesh is controlled by the solution thus far, have been the subject of extensive research in various fields over recent years.

Experience of the authors indicate that sometimes alignment of the mesh via an oriented stretching of the finite elements can be a very powerful approach to the problem of capturing localised failure, see also [1]. However, the method appears to be less amenable for general cases, and can be dependent on both the specific problem and the mesh generation tools available.

Here, an alternative strategy for the alignment of the mesh will be explored, whereby smooth internal boundaries will be constructed along which the finite element mesh is aligned. Similar approaches were suggested in the past and the process by which these apparent discontinuity lines may be constructed varies. Some researchers, [2, 3], have utilised the acoustic tensor analysis to predict the direction of possible discontinuity and therefore construct internal 'slip lines'. A more simple approach is explored here, whereby the internal boundary is formed via tracing the ridge of an appropriate scalar control field.

All analyses presented use the fracture energy controlled softening Rankine plasticity model in a generalised three dimensional form. This model and associated algorithmic issues have recently been reported in [4].

An exponential softening law is adopted and to regularise the inherent problem of mesh sensitivity with respect to the energy release, the fracture toughness G_f is introduced as a material parameter where the local softening slope becomes a function of a length scale δ, which in turn is related to the mesh size [5]. The fracture energy release regularisation concept operates well for aligned 'well-behaved' cases where localisation occurs across the width of one element, whereas its performance is less successful for more general cases where more diffuse failure is identified. It is believed that correct mesh alignment will reduce the diffusivity of the localised failure mode and thereby help the regularisation effect of the fracture energy concept. Moreover, it will be seen that the length scale, δ, can now be more suitably defined due to the introduction of double internal boundaries.

2. Alignment and Construction Of Internal Boundaries

As already stated, in an attempt to improve localised failure predictions, finite elements are aligned along internal boundaries, where the position of these internal boundaries is controlled by the propagation of the localisation zone. In order to determine the extent and orientation of this localised failure zone, the 'ridge' of a suitable scalar control field ζ is topologically 'traced'.

2.1. SUITABLE SCALAR CONTROL FIELD

Scalar fields associated directly or indirectly with state variables (e.g. effective plastic strain) as well as scalar fields associated with the norms of some incremental quantity (e.g. incremental total strain or displacement) have been studied. In addition, scalar fields associated with the predominant eigenvector have been investigated, to help filter out the eventual failure mode. In this case, the incremental displacement is expanded in terms of the eigenvectors of the tangential stiffness matrix, and the strain norm associated with the eigenvector which has the greatest participation factor can be utilised as a suitable scalar control field.

For the problem outlined in Section 3, all of the above scalar control fields performed in a similar way, so in the remainder of this paper the intensity of the incremental strain tensor $\zeta = \Delta\boldsymbol{\epsilon} : \Delta\boldsymbol{\epsilon}$ is adopted.

2.2. RIDGE TRACING

The process of ridge tracing starts from the node corresponding to the maximum value of the scalar control field ζ; the local minimum curvature of the ridge is then numerically obtained. From a 'patch' of the surrounding nodes a quadratic surface of the scalar field is constructed via a least squares fit; the principal curvatures of this surface are consequently determined and the eigenvector $\boldsymbol{n}'$ corresponding to the minimum principal curvature is taken as the local direction of the ridge.

The next step is to search along this direction $\boldsymbol{n}'$ for the nearest node p_i within the patch which is in some sense closest to the search direction. Here the new node along the ridge corresponds to the maximum of

$$e_i = \frac{1}{(\boldsymbol{t}_i \cdot \boldsymbol{n})\,(\boldsymbol{t}_i \cdot \boldsymbol{n}')} \tag{1}$$

where $\boldsymbol{t}_i$ is the direction vector of the new node p_i with respect to the previous node p_{i-1} and $\boldsymbol{n}$ is the unit eigenvector corresponding to the maximum principal curvature at node p_{i-1}, locally orthogonal to $\boldsymbol{n}'$, see Fig 1(a).

Once the next node is located the process is repeated until no new node can be identified. The resulting set of nodes which comprise the ridge are

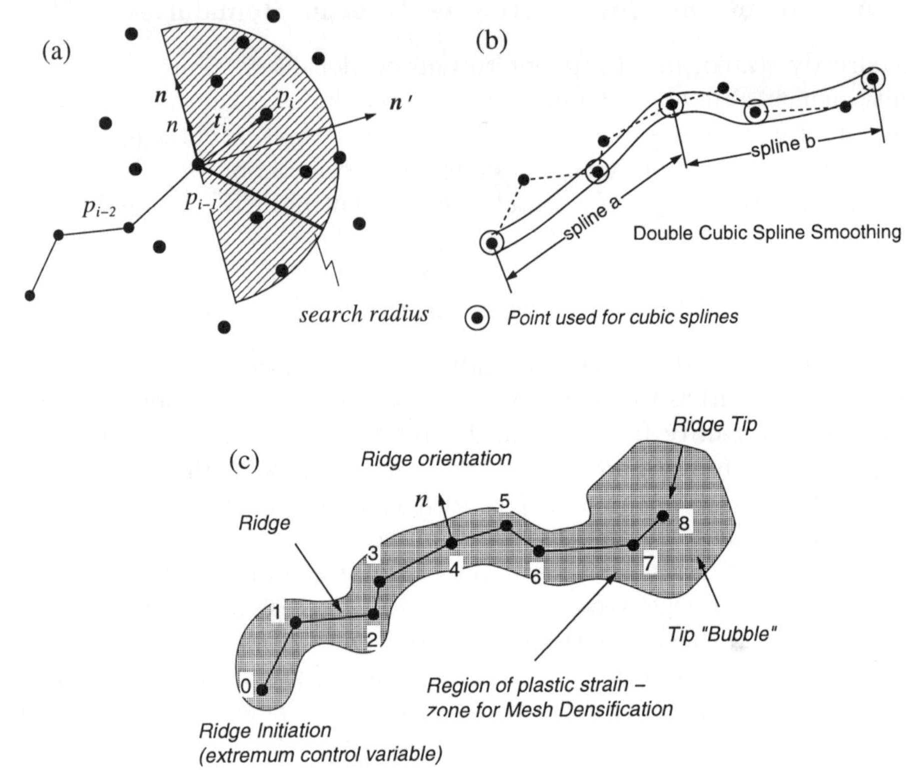

Figure 1. (a) Ridge Tracing. Searching for the next node which lies within the search radius along the direction of the minimum principal curvature. (b) Discontinuity band constructed from a double cubic spline smoothing. (c) Numerically determined ridge and associated zone of mesh densification.

then used to construct a smooth internal boundary; within this work the boundary was restricted to a double cubic spline fit. A parallel line was also constructed in order to produce an internal 'discontinuity band' of a given thickness, see Fig 1(b) and the thickness of this band can now be used directly as the internal length scale δ for the fracture energy regularisation.

In reality, often multiple localisation paths may develop, but it is specifically the propagation of the dominant fracture to which attention must be paid. The proposed procedure is currently limited to the construction of a single 'discontinuity band', corresponding to the propagation of the dominant failure mode.

2.3. ADDITIONAL MESH DENSIFICATION

The aim of the construction and the alignment of the discontinuity band is to reflect the current, dominant failure mode of the analysis. However, any such mesh should not overly influence further analysis by favouring only a single mode of failure. It is therefore suggested that in order to prevent the alignment bias introduced and to adequately provide further freedom for the analysis to reflect other active or previously active localisation paths, mesh densification should in addition allow for any alternative paths to develop. This is achieved by inversely relating the new mesh size to the intensity of the accumulated plastic strain in all damaged regions, Fig 1(c).

The mesh generation is performed using an advancing front mesh generator NeGe [6], an unstructured mesh generator, where the mesh density and orientation are controlled by the background mesh. In algorithmic terms the old mesh is passed to the mesh generator as a background mesh with the corresponding information about mesh size.

3. Illustrative Example - Rilem Round Robin Anchor Pull-Out

This example broadly follows the problem described by Feenstra [7] of the pull-out of an anchor from mass concrete, Fig 2(a), (some corresponding experimental results have been reported by Vervuurt et al. [8]). A plain strain analysis under displacement control has been undertaken as opposed to the plane stress, load control analysis of Feenstra. The initial mesh and the geometry are shown in Fig 2(a) and it should be noted that the anchor itself has been omitted and controlled displacement applied along the upper edge of the notch. The meshes are composed of three noded, constant strain, plane strain elements.

Several aligned meshes have been analysed in the context of an incremental remeshing strategy, i.e. internal discontinuity band and associated mesh alignment is produced at various load increments, followed by a complete re-analysis with a newly aligned mesh, see Fig 3. Note that in addition to the presence of an aligned internal boundary the mesh densification of constant element sizes is adopted in all zones including a ‘tip bubble’, see Fig 1(c).

Aligned meshes and associated scalar field (effective plastic strain presented here) clearly indicate the effectiveness of the proposed procedure to channel localised failure into a narrow zone, Fig 4. However, it is worth noting that the lower, secondary localisation band is modelled quite well while relying on mesh densification only (only a single, dominant localised failure mode is traced by the above strategy, in this case clearly the upper branch). Finally, force displacement diagrams, Fig 2(b), indicate that both

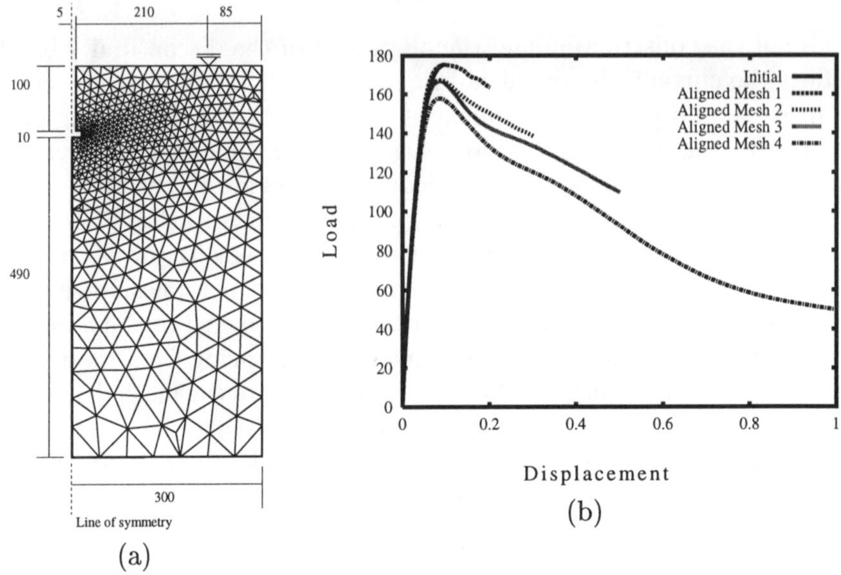

Figure 2. Anchor Pull-Out. (a) Initial mesh and geometry (dimensions in mm). (b) Load-displacement curves for various aligned meshes. Material properties: $E = 37000N/mm^2$, $\nu = 0.3$, $f_t = 2.5N/mm^2$, $G_f = 0.09N/mm$.

the peak load as well as predictions of the softening branch response are influenced by mesh alignment, and that an improved alignment allows for a better control of discretisation effects. Subsequent mesh alignment should eventually lead to a unique force displacement diagram. As the localisation becomes increasingly more and more concentrated within the aligned band there should also be an improvement in the regularisation effect with respect to mesh objectivity provided by the fracture energy control in the softening Rankine model, and in this context the actual thickness of the band becomes immaterial and it does not affect the solution.

4. Conclusions

A simple alignment meshing strategy for localised failure analysis of concrete is presented, whereby a discontinuity band is constructed by tracing a ridge of a suitable scalar control field. The present format includes staggered remeshing, where every mesh alignment is followed by a complete reanalysis to a higher load level than previously achieved, thereby avoiding difficulties in mapping the stress state and internal variables incrementally from one mesh to another. The adopted alignment strategy allowed for a

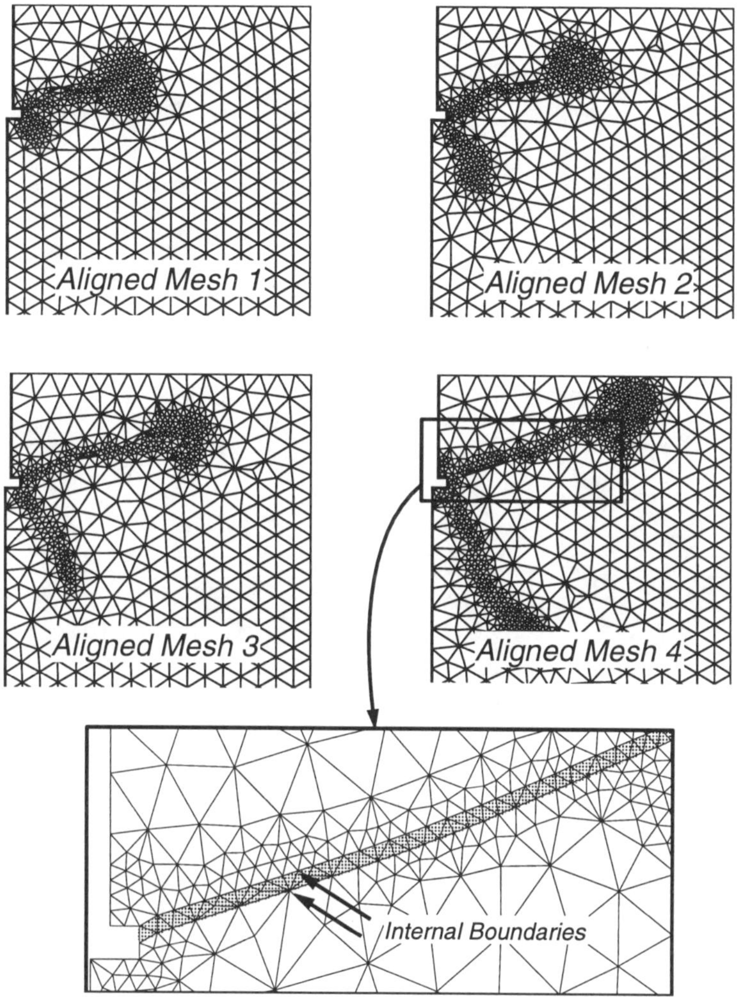

Figure 3. Consecutive meshes showing mesh densification and aligned discontinuity band (including close up of the fourth aligned mesh).

highly localised resolution of a pull-out model problem failure mode.

Further work will introduce multiple internal boundaries to enable the modelling of any number of crack paths through extension of the ridge tracing strategy. Moreover, the concept of filtering out the final failure mode by considering scalar fields associated with the predominant eigenvector is to be investigated further.

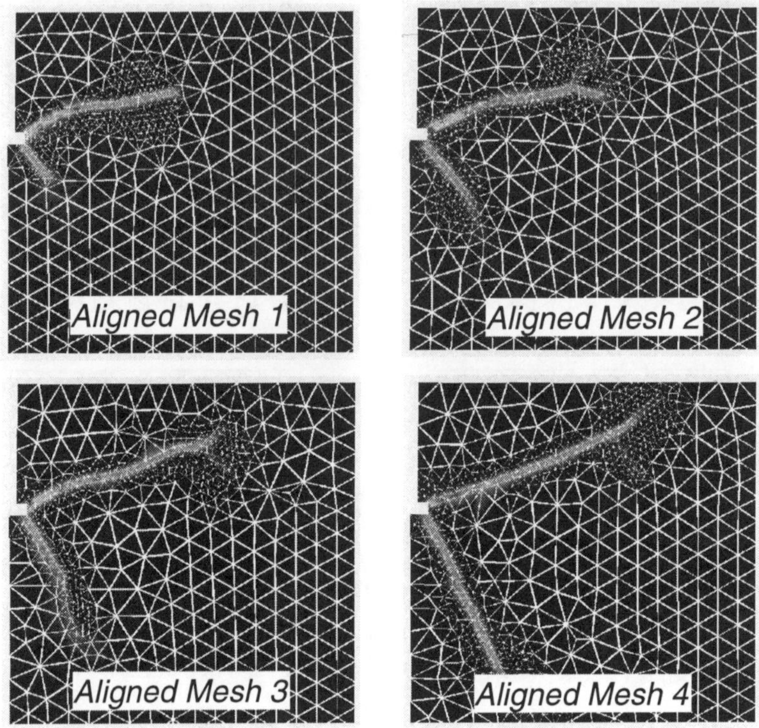

Figure 4. Consecutive meshes showing plots of effective plastic strain.

References

1. Zienkiewicz, O. C., Huang, M., and Pastor, M.: Localization problems in plasticity using finite elements with adaptive remeshing, *Int. J. Num. Meth. Engng.* **19** (1995) 127–148.
2. Larsson, R.: *Numerical simulation of plastic localization*, Ph.D. thesis, Department of Structural Engineering, Chalmers University of Technology (1990).
3. Pearce, C. J.: *Computational plasticity in concrete failure mechanics*, Ph.D., Department of Civil Engineering, University Of Wales Swansea (1996).
4. Pearce, C. J. and Bićanić, N.: On multi-surface plasticity and Rankine model, D. R. J. Owen, E. Oñate, and E. Hinton, eds., *Computational Plasticity: Fundamentals and Applications*, CIMNE (1997), 957–964.
5. Rots, J. G.: *Computational modeling of concrete fracture*, Ph.D., Civil Engineering Dept., Delft University of Technology, Netherlands (1988).
6. Sienz, J.: *NeGe Reference Manual*, Department of Civil Engineering, University of Wales Swansea (1992).
7. Feenstra, P.: *Computational aspects of biaxial stress in plain and reinforced concrete*, Ph.D., Civil Engineering Dept., Delft University of Technology, Netherlands (1993).
8. Vervuurt, A., Schlangen, E., and Mier, J. G. M. V.: A numerical and experimental analysis of the pull-out behaviour of steel anchors embedded in concrete, Technical Report Report 25.5-93-1/VFL, Faculty of Civil Engineering, Delft University of Technology (1993).

FINITE STRAIN ELASTOPLASTICITY WITH FINITE ELEMENTS APPLIED TO INDUSTRIAL FORMING-MODELLING AND DISCRETIZATION PROBLEMS

A. SAMUELSSON, K. MATTIASSON, U. WENDT
Department of Structural Mechanics
Chalmers University of Technology
S-412 96 Göteborg, Sweden

Abstract – Finite strain elastoplasticity is applied to industrial forming processes – sheet metal forming by stamping and by fluid cell pressure. The objective of the work is to get an efficient code for simulation of the processes. The code should be used in the industry for design of the manufacturing tools and for steering of the process. Typical for our approach is the use of exaggerated inertia terms and an explicit integration method. The positive effects are numerical stability and shorter computer time compared to a quasistatic analysis. Another advantage with this method is that instability phenomena like wrinkling and buckling can be simulated efficiently. The method works well for displacement driven processes as stamping but can not be used directly for force driven processes such as fluid cell pressure. We show an adaptive method which seems to be promising. In order to get a realistic simulation of the springback of the sheet when the load is removed the stress history must be simulated with high accuracy. The simulation time will then be high. We suggest some remedies: A new explicit code is developed based on Kirchhoff-Love shell theory, facet triangular elements and constant strain constant curvature discretization. It is tested against experiments and compared to a well established code.

1. Introduction

Cold punching, Fig 1a, and hydroforming, Fig 1b, of sheet metal are two processes of great importance in automotive industries. Design and manufacturing of pressure tools are so time-consuming that computer simulations are of great importance for speeding up this process. The formed parts are required to have exact form without thin parts or wrinkles. The amount of material used should also be kept as low as possible.

Computer simulation of cold punching with the finite element method was first proposed in 1978 by Wang and Budiansky [1], by Kobayashi and Kim [2] and by Oñate, Zienkiewicz and Heinrich [3]. In the first paper a quasi static membrane

H. A. Mang and F. G. Rammerstorfer (eds.),
IUTAM Symposium on Discretization Methods in Structural Mechanics, 331–338.

formulation and an elasto-plastic formulation for large deformations according to Hutchinson [4] was used.

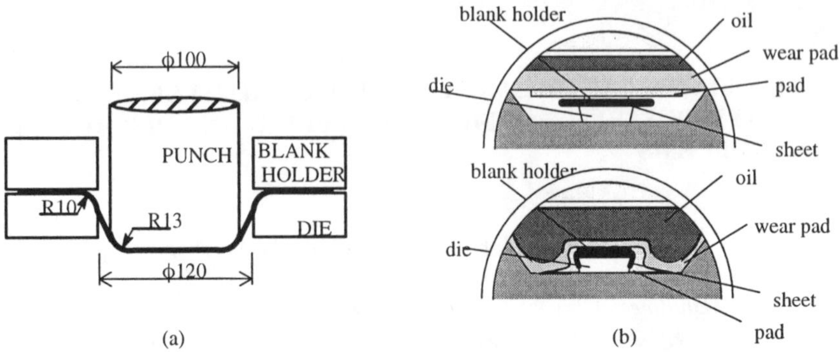

Figure 1. (a):Cold punching; (b):Hydroforming.

The pioneering work was followed by a lot of research and development, at universities and at automotive manufacturers. After a decade of development the first workshop application was made by Ford. They used a code developed by Wang and Tang [5]. At our department in Gothenburg we tried to improve the Wang and Budiansky algorithm with regard to accuracy and efficiency. We compared our simulation results with results from experiments made at the Swedish Institute for Metals Research in Stockholm [6]. In [2] and [3] a plastic flow formulation was used. The disregardence of elasticity was the main disadvantage.

In 1989 Honecker and Mattiasson [7] presented the idea to use a dynamic approach in order to speed up the computer time. The punching process is a relatively slow process. The speed is 0.1-1.0 m/s, so the inertia effect can be disregarded. A dynamic approach can be made more efficient than the quasi static. The reason is that with an explicit time stepping procedure no equation systems have to be solved. Explicit dynamic codes had in 1989 been used since some time with great success for simulation of car collisions.

Explicit time stepping is, however, an only conditionally stable procedure. The size of the critical time step is proportional to the size of the smallest finite element. In order to get required accuracy very small elements are needed in parts of the sheet so the number of time steps will in general be prohibitively large if the process is run with the real velocity. Since the inertia effects are negligible the same solution will be obtained even if the velocity is somewhat increased. Numerical experiments can tell how much the velocity can be increased without destroying the solution by introducing unphysical inertia. Punching in Fig 1 has been simulated with a dynamic approach with the velocity increased with a factor 100. As can be seen from Fig 2, where this solution is compared to a quasi static solution as well as experiments, the dynamic solution gives an average behavior very close to the quasi static as well as the experimental results. The CPU time for the explicit one is, however, only a small fraction of the quasi static one.

The problem solved is displacement driven so the inertia forces are under strict control. The hydroforming process is, however, force driven and can not be analyzed directly with a dynamic approach with higher velocity than the real. We have come around this problem, see [8], by steering the loading adaptively in such a way that the velocity becomes almost constant, see below.

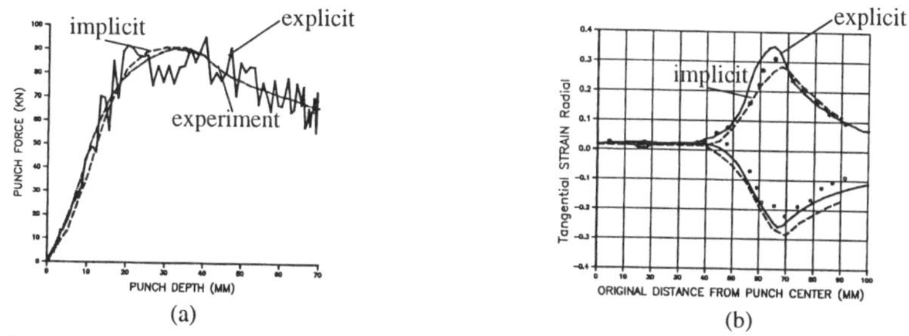

Figure 2. (a):Punch force; (b):Tangential and radial strains.

Another phenomenon in simulation of sheet metals is the springback after the pressure tool is released. The understanding and handling of this problem is very important as the effect on the form of the workpiece is very substantial. We have investigated it in detail [9] and advised ways of finding an efficient simulation procedure.

A main problem with simulation of the springback is the difficulty to calculate correctly the bending stresses obtained when the sheet is drawn over the die corners. Very small elements are needed in this region in order to obtain a useful result. In order to be able to solve this problem efficiently a new adaptive finite deformation shell code is under development, see below.

2. Force driven problems. an adaptive algorithm

A dynamic approach to quasi-static elastic-plastic problems is easy and efficient for displacement driven problems. The unphysical dynamic forces are much more difficult to control when the problem is force driven. Hydroforming in which the load is transferred to the sheet by oil pressure is such a force driven problem.

The success with the dynamic approach is due to the use of explicit time integration method with lumped masses and neglected damping. Further, to make the algorithm efficient the fictitious time for the process should be kept as short as possible but still so long that the inertia (unphysical) effects can be neglected. This is possible for displacement-driven problems for which the inertia effects are well controlled. For force-driven problems we advice an adaptive procedure aiming at getting a loading function that corresponds to a prescribed velocity norm.

We have chosen this norm to be

$$\| v \| = \left(\sum_{m} \frac{f_i \, v_i}{| f_i |} \right) \Big/ m \tag{1}$$

where f_i is a force component, m is the number of non-zero such components and v_i are the corresponding velocity components.

The nominal load increment is

$$\Delta f_0 = \frac{k\, f_{\max}\, \Delta t}{T_{\max}} \tag{2}$$

where $f_{\max}$ and $T_{\max}$ are given quantities – the maximum load factor and the process time.

Further, Δt is the critical time step for numerical stability of the explicit integration method.

In order to get $f_{\max}$ to be reached at $T_{\max}$ a factor k has to be given a positive value determined by experiments.

We define an incremental load factor Δp^n at time n in such a way that the velocity norm is kept within a band around a given target function $\| v \|^*$:

$$(1 - \alpha) \| v \|^* < \| v \|^n < (1 + \alpha) \| v \|^* \tag{3}$$

where α is taken to 0.25.

The load factor is taken as

$$\Delta p^n = - \Delta f_0 \tag{4}$$

if the nominal velocity is over the band or in the band but over the target velocity and tending upwards.

Finally, the load factor is taken as

$$\Delta p^n = 0 \tag{5}$$

if the nominal velocity is within the band and tending towards the target velocity.

When the velocity is under the band or in the band under the target and tending out from the band, the load factor Δp^n is taken as Δf^n where

$$\begin{aligned} \Delta p^n &= \Delta f^n = (1 + \varepsilon)\Delta f^{n-1} && \text{if} \quad \Delta p^{n-1} > 0 \\ \Delta p^n &= \Delta f^n = \Delta f_0 && \text{if} \quad \Delta p^{n-1} = 0 \end{aligned} \tag{6}$$

where ε is a small acceleration factor (taken as $2 \cdot 10^{-3}$) and the nominal velocity is below the band or in the band but below the target velocity and tending downwards.

The algorithm is illustrated in Fig 3.

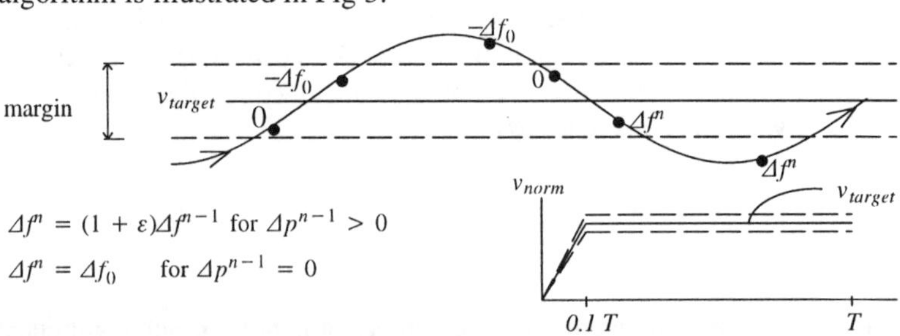

Figure 3. Illustration of the incremental load factor Δp^n.

3. Calculation of springback effects

A considerable springback takes place when the sheet is removed from the tool. A correct simulation of the geometrical effects is of great interest.

The springback is mainly an elastic phenomenon and a real time scale must be used if treated dynamically. The most efficient way is however to change from a dynamic, explicit method to a static method at the end of the forming. Even so, it is not an easy problem to get accurate springback results. The reason is that the stress field at the end of the forming must have high accuracy as the strain release at unloading is proportional to this stress field and the strain release causes the springback deformations.

There are several reasons for the difficulty to get accurate stresses. The main reason seems to be discretization errors at the die corners.

At the NUMISHEET conference in 1995 [9] a simple two dimensional sheet metal problem was studied, Fig 4. A number of research groups had been invited to calculate the springback to be expected and also to present experimental evidence.

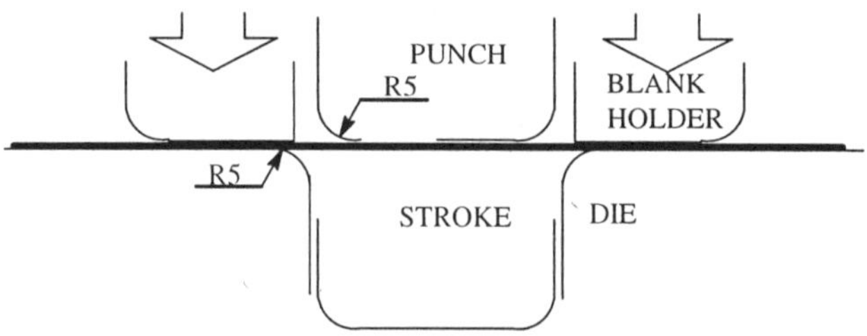

Figure 4. Illustration of the 2D draw bending benchmark problem.

In Fig 5 results from experiments and simulations for two springback angles are compared. The simulations are all made with a dynamic explicit method. The angles are all underestimated.

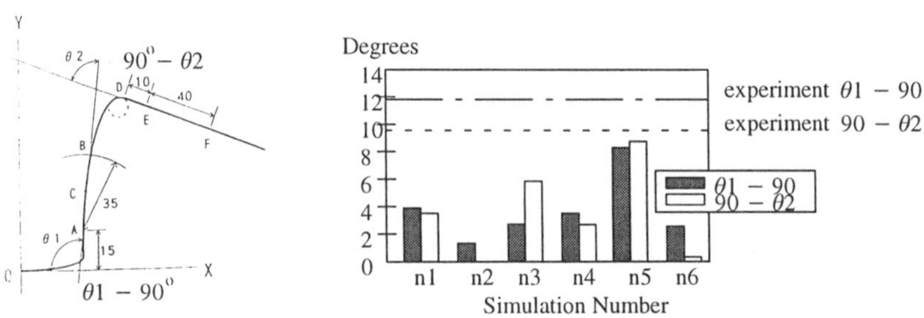

Figure 5. (a):Definition of spring back parameters; (b):Benchmark results from prediction of springback angles.

The longitudinal strains and stresses in the two points at the upper surface of the sheet are studied when formed round a corner, Fig 6, see [9].

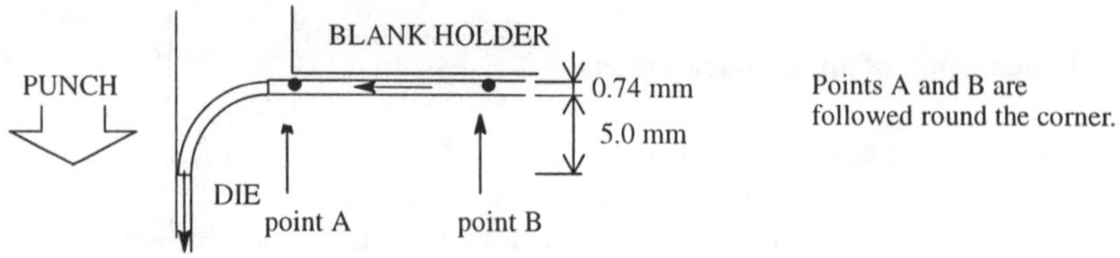

Figure 6. Two points A and B on the outer surface of the sheet strip.

The effective plastic strain increases rapidly when the point reaches the corner and the sheet is bent. It is then constant round the corner after which it again increases rapidly when it is bent back. No further plastic strain is obtained after the corner, Fig 7.

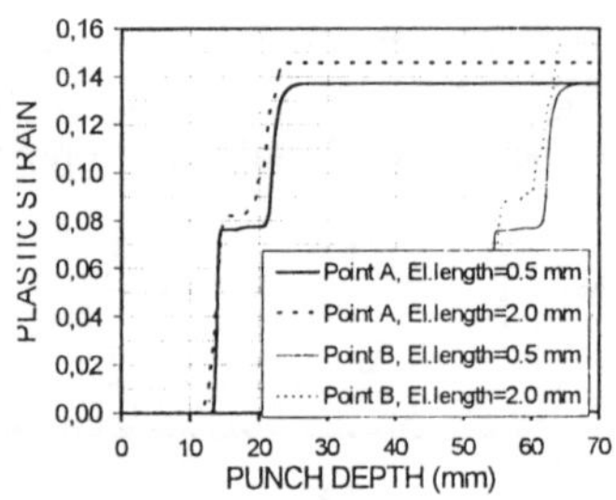

Figure 7. Plastic strain histories for two different element sizes for points A and B illustrated in Fig 6.

The stresses at the same points reach a peak after the bending and falls to negative values when the sheet is bent back. The final stress is very sensitive for the size of the element. First for very small elements, see Fig 8 a constant stress is obtained after the corner.

It was also shown in [9] that the stress relaxation is influenced by the choice of hardening model, representation of the die surface and inertia effects. Kinematic hardening, smooth surface representation and not too fast simulation are remedies.

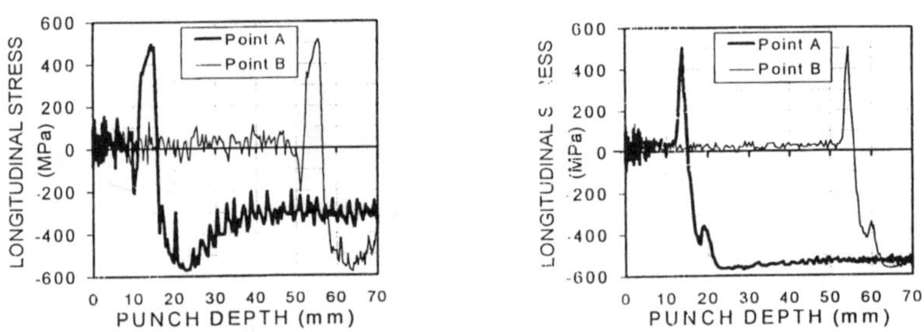

Figure 8. Stress history in points A and B. (a):Element size = 2.0 mm; (b):Element size = 5.0 mm.

4. A new explicit code

An explicit code using the 12 degrees of freedom Morley shell element has been developed, see [10]. The program takes large deformations, large displacements and material non-linearities into account.

In Fig 9 an experiment carried out by Balmer and Witmer [11] is illustrated. The shell is submitted to an impact load in the shaded area of the shell in Fig 9. The numerical simulations are carried out using the four meshes. The displacement of point A using the new explicit code are compared to simulation using DYNA3D and to the experimental result given by Balmer and Witmer, Fig 10.

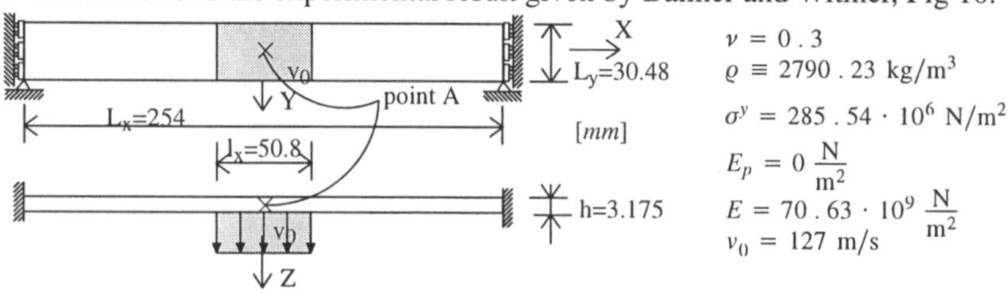

Figure 9. The Balmer and Witmer experiment, geometry and material properties

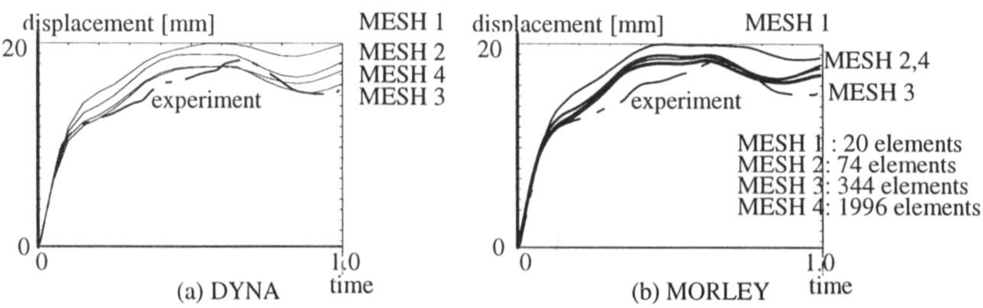

Figure 10. Displacement of point A. (a): DYNA3D; (b) new explicit formulation.

In Fig 11 a buckling test where a thin square plate subjected to tension is described, see [12]. The plate is loaded by a prescribed velocity $v(t)$ in the x-direction. When the plate is submitted to this load wrinkles occur in the y-direction. For the three meshes shown in Fig 12 these wrinkels are viewed in the x-direction at the end of the calculation (t=1 ms).

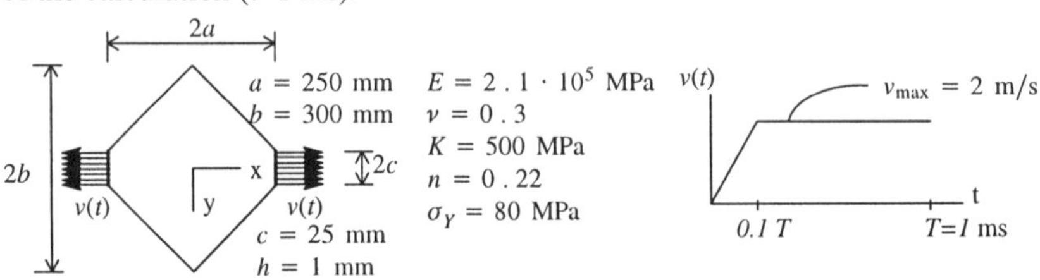

Figure 11. Geometry, material data and load data for the buckling test.

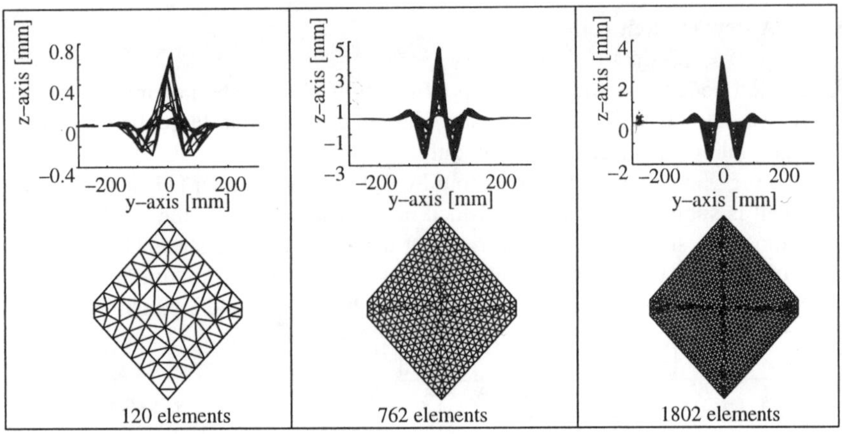

Figure 12. Wrinkles occurring in the buckling test for three meshes.

5. References

[1] N.-M. Wang and B. Budiansky: Analysis of sheet metal stamping by the finite element method', Journal of Applied Mechanics 45, 73-82, 1978.

[2] S. Kobayashi and J. H. Kim: 'Deformation Analysis of Axisymmetric Sheet Metal Forming Processes by the Rigid-Plastic Finite Element Method', Mechanics of Sheet Metal Forming, Plenum Press, New York, 1978.

[3] E. Oñate, O. C. Zienkiewicz and J. C. Heinrich: 'Plastic Flow in Metal Forming. I. Coupled Thermal Behavior in Extrusion. II. Thin Sheet Forming', Applications of Numerical Methods to Forming Process, AMD-Vol. 28, ASME, New York, 1978.

[4] J. W. Hutchinson: Finite solution of nonlinear structural problems, AMD–Vol6, ASME, New York, (1973)

[5] N.-M. Wang and S. C. Tang: 'Analysis of bending effects in sheet forming operations, NUMIFORM86, Balkema, 1986.

[6] M. Saran, E. Schedin, A. Samuelsson, A. Melander and C. Gustafsson:'Numerical and Experimental investigations of deep drawing of metal sheets', J. Materials Shaping Tech. 8:2, 1990.

[7] A. Honecker and K. Mattiasson:'Finite element procedures for 3D sheet forming simulation', NUMIFORM89, Balkena, 1980.

[8] K. Mattiasson, L. Bernspång and A. Samuelsson:'Solution of Quasi-Static, Force-Driven Problems by Means of a Dynamic, Explicit Approach and and Adaptive Loading Procedure', Chalmers University of Technology, Göteborg, Sweden, 1995.

[9] K. Mattiasson, A. Strange and A. Samuelsson:'Simulation of springback in sheet metal forming', NUMIFORM95, 1995.

[10] U. Wendt: 'Linear and non-linear analysis of plates and shells by use of Morley finite elements', Dept. Structural Mechanics, Chalmers University of Technology, Publ. 97:4, 1997.

[11] H. A. Balmer and E. A. Witmer: 'Theoretical-experimental correlation of large dynamic and permanent deformation of impulsively loaded simple structures', Air Force Flight Dynamics Laboratory, Rept. FDP-TDR-64-108, Wright Patterson AF Base, Ohio, 1965.

[12] Y. Tomita and A. Shindo:'Onset and Growth of Wrinkles in Thin Square plates Subjected to diagonal Tension', Int. J. Mech. Sci., Vol. 30, No. 12, 921-931, 1988.

NUMERICAL SIMULATION OF INELASTIC STRUCTURAL CRACK-DAMAGE

W.B. KRÄTZIG and C. KÖNKE

Institute for Statics and Dynamics, Ruhr-University
Universitätsstraße 150, D-44780 Bochum, Germany

1. Introduction

Cracking processes damage a structure in local areas, but their effects can be – apart from very small cracks – not at all local. Cracks can influence the overall load-bearing capacity considerably and thus modify safety and reliability of the structure. The simulation of cracking phenomena consequently requires a computational treatment in a 3D environment in which inelastic phenomena are formulated. Nevertheless, the application of classical structural models of reduced internal dimensions, such as rods or shells, leads to considerable advantages, because the kinematic reduction in general emphasizes the main deformation modes and always saves computing-time. Since for exactness the material description should remain in the E3, the computational evaluation calls for rather sophisticated multi-level simulation strategies from 3D material points over reduced structural models, different finite elements to the final structural level, as described in the following paper.

2. Nonlinear Structural Model Theories

2.1. GENERAL FORMULATION

The computer-simulation of inelastic deformation processes requires, apart from approximative results, a treatment on 3D material point level $\{\sigma^{ij}, \gamma_{ij}; i,j = 1,2,3\}$ on which the respective constitutive relations are formulated. On the other hand, engineering structural responses are often evaluated on the basis of classical

H. A. Mang and F. G. Rammerstorfer (eds.),
IUTAM Symposium on Discretization Methods in Structural Mechanics, 339–348.

mechanical models of reduced internal dimensions, mainly for easier interpretation of the results and for saving of computing-time.

All dimensionally reduced structural models as well as 3D continua posses the same **mathematical structure** as follows. Loads **p** and stress variables $\boldsymbol{\sigma}$ are connected by the **dynamic equilibrium conditions**

$$-\mathbf{p} = \mathbf{D}_e \cdot \boldsymbol{\sigma} = (\mathbf{D}_{el} + \mathbf{D}_{eN}(\mathbf{u})) \cdot \boldsymbol{\sigma}, \quad \{\mathbf{p}, \boldsymbol{\sigma}\} \in V \tag{1}$$

in which $\mathbf{D}_{el}$ describes a linear, $\mathbf{D}_{eN}$ a nonlinear partial differential operator and linear functional of **u**. Displacement variables **u** and strain measures $\boldsymbol{\varepsilon}$ are linked by the **kinematic relations**

$$\boldsymbol{\varepsilon} = \mathbf{D}_k \cdot \mathbf{u} = (\mathbf{D}_{kL} + \mathbf{D}_{kN}(\mathbf{u})) \cdot \mathbf{u}, \quad \{\mathbf{u}, \boldsymbol{\varepsilon}\} \in V \tag{2}$$

with the kinematic operator $\mathbf{D}_k$ consisting of the linear part $\mathbf{D}_{kL}$ and the nonlinear one $\mathbf{D}_{kN}$, the latter also a linear functional of **u** and/or its covariant derivatives. $\mathbf{D}_e$ and $\mathbf{D}_k$ are formally adjoint differential operators.

The given **field equations** (1, 2), valid in the reference volume V of the structure, are completed by the **boundary conditions** on its surface S. We confirm once again that the described mathematical structure (1, 2) holds for all solid mechanical models (Krätzig 1990), (Basar 1989), if energetically corresponding sets of internal variables are applied.

For general nonlinear material behaviour the standard **constitutive laws** are formulated as initial-value problems in the E3 yielding

$$\dot{\boldsymbol{\sigma}} = \mathbf{C} \cdot \dot{\boldsymbol{\varepsilon}} \;\rightarrow \boldsymbol{\sigma} = \int \dot{\boldsymbol{\sigma}}\, dt = \int \mathbf{C} \cdot \dot{\boldsymbol{\varepsilon}}\, dt \,,$$

$$\boldsymbol{\varepsilon} = \int \dot{\boldsymbol{\varepsilon}}\, dt = \int \mathbf{C}^{-1} \cdot \dot{\boldsymbol{\sigma}}\, dt = \int \mathbf{I} \cdot \dot{\boldsymbol{\sigma}}\, dt \,, \; \{\dot{\boldsymbol{\sigma}}, \dot{\boldsymbol{\varepsilon}}\} \in V3 \,. \tag{3}$$

Herein, $\dot{\boldsymbol{\sigma}}, \dot{\boldsymbol{\varepsilon}}$ describe objective stress-rate and strain-velocity tensors. **I** denotes the material compliance tensor and **C** the corresponding material stiffness.

Using structural models of reduced internal dimensions, $\dot{\boldsymbol{\sigma}}$ and $\dot{\boldsymbol{\varepsilon}}$ have to be integral-transformed into the respective model space V2 of E2 or V1 of E1, generally by numerical integrations.

2.2. VIRTUAL WORK

Introducing the incremental forms of all fields of response variables into the **principle of virtual work**, which describes in a weak formulation dynamic equilibrium (1, 2), we obtain after some tedious transformations explained in detail in (Krätzig 1990) the following **incremental principle of virtual work**:

$$\int_{\overset{\circ}{V}} \overset{\circ}{\rho}\, \overset{+}{\ddot{\mathbf{u}}}^T \delta\mathbf{u}\, d\overset{\circ}{V} + \int_{\overset{\circ}{V}} \left[(\overset{\circ}{\boldsymbol{\sigma}} + \bar{\boldsymbol{\sigma}})^T \frac{1}{2} \delta \overset{++}{\boldsymbol{\varepsilon}} + \overset{+}{\boldsymbol{\sigma}}^T \delta \overset{+}{\boldsymbol{\varepsilon}} \right] d\overset{\circ}{V}$$

$$= \delta\left[\int_{\overset{\circ}{V}} \mathbf{p}^T \bar{\delta}\bar{\mathbf{u}}\, d\overset{\circ}{V} + \int_{\overset{\circ}{S}_t} \bar{\mathbf{t}}^T \bar{\delta}\bar{\mathbf{r}}\, d\overset{\circ}{S}_t - \int_{\overset{\circ}{V}} \overset{\circ}{\rho}\, \ddot{\bar{\mathbf{u}}}^T \bar{\delta}\bar{\mathbf{u}} - \int_{\overset{\circ}{V}} (\overset{\circ}{\boldsymbol{\sigma}} + \bar{\boldsymbol{\sigma}})^T \bar{\delta}\bar{\boldsymbol{\varepsilon}}\, d\overset{\circ}{V} \right]$$

$$+ \int_{\overset{\circ}{V}} \overset{+}{\mathbf{p}}^T \delta\overset{+}{\mathbf{u}}\, d\overset{\circ}{V} + \int_{\overset{\circ}{S}_t} \overset{+}{\mathbf{t}}^T \delta\overset{+}{\mathbf{r}}\, d\overset{\circ}{S}_t = 0 \tag{4}$$

with $\boldsymbol{\varepsilon} = \boldsymbol{\varepsilon} + \bar{\delta}\bar{\boldsymbol{\varepsilon}} + \frac{1}{2!}\bar{\delta}^2\bar{\boldsymbol{\varepsilon}} = \bar{\boldsymbol{\varepsilon}} + \overset{+}{\boldsymbol{\varepsilon}} + \frac{1}{2}\delta\overset{++}{\boldsymbol{\varepsilon}}$,

$$\boldsymbol{\sigma} = \overset{\circ}{\boldsymbol{\sigma}} + \bar{\boldsymbol{\sigma}} + \overset{+}{\boldsymbol{\sigma}} + \overset{++}{\boldsymbol{\sigma}}\ , \tag{5}$$

$$\delta\boldsymbol{\varepsilon} = \delta(\bar{\delta}\bar{\boldsymbol{\varepsilon}}) + \frac{1}{2!}\delta(\bar{\delta}^2\bar{\boldsymbol{\varepsilon}}) = \delta\overset{+}{\boldsymbol{\varepsilon}} + \frac{1}{2}\delta\overset{++}{\boldsymbol{\varepsilon}}$$

$$= (\mathbf{D}_{kL} + \mathbf{D}_{kN}(\overset{\circ}{\mathbf{u}} + \bar{\mathbf{u}}))\cdot\delta\overset{+}{\mathbf{u}} + \mathbf{D}_{kN}(\overset{+}{\mathbf{u}})\cdot\delta\overset{+}{\mathbf{u}} \tag{6}$$

with $\overset{\circ}{...}$ as imperfections or prestresses, $\overline{...}$ as variables of the fundamental state and $\overset{+}{...}$ as incremental changes from the fundamental to the adjacent (neighbouring) state. Like its origin, the incremental principle of virtual work holds independently of the material law applied.

3. Multi-level Simulation Strategies

3.1. DISCRETIZATION AND STRUCTURAL LEVEL

The complete incremental principle of virtual work (4) contains all necessary physical informations for the determination of arbitrary neighbouring states, starting from a known fundamental state in dynamic equilibrium. To obtain the discretized functional (4), we approximate the displacement field $\mathbf{u}^p$ of an arbitrary p-th element by

$$\mathbf{u}^p = \boldsymbol{\Omega}^p \cdot \mathbf{v}^p\ , \tag{7}$$

with $\boldsymbol{\Omega}^p$ as matrix of shape functions and $\mathbf{v}^p$ the column of the element degrees of freedom.

After specification of the intended constitutive behaviour, we insert (7) into the incremental principle of virtual work (6). Applying the variations to the nodal degrees of freedom of the p-th element, we finally derive the **element tangential equation of motion**

$$\mathbf{m}^{\mathrm{p}} \cdot \ddot{\overset{+}{\mathbf{v}}}{}^{\mathrm{p}} + \mathbf{c}_{\mathrm{T}}^{\mathrm{p}} \cdot \dot{\overset{+}{\mathbf{v}}}{}^{\mathrm{p}} + \mathbf{k}_{\mathrm{T}}^{\mathrm{p}} \cdot \overset{+}{\mathbf{v}}{}^{\mathrm{p}} = \mathbf{p}^{\mathrm{p}} - \mathbf{f}_{\mathrm{I}}^{\mathrm{p}} , \tag{8}$$

in which $\mathbf{m}^{\mathrm{p}}$ abbreviates its mass matrix, $\mathbf{c}_{\mathrm{T}}^{\mathrm{p}}$ its tangential damping matrix and $\mathbf{k}_{\mathrm{T}}^{\mathrm{p}}$ its tangential stiffness matrix. $\mathbf{p}^{\mathrm{p}}$ denotes the total element load vector and $\mathbf{f}_{\mathrm{I}}^{\mathrm{p}}$ the internal nodal forces. All coefficient matrices and vectors in (8) are to evaluate from integrals over the element domain. Accomplishing the discretization process for all elements of a certain structure and assembling then all elements together, the **global equation of motion** is received.

3.2 FROM STRUCTURAL LEVEL TO MATERIAL POINT LEVEL

As pointed out, the final simulation strategy is controlled on the **highest level**, the structural one, by the **global tangential equation of motion**. On Fig. 1 the

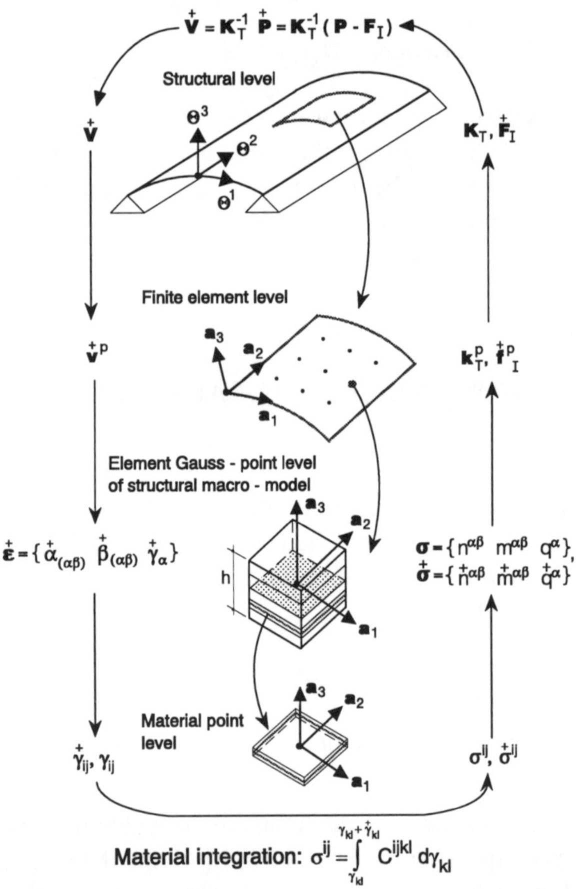

Figure 1. Multi-level simulation strategy

simulation concept is explained by example of a shell response, for simplicity assumed as **time-invariant**. An **incremental-iterative strategy** then is applied to the global tangential stiffness equation.

We start from a certain equilibrated fundamental state ($\overset{+}{\mathbf{P}} = \mathbf{0}$) with a set of internal nodal forces $\mathbf{F}_I$, at which then a new load level $\mathbf{P}$ is chosen. With a new increment $\overset{+}{\mathbf{V}}$ of the global degrees of freedom, we enter at each **finite element** on all **GAUSS-points** and proceed with the aid of the kinematic model $\{\alpha_{\alpha\beta}, \beta_{\alpha\beta}, \gamma_\alpha\}$ to the strain increments at each **material sub-point** of each of the L sub-layers. Here the constitutive law (3) is evaluated, and its results are integrated numerically over all sub-layers of the shell-thickness h arriving back at the original GAUSS-point of the layered finite element level.

The material tensor C^{ijkl} stems from the original material law

$$\dot{\sigma}^{ij} = C^{ijkl}\,\dot{\gamma}_{kl} \quad \rightarrow \quad \overset{+}{\sigma}^{ij} = \int_{\gamma_{kl}}^{\gamma_{kl}+\overset{+}{\gamma}_{kl}} C^{ijkl}\,\dot{\gamma}_{kl}\,dt\,. \tag{9}$$

This step (9) requires the integration of the **constitutive initial-value problem**, embedded into the most inner level, solving for the stress increments, then for the total strains and total stresses. From here we proceed further to the element level with $\mathbf{k}_T^p, \mathbf{f}_I^p$ and then back to the structural level with $\mathbf{K}_T, \mathbf{F}_I$ for the evaluation of the next increment $\overset{+}{\mathbf{V}}$ in the sense of a NEWTON-iteration. In case of time-invariant process simulations, the solution (9) of the constitutive initial-value problem can be carried out using non-physical (endochronic) time t*, or just by multiplying (9) with dt.

4. Elasto-plastic Constitutive Equations

4.1. STANDARD FORMULATION

The described multi-level simulations of elasto-plastic crack damage problems are based on a standard form of constitutive laws. Due to (Bergander 1996) each (time-dependent) material description can be formulated as a system of 1st-order differential equations as follows:

$$\begin{bmatrix} \dot{\boldsymbol{\varepsilon}} \\ \dot{\mathbf{h}} \end{bmatrix} = \begin{bmatrix} \mathbf{I} \\ \mathbf{A} \end{bmatrix} \cdot \dot{\boldsymbol{\sigma}} + \begin{bmatrix} \mathbf{b}_T \\ \mathbf{c}_T \end{bmatrix} \cdot \dot{T} + \begin{bmatrix} \mathbf{b} \\ \mathbf{c} \end{bmatrix}. \tag{10}$$

Herein $\dot{\boldsymbol{\varepsilon}}$ abbreviates the strain-velocities, $\dot{\boldsymbol{\sigma}}$ corresponding stress-rate variables and $\dot{T}$ temperature rates. $\mathbf{I}$ describes the material compliance matrix: quadratic, in general regular and positive-definite; $\mathbf{b}_T$ contains the thermal expansion coefficients. $\mathbf{h}$ denotes a vector of so-called **internal thermodynamic varia-**

bles, which appear in the constitutive description without contributing to the rate of internal work:

$$W_i = \boldsymbol{\sigma}^T \cdot \dot{\boldsymbol{\varepsilon}} \tag{11}$$

Because of the regularity of the compliance matrix **I**, also the inverse form of (10) can be formulated. All matrices in (10) are admitted to depend on the states of stress $\boldsymbol{\sigma}$, strain $\boldsymbol{\varepsilon}$ and temperature T as well as on time t. Matrices and variables $\dot{\boldsymbol{\sigma}}, \dot{\boldsymbol{\varepsilon}}, \mathbf{h}$ characterize in size and content a specific material. The larger the vector **h** is, the more complicated is the nonlinearity or path-dependence of the material considered.

4.2 NON-ASSOCIATED THERMO-PLASTICITY

Exemplifying the general form (10) of the constitutive law for elasto-plastic materials we cite the derived rate formulation for a general thermo-elasto-plastic solid. We start with the thermo-elastic stress-strain-law

$$\dot{\boldsymbol{\sigma}} = \mathbf{E}^e (\dot{\boldsymbol{\varepsilon}} - \dot{\boldsymbol{\varepsilon}}^p) + \mathbf{c}_T^e \cdot \dot{T} = \mathbf{E}^e \cdot \dot{\boldsymbol{\varepsilon}}^e + \mathbf{c}_T^e \cdot \dot{T} \ , \tag{12}$$

given again in the symbolic notation previously used. We further assume the yield surface as

$$F = F(\boldsymbol{\sigma}, \mathbf{h}, T) = 0 \tag{13}$$

with the stress tensor $\boldsymbol{\sigma}$, the temperature change T and the set **h** of **internal hardening parameters**, for which evolution laws can be formulated as

$$\dot{\mathbf{h}} = \dot{\lambda}\, \mathbf{q}(\boldsymbol{\sigma}, \mathbf{h}, T) \ . \tag{14}$$

h shall be specified in a given order by parameters for isotropic, anisotropic, kinematic and cyclic hardening. In (14) $\dot{\lambda}$ describes the well-known plastic multiplier for the plastic flow

$$\dot{\boldsymbol{\varepsilon}}^p = \dot{\lambda}\, \mathbf{m} \ , \tag{15}$$

where **m** illustrates its direction. Finally we abbreviate the normal to the yield surface in stress space as usual by

$$\mathbf{n} = \partial F / \partial \boldsymbol{\sigma} \tag{16}$$

and introduce the matrix product:

$$\frac{\partial F}{\partial \mathbf{h}} \cdot \mathbf{q} = -V \ . \tag{17}$$

We now are able to derive the standard rate form (10) of the constitutive equation for general non-associated thermo-plasticity which reads as follows:

$$\begin{bmatrix} \dot{\boldsymbol{\sigma}} \\ \\ \dot{\mathbf{h}} \end{bmatrix} = \begin{bmatrix} \mathbf{E}^e - \dfrac{\mathbf{E}^e \cdot \mathbf{m} \cdot \mathbf{n}^T \cdot \mathbf{E}^e}{D} \\ \dfrac{\mathbf{q} \cdot \mathbf{n}^T \cdot \mathbf{E}^e}{D} \end{bmatrix} \cdot \dot{\boldsymbol{\varepsilon}} + \begin{bmatrix} \mathbf{c}^{T^e} - \dfrac{\mathbf{E}^e \cdot \mathbf{m} \cdot (\mathbf{n}^T \cdot \mathbf{c}_T^e + F_{,T})}{D} \\ \dfrac{\mathbf{q}^T \cdot (\mathbf{n}^T \cdot \mathbf{c}_T^e + F_{,T})}{D} \end{bmatrix} \cdot \dot{T} \tag{18}$$

using the abbreviated denominator:

$$D = \mathbf{n}^T \cdot \mathbf{E}^e \cdot \mathbf{m} + V \ . \tag{19}$$

All future simulations will be based on simplified material descriptions of the type (18).

5. Damage Evolution

5.1 MATERIAL MICRO-DAMAGE

In continuum damage theories, the effects of discrete **micro-voids** and **micro-cracks** are considered in a smeared sense. This results in an easy way of incorporating damage into existing continuum models. Nevertheless it remains necessary to recognize the limits of the various models used for damage description. The validity range of continuum damage mechanics terminates, if neighbouring voids start to coalescence, forming the first macro-cracks. From thereon, **fracture mechanics** concepts should substitute the continuum damage model. Combining continuum damage models with fracture mechanics models enables studies of complete damage evolution processes from nucleation and growth of first micro defects to the range of macro-damage.

In this course we investigate the damage evolution for ductile, polycrystalline materials, under quasi-static loading. The phase of micro-damage is described by the Cocks/Ashby (Cocks et al. 1980) void-growth model. This model is coupled with a macro-crack model for the description of macro-damage. Increments and directions of crack propagation will be controlled by the micro-damage evolution. Two fundamental advantages of this concept compared with continuum-damage models used to simulate macro-cracking can be observed:

- the physical process of macro-cracking, evolution of new free crack surfaces within the solid, is modelled in an appropriate way (Wawrzynek et al. 1991, Cervenka 1994)
- the mesh depencies of the solution, which have been observed in smeared crack models (de Borst et al. 1992, Oliver et al. 1997) can be avoided. No

regularization techniques have to be applied, because the macro-crack is modelled as a real crack in the numerical model.

The void-growth model is based upon micromechanical material considerations (Cocks et al. 1980). Cocks/Ashby assume that microvoids will nucleate at grain boundaries, inclusions or segregations. In this theory the microvoids are assumed as spherical holes, which will grow due to inelastic deformation of the surrounding matrix material until neighbouring voids coalescence.

The following void-growth evolution equation can be obtained for future use:

$$\dot{f} = \beta(n, p, T_e)\left(\frac{1}{(1-f)^n} - (1-f)\right)d_{ss}\ , \quad \beta = \sin h\left\{2\frac{n-\frac{1}{2}}{n+\frac{1}{2}}\left(\frac{p}{T_e}\right)\right\}. \tag{20}$$

with d_{ss} the equivalent plastic strain rate, p the hydrostatic stress, T_e the equivalent stress and n as a material parameter. Cocks/Ashby assign the value of the power law creep exponent to n. This value has to be determined from experimental results.

5.2 MACRO-DAMAGE SIMULATION

In experimental and numerical investigations it has been found that a critical void-volume fraction f_c of approximately 20% can be defined (Tjotta 1992) after which a macro-crack is initiated. The change from micro- to macro-damage is progressive, depending on material, shape of the test specimen and loading history. The concept applied will include micro- and macro-damage effects in one model (Könke 1994). After passing the limit value of the void-volume fraction f_c, a macro crack is established in the model. Crack propagation is persecuted incrementally, letting the crack grow in finite crack increments. Crack propagation, which depends on the direction of the crack growth and the size of the increment, is controlled by the distribution of the micro-damage variable f in the crack-tip region. With this technique the crack-path remains independent of the Finite Element mesh. Adaptively and automatically, respective new meshes, fitted to the latest geometry, are designed for every new crack increment. All calculated history dependent data-displacements in nodes, stresses, equivalent plastic strains and damage variables in GAUSSpoints – are interpolated over the old mesh and transformed to the new one.

The critical void-volume fraction f_c limits the range of validity for the micro-damage model. After exceedance of f_c, the material fails as a result of void coalescence, leading to local stiffness cancellations and geometry as well as boundary condition changes. Two crack faces must be introduced as free surfaces into the solid. Till today macro-crack propagation in most Finite Element simulation have been simulated with the **Nodal Release Approach** or the **Smeared Crack Approach**, both methods describing the crack trajectories approximately. In this context, a more realistic approach is applied as explained in the next section.

5.3. DELETE-AND-FILL-STRATEGY

In the early 1980's, Saouma and Ingraffea introduced a simulation method for discrete crack propagation without regarding the existing mesh (Wawrzynek et al. 1991). The strategy of this delete-and-fill process is demonstrated in Figure 2, as follows. A group of elements in a region around the new crack-tip is deleted (Figures 2a, b). Then the crack is extended into this region (Figure 2c) and elements are placed around the new crack-tip (Figure 2d). These crack-tip elements i.e. are elements with special features, such as stress singularities at the crack-tip. In the final step, the area left between crack-tip elements and the original mesh is suitable remeshed (Figure 2e). The last picture shows the new open crack, demonstrating that not only the meshing, but also the geometry and the boundary conditions of the system had to be changed (Figure 2f).

In this paper the coupling between micro-damage and macro-crack model is applied as just described. The scalar damage parameter f is calculated in every element at the GAUSSpoints. If the damage value f in one GAUSSpoint reaches

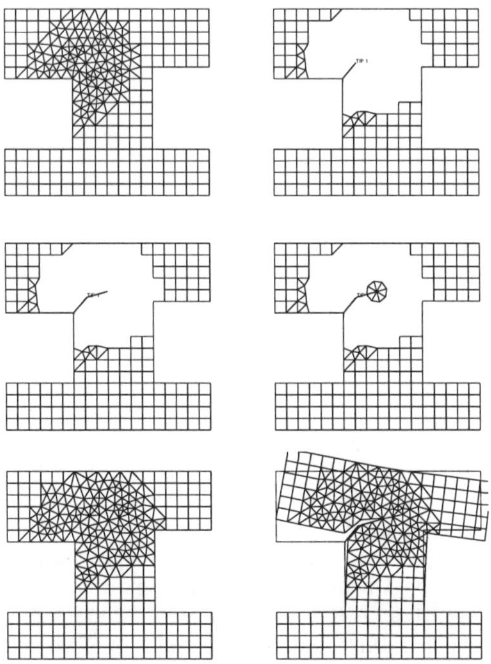

Figure 2. Macro-crack propagation with delete-and-fill method

the critical limit f_c, a macro-crack is initiated from the nearest element corner node - the node with the maximum value for f. A circular shaped region is then generated with its central point in the crack initiation point. With the next step the new crack-tip is positioned in the direction of maximum void volume fraction. In this way crack propagation direction and crack increment are determined. In the last step, geometry and mesh are matched again according to the delete-and-fill method (see Figures 2d – f). The propagation of an existing crack is driven by the distribution of the damage parameter around the crack tip. For further details see (Könke 1995).

Acknowledgement: Financial support of the DFG–National German Science Foundation within the SFB 398 and of the European Union within the HCM Project ERBCHRXCT 940629 is gratefully acklowledged.

6. References

Basar, Y., Krätzig, W.B. 1989: A consistent shell theory for finite deformations, Acta Mechanica, 76, 73-87.

Bergander, H. 1996: Standard Formulations of Constitutive Laws in Structural Dynamics, In: W.B. Krätzig, H.-J. Niemann (Eds.): Dynamics of Civil Engineering Structures, A.A. Balkema, Rotterdam, 125-136.

Cervenka, J. 1994: Discrete Crack Modelling in Concrete Structures, Ph.D. Thesis, Department of Civil Engineering, University of Colorado, Boulder.

Cocks, A.C.F., Ashby, M.F. 1980: Intergranular fracture during powerlaw creep under mulrtiaxial stresses, Metal Science, August-September 1980: 395-402.

de Borst, R., Mühlhaus, H.-B., Pamin, J., Sluys, L.J. 1992: Computational Modelling of Localisation of Deformation, Computational Plasticity – Fundamentals and Applications, Owen, D.R.J., Onate, E., Hinton, E. (eds.), CIMNE.

Könke, C. 1995: Damage evolution in ductile materials: from micro- to macro-damage, Computational Mechanics 15, 497-510.

Könke, C. 1994: Coupling of a micromechanically based void-growth model and a macro-crack model for the simulation of damage evolution in ductile materials, IKIB Techn.-wiss. Mitt. Nr. 94-4, Ruhr-Universität Bochum.

Krätzig, W.B. 1990: Fundamentals of Numerical Algorithms for Static and Dynamic Instability Phenomena, In: W.B. Krätzig, E. Onate (Eds.): Computational Mechanics of Nonlinear Response of Shells, Springer-Verlag, 101-124.

Mathur, K.K., Dawson, P.R. 1987: On Modeling Evolution during the Drawing of Metals, Mechanics of Materials 6: 179-186.

Oliver, J., Cervera, M., Manzoli, O. 1997: On the use of 72 Plasticity Models for the Simulation of 2D Strong Discontinuities in Solids, Computational Plasticity – Fundamentals and Applications, Owen, D.R.J., Onate, E., Hinton, E. (eds.), CIMNE.

Tjotta, St. 1992: Formability and the growth of damage, Numerical Methods in Industrial Forming Processes, Chenot, Wood and Zienkiewicz (eds), Balkema 187-192.

Wawrzynek, P.A., Ingraffea, A.R. 1991: Discrete Modeling of Crack Propagation. Theoretical Aspects and Implementation Issues in Two and Three Dimensions, Department of Structural Engineering, Report Number 91-5, Cornell University, Ithaca, New York.

ASPECTS OF LARGE–STRAIN THERMOPLASTIC ANALYSIS OF SHELL–LIKE STRUCTURES

C. MIEHE AND S. SCHLEY
Institut für Mechanik (Bauwesen) Lehrstuhl I
Universutät Stuttgart
D–70550 Stuttgart, Pfaffenwaldring 7

1. Introduction

The paper presents aspects of non–isothermal elastoplastic analysis of shells at large strains. The two main topics are the construction of a new 8–node brick–type finite shell element for coupled thermomechanical analysis and the setup of a constitutive framework of large–strain thermo–elasto–plasticity in shells based on the notion of a plastic metric. The development is in line with recent works [7], [5], [12], [17], [16], [10] and others on finite elastoplastic analysis of shell–like structures.

On the side of structural modelling, we focus on a continuum based shell parametrization based on *pure displacement degrees* at the top and bottom surface of a shell–like continuum, see e.g. Schoop [13], Parisch [11], Miehe [10] and the references therein. The spatial discretization of this type of shell parametrization can be performed in a straightforward manner on the basis of a nonlinear 8–node brick–type finite shell element proposed by Miehe [10] and denoted there as the SQ1E5A8 element. The element bases on a trilinear interpolation (SQ1) of the displacement field and is equipped with five (E5) gradient–type enhanced strain modes as well as shell–typical assumed strain modifications on the basis of 8 (A8) assembling points. The enhanced strain modifications are based on the *gradient–type enhancing* methodology originally proposed by Simó, Armero & Taylor [15] and affects with 5 additional internal modes the *membrane strains* and the *thickness strains*. Here, we enhance the Jacobian of the Eulerian parameter map of the brick–type shell element by taking into account deformation–dependent *objective* enhanced strains. To the experience of the authors, this approach results in a more robust formulation in the large strains range than the alternative *metric–type enhancing* strategies used in [3] and [2] which extend the small–strain methodology originally proposed in [14] to the Green–Lagrangian strains of the geometric nonlinear theory. The proposed additional assumed strain modification are the shell-typical assumed interpolations of the *transverse shear strains* and the *thickness strains* as introduced for alternative shell element formulations by Dvorkin & Bathe [4] and Betsch & Stein [1]. They are obtained based on the introduction of 8 assembling points for discrete assumed strains. The brick–type shell element is

H. A. Mang and F. G. Rammerstorfer (eds.),
IUTAM Symposium on Discretization Methods in Structural Mechanics, 349–360.

extended to non–isothermal applications by introducing mechanical displacement–type and thermal temperature–type discrete nodal degrees. It has a well–defined strain–driven interface to a three–dimensional thermoplastic constitutive model.

On the side of the constitutive modelling, we extend the isothermal framework of finite strain plasticity in shells proposed by Miehe [10] to the non–isothermal application under consideration. The constitutive equations are exclusively formulated in a *geometric setting relative to the parameter manifold* associated with the shell–typical curvilinear coordinate space. The framework is based on the introduction of an a priori six–dimensional *plastic metric*, which is assumed to describe locally the history–dependent inelastic material response in the sense of an internal variable formulation, and uses elastic domains in the space of the *plastic force* conjugate to the plastic metric. A key ingredient of the framework presented here is the representation of the constitutive response functions in terms of principal strains and principal stresses associated with *dual co– and contra–variant eigenvalue triads defined on the parameter manifold* of the shell. The associated dual vector triads are left and right eigenvectors of a mixed–variant elastic strain tensor and form a natural basis of the tangent and cotangent spaces of the parameter manifold. As proposed by Miehe [9, 10] the eigenvectors are *normalized with respect to the plastic metric* which results in an extremely simple representation of the partial thermoelastic response in spectral form.

2. Parametrization of the Shell Continuum

We consider a shell as a standard continuum where one dimension – the one in the thickness direction – is typically of lower order than the other dimensions. As a consequence, standard local parametrizations of the shell manifolds introduce independent parametrizations of a *reference surface* and a *shell fiber*. This parametrization is typically carried out in terms of curvilinear coordinates. Therefore, we consider a domain $\mathcal{A} \subset \mathbb{R}^3$ which we call the *parameter manifold* of the shell. This domain has the particular cartesian structure $\mathcal{A} = \mathcal{M} \times \mathcal{H}$ where $\mathcal{M} \subset \mathbb{R}^2$ is the *parameter space of the reference surface* and $\mathcal{H} \subset \mathbb{R}$ the *parameter space of the shell fiber*. Elements $\boldsymbol{\theta} \in \mathcal{A}$ are called the *local curvilinear coordinates* of the shell. The geometric relationships are visualized in Figure 1.

The local parametrization of the reference configuration $\mathcal{B} \subset \mathbb{R}^3$ of the shell is determined by the map $\hat{\boldsymbol{X}} : \mathcal{A} \to \mathcal{B}$. This map is assumed to have the particular form

$$\hat{\boldsymbol{X}} = \hat{\boldsymbol{M}}(\theta^1, \theta^2) + \theta^3 \hat{\boldsymbol{D}}(\theta^1, \theta^2) \tag{1}$$

where $\hat{\boldsymbol{M}}$ and $\hat{\boldsymbol{D}}$ are the parametrizations of the Lagrangian reference surface and the Lagrangian director field, respectively. Now let $\boldsymbol{J} : T_{\theta}\mathcal{A} \to T_X\mathcal{B}$ with components $J^A{}_i := \partial \hat{X}^A / \partial \theta^i$ denote the tangent map of $\hat{\boldsymbol{X}}$ and $\boldsymbol{J}^{-T} : T^*_{\theta}\mathcal{A} \to T^*_X\mathcal{B}$ the dual normal map. We denote

$$\bar{\boldsymbol{G}} := \boldsymbol{J}^T \boldsymbol{G} \boldsymbol{J} = \hat{\boldsymbol{X}}_{,i} \cdot \hat{\boldsymbol{X}}_{,j} \tag{2}$$

at $\boldsymbol{\theta} \in \mathcal{A}$ as the *covariant reference metric* in the representation with respect to the parameter manifold. Using geometric terminologies, we view $\bar{\boldsymbol{G}}$ as the *pull*

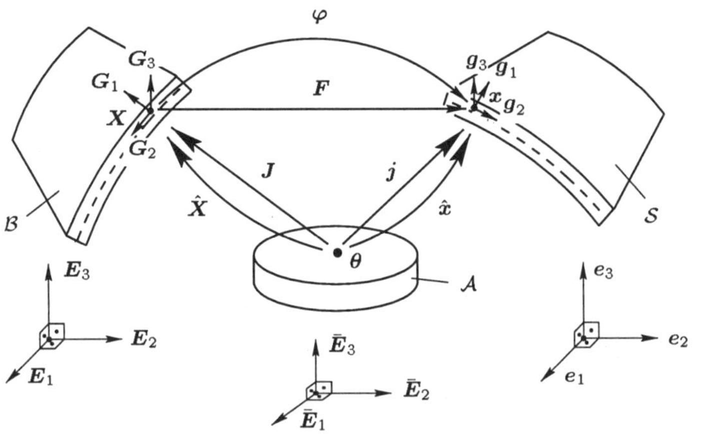

Figure 1. Geometry of a Shell. We denote $\mathcal{B}$, $\mathcal{S}$ and $\mathcal{A}$ as the *Lagrangian*, *Eulerian* and *parameter manifold* of the shell, where $\mathcal{B}$ is associated with an initial configuration and $\mathcal{S}$ with a current configuration of a local patch of the shell. $\hat{\boldsymbol{X}}$, $\hat{\boldsymbol{x}}$ and φ are *nonlinear point maps* which relate *points* $\boldsymbol{X} \in \mathcal{B}$, $\boldsymbol{x} \in \mathcal{S}$ and $\boldsymbol{\theta} \in \mathcal{A}$. The associated *linear tangent maps* $\boldsymbol{J}$, $\boldsymbol{j}$ and $\boldsymbol{F}$ relate *vectors* defined on $\mathcal{B}$, $\mathcal{S}$ and $\mathcal{A}$, respectively. $\{\boldsymbol{E}_i\}_{i=1,3}$, $\{e_i\}_{i=1,3}$ and $\{\bar{\boldsymbol{E}}_i\}_{i=1,3}$ are standard *cartesian* covariant bases associated with $\mathcal{B}$, $\mathcal{S}$ and $\mathcal{A}$. $\{\boldsymbol{G}_i\}_{i=1,3}$ and $\{\boldsymbol{g}_i\}_{i=1,3}$ are *curvilinear* covariant bases at $\boldsymbol{X} \in \mathcal{B}$ and $\boldsymbol{x} \in \mathcal{S}$, respectively.

back of the standard cartesian metric $\boldsymbol{G} = \delta_{AB}$ of the Lagrangian manifold to the parameter manifold.

The local parametrization of the current configuration $\mathcal{S} \subset \mathbb{R}^3$ of the shell *at a given time* $t \in \mathbb{R}_+$ is obtained in a completely analogous way and determined by the map $\hat{\boldsymbol{x}} : \mathcal{A} \rightarrow \mathcal{S}$. Again, this map is assumed to have the particular form

$$\hat{\boldsymbol{x}} = \hat{\boldsymbol{m}}(\theta^1, \theta^2; t) + \theta^3 \hat{\boldsymbol{d}}(\theta^1, \theta^2; t) \tag{3}$$

where $\hat{\boldsymbol{m}}$ and $\hat{\boldsymbol{d}}$ are the parametrizations of the Eulerian reference surface and the Eulerian director field, respectively. Now let $\boldsymbol{j} : T_\theta \mathcal{A} \rightarrow T_x \mathcal{S}$ with components $j^a{}_i := \partial \hat{x}^a / \partial \theta^i$ denote the tangent map of $\hat{\boldsymbol{x}}$ and $\boldsymbol{j}^{-T} : T^*_\theta \mathcal{A} \rightarrow T^*_x \mathcal{S}$ the dual normal map. We denote

$$\bar{\boldsymbol{C}} := \boldsymbol{j}^T \boldsymbol{g} \boldsymbol{j} = \hat{\boldsymbol{x}}_{,i} \cdot \hat{\boldsymbol{x}}_{,j} \tag{4}$$

at $\boldsymbol{\theta} \in \mathcal{A}$ as the *covariant current metric* in the representation with respect to the parameter manifold. $\bar{\boldsymbol{C}}$ is the *pull back* of the standard cartesian metric $\boldsymbol{g} = \delta_{ab}$ from the Eulerian manifold to the parameter manifold.

In order to get a strain–like interface with independent non–constant terms of (4) to three–dimensional constitutive models for the stress response in the shell continuum, we need a parametrizations of the *shell deformation* with *at least six local parameters*. We use here a parametrization which introduces two displacement fields $\hat{\boldsymbol{u}}$ and $\hat{\boldsymbol{v}}$ for the description of a pair of associated material points at the top and bottom surface of the shell continuum, see also Schoop [13] and Parisch [11]. The formulation results in the *linear* relationships

$$\hat{\boldsymbol{m}} = \hat{\boldsymbol{M}} + (\hat{\boldsymbol{u}} + \hat{\boldsymbol{v}})/2 \qquad \text{and} \qquad \hat{\boldsymbol{d}} = \hat{\boldsymbol{D}} + (\hat{\boldsymbol{u}} - \hat{\boldsymbol{v}}) \tag{5}$$

between the Lagrangian and Eulerian reference surfaces and director fields in (1) and (3), respectively, when $H = |\boldsymbol{D}|$ is the initial thickness of the shell. Note that (5) induces a *generalized shell displacement vector* $(\theta^1, \theta^2; t) \mapsto \underline{\boldsymbol{s}} := [\hat{\boldsymbol{u}}, \hat{\boldsymbol{v}}]^T \in \mathbb{R}^6$.

The current absolute temperature field is parametrized on the parameter manifold $\mathcal{A}$ of the shell by the map $\hat{\vartheta} : \mathcal{A} \to \mathbb{R}_+$. This map is assumed to have the particular form

$$\hat{\vartheta} = \hat{\vartheta}_m(\theta^1, \theta^2; t) + \theta^3 \hat{\vartheta}_d(\theta^1, \theta^2; t) \tag{6}$$

in analogy to (1) and (3), where $\hat{\vartheta}_m$ and $\hat{\vartheta}_d$ are denoted as the membrane temperature and the bending temperature gradient, respectively.

3. Thermomechanical Initial Boundary Value Problem

With regard to the construction of algorithms for the solution of the initial boundary problem we consider first the field equations of coupled thermoplasticity for the standard three–dimensional continuum, which can be written as a first order system

$$\left.\begin{aligned} \dot{\boldsymbol{\varphi}} &= \boldsymbol{v} \\ \dot{\boldsymbol{v}} &= \tfrac{1}{\rho_0} j \operatorname{div}[\boldsymbol{\tau}/j] \\ \dot{\vartheta} &= \tfrac{1}{c} j \operatorname{div}[-\boldsymbol{q}/j] + \tfrac{1}{c}\mathcal{D} \end{aligned}\right\} \tag{7}$$

with respect to the Eulerian manifold $\mathcal{S}$. These equations cover essentially the definition of the spatial velocity field and the field equations of the balance of momentum and energy, here written in terms of the Kirchhoff stress field $\boldsymbol{\tau}$ and the Kirchhoff–type heat flux $\boldsymbol{q}$ on the Eulerian manifold. In the $(7)_3$ we have neglected terms associated with the latent thermoelastic heating. $\mathcal{D}$ is the mechanical dissipation field, c the heat capacity and $j := \det[\boldsymbol{j}\boldsymbol{J}^{-1}]$ the determinant of the local deformation gradient of the shell continuum. We regard the right hand side of (7) as a function of the *primary variables* $\{\boldsymbol{\varphi}, \boldsymbol{v}, \vartheta\}$, the deformation field, the velocity field and the temperature field. Note that the local deformation $\boldsymbol{\varphi}$ and velocity $\boldsymbol{v}$ of material points of the shell continuum are determined by the generalized shell deformation vector $\underline{\boldsymbol{s}}$ defined in (5).

We now consider a time interval $[t_n, t_{n+1}] \subset \mathbb{R}$ where the solution at time t_n is assumed to be known. Then the initial-boundary-value problem of coupled thermoplasticity is completed by the initial conditions $\{\boldsymbol{\varphi}, \boldsymbol{v}, \vartheta\}|_{(t=t_n)} = \{\boldsymbol{\varphi}_n, \boldsymbol{v}_n, \vartheta_n\}$ for the primary variables. The essential boundary conditions for the deformation field and the temperature field are $\boldsymbol{\varphi} = \bar{\boldsymbol{\varphi}}$ on Γ_φ and $\vartheta = \bar{\vartheta}$ on Γ_ϑ. Finally, one has to take into account boundary conditions for the tractions $\boldsymbol{t}$ and the heat exchange q, i.e. $\boldsymbol{t} := \boldsymbol{\tau} \cdot \boldsymbol{n} = \bar{\boldsymbol{t}}$ on $\boldsymbol{\varphi}(\Gamma_t)$ and $q := \boldsymbol{q} \cdot \boldsymbol{n} = \bar{q}$ on $\boldsymbol{\varphi}(\Gamma_q)$ with the decompositions of the surface of the Lagrangian manifold $\Gamma = \Gamma_\varphi \cup \Gamma_t = \Gamma_\vartheta \cup \Gamma_q$ and $\Gamma_\varphi \cap \Gamma_t = \emptyset$, $\Gamma_\vartheta \cap \Gamma_q = \emptyset$ of the shell continuum.

An operator split algorithm for the solution of the coupled system (7), see e.g. Miehe [8] and references therein, then yields for the *quasi–static case* by applying a backward Euler scheme to $(7)_3$ the following global solution procedure for the coupled problem. We first solve within an *isothermal predictor* the balance of momentum

$$G_I := \int_{\mathcal{A}} \tfrac{1}{2}\delta\bar{\boldsymbol{C}} : \bar{\boldsymbol{S}}\, JdV - \int_{\boldsymbol{\varphi}(\Gamma_t)} \delta\boldsymbol{\varphi} \cdot \bar{\boldsymbol{t}} dA = 0 \tag{8}$$

for the mechanical variables $\{\varphi, \boldsymbol{v}\}$ at time t_{n+1} with frozen temperature $\{\vartheta_n\}$. We then solve within a *heat conduction corrector* the balance of energy

$$G_{II} := \int_{\mathcal{A}} \{\delta\vartheta [\frac{c}{\Delta t}(\vartheta - \vartheta_n) - \mathcal{D}] - \delta\bar{\boldsymbol{T}} \cdot \bar{\boldsymbol{Q}}\} J dV + \int_{\varphi(\Gamma_q)} \delta\vartheta \bar{q} dA = 0 \quad (9)$$

for the current temperature $\{\vartheta\}$ at frozen deformation $\{\varphi, \boldsymbol{v}\}$.

Observe that the domain parts of the weak forms (8) and (9) are written in terms of a representation relative to the parameter manifold of the shell continuum in terms of the convected stresses $\bar{\boldsymbol{S}} := \boldsymbol{j}^{-1}\boldsymbol{\tau}\boldsymbol{j}^{-T}$ and the convected heat flux $\bar{\boldsymbol{Q}} := \boldsymbol{j}^{-1}\boldsymbol{q}$. Furthermore, we have set $\frac{1}{2}\delta\bar{\boldsymbol{C}} := \text{sym}[\nabla_\theta^T(\delta\varphi)\boldsymbol{g}\boldsymbol{j}]$, $\delta\bar{\boldsymbol{T}} := \nabla_\theta(\delta\vartheta)$ and $J := \det[\boldsymbol{J}]$.

4. Brick–type Finite Shell Element SQ1E5A8

The spatial discretization of the weak forms (8) and (9) is performed by a coupled thermomechanical finite shell element. We here use a brick–type shell element proposed by Miehe [10], which meets in a natural format the geometric constraints of the shell geometry given in (1), (3), (5), (6) above. The element bases on the classical 3D brick with trilinear interpolation functions where the θ^3–parameter is assumed to be associated with the thickness direction of the shell. The element, visualized in Figure 2, is equipped with 5 enhanced strain modes and 8 assembling points for assumed strain interpolations.

Within the basic underlying displacement approach we first discretize the geometry

$$\boldsymbol{X} = \textstyle\sum_{I=1}^n N^I \boldsymbol{X}_I \qquad \text{and} \qquad \boldsymbol{x} = \textstyle\sum_{I=1}^n N^I \boldsymbol{x}_I \quad (10)$$

with the trilinear interpolation functions $N^I(\theta^1, \theta^2, \theta^3) = \overset{0}{N}{}^I(\theta^1, \theta^2) + \theta^3 \overset{1}{N}{}^I(\theta^1, \theta^2)$. Observe that these interpolation functions are partitioned in view to a *numerical pre–integration* in the thickness direction. This yields Lagrangian and Eulerian Jacobian

$$\boldsymbol{J} = \textstyle\sum_{I=1}^n \boldsymbol{X}_I \otimes \nabla N^I \qquad \text{and} \qquad \boldsymbol{j} = \textstyle\sum_{I=1}^n \boldsymbol{x}_I \otimes \nabla N^I \quad (11)$$

of the displacement model and therefore the current metric and reference metric $\bar{\boldsymbol{C}} = \boldsymbol{j}^T\boldsymbol{g}\boldsymbol{j}$ and $\bar{\boldsymbol{G}} = \boldsymbol{J}^T\boldsymbol{G}\boldsymbol{J}$.

We then introduce an enhanced strain modifications of the membrane and thickness strains based on the methodology conceptually described in Simó, Armero & Taylor [1992]. With respect to the curvilinear coordinates of the shell continuum we consider an *enhanced Eulerian parameter map* with associated enhanced current metric

$$\tilde{\boldsymbol{j}} = \boldsymbol{j}_C + \boldsymbol{j}_E \qquad \text{and} \qquad \tilde{\bar{\boldsymbol{C}}} = \tilde{\boldsymbol{j}}^T\boldsymbol{g}\tilde{\boldsymbol{j}} \quad (12)$$

The non–conforming part is governed by the ansatz $\boldsymbol{j}_E = \boldsymbol{j}_0\boldsymbol{J}_E$ with $\boldsymbol{J}_E := \sum_{e=1}^m \boldsymbol{a}_e \otimes \nabla\tilde{M}^e$ based on the modified gradients $\nabla\tilde{M}^e := \frac{J_0}{J}\nabla M^e(\boldsymbol{J}_0^{-1}\boldsymbol{J})$ of Wilson modes $M^e(\theta^1, \theta^2, \theta^3) = \frac{1}{2}[(\theta^e)^2 - 1]$ and the a priori choosen internal degrees

$$\boldsymbol{a}_1 = \begin{bmatrix} \alpha_1 \\ \alpha_2 \\ 0 \end{bmatrix} \quad ; \quad \boldsymbol{a}_2 = \begin{bmatrix} \alpha_3 \\ \alpha_4 \\ 0 \end{bmatrix} \quad ; \quad \boldsymbol{a}_3 = \begin{bmatrix} 0 \\ 0 \\ \alpha_5 \end{bmatrix} \quad (13)$$

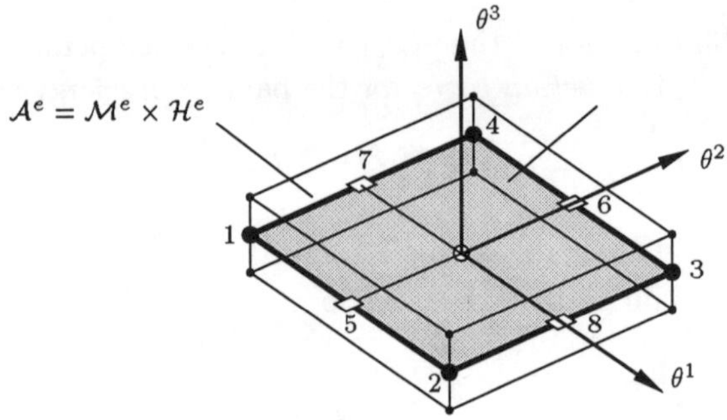

Figure 2. 8–Node Brick–Type Shell Element SQ1E5A8 in the parameter space. $A = 1, ..., 4$ with (-1,-1,0), (+1,-1,0), (+1,+1,0) and (-1,+1,0) denote assumed strain points associated with the thickness strain. $A = 5, ..., 8$ with (0,-1,0), (0,+1,0), (-1,0,0) and (+1,0,0) denote assumed strain points for the transverse shear strain interpolation. The element has *five* internal degrees which govern enhanced strain modes and *eight* collocation points for assumed strain modes.

which can be eliminated by a typical static condension procedure.

As a third step, we introduce assumed strain interpolations of the transverse shear and thickness strains as suggested by Dvorkin & Bathe [4] and Betsch & Stein [1]. The application to the brick–type element yields modification of reference metric and the current metric

$$\tilde{\tilde{C}} = \tilde{C} + [\, \overset{0}{\tilde{\tilde{C}}}{}^{ass} - \overset{0}{\tilde{C}} \,] \qquad \text{and} \qquad \tilde{G} = \bar{G} + [\, \overset{0}{\tilde{G}}{}^{ass} - \overset{0}{\bar{G}} \,] \tag{14}$$

of the form

$$\left.\begin{aligned} \overset{0}{\tilde{\tilde{C}}}{}^{ass}_{33} &:= \textstyle\sum_{A=1}^{4} \frac{1}{4}(1+\theta^1\theta^1_A)(1+\theta^2\theta^2_A)\overset{0}{\tilde{\tilde{C}}}{}^{A}_{33} \\ \overset{0}{\tilde{C}}{}^{ass}_{13} &:= \textstyle\sum_{A=5}^{6} \frac{1}{2}(1+\theta^2\theta^2_A)\overset{0}{\tilde{C}}{}^{A}_{13} \\ \overset{0}{\tilde{C}}{}^{ass}_{23} &:= \textstyle\sum_{A=7}^{8} \frac{1}{2}(1+\theta^1\theta^1_A)\overset{0}{\tilde{C}}{}^{A}_{23} \end{aligned}\right\} \ . \tag{15}$$

The modified metric tensors (14) provide the strain–driven interface to the constitutive model of thermoplasticity discussed in section 5 below. The modified Eulerian and Lagrangian tangent maps can be recovered based on the polar decompositions

$$\tilde{\tilde{j}} = [\tilde{j}\tilde{C}^{-1/2}]\tilde{\tilde{C}}^{+1/2} \qquad \text{and} \qquad \tilde{J} = [J\bar{G}^{-1/2}]\tilde{G}^{+1/2} \tag{16}$$

similar to a proposal of Dvorkin, Pantuso & Repetto [5].

The discretization of the current temperature on the shell continuum is assumed to have the form

$$\vartheta = \textstyle\sum_{I=1}^{n} N^I \vartheta_I \tag{17}$$

in terms of the trilinear interpolation functions defined above. Following a proposal of Glaser [6] we assume elementwise constant bending–type temperature to avoid overshooting phenomena be redefining $\overset{1}{N}{}^I(\theta^1, \theta^2) = \frac{1}{8}\theta_I^3 = const.$ This postulates a decoupled membrane–thickness heat conduction response of the shell continuum.

5. A Model of Finite Thermoplasticity in Shells

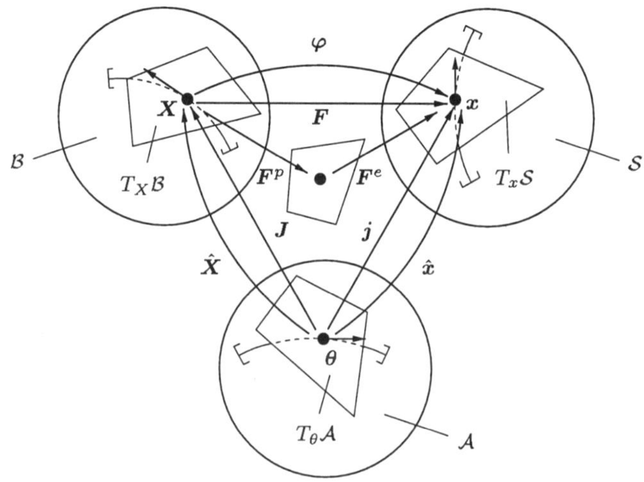

Primary Variables:	$[\cdot]^\flat$	—	$\{\cdot\}^\flat$	—	$(\cdot)^\flat$
Current Metric	$\boldsymbol{C}$	—	$\bar{\boldsymbol{C}}$	—	$\boldsymbol{g}$
Plastic Metric	$\boldsymbol{G}^p$	—	$\bar{\boldsymbol{G}}^p$	—	$\boldsymbol{c}^p$
Conjugate Variables:	$[\cdot]^\sharp$	—	$\{\cdot\}^\sharp$	—	$(\cdot)^\sharp$
Stress	$\boldsymbol{S}$	—	$\bar{\boldsymbol{S}}$	—	$\boldsymbol{\tau}$
Plastic Force	$\boldsymbol{S}^p$	—	$\bar{\boldsymbol{S}}^p$	—	$\boldsymbol{\tau}^p$

Figure 3. Primary and Work–Conjugate Mechanical Dual Variables. The Eulerian variables $(\cdot)$ on $\mathcal{S}$ are connected with their counterparts $\{\cdot\}$ on the parameter manifold $\mathcal{A}$ by compositions with the Eulerian parameter map $\hat{\boldsymbol{x}}$ and the associated linear tangent and normal maps $\boldsymbol{j}$ and $\boldsymbol{j}^{-T}$, respectively. That is $(\cdot)^\flat = \boldsymbol{j}^{-T}\{\cdot\}^\flat\boldsymbol{j}^{-1} \circ \hat{\boldsymbol{x}}^{-1}$ for co-variant fields and $(\cdot)^\sharp = \boldsymbol{j}\{\cdot\}^\sharp\boldsymbol{j}^T \circ \hat{\boldsymbol{x}}^{-1}$ for contra-variant fields. The relationships between Lagrangian variables $[\cdot]$ on $\mathcal{B}$ and counterparts on $\mathcal{A}$ are governed by analogous transformation, determined by the Lagrangian parameter map $\hat{\boldsymbol{X}}$ and associated tangent and normal maps $\boldsymbol{J}$ and $\boldsymbol{J}^{-T}$.

In the subsequent development we describe the history dependence of the inelastic material response by means of additional local variables. The evolution of these additional variables is then governed by constitutive evolution equations. Following the treatments of Miehe [9, 10], we denote by $\bar{\boldsymbol{G}}^p$ with the *initial condition*

$$\bar{\boldsymbol{G}}^p(\boldsymbol{\theta}; t_0) = \bar{\boldsymbol{G}} := \boldsymbol{J}^T\boldsymbol{G}\boldsymbol{J} \tag{18}$$

the covariant *plastic metric on the parameter manifold* $\mathcal{A}$. $\bar{\boldsymbol{G}}^p$ is assumed to be a symmetric, positive definite tensor field for the phenomenological description of the local plastic deformation. Its evolution is determined by a plastic flow rule. Observe that the initial condition in (18) simply states that at the initial time t_0 of the process the plastic metric is assumed to be identical to the reference metric $\bar{\boldsymbol{G}}$ in the representation relative to the parameter manifold $\mathcal{A}$. Note furthermore that the plastic metric can be motivated by a mutliplicative decomposition of the local deformation gradient $\boldsymbol{F} = \boldsymbol{F}^e \boldsymbol{F}^p$ into plastic and elastic parts as indicated in Figure 3 in the way $\bar{\boldsymbol{G}}^p = \boldsymbol{J}^T \boldsymbol{C}^p \boldsymbol{J}$ with $\boldsymbol{C}^p := \boldsymbol{F}^{pT} \boldsymbol{F}^p$.

We now introduce the thermodynamic variables work–conjugate to the current metric $\bar{\boldsymbol{C}}$ defined in (4) and the plastic metric $\bar{\boldsymbol{G}}^p$ introduced above. Let $\boldsymbol{\tau}$ denote the contravariant Eulerian Kirchhoff *stress tensor* work–conjugate to the current metric $\boldsymbol{g}$ in (4). Then the field $\bar{\boldsymbol{S}} := \boldsymbol{j}^{-1} \boldsymbol{\tau} \boldsymbol{j}^{-T}$ is the symmetric Piola *convected stress tensor* on the parameter manifold conjugate to the convected current metric $\bar{\boldsymbol{C}}$ in (4). Furthermore, let the symmetric field $\bar{\boldsymbol{S}}^p$ be the thermodynamic stress–like variable conjugate to the plastic metric $\bar{\boldsymbol{G}}^p$ introduced above and in what follows denoted as the *plastic force*. Then the Eulerian field $\boldsymbol{\tau}^p$, defined by $\boldsymbol{\tau}^p := \boldsymbol{j} \bar{\boldsymbol{S}}^p \boldsymbol{j}^T$ is the *convected plastic force* work–conjugate to the convected Eulerian plastic metric $\boldsymbol{c}^p$. Figure 3 visualizes the geometric relationships between associated variables on the the manifolds under consideration.

Based on the variables introduced above, a canonical set of constitutive equations for large–strain elastoplasticity relative to the parameter manifold has been derived by Miehe [10] on the basis of standard thermodynamical arguments. The structure is summarized in Table 1. Observe that the free energy function $\hat{\psi}$ in Table 1 is formulated in terms of the current metric $\bar{\boldsymbol{C}}$ and the plastic metric $\bar{\boldsymbol{G}}^p$ in a completely general context. We restrict this general framework by considering a particular definition of a mixed–variant (covariant–contravariant) elastic strain measure

$$\bar{\boldsymbol{E}} := \bar{\boldsymbol{C}} \bar{\boldsymbol{G}}^{p-1} \tag{19}$$

which were introduced by Miehe [8] and denoted there as metric transformation tensor. This tensor constitutes an a priori relationship between the current metric and the plastic metric which is assumed to enter the isotropic free energy function, i.e. we consider the particular form $\psi = \hat{\psi}(\bar{\boldsymbol{E}}, A, \vartheta)$. A straightforward exploitation of the constitutive function for the plastic forces in Table 1 then yields the relation $\bar{\boldsymbol{S}}^p := \partial_{\boldsymbol{G}^p} \hat{\psi} = \bar{\boldsymbol{G}}^{p-1} \bar{\boldsymbol{C}} \bar{\boldsymbol{S}}$ to the stress $\bar{\boldsymbol{S}}$. Recall furthermore that the yield criterion function $\hat{\phi}$ has been formulated in terms of the plastic force $\bar{\boldsymbol{S}}^p$ and the plastic metric $\bar{\boldsymbol{G}}^p$. We restrict the possible settings to a particular dependence of the mixed–variant *driving stress tensor* $\bar{\boldsymbol{\Sigma}} := \bar{\boldsymbol{G}}^p \bar{\boldsymbol{S}}^p$ and consider the form $\phi = \hat{\phi}(\bar{\boldsymbol{\Sigma}}, B, \vartheta)$. Taking into account the relationship of the plastic force to the stress, we observe that the driving stress is identical to the mixed–variant Mandel–type stress $\bar{\boldsymbol{C}} \bar{\boldsymbol{S}}$.

Based on the definitions introduced above, we now specify the set of constitutive elastoplastic equations for isotropic response formulated in terms of elastic principal stretches and principal stresses. We therefore consider the eigenvalue problems

$$\bar{\boldsymbol{N}}^i \bar{\boldsymbol{E}} = \lambda_i^2 \bar{\boldsymbol{N}}^i \qquad \text{and} \qquad \bar{\boldsymbol{E}} \bar{\boldsymbol{N}}_i = \lambda_i^2 \bar{\boldsymbol{N}}_i \tag{20}$$

TABLE 1. Thermoplasticity on Parameter Space in Spectral Form

free energy	ψ	$=$	$\hat{\psi}(\varepsilon_1, \varepsilon_2, \varepsilon_3, A, \vartheta)$
stresses	$\bar{\boldsymbol{S}}$	$=$	$\sum_{i=1}^{3} \hat{\psi}_{,i}/\lambda_i^2 \bar{\boldsymbol{N}}^i \otimes \bar{\boldsymbol{N}}^i$
internal force	B	$=$	$-\partial_A \hat{\psi}$
yield function	ϕ	$=$	$\hat{\phi}(\tau_1, \tau_2, \tau_3, B, \vartheta)$
flow rule	$\partial_t \bar{\boldsymbol{G}}^p$	$=$	$\lambda \sum_{i=1}^{3} 2\hat{\chi}_{,i} \bar{\boldsymbol{N}}_i \otimes \bar{\boldsymbol{N}}_i$
evolution	$\partial_t A$	$=$	$\lambda\, \partial_B \hat{\phi}$
dissipation	$\mathcal{D}$	$=$	$\lambda[\sum_{i=1}^{3} (\tau_i \hat{\phi}_{,i}) - B\hat{\phi}_{,B}]$
loading			$\lambda \geq 0\ ;\ \hat{\phi} \leq 0\ ;\ \lambda\, \hat{\phi} = 0$

associated with the mixed–variant elastic strain tensor introduced in (19). Here $\{\lambda_i\}_{i=1\ldots3}$ are the *elastic principal stretches*. $\{\bar{\boldsymbol{N}}^i\}_{i=1\ldots3} \in T_\theta \mathcal{A}$ and $\{\bar{\boldsymbol{N}}_i\}_{i=1\ldots3} \in T_\theta^* \mathcal{A}$ are the dual sets of right and left eigenvectors relative to the parameter manifold $\mathcal{A}$. Note that the dual eigenvectors are related via $\bar{\boldsymbol{N}}^i \cdot \bar{\boldsymbol{N}}_j = \delta^i{}_j$ where $\delta^i{}_j$ denotes the Kronecker symbol. We assume in addition a *normalization of the eigenvectors with respect to the plastic metric*, which plays the role of a *reference metric for the elastic response*, i.e.

$$\bar{\boldsymbol{N}}^i \cdot \bar{\boldsymbol{G}}^p \bar{\boldsymbol{N}}^i = 1 \tag{21}$$

This induces the relationship $\bar{\boldsymbol{N}}_i = \bar{\boldsymbol{G}}^p \bar{\boldsymbol{N}}^i$. The normalization (21) with respect to the plastic metric generates the alternative normalization $\bar{\boldsymbol{N}}^i \cdot \bar{\boldsymbol{C}} \bar{\boldsymbol{N}}^i = \lambda_i^2$ with respect to the current metric. This induces the relationship $\lambda_i^2 \bar{\boldsymbol{N}}_i = \bar{\boldsymbol{C}} \bar{\boldsymbol{N}}^i$. The eigenvalue problems (20) yield the spectral representation $\bar{\boldsymbol{C}} = \sum_{i=1}^{3} \lambda_i^2 \bar{\boldsymbol{N}}_i \otimes \bar{\boldsymbol{N}}_i$ of the current metric and $\bar{\boldsymbol{G}}^p = \sum_{i=1}^{3} \bar{\boldsymbol{N}}_i \otimes \bar{\boldsymbol{N}}_i$ of the plastic metric, respectively, in terms of the *covariant* eigenvector triads. We now assume a particular dependence of the isotropic free energy

$$\psi = \hat{\psi}(\varepsilon_1, \varepsilon_2, \varepsilon_3, A, \vartheta) \tag{22}$$

on the *logarithmic elastic principal strains* $\varepsilon_i := \frac{1}{2}\ln[\lambda_i^2]$ and the generic scalar internal variable A. This results in the spectral representation $\bar{\boldsymbol{S}} = \sum_{i=1}^{3} \tau_i/\lambda_i^2 \bar{\boldsymbol{N}}^i \otimes \bar{\boldsymbol{N}}^i$ of the stresses and $\bar{\boldsymbol{S}}^p = \sum_{i=1}^{3} \tau_i \bar{\boldsymbol{N}}^i \otimes \bar{\boldsymbol{N}}^i$ of the plastic force relative to the parameter manifold in terms of the *contravariant* eigenvector triads and the *principal stresses* $\tau_i := \hat{\psi}_{,i}$. Assuming furthermore the dependence

$$\phi = \hat{\phi}(\tau_1, \tau_2, \tau_3, B, \vartheta) \tag{23}$$

of the isotropic yield function on the principal stresses τ_i and the conjugate internal variable B, we obtain by exploition the evolution equation in Table 1 the representation $\partial_t \bar{\boldsymbol{G}}^p = \lambda \sum_{i=1}^{3} 2\hat{\phi}_{,i} \bar{\boldsymbol{N}}_i \otimes \bar{\boldsymbol{N}}_i$ of the flow rule in spectral form.

A fully implicit exponential–shifted integration of the flow rule in Table 1 within a typical time interval $[t_n, t_{n+1}]$ results a stress update algorithm relative to the parameter manifold of the shell, which can methodically be organized in a geometric pre– and postprocessing and an algorithmic box in the eigenvalue space as outlined in Table 2. The algorithmic box in the eigenvalue space typically consists of a general return mapping scheme which covers a wide range of possible applications. Details of the stress update algorithm are given in Miehe [9, 10].

TABLE 2. Tensorial Stress Update Algorithm on Parameter Manifold

1. *Preprocessing.* The current metric $\bar{\boldsymbol{C}}$ and the *trial* plastic metric $\bar{\boldsymbol{G}}^{p-1*} := \bar{\boldsymbol{G}}_n^{p-1}$ at time t_{n+1} are given. Solve general eigenvalue problem

$$\bar{\boldsymbol{N}}^{i*}(\bar{\boldsymbol{C}}\bar{\boldsymbol{G}}^{p-1*}) = \lambda_i^{2*}\bar{\boldsymbol{N}}^{i*} \qquad \text{with} \qquad \bar{\boldsymbol{N}}^{i*}\cdot\bar{\boldsymbol{G}}^{p*}\bar{\boldsymbol{N}}^{i*} = 1$$

and compute trial logarithmic elastic strains $\varepsilon_i^* := \ln[\lambda_i^*]$.

2. *Constitutive Box.* Compute elastic strains ε_i, principal stresses τ_i and algorithmic moduli $\mathbf{E}_{ij}^{ep}$ in the eigenvalue space by means of a general return mapping scheme.

3. *Postprocessing.* Compute current elastic principal stretches $\lambda_i = \exp[\varepsilon_i]$, update eigenvectors $\bar{\boldsymbol{N}}^i = (\lambda_i/\lambda_i^*)\bar{\boldsymbol{N}}^{i*}$, compute plastic metric

$$\bar{\boldsymbol{G}}^{p-1} = \sum_{i=1}^{3}\bar{\boldsymbol{N}}^i\otimes\bar{\boldsymbol{N}}^i$$

and get stresses and algorithmic tangent moduli

$$\begin{aligned}
\bar{\boldsymbol{S}} &= \sum_{i=1}^{3}\tau_i/\lambda_i^{2*}\bar{\boldsymbol{N}}^{i*}\otimes\bar{\boldsymbol{N}}^{i*}\\
\bar{\mathbb{C}} &= \sum_i^3\sum_j^3[\mathbf{E}_{ij}^{ep} - 2\tau_i\delta_{ij}]/\lambda_i^{2*}\lambda_j^{2*}\,\bar{\boldsymbol{N}}^{i*}\otimes\bar{\boldsymbol{N}}^{i*}\otimes\bar{\boldsymbol{N}}^{j*}\otimes\bar{\boldsymbol{N}}^{j*}\\
&+ \sum_i^3\sum_{j\neq i}^3\frac{\tau_i/\lambda_i^{2*}-\tau_j/\lambda_j^{2*}}{\lambda_i^{2*}-\lambda_j^{2*}}\bar{\boldsymbol{N}}^{i*}\otimes\bar{\boldsymbol{N}}^{j*}\otimes(\bar{\boldsymbol{N}}^{i*}\otimes\bar{\boldsymbol{N}}^{j*}+\bar{\boldsymbol{N}}^{j*}\otimes\bar{\boldsymbol{N}}^{i*})
\end{aligned}$$

Finally the heat flux of the shell continuum is assumed to be governed by the Fourier–type law of the form

$$\bar{\boldsymbol{Q}} = -k\,\bar{\boldsymbol{C}}^{-1}\bar{\boldsymbol{T}} \qquad \text{with} \qquad \bar{\boldsymbol{T}} := \nabla_\theta(\vartheta) \tag{24}$$

in a representation relative to the parameter manifold of the shell continuum.

5. Example

The numerical example outlined in the Figures 4 and 5 is concerned with a typical thermoplastic deformation process at large strains, the thermoplastic necking of a thin–walled tube.

Figure 4a depicts the initial geometry of the half of the tube where a quarter has been discretized by 40 x 25 brick–type shell elements as described in section 4 above. The shell intersection is performed by a particular interface element. The tube is clamped at both ends and then stretched in axial direction within a deformation–controlled process. We consider adiabatic thermal boundary conditions. A thermoplastic material model based on the von Mises plasticity model has been applied as outlined in Miehe [8]. Figure 5 shows deformed configurations of the tube where a necking develops in the middle of the specimen. The distributions of the equivalent plastic strain and the temperature on the surface of the shell are displayed.

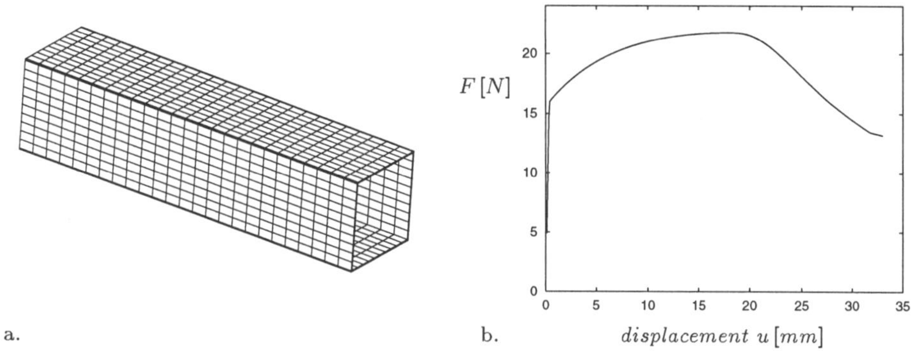

Figure 4. a.) Discretization with 40 × 25 shell elements. b.) Load-deflection curve. (F_{max} = $21.76 N$ at $u = 17.9 mm$)

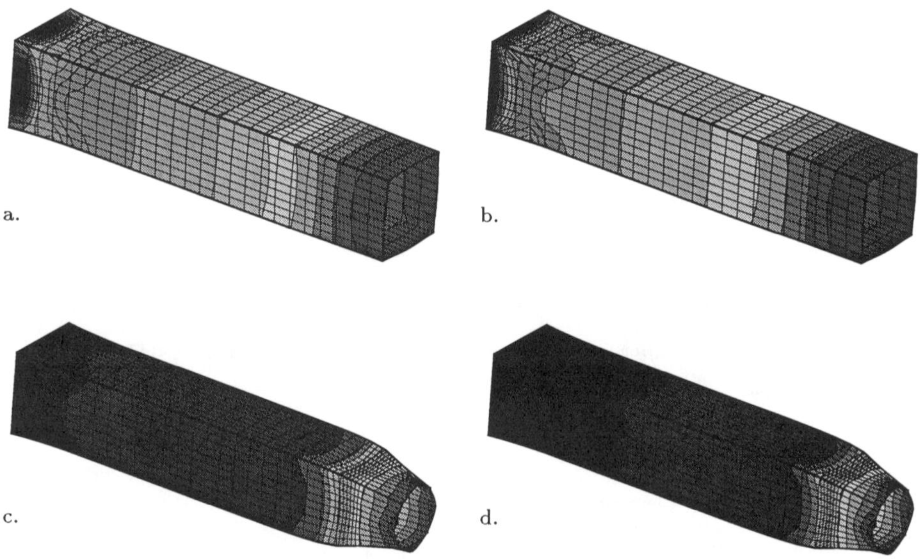

Figure 5. a.) Equivalent plastic strains at displacement $u = 20\,mm$. ($e^p = 0.032 - 0.178$) b.) Temperature distribution at displacement $u = 20\,mm$. ($\theta = 0.0° - 2.804°$) c.) Equivalent plastic strains at displacement $u = 33\,mm$. ($e^p = 0.032 - 1.163$) d.) Temperature distribution at displacement $u = 33\,mm$. ($\theta = 0.0° - 28.52°$)

Acknowledgement

Support for this research was provided by the Deutsche Forschungsgemeinschaft (DFG) under grant Mi 295/4.

References

[1] Betsch, P. and Stein, E. (1996) An Assumed Strain Approach Avoiding Artificial Thickness Straining for a Nonlinear 4–Node Shell Element, *Communications in Applied Numerical Methods* **11**, 899–909.

[2] Betsch, P., Gruttmann, F. and Stein, E. (1996) A 4–node Finite Shell Element for the Implementation of General Hyperelastic 3D–Elasticity at Finite Strains, *Computer Methods in Applied Mechanics and Engineering* **130**, 57–79.

[3] Büchter, N., Ramm, E. and Roehl, D. (1994) Three–Dimensional Extension of Non–Linear Shell Formulation Based on the Enhanced Assumed Strain Concept, *International Journal of Numerical Methods in Engineering* **37**, 2551–2568.

[4] Dvorkin, E.N. and Bathe, K.J. (1984) A Continuum Mechanics Based Four–Node Shell Element for General Nonlinear Analysis, *Engineering Computations* **1**, 77–88.

[5] Dvorkin, E.N., Pantuso, D. and Repetto, E.A. (1995) A Formulation of the MITC4 Shell Element for Finite Strain Elasto–Plastic Analysis, *Computer Methods in Applied Mechanics and Engineering* **125**, 17–40.

[6] Glaser, S. (1996) Gekoppelte Thermomechanische Berechnung dünnwandiger Strukturen mit der Methode der Finiten Elemente, Habilitationsschrift, Institut für Statik und Dynamik der Luft– und Raumfahrtkonstruktionen, Universität Stuttgart.

[7] Ibrahimbegović, A. (1994) Finite Elastoplastic Deformations of Space–Curved Membranes, *Computer Methods in Applied Mechanics and Engineering* **119**, 371–394.

[8] Miehe, C. (1995) A Theory of Large–Strain Isotropic Thermoplasticity Based on Metric Transformation Tensors, *Archive of Applied Mechanics* **66**, 413–427.

[9] Miehe, C. (1996) A Formulation of Finite Elastoplasticity Based on Dual Co– and Contra–Variant Eigenvector Triads Normalized with Respect to a Plastic Metric, *Computer Methods in Applied Mechanics and Engineering*, in press.

[10] Miehe, C. (1996) A Theoretical and Computational Model for Isotropic Elastoplastic Stress Analysis in Shells at Large Strains, *Computer Methods in Applied Mechanics and Engineering*, in press.

[11] Parisch, H. (1995) A Continuum–Based Shell Theory for Non–Linear Applications, *International Journal of Numerical Methods in Engineering* **38**, 1855–1883.

[12] Roehl, D. and Ramm, E. (1996) Large Elasto–Plastic Finite Element Analysis of Solids and Shells with the Enhanced Assumed Strain Concept, *International Journal of Solids and Structures* **33**, 3215–3237.

[13] Schoop, H. (1986) Oberflächenorientierte Schalentheorien endlicher Verschiebungen, *Ingenieur Archiv* **56**, 427- 437.

[14] Simó, J.C. and Rifai, M.S. (1990) A class of mixed assumed strain methods and the method of incompatible modes, *International Journal of Numerical Methods in Engineering* **29**, 1595-1638.

[15] Simó, J.C., Armero, F. and Taylor, R.L. (1993) Improved Versions of Assumed Enhanced Strain Tri–Linear Elements for 3D Finite Deformation Problems, *Computer Methods in Applied Mechanics and Engineering,* **110**, 359–386.

[16] Seifert, B. (1996) Zur Theorie und Numerik finiter elastoplastischer Deformationen von Schalenstrukturen, *Bericht Nr. I–2 (1996) des Instituts für Mechanik (Bauwesen), Universität Stuttgart.*

[17] Wriggers, P., Eberlein, R. and Reese, S. (1996) A Comparison of Three–Dimensional Continuum and Shell Elements for Finite Plasticity, *International Journal of Solids and Structures* **33**, 3309–3326.

FINITE ELEMENT MODELS OF MULTIFACTORAL PHYSICO MECHANICAL PROBLEMS

A. G. UGODCHIKOV and N. A. UGODCHIKOV
Nizhny Novgorod State University
Gagarin ave., 23, Nizhny Novgorod, Russia

1. Introduction

The problems of strength and reliability of elements and constructions are very important in development and design of apparatuses and devices for new equipment. Strain-stress analysis is impossible without modeling of real operating conditions to determine properties of materials and loading modes. To deal with such problems it is necessary to have an opportunity to model connected physical and mechanical fields.

Figure 1 gives a schematic diagram of interaction between the physical fields and quasistatic deformations in constructions. Changes in temperature, concentration of doping admixtures, neutron radiation are known to lead to partially irreversible volumetric strain as well as changes in strain-strength properties, conductivity and other properties. Electric current in conductive structures causes non-uniform and non-stationary heating and panderomotoric volume forces. Radiation action also brings about non-uniform and non-stationary heating. Gradients of the spherical component of stress tensor, temperature and electrical potential determine the intensity of drift flows in diffusion processes.

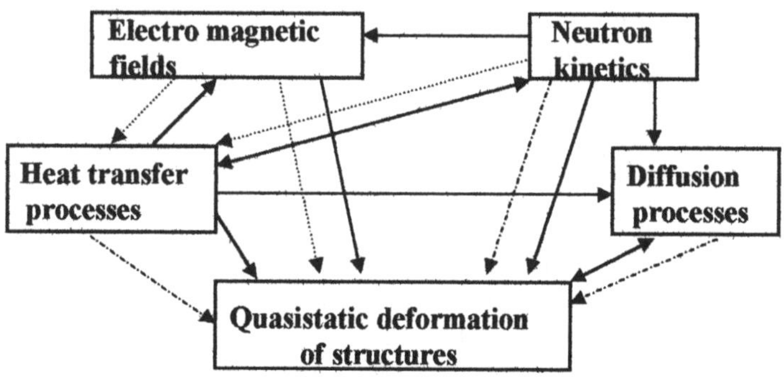

→ - changing properties; ⋯→ - volume loads; -·-→ -volumetric strain.

Figure 1. A schematic diagram of interaction between the physical fields.

H. A. Mang and F. G. Rammerstorfer (eds.),
IUTAM Symposium on Discretization Methods in Structural Mechanics, 361–366.

A method of computer modeling of interrelated physical and mechanical problems in various combinations, and connections between them, is therefore necessary for solving the problems of strength in a number of structures. These are structures like nuclear and thermonuclear plant components or constructions subjected to aggressive action of hydrogen containing media.

2. Generalized variational formulation

It is shown [1] that for the stationary case in the majority of the problems the generalized variational formulation can be used.

$$\chi(\psi)=\frac{1}{2}\int_V([A]\{\psi\})^T[D][A]\{f\}dV-\int_V\{\psi\}^T\{Q\}dV-$$
$$-\int_{S_2}\{\psi\}^T\{f\}dS++\frac{1}{2}\int_{S_3}\{[\eta](\{\psi\}-\{\psi\}_0)\}^T(\{\psi\}-\{\psi\}_0)dS \tag{1}$$

Where $\{\psi\}$ is the vector of unknowns, $[A]$ is the differentiation matrix, $[D]$ is the matrix of media properties, $\{Q\}$ is the vector of volumetric loads, τ is the. time constant, $\{f\}$ is the vector of surface loads, $[A_S]$ is the matrix of directions cosines, $[\eta]$ is the matrix of surface interaction.

Models of the following physical fields can be brought to this form: quasistatic deformation of structures, heat transfer processes, electro-magnetic field for nonferromagnetic conducting solids in quasistatic approximation, electric potential, diffusional doping of semiconducting devices, hydrogen diffusion in metals, neutron kinetics.

To bring the above models to a generalized form three types of differentiation matrices and matrices of directing cosines are used [1,2].

It may also be shown that each summand of the functional is energy type

For the problem of quasistatic deformation of structures the functional 6 coincides with the Lagrangean functional.

If the intensity of magnetic field is the main unknown, the first summand is half the power of Joule's losses; the second and the fourth summands are equal to zero ; the third summand is the power flux across the boundary (integrand is Poynting vector).

If the vector magnetic potential is the main unknown, the first summand is the energy of the electromagnetic field; the second summand is the work of magnetic field on the system of prescribed currents; the third summand is the energy flow across the boundary and the forth one is equal to zero

For diffusion and heat transfer problems functional 6 coincides with the Mikhlin energy functional [3].

The energy sense of the functional provides solution of the problems in curvilinear coordinates or allows to use simplified schemes. It is only necessary to change the differentiation matrix and the matrix of direction cosines.

The peculiarity of the diffusion problems is the necessity of modeling drift flows, which are determined by gradients of pressure, temperature and electric potential. The full flux of particles is the sum of diffusion and drift fluxes in a number of cases it can be written as

$$\bar{J} = \bar{J}_{diff} + \bar{J}_{dr} = -\tilde{D}\left(\nabla C + C\nabla\gamma\right),$$

where $\tilde{D}$ is the diffusion coefficient, C is particle concentration, γ is a function of the coordinates and of time.. In the problem hydrogen diffusion in metals, for example, if the drift flow is dependent on the temperature only $\gamma = -\frac{Q^*}{kT}$, where k is the Boltzmann constant, Q^* is the experimentally determined diffusion heat. After substituting for $V = C\exp\gamma$

$$\bar{J} = D_{eff}\nabla V\text{, where } D_{eff} = \tilde{D}\exp(-\gamma)$$

3. Forming the relations for the solution of pertinent problems

After transformation of the functional (1), presentation the solution for each finite element in the standard form

$$\chi(\{\psi\}) = \sum_{l=1}^{M}\chi_i(\{\psi\}); \quad \{\psi\} = [N]\{\delta\}$$

and variation of the nodal variables the system of linear algebraic equations can be obtained for the solution of linear stationary problems (of all the types considered above).

$$[K]\{\delta\} + \{\tilde{Q}\} = 0$$

A step-by-step linearization is applied for solving non-linear problems like the initial deformation method in case of the problems of elastic-plastic deformation in structures. Here the problem is reduced to the solution of linear algebraic equations at the iteration step.

Non-stationary problems are reduced by the finite element procedure, to a system of ordinary differential equations. Minimization of the functional yields the equations containing time derivatives that are treated as generalized loads, analogous to d'Alambert forces in mechanical problems. The same result can be obtained through direct application of Galerkin's method with the system of piece-wise functions to the respective hyperbolic equation. The same method with the system of piece-wise functions approximating the solution at the time step [4] is applied to the solution of the system of ordinary differential equations. In this case the solution of the problem at the time step is also reduced to the solution of linear algebraic equations.

Thus, the solution of the linear stationary problems, the non-linear problems on

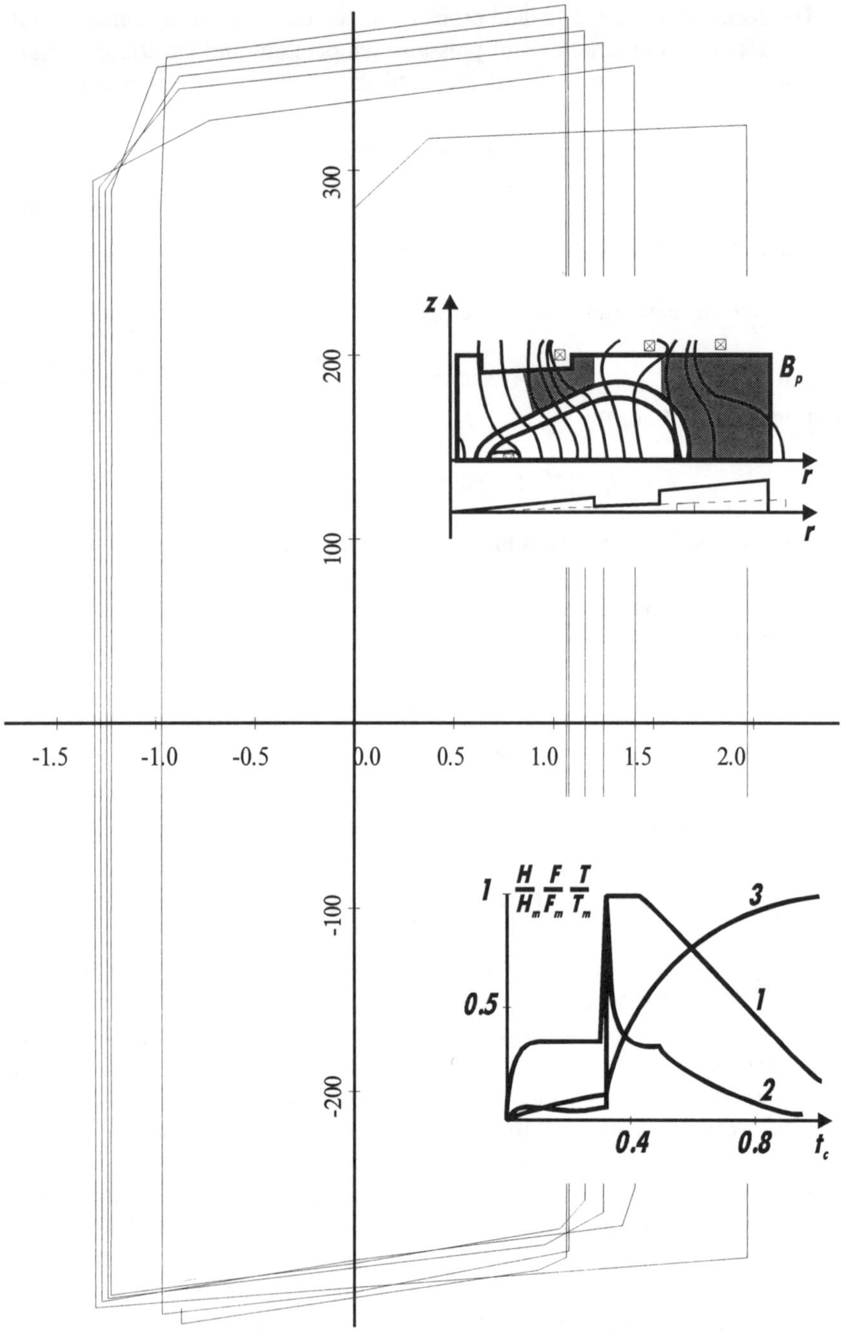

Figure 2. Numerical result of the research of physical-mechanical fields in toroidal magnet bending.

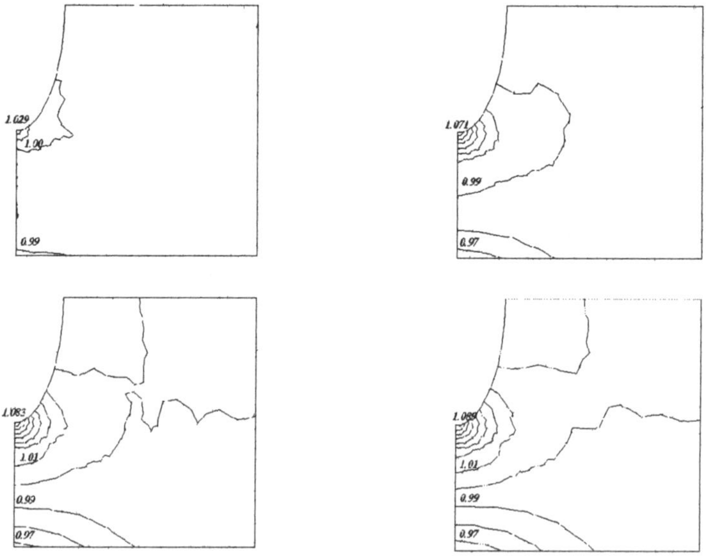

Figure 3. Lines of equal value of hydrogen concentration in different point time.

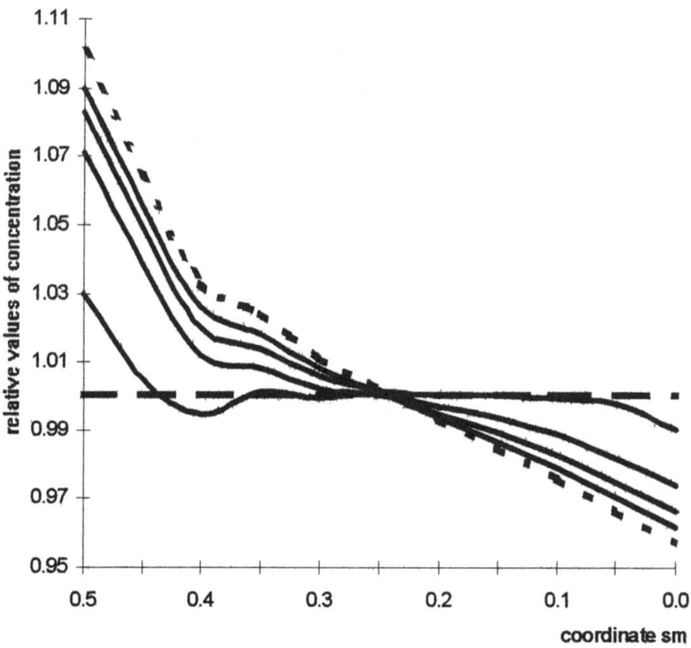

Figure 4. Hydrogen distribution in specimen cross under concentrator in different point time.

iteration step, non-stationary problems at a time step is reduced to a unified discrete model - a system of linear algebraic equations formed according to standard rules.

This methodology allows organization of the procedure solution of all the problems considered in a well-developed FE calculation system by means of widening the library of finite elements and through translation of information from one problem to the other one with respective transformations.

This was the basis for the numerical solution of the problems of magneto-thermo-elasticity, magneto-thermo-plasticity, diffusion doping of semi-conducting devices and hydrogen diffusion in metals under non-uniform stress field and some other.

4. Examples

In the first example the problems of deformation the toroidal magnet field bending is investigated. Coupled non-stationary problems of electro-magnetic fields and heat transfer (the conductivity is temperature dependent) was solved for to determine time and space dependent fields of volume forces and temperature. Some numerical results of the solution of these problems and the results of the research of nonlinear quasistatic deformation of the structure are presented in Figure 2

The upper fragment of Figure 2 depicts the structural circuit diagram of the toroidal windings, consisting of steel straining structure and a bronze current-carring bus, and the poloidal field lines. The lower fragment of Figure 2 shows the time dependans nature of the toroidal magnetic field, of the pondermotive loads and the temperature (curves 1,2 and 3) on surface of the current-carring bus. The main fragment of Figure 2 shhows thhe deformation diagram for the most straight part of the bus under cycle loading followed by a steadily repeating cycle. The ordinat axis shows stress in megapascals, the absciss axis shows deformation in percent. Using a Coffin-Manson-type theory, the strength of the structure can be assessed based on the amplitude of plastic deformation in the cycle.

In the second example the solution of the non-stationary problem of hydrogen diffusion under the condition of a non-uniform state of stress was investigated. The specimen with an elliptic concentrator and a uniform distribution of hydrogen at the initial moment was loaded with a three point bending load. The computed curves of the equal hydrogen concentration of different time points obtained from sequential analyses of deformation and non-stationary diffusion is presented in Figure 3. The Figure 4 shows the hydrogen concentration distribution in the specimen cross under the concentrator in a different time instants. With the dotted lines is shown the initial and the stationary hydrogen distribution.

Our investigation are backed by Russian Fundamental Research Fond (prog. numb. 96-01-00576)

5. References

[1.] Ugodchikov A.G., Ugodchikov N.A. (1995) Generalized Variational Formulation and Computer-Aided Simulation of Physical Mechanical Problems. Proc. of the Internat. Conference on Computational Engineering Science ICES'95 July 30 - Aug.3 Hawaii USA

[2.] Mikhlin S. G. (1970) Variational Methods in Mathematical Physics. Moscow: Nauka,.

[3.] Rozin L.A. (1977) The Finite Element Method as Applied to Elastic Systems. Moscow: Stroyizdat,.

[4.] Zienkewicz O.(1971) The Finite Element Method, McGraw-Hill, New-York.

CONCLUDING REMARKS

Erwin Stein
Institute for Structural and Computational Mechanics
University of Hannover
Appelstraße 9 A, D-30167 Hannover, Germany

Ladies and Gentlemen,

It is surely not an easy task to summarize the topics and the output of our Symposium, not to talk of evaluating this output. Of course, this attempt is subjective and incomplete. In general, I shall follow the sequence of our program.

1. Error-controlled adaptivity of FEM for linear and non-linear problems

We have the competition between residual-based error indicators and those derived from local recovery methods, namely 1. smoothing plus superconvergent techniques as well as 2. the postcomputing of local Dirichlet problems with higher test functions or 3. of local Neuman-problems with higher test function, in order to get anisotropic estimates of the stability constants and the interpolation constants such that at least upper bounds of the global and local relative errors can be computed. We know that the error estimators derived from the local dual problem, namely the local equilibrium, are not subjected to locking and, hence, they are more robust than displacement-based indicators. Furthermore, they are more effective because they are acting more locally. The above mentioned error indicators are derived from global error analysis of a problem.

A second rather new but important approach is local error estimation in order to get quantitative error bounds for single values of displacements, strains, stresses or functionals of them such as the stress intensity factor of a crack, or the integral of the stresses at a certain boundary. What is needed is a guaranteed quantitative error bound at minimum cost for the adaptive process. We have heard contributions to this topic by Dr. Cirak and by myself.

I should like to comment on efforts for error analysis and mesh adaptation. It is not the primal demand to make this postprocessing as cheap as possible but to make it as reliable as possible and to get bounds for quantitative errors. If there is no method available to guarantee the accuracy of, e.g., a stress intensity factor, one can only solve the problem with very big systems of equations, which may be an overkill. Therefore, the rigorous adaptivity is much cheaper. Another subject is the coupled treatment of the truncation error of solution methods and the discretization error, i.e., the coupling of establishing FE-systems and of their solution. This topic was touched by J. Orkisz.

H. A. Mang and F. G. Rammerstorfer (eds.),
IUTAM Symposium on Discretization Methods in Structural Mechanics, 367–371.

So far, there are only few versions of general-purpose adaptive FE-codes available because the data base has to be arranged on the level of the system topology and the geometry. I believe that some adaptive codes will appear in the near future.

It is my conviction that we have to integrate the dimensional-adaptivity and the model-adaptivity with solution-adaptivity in order to consider boundary layers, stiffeners and other restrictions of usually simple mathematical models in undisturbed subdomains, following the engineering rule: "As simple as possible but as complicated as necessary". I am convinced that this overall adaptivity will provide a quality jump of the Finite-Element-Method.

2. Large deformations and buckling including inelastic deformations

This subject as well as local damage and localization of inelastic deformations, followed by global failure modes, is a challenging topic of research.

About 70 % of all FE-applications are still in the linear elastic range, solved by direct solvers. However, coupled problems with different models also play an important role.

Nonlinear codes usually are needed in research-environments and development-areas in engineering. They will be more important in the future, especially in structural and aircraft engineering, because e. g. the Euro-Codes for structural engineering demand a stochastically based safety analysis where the ultimate load states or failure processes of systems have to be analyzed.

On the other hand, most of the available nonlinear codes, especially those implementing inelastic damage and failure, are not yet suitable for less trained people in consulting or industrial companies. The susceptibility to severe errors caused by incompetent use is great.

What can be done to increase the reliability? I believe we need a new type of postprocessing with competitive codes which enable the user to have an intelligent and critical interaction with the computer on the basis of engineering expertise of the user who does not know the complexity of the Finite-Element-Theory and of Finite-Element-Codes. He likes to pose simple but crucial questions concerning equilibrium as well as local and global deformation behavior of a mechanical system which he wishes to control independently. He is not interested in the error in the energy norm but in the maximum norm and so on. I believe that within the adaptive strategy we need a new type of research and software development making programs mentally acceptable, reliable and controllable. The force method will play again a more important role for postprocessing purposes. In general, reliability and efficiency always have to be combined.

3. Inelastic brittle or ductile localization, phase transition and system failure, resulting from monotonic, cyclic or impact loading

Pertinent theories, algorithms and codes are still being developed at a fast rate. These theories have to consider meso- or micromechanical properties and must be strictly goal - oriented in order to be operable. As shown by F.D. Fischer, aging processes, developing local anisotropy with change of material properties and phase transitions have to be regarded. Homogenization of inelastic deformations from micro- or meso- to macro-levels of the material structure are of particular interest in material science and mathematics. Obviously, computational material mechanics plays an important role in connection with computational structural mechanics.

At present, the value of almost all nonlinear computations is the one of case studies. Comparative investigations with disturbed parameters which are stochastic in nature and which can produce large changes of results are needed. This holds for the failure of concrete, metals, composites, and so on.

This brings me to the 4th topic, namely

4. Sensitivity analysis and inverse problems, especially parameter identification of material parameters

As was shown by I. Babuska and M. Kleiber in their lectures, this field of research is still at an early stage of development concerning the quantitative assessment of the quality and reliability of chosen material models (constitutive equations) in the inelastic range in connection with the availability of trustworthy experimental data.

5. Development of linear and nonlinear finite elements for thin-walled structures and composites:

Two research directions can be observed:

a) Improvement of the robustness of elements against locking phenomena - and, of course, with optimal convergence rate with respect to the computational effort
b) Formulations of elements within 3D-kinematics in order to admit 3D-material equations without constraints which usually produce locking

I should like to emphasize that one can show that the Babuska-Brezzi stability condition for the Enhanced Assumed Strain Concept holds only in the geometrically linear case, and we computed counterexamples for shell buckling. The research work of F. Armero and P. Wriggers aiming at stable enhanced nonlinear shell elements have not yet come to a final result. I personally have the impression that new stable mixed elements can have better convergence rates than the EAS-elements, and they should be open for hierarchical generalization.

Similar thoughts hold for the p-version of the FEM, as realized in the Probe-program of I. Szabo.The p-elements are very appealing in elasticity because of their superior convergence, but, of course, the software is not easy, especially the equivalent representation of the geometry of a system. In elastoplasticity higher-order elements are also superior in smooth areas, but not at the yield front of plastic flow processes.

6. Implicit integration schemes in nonlinear dynamics, especially for complex systems.

Satisfaction of conservation principles within time integration in combination with optimal convergence and stability requirements were treated in the lectures by M. Crisfield, E. Wiberg and others. There is room for further mathematical and engineering developments. The enhanced midpoint rule enforcing the conservation of energy can become unstable. Weak forms of conservation of energy by Lagrangian multipliers can transfer erroneous high frequencies into low frequencies and such deteriorate the important results for low frequencies. It seems to me that the algorithms should have diagnosis tools to recognize what type of process is predominant in order to select the appropriate time integration algorithm in a purpose-oriented way.

7. Couplings of rigid- and soft-elastic structures, fluid-structure and acoustic-structure interactions

These fields as well as thermo-plastic-coupling became increasingly important in technology. We heard pertinent lectures by B. Taylor and R. Ohayon. They demonstrated the progress of solving complicated problems by new algorithms and program structures, e. g. for the acoustics of motor cars, aircrafts, ships, etc.. Existing program systems are not sufficiently advanced to cover all the challenging requirements of actual research projects in this field.

8. Competition of FEM, BEM and couplings

BEM is strongly moving to nonlinear problems and adaptivity, benefiting from the dimensional reduction. Contact problems and elastoplastic deformations are treated. Coupling strategies with FEM attract increasing interest, using additive and multiplicative Schwarz methods as solution algorithms.

9. Identification of material and structural data

The development of traditional identification methods with evolution and genetic algorithms, using learning neural networks, or with gradient methods is also showing fast progress. Programs are available which directly image the implicit connection of input - output functions and the mathematical theory. Many questions of convergence and robustness are still open.

I apologize that I did not mention all the challenging problems which were treated during the symposium.

Altogether, it can be observed that controllable adequate modeling with computable error bounds for the parameters becomes more and more important in computational mechanics.

Object-oriented micro- and meso-considerations of materials with anisotropic micro-structure bring us nearer to modern material physics.

In the computational area itself we need deeper mathematical insight into complex finite elements, where mathematical programs like Mathematica or Maple can be very helpful, see also the lecture by P. Wriggers. The same holds for complicated deformation processes like phase transition, inelastic localization a.s.o.. This keeps computational mechanics close to mathematics. Concerning the continuously fast developing hardware one can state, in general, that the more powerful the hardware becomes, the more mathematics combined with deep physical and engineering insight is needed to operate powerful computers effectively.

I got the impression that our symposium reflected these challenges for the future.

Hence, in total, the symposium met the expectations of IUTAM and IACM to be a high-level meeting at the front of computational mechanics.

At the end it is my pleasant and honorable duty to thank Professor Mang and Professor Rammerstorfer, and all the nice people behind them for their great efforts to make this symposium successful, both from the scientific and the social point of view.

It needs really tremendous efforts to prepare all the necessary things - and then to see that some of the eminent people do not appear. This is real life, and you also overcame this in a smooth way.

So, thank you very much, you did a great job to promote computational mechanics and, thus, to strengthen IUTAM and IACM.

Mechanics

FLUID MECHANICS AND ITS APPLICATIONS

Series Editor: R. Moreau

Aims and Scope of the Series

The purpose of this series is to focus on subjects in which fluid mechanics plays a fundamental role. As well as the more traditional applications of aeronautics, hydraulics, heat and mass transfer etc., books will be published dealing with topics which are currently in a state of rapid development, such as turbulence, suspensions and multiphase fluids, super and hypersonic flows and numerical modelling techniques. It is a widely held view that it is the interdisciplinary subjects that will receive intense scientific attention, bringing them to the forefront of technological advancement. Fluids have the ability to transport matter and its properties as well as transmit force, therefore fluid mechanics is a subject that is particularly open to cross fertilisation with other sciences and disciplines of engineering. The subject of fluid mechanics will be highly relevant in domains such as chemical, metallurgical, biological and ecological engineering. This series is particularly open to such new multidisciplinary domains.

1. M. Lesieur: *Turbulence in Fluids*. 2nd rev. ed., 1990 ISBN 0-7923-0645-7
2. O. Métais and M. Lesieur (eds.): *Turbulence and Coherent Structures*. 1991 ISBN 0-7923-0646-5
3. R. Moreau: *Magnetohydrodynamics*. 1990 ISBN 0-7923-0937-5
4. E. Coustols (ed.): *Turbulence Control by Passive Means*. 1990 ISBN 0-7923-1020-9
5. A.A. Borissov (ed.): *Dynamic Structure of Detonation in Gaseous and Dispersed Media*. 1991 ISBN 0-7923-1340-2
6. K.-S. Choi (ed.): *Recent Developments in Turbulence Management*. 1991 ISBN 0-7923-1477-8
7. E.P. Evans and B. Coulbeck (eds.): *Pipeline Systems*. 1992 ISBN 0-7923-1668-1
8. B. Nau (ed.): *Fluid Sealing*. 1992 ISBN 0-7923-1669-X
9. T.K.S. Murthy (ed.): *Computational Methods in Hypersonic Aerodynamics*. 1992 ISBN 0-7923-1673-8
10. R. King (ed.): *Fluid Mechanics of Mixing*. Modelling, Operations and Experimental Techniques. 1992 ISBN 0-7923-1720-3
11. Z. Han and X. Yin: *Shock Dynamics*. 1993 ISBN 0-7923-1746-7
12. L. Svarovsky and M.T. Thew (eds.): *Hydroclones*. Analysis and Applications. 1992 ISBN 0-7923-1876-5
13. A. Lichtarowicz (ed.): *Jet Cutting Technology*. 1992 ISBN 0-7923-1979-6
14. F.T.M. Nieuwstadt (ed.): *Flow Visualization and Image Analysis*. 1993 ISBN 0-7923-1994-X
15. A.J. Saul (ed.): *Floods and Flood Management*. 1992 ISBN 0-7923-2078-6
16. D.E. Ashpis, T.B. Gatski and R. Hirsh (eds.): *Instabilities and Turbulence in Engineering Flows*. 1993 ISBN 0-7923-2161-8
17. R.S. Azad: *The Atmospheric Boundary Layer for Engineers*. 1993 ISBN 0-7923-2187-1
18. F.T.M. Nieuwstadt (ed.): *Advances in Turbulence IV*. 1993 ISBN 0-7923-2282-7
19. K.K. Prasad (ed.): *Further Developments in Turbulence Management*. 1993 ISBN 0-7923-2291-6
20. Y.A. Tatarchenko: *Shaped Crystal Growth*. 1993 ISBN 0-7923-2419-6

Kluwer Academic Publishers – Dordrecht / Boston / London

Mechanics

FLUID MECHANICS AND ITS APPLICATIONS

Series Editor: R. Moreau

21. J.P. Bonnet and M.N. Glauser (eds.): *Eddy Structure Identification in Free Turbulent Shear Flows.* 1993 ISBN 0-7923-2449-8
22. R.S. Srivastava: *Interaction of Shock Waves.* 1994 ISBN 0-7923-2920-1
23. J.R. Blake, J.M. Boulton-Stone and N.H. Thomas (eds.): *Bubble Dynamics and Interface Phenomena.* 1994 ISBN 0-7923-3008-0
24. R. Benzi (ed.): *Advances in Turbulence V.* 1995 ISBN 0-7923-3032-3
25. B.I. Rabinovich, V.G. Lebedev and A.I. Mytarev: *Vortex Processes and Solid Body Dynamics.* The Dynamic Problems of Spacecrafts and Magnetic Levitation Systems. 1994 ISBN 0-7923-3092-7
26. P.R. Voke, L. Kleiser and J.-P. Chollet (eds.): *Direct and Large-Eddy Simulation I.* Selected papers from the First ERCOFTAC Workshop on Direct and Large-Eddy Simulation. 1994 ISBN 0-7923-3106-0
27. J.A. Sparenberg: *Hydrodynamic Propulsion and its Optimization.* Analytic Theory. 1995 ISBN 0-7923-3201-6
28. J.F. Dijksman and G.D.C. Kuiken (eds.): *IUTAM Symposium on Numerical Simulation of Non-Isothermal Flow of Viscoelastic Liquids.* Proceedings of an IUTAM Symposium held in Kerkrade, The Netherlands. 1995 ISBN 0-7923-3262-8
29. B.M. Boubnov and G.S. Golitsyn: *Convection in Rotating Fluids.* 1995 ISBN 0-7923-3371-3
30. S.I. Green (ed.): *Fluid Vortices.* 1995 ISBN 0-7923-3376-4
31. S. Morioka and L. van Wijngaarden (eds.): *IUTAM Symposium on Waves in Liquid/Gas and Liquid/Vapour Two-Phase Systems.* 1995 ISBN 0-7923-3424-8
32. A. Gyr and H.-W. Bewersdorff: *Drag Reduction of Turbulent Flows by Additives.* 1995 ISBN 0-7923-3485-X
33. Y.P. Golovachov: *Numerical Simulation of Viscous Shock Layer Flows.* 1995 ISBN 0-7923-3626-7
34. J. Grue, B. Gjevik and J.E. Weber (eds.): *Waves and Nonlinear Processes in Hydrodynamics.* 1996 ISBN 0-7923-4031-0
35. P.W. Duck and P. Hall (eds.): *IUTAM Symposium on Nonlinear Instability and Transition in Three-Dimensional Boundary Layers.* 1996 ISBN 0-7923-4079-5
36. S. Gavrilakis, L. Machiels and P.A. Monkewitz (eds.): *Advances in Turbulence VI.* Proceedings of the 6th European Turbulence Conference. 1996 ISBN 0-7923-4132-5
37. K. Gersten (ed.): *IUTAM Symposium on Asymptotic Methods for Turbulent Shear Flows at High Reynolds Numbers.* Proceedings of the IUTAM Symposium held in Bochum, Germany. 1996 ISBN 0-7923-4138-4
38. J. Verhás: *Thermodynamics and Rheology.* 1997 ISBN 0-7923-4251-8
39. M. Champion and B. Deshaies (eds.): *IUTAM Symposium on Combustion in Supersonic Flows.* Proceedings of the IUTAM Symposium held in Poitiers, France. 1997 ISBN 0-7923-4313-1
40. M. Lesieur: *Turbulence in Fluids.* Third Revised and Enlarged Edition. 1997 ISBN 0-7923-4415-4; Pb: 0-7923-4416-2

Kluwer Academic Publishers – Dordrecht / Boston / London

Mechanics

FLUID MECHANICS AND ITS APPLICATIONS

Series Editor: R. Moreau

41. L. Fulachier, J.L. Lumley and F. Anselmet (eds.): *IUTAM Symposium on Variable Density Low-Speed Turbulent Flows.* Proceedings of the IUTAM Symposium held in Marseille, France. 1997 ISBN 0-7923-4602-5
42. B.K. Shivamoggi: *Nonlinear Dynamics and Chaotic Phenomena.* An Introduction. 1997 ISBN 0-7923-4772-2
43. H. Ramkissoon, *IUTAM Symposium on Lubricated Transport of Viscous Materials.* Proceedings of the IUTAM Symposium held in Tobago, West Indies. 1998 ISBN 0-7923-4897-4
44. E. Krause and K. Gersten, *IUTAM Symposium on Dynamics of Slender Vortices.* Proceedings of the IUTAM Symposium held in Aachen, Germany. 1998 ISBN 0-7923-5041-3
45. A. Biesheuvel and G.J.F. van Heyst (eds.): *In Fascination of Fluid Dynamics.* A Symposium in honour of Leen van Wijngaarden. 1998 ISBN 0-7923-5078-2
46. U. Frisch (ed.): *Advances in Turbulence VII.* Proceedings of the Seventh European Turbulence Conference, held in Saint-Jean Cap Ferrat, 30 June–3 July 1998. 1998 ISBN 0-7923-5115-0
47. E.F. Toro and J.F. Clarke, *Numerical Methods for Wave Propagation.* Selected Contributions from the Workshop held in Manchester, UK. 1998 ISBN 0-7923-5125-8
48. A. Yoshizawa: *Hydrodynamic and Magnetohydrodynamic Turbulent Flows.* Modelling and Statistical Theory. 1998 ISBN 0-7923-5225-4
49. T.L. Geers (ed.): *IUTAM Symposium on Computational Methods for Unbounded Domains.* 1998 ISBN 0-7923-5266-1
50. Z. Zapryanov and S. Tabakova: *Dynamics of Bubbles, Drops and Rigid Particles.* 1999 ISBN 0-7923-5347-1
51. A. Alemany, Ph. Marty and J.P. Thibault (eds.): *Transfer Phenomena in Magnetohydrodynamic and Electroconducting Flows.* 1999 ISBN 0-7923-5532-6

Kluwer Academic Publishers – Dordrecht / Boston / London

Mechanics

SOLID MECHANICS AND ITS APPLICATIONS

Series Editor: G.M.L. Gladwell

20. W. Schiehlen (ed.): *Advanced Multibody System Dynamics.* Simulation and Software Tools. 1993 ISBN 0-7923-2192-8
21. C.-W. Lee: *Vibration Analysis of Rotors.* 1993 ISBN 0-7923-2300-9
22. D.R. Smith: *An Introduction to Continuum Mechanics.* 1993 ISBN 0-7923-2454-4
23. G.M.L. Gladwell: *Inverse Problems in Scattering.* An Introduction. 1993 ISBN 0-7923-2478-1
24. G. Prathap: *The Finite Element Method in Structural Mechanics.* 1993 ISBN 0-7923-2492-7
25. J. Herskovits (ed.): *Advances in Structural Optimization.* 1995 ISBN 0-7923-2510-9
26. M.A. González-Palacios and J. Angeles: *Cam Synthesis.* 1993 ISBN 0-7923-2536-2
27. W.S. Hall: *The Boundary Element Method.* 1993 ISBN 0-7923-2580-X
28. J. Angeles, G. Hommel and P. Kovács (eds.): *Computational Kinematics.* 1993 ISBN 0-7923-2585-0
29. A. Curnier: *Computational Methods in Solid Mechanics.* 1994 ISBN 0-7923-2761-6
30. D.A. Hills and D. Nowell: *Mechanics of Fretting Fatigue.* 1994 ISBN 0-7923-2866-3
31. B. Tabarrok and F.P.J. Rimrott: *Variational Methods and Complementary Formulations in Dynamics.* 1994 ISBN 0-7923-2923-6
32. E.H. Dowell (ed.), E.F. Crawley, H.C. Curtiss Jr., D.A. Peters, R. H. Scanlan and F. Sisto: *A Modern Course in Aeroelasticity.* Third Revised and Enlarged Edition. 1995 ISBN 0-7923-2788-8; Pb: 0-7923-2789-6
33. A. Preumont: *Random Vibration and Spectral Analysis.* 1994 ISBN 0-7923-3036-6
34. J.N. Reddy (ed.): *Mechanics of Composite Materials.* Selected works of Nicholas J. Pagano. 1994 ISBN 0-7923-3041-2
35. A.P.S. Selvadurai (ed.): *Mechanics of Poroelastic Media.* 1996 ISBN 0-7923-3329-2
36. Z. Mróz, D. Weichert, S. Dorosz (eds.): *Inelastic Behaviour of Structures under Variable Loads.* 1995 ISBN 0-7923-3397-7
37. R. Pyrz (ed.): *IUTAM Symposium on Microstructure-Property Interactions in Composite Materials.* Proceedings of the IUTAM Symposium held in Aalborg, Denmark. 1995 ISBN 0-7923-3427-2
38. M.I. Friswell and J.E. Mottershead: *Finite Element Model Updating in Structural Dynamics.* 1995 ISBN 0-7923-3431-0
39. D.F. Parker and A.H. England (eds.): *IUTAM Symposium on Anisotropy, Inhomogeneity and Nonlinearity in Solid Mechanics.* Proceedings of the IUTAM Symposium held in Nottingham, U.K. 1995 ISBN 0-7923-3594-5
40. J.-P. Merlet and B. Ravani (eds.): *Computational Kinematics '95.* 1995 ISBN 0-7923-3673-9
41. L.P. Lebedev, I.I. Vorovich and G.M.L. Gladwell: *Functional Analysis.* Applications in Mechanics and Inverse Problems. 1996 ISBN 0-7923-3849-9
42. J. Menčik: *Mechanics of Components with Treated or Coated Surfaces.* 1996 ISBN 0-7923-3700-X
43. D. Bestle and W. Schiehlen (eds.): *IUTAM Symposium on Optimization of Mechanical Systems.* Proceedings of the IUTAM Symposium held in Stuttgart, Germany. 1996 ISBN 0-7923-3830-8

Kluwer Academic Publishers – Dordrecht / Boston / London

Mechanics

SOLID MECHANICS AND ITS APPLICATIONS

Series Editor: G.M.L. Gladwell

44. D.A. Hills, P.A. Kelly, D.N. Dai and A.M. Korsunsky: *Solution of Crack Problems*. The Distributed Dislocation Technique. 1996 ISBN 0-7923-3848-0
45. V.A. Squire, R.J. Hosking, A.D. Kerr and P.J. Langhorne: *Moving Loads on Ice Plates*. 1996 ISBN 0-7923-3953-3
46. A. Pineau and A. Zaoui (eds.): *IUTAM Symposium on Micromechanics of Plasticity and Damage of Multiphase Materials*. Proceedings of the IUTAM Symposium held in Sèvres, Paris, France. 1996 ISBN 0-7923-4188-0
47. A. Naess and S. Krenk (eds.): *IUTAM Symposium on Advances in Nonlinear Stochastic Mechanics*. Proceedings of the IUTAM Symposium held in Trondheim, Norway. 1996 ISBN 0-7923-4193-7
48. D. Ieşan and A. Scalia: *Thermoelastic Deformations*. 1996 ISBN 0-7923-4230-5
49. J. R. Willis (ed.): *IUTAM Symposium on Nonlinear Analysis of Fracture*. Proceedings of the IUTAM Symposium held in Cambridge, U.K. 1997 ISBN 0-7923-4378-6
50. A. Preumont: *Vibration Control of Active Structures*. An Introduction. 1997 ISBN 0-7923-4392-1
51. G.P. Cherepanov: *Methods of Fracture Mechanics: Solid Matter Physics*. 1997 ISBN 0-7923-4408-1
52. D.H. van Campen (ed.): *IUTAM Symposium on Interaction between Dynamics and Control in Advanced Mechanical Systems*. Proceedings of the IUTAM Symposium held in Eindhoven, The Netherlands. 1997 ISBN 0-7923-4429-4
53. N.A. Fleck and A.C.F. Cocks (eds.): *IUTAM Symposium on Mechanics of Granular and Porous Materials*. Proceedings of the IUTAM Symposium held in Cambridge, U.K. 1997 ISBN 0-7923-4553-3
54. J. Roorda and N.K. Srivastava (eds.): *Trends in Structural Mechanics*. Theory, Practice, Education. 1997 ISBN 0-7923-4603-3
55. Yu. A. Mitropolskii and N. Van Dao: *Applied Asymptotic Methods in Nonlinear Oscillations*. 1997 ISBN 0-7923-4605-X
56. C. Guedes Soares (ed.): *Probabilistic Methods for Structural Design*. 1997 ISBN 0-7923-4670-X
57. D. François, A. Pineau and A. Zaoui: *Mechanical Behaviour of Materials*. Volume I: Elasticity and Plasticity. 1998 ISBN 0-7923-4894-X
58. D. François, A. Pineau and A. Zaoui: *Mechanical Behaviour of Materials*. Volume II: Viscoplasticity, Damage, Fracture and Contact Mechanics. 1998 ISBN 0-7923-4895-8
59. L. T. Tenek and J. Argyris: *Finite Element Analysis for Composite Structures*. 1998 ISBN 0-7923-4899-0
60. Y.A. Bahei-El-Din and G.J. Dvorak (eds.): *IUTAM Symposium on Transformation Problems in Composite and Active Materials*. Proceedings of the IUTAM Symposium held in Cairo, Egypt. 1998 ISBN 0-7923-5122-3
61. I.G. Goryacheva: *Contact Mechanics in Tribology*. 1998 ISBN 0-7923-5257-2
62. O.T. Bruhns and E. Stein (eds.): *IUTAM Symposium on Micro- and Macrostructural Aspects of Thermoplasticity*. Proceedings of the IUTAM Symposium held in Bochum, Germany. 1999 ISBN 0-7923-5265-3

Kluwer Academic Publishers – Dordrecht / Boston / London

Mechanics

SOLID MECHANICS AND ITS APPLICATIONS

Series Editor: G.M.L. Gladwell

63. F.C. Moon: *IUTAM Symposium on New Applications of Nonlinear and Chaotic Dynamics in Mechanics*. Proceedings of the IUTAM Symposium held in Ithaca, NY, USA. 1998 ISBN 0-7923-5276-9
64. R. Wang: *IUTAM Symposium on Rheology of Bodies with Defects*. Proceedings of the IUTAM Symposium held in Beijing, China. 1999 ISBN 0-7923-5297-1
65. Yu. I. Dimitrienko: *Thermomechanics of Composites under High Temperatures*. 1999 ISBN 0-7923-4899-0
66. P. Argoul, M. Frémond and Q.S. Nguyen (eds.): *IUTAM Symposium on Variations of Domains and Free-Boundary Problems in Solid Mechanics*. Proceedings of the IUTAM Symposium held in Paris, France. 1999 ISBN 0-7923-5450-8
67. F.J. Fahy and W.G. Price (eds.): *IUTAM Symposium on Statistical Energy Analysis*. Proceedings of the IUTAM Symposium held in Southampton, U.K. 1999 ISBN 0-7923-5457-5
68. H.A. Mang and F.G. Rammerstorfer (eds.): *IUTAM Symposium on Discretization Methods in Structural Mechanics*. Proceedings of the IUTAM Symposium held in Vienna, Austria. 1999 ISBN 0-7923-5591-1

Kluwer Academic Publishers – Dordrecht / Boston / London